T0198002

International Space Station
Architecture beyond Earth

International Space Station
Architecture beyond Earth

David Nixon

C I R C A

To Kathryn and Anya

Contents

Foreword

One of the pleasures of great works of architecture and engineering is that you can visit them. You can fly into Washington-Dulles Airport, drive across the Golden Gate Bridge, climb the stairs of the Eiffel Tower and sail past the Sydney Opera House. Such visits are not possible with the International Space Station. Only highly trained astronauts and cosmonauts or individuals with very deep pockets can travel to it. The rest of us who have to stay on the ground must rely on films, videos, photographs, articles and books to get a feeling for what the International Space Station is like, up close and personal. It is all the more surprising, then, that so little has been published on it. Not much more than a few paperback books, now out of date, and a pictorial reference guide from the National Aeronautics and Space Administration (henceforth referred to in this book as NASA) are available for an achievement that is widely regarded as the engineering and construction masterpiece of modern times. I hope this book will begin to fill the gap.

I have attempted in the narrative to follow a winding path between two conflicting approaches. On the one hand, it was essential to describe carefully and chronicle accurately the eventful development of an extremely complex piece of design and engineering. In other words, to stick to the technical script. This was important for the book to be useful as a resource and a reference. On the other hand, it was vital to liberate the story from the heavy technical jargon and countless acronyms that saturate the field

International Space Station solar transit
This composite image by Bill Ingalls is made from five frames and shows the Station silhouette as it transits the Sun at roughly five miles per second relative to the Earth's surface. Image: NASA/Bill Ingalls.

of astronautics. This was important for the story to hold the attention of readers, far and wide, without them dozing off.

Because NASA has become a household acronym – a worldwide acronym even – it is the only one that appears regularly throughout the book. I have expelled almost all others and described what they signify in words. Purists may grumble that I have contracted Environmental Control and Life Support System (ECLSS) into life support system, demystified Extra-Vehicular Activity (EVA) into a spacewalk and demoted an Orbital Replacement Unit (ORU) into a spare part. An Extravehicular Mobility Unit (EMU) becomes a spacesuit and an Expendable Launch Vehicle (ELV) becomes a rocket. Weightlessness on the Station is described in the space field as microgravity because a minuscule gravitational pull occurs that can affect sensitive experiments. Most of the time I have used weightlessness as everyone understands what it means. I make no apologies for these and other simplifications. The complex tale of the International Space Station deserves to be told in clear and straightforward language. There is a Glossary at the back that will help to shed light on technical words that are not in common use.

There are individuals and organisations that I wish to acknowledge and thank sincerely for their contributions, interest and support. First, astronaut Nicole Stott, now retired from NASA, for sharing some of her fond memories of the Station in her contribution entitled 'A Home in Space'. Nicole paints a word picture of what the Station is like to live in. Nicole flew to the Station on Shuttle flight STS-128 in August 2009, spent three months on board as part of Expeditions 20 and 21, and returned on Shuttle flight STS-129 in November 2009. She went to the Station again on STS-133, the final flight of the Orbiter *Discovery* in February and March 2011.

Marc Cohen, formerly with NASA Ames Research Center, guided me on the chronology of Space Station studies from the early 1980s. It was he who beckoned me into the space field in 1985. At Boeing, Ed Memi, Jim Waterman and Steve Ernst provided descriptions of the fabrication of the modules and truss pieces. At NASA Johnson Space Center, Rod Jones and Michael Berdich helped with contacts and information and Kriss Kennedy gave me the background of the TransHab inflatable module. Martin Mikulas, formerly of NASA Langley Research Center, reminisced about the Station's truss structure in the late 1980s. Alan Thirkettle, formerly head of human spaceflight at the European Space Agency, recalled some history of the Columbus programme. Brand Griffin contributed some images and notes from the years he spent at NASA and Boeing as an architect and I have included some of his clever sketches and drawings. MacDonald Dettwiler and Associates of Brampton, Ontario provided me with descriptions and illustrations of their acrobatic Canadarm2 robotic arm and Dextre manipulator and Lonnie Cundieff at Barrios Technology in Houston briefed me on how astronauts operate the arm on the Station.

The University of Calgary, Alberta, supplied information on their pioneering NeuroArm medical spin-off programme. RSC Energia in Russia sent me material on the Proton launch vehicle as did the European Organization for Nuclear Research (CERN) in Switzerland for the Large Hadron Collider. At NASA Headquarters in Washington, Bill Barry and Stephen Garber of the History Office offered encouragement and helped with publisher suggestions and reference sources while Robyn Gatens, manager for technology demonstrations on the Station, provided an update on life support system plans and Joel Kowsky tracked down and sent me an important photograph. From the European Space Agency Headquarters in Paris, Nathalie Tinjod generously sent me an important reference.

I owe a debt of gratitude to Paul Clancy, formerly head of planning for European use of the International Space Station at the European Space Agency, for his technical review of the manuscript and his recollections of early NASA meetings. I owe a similar debt to Carrie Paterson of DoppelHouse Press in California for her thoughtful comments and her interest and support. Richard Horden gave me some publishing suggestions. My daughter Anya helped to iron out the syntax. This book would not have been possible without access to the extensive archives of reports, documents, press kits and photographs that NASA and the European Space Agency make publicly available through their various excellent websites. These are truly valuable resources. I was also glad to be able to include Dylan O'Donnell's extraordinary photograph of the International Space Station transiting the Moon. Last but not least, David Jenkins of Circa Press in London had the courage and vision to take the book on and publish it. Jean-Michel Dentand did a superb job on its design and Jayson Young worked magic on its 3D images. Julia Dawson did the final proofreading with a sharp eye and Vanessa Bird carefully assembled the Index. I am very grateful to them all.

I am conscious that in the seven years, on and off, that I have spent working on this book, I have just scratched the surface of its complex and compelling subject. For every source of information in the text there were several more, waiting in the wings, that were relevant but I had to stop somewhere and I wanted the book to be of manageable length. There is a need for someone to write a book about the vast amount of excellent research work taking place on the Station and how valuable it is to mankind and the planet. This book is a starting point and I hope that it will stimulate others into delving more deeply into the Station's fascinating story before the trail begins to grow cold. The International Space Station's life may be secure until 2024 but its future after that is murky and sooner or later it will cease to exist.

David Nixon
November 2015

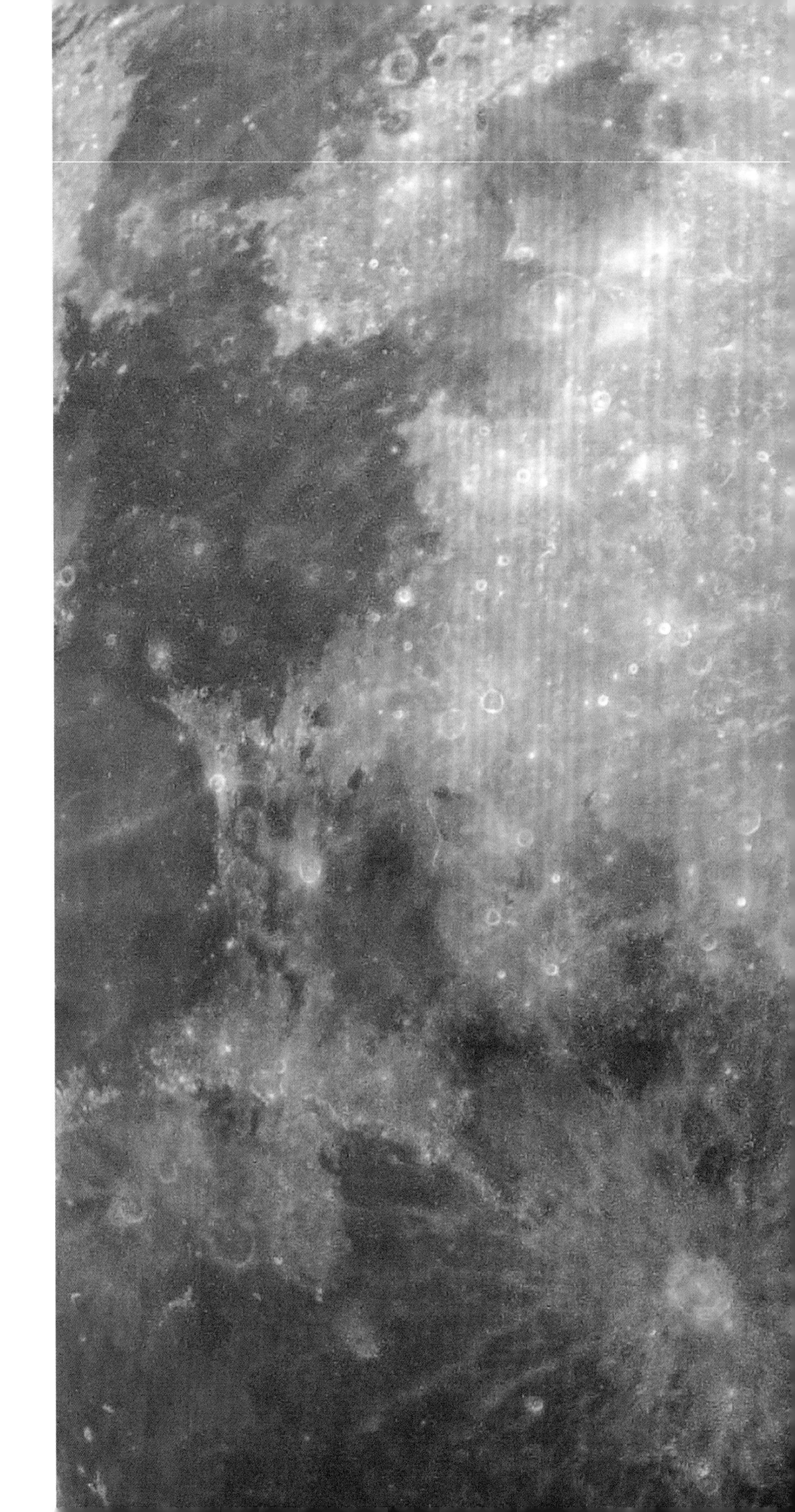

International Space Station lunar transit Photograph by Dylan O'Donnell of the International Space Station in transit across the Moon. He took the photograph on 30 June 2015 at 19:54 local time in Byron Bay, New South Wales, at the easternmost point of Australia where there is minimal light pollution of the night sky. Dylan O'Donnell used a Canon 70D camera attached to the rear cell of a Celestron 9.25" telescope (2350mm / f10). At a distance of about 400km and with such magnification, the Station's transit time across the Moon was about one-third of a second. He set the shutter speed at 1/1650th of a second at ISO 800. He added about a second of further exposures on either side of the transit to strengthen the lunar surface detail using AutoStakkert!2. He later increased the saturation to add some colour enhancement of the lunar surface.

American Eastern Seaboard at night from the International Space Station
This photograph was taken on the Station during Expedition 30 between November 2011 and April 2012. It shows a stretch of the United States Atlantic coast. In the foreground, the curving line of six bright clusters of light from left are Norfolk-Hampton, Richmond, Washington DC, Baltimore, Philadelphia and New York-Brooklyn. In the left foreground are two Russian vehicles docked at the Station.

Study in grey and gold The Sun reflects off a golden solar array on the International Space Station in this photograph taken by an Expedition 38 crew member in January 2014. In the top foreground in grey is one of the Station's radiators that is used to dump the Station's internal heat into space.

A home in space

I was very pleased when David Nixon asked me to write a memoir for his book about the International Space Station. I was intrigued by the perspective he shares in presenting the history of space station architecture, the different architectural concepts that were proposed and, after all political and technical challenges, how we ended up with the incredible Space Station we have on orbit today.

The implication of the architecture of our spacecraft, of our space stations, is significant not only from physical and aesthetic viewpoints, but also with respect to the utility and habitability of the 'space' within the place. This is true of architecture in general, of course, but it is particularly interesting when we consider it from the standpoint of the unique environment of space.

I have been blessed to live and work on the Space Station for over three months. It makes me smile when I think about my time there, the experiences I shared with my crew mates and our support team on the ground. I am privileged to be able to call the Station one of my homes.

When I go to schools and talk to students about the Space Station, one of the first things I do is ask them how old they are. For those in high school and younger, the fact is, that for as long as they have been alive, people have been living and working continuously and peacefully together in space on the Space Station. That fact alone is compelling. To bring it

Nicole Stott who wrote this piece, was flight engineer on Space Station Expeditions 20 and 21. This photograph shows her during Shuttle Mission STS-128 on a six-and-a-half hour spacewalk that she carried out with astronaut Danny Olivas. They performed maintenance tasks along the Station's truss structure and transferred an external experiment from the European Columbus Module to the Shuttle Orbiter *Discovery* for return to Earth. Though Nicole may appear to be floating freely in this view, she is firmly attached to the module in the top-left corner by a tether line that is shackled to the gold-coloured grabrail at the end of her right arm. The exterior of the Station is covered in grabrails such as these to help with spacewalk mobility and safety.

closer to home for me, for as long as my son has been alive there have been people living and working there — people like his mom, and his godfathers (he has both an American astronaut and a Russian cosmonaut godfather). Some of his close family friends and acquaintances have been there too. I am hopeful that these first fifteen years of the Space Station represent just the beginning of humanity's uninterrupted, continuous presence in space.

For those who know about the Space Station (young and old alike), one of the first things they want to know is what it looks like, physically there in space. Some people have mental images inspired by science fiction books and movies. Again, it illustrates why architecture plays such an important role in how we envision life in space. It is why the title of this book was so interesting to me: *International Space Station: Architecture Beyond Earth*. It implies very simply that there are, and will continue to be, people living beyond Earth, and that is of huge significance.

Reading this book, I was reminded of the time I spent on the Space Station. But I also found it interesting to think about how my experiences there might have been different if the shape or size or international partnerships or system architecture of the Station had been different. It made me understand just how much of the Station's development story I had either forgotten or maybe never knew.

Based on my time there, I can tell you that the Station is awesome, and living and working there is incredible. All expectations I had before flying were exceeded once I got there. Everything from spacewalking to routine maintenance is part of the amazing adventure of living in space.

For the human body, living and working in the weightless environment of space is liberating. To float, to literally fly from one place to another, is like a childhood dream come to life. Just as remarkable, is the fact that our brains and bodies very quickly adjust to this new environment and help us navigate and move through it gracefully. Because our brains and bodies don't give us any physical cues of up or down, the Station's designers very thoughtfully incorporated visual cues. To help us better communicate with each other about where we are and where we're going, there is simple signage for location coding and orientation (e.g. port/starboard, overhead/deck).

This liberating feeling that our bodies have in weightlessness is also true for our 'stuff'. Organisation is key in zero-g. Hence the prolific use of Velcro and bungees. If you don't consciously manage your belongings – secure them with Velcro or otherwise – they are likely to float away and not be seen again for a very long time, if ever. There are many stories of astronauts letting go of their glasses or tools 'just for a second' and not seeing them again until they float up out of nowhere months later.

The downside of our brains and bodies adjusting to a zero-g environment is that they figure out that you don't need bones or muscles any more to survive there. We essentially start experiencing an accelerated osteoporosis.

So astronauts exercise two hours a day – both resistive and aerobic exercise. Thanks to the resistive exercise device we have on orbit now, crew members are coming back to Earth with statistically insignificant bone and muscle loss that's fully recoverable. As a result, we are learning a lot about how our bodies respond to zero-g and also how we should be preventing and managing osteoporosis here on Earth. This is just one example of the kind of science that's happening on the Space Station, which is why I love our motto – 'Off the Earth, For the Earth'.

The Space Station is physically set up to function as an off-Earth, microgravity laboratory. All the 'walls' of the Station modules are usable surfaces that incorporate laboratory or systems equipment, or habitation or stowage spaces. From the laboratory standpoint, the research that's going on everyday is impressive. Every area of scientific research is represented in some way on the Space Station; and everything about that research, as well as the way the Station systems operate, is helping us to explore further off our planet and simultaneously improve life here on Earth.

From the habitation standpoint, the shell of the Space Station modules is like a comfortable cocoon around the crew. It is a place with its own wonderful character, its own sights and sounds and smells. To share a little of what it's like, I offer the following as a sample of the subtleties of living in this unique place.

Our day-to-day activities are managed via a schedule prepared by the mission control team on the ground and presented to us on our computers. Our 'time zone' is set to GMT. Travelling around the Earth at 17,500mph, orbiting sixteen times a day, means you experience a sunrise and sunset about every 45 minutes (very cool by the way!). So we are not able to use light and dark as seen through the windows to give us any indication of 'day' or 'night' or the passing of time. It is natural at first to use your schedule, or your watch to indicate the time of day. But it soon becomes more manageable when you pay attention to your own physical cues, such as being hungry or tired.

Due to the large temperature changes that take place outside the Station, as it is exposed repeatedly in orbit to extreme heat while in the sun and extreme cold when not, the metal of the Station expands and contracts and makes a creaking sound. At first it is slightly alarming, but soon becomes familiar. Because of the unique environment of space, 'at first alarming, but soon familiar' is true for a lot of things experienced on the Station.

The atmosphere inside the Station is maintained at sea level pressure and the air is cool and dry and very comfortable, so no special clothing or equipment is required. (Very different to the complex equipment that's required in order to leave the Station's airlock for a spacewalk.) The big difference from working on Earth is the floating, mentioned above, but even that becomes 'normal'. To 'hold on' to do our work, we slip our socked feet under metal rails for stability and leverage. As a result, the soles of the feet become softer, while the tops develop callouses. (Most crew members don't wear shoes during the day, except while they're using the exercise equipment.)

Food and mealtimes on board the Station are as important as they are on Earth. The variety of the food is quite good, although the packaging and presentation are very different to what we would normally expect. Most items are individually packed, dehydrated or ready to eat. The only methods of preparing food are to rehydrate it with hot or cold water, or to heat it up in a simple food warmer. The only utensil required is a long-handled spoon. It is amazing how nice a meal you can have, given such limited means. The smell of food is welcome too. Not quite like the smells that come wafting from the kitchen at home, but nonetheless comforting and pleasant. Some of my best memories of spaceflight are from mealtimes, floating around the table with my crew mates – sharing stories from our day, solving the world's problems, with bags of food stuck to the wall or to the Velcro on our pants.

Each crew member has his or her own sleeping compartment. About the size of a phone booth, it is just right for hanging your sleeping bag on one wall and distributing the rest of your personal items across the others. Everything is within easy reach of the comfort of your floating sleeping bag. The compartments are quiet and dark and have good air circulation. Without the pressure points of a mattress, it was the best sleep I have ever had.

There is a difference in the interior design and 'feel' of the US-built and Russian-built modules, although both are comfortable and technically sound. The US modules are somewhat sterile, with a lot of white panels and exposed cables and equipment, while the Russian modules are what I would describe as 'cosy', with a plush tan fabric covering the major surfaces. The smell of the modules is distinctive too – neither smells bad, just different. You can close your eyes and float from one module to another and tell where you are just from the smell or feel of the place.

A common factor throughout the Station is the background noise. There is constant hum of pumps and fans and surrounding equipment. You almost forget it's there, rather like the noise of the refrigerator at home. You mostly notice it when it's gone. Because of that, our very creative engineers designed the layout and location of different pieces of equipment so that we can troubleshoot system failures just by the loss of the associated background noise. It is a very helpful and simple strategy for the crew.

Another noise, which I had not expected, comes when crew mates are outside on a spacewalk. If they are moving around one of the pressurised modules, you can hear where they are the entire time and follow them.

A big surprise to me was the 'smell of space'. By that, I mean what something smells like after being exposed to the vacuum of space. The best example is the suit used for a spacewalk right after a crew member re-enters the Station from the airlock. I've heard different descriptions, but to me space has a sweet, metallic smell similar to that of an overheating car radiator.

This account would not be complete without saying something about the interior size of the Space Station. To use a word favoured by my young son, it's 'ginormous'. It has an interior volume greater than that of a Boeing 747. Six crew members can spread out very nicely and not see each other for hours, depending on where their work is being done. There is nothing claustrophobic or small about the Station.

As human beings on the Space Station, beyond the adventure of the day-to-day activities on board, having a connection to our home planet is really important. The single most impressive connection we have to Earth from the Station is the view through our windows. I give huge thanks to the designers who convinced NASA to include windows – a fairly significant structural problem to solve when building spacecraft. I could devote this whole piece to the view, about the philosophical discussions it generates, about its overwhelming beauty and seemingly endless surprises, and never ever tiring of seeing it. Every glimpse through a window provides you with something new to consider about our planet and the space around it. Seeing a shooting star below you is one of the surprises that stands out for me. Whether looking through the port windows of the Japanese Experiment Module or the horizon-to-horizon views presented through the Cupola windows or through your own tiny visor during a spacewalk, the only word to even begin to describe the experience is awe!

In closing, one of the central messages of this book is the significance of the word 'international' in International Space Station. I believe that the Space Station, because of the overwhelmingly successful international partnership it is built upon, will always be *the* shining example of how we can work together peacefully as a global community to solve the world's most complex problems.

Many designs were proposed before the final one for the International Space Station was chosen. Each of these designs would have had its pluses and minuses in comparison, and quite honestly I wish we could have built them all. But I have to admit that I'm partial to the design that was chosen and grateful to have had the opportunity to call the International Space Station, our Space Station, my home.

David Nixon has done an outstanding job bringing the Space Station to life for us. I look forward to a future in which human beings continue to live and work in space, and do so based on the amazing place that the International Space Station has given us. Thank you David for sharing the wonderful story of the International Space Station and mankind's history in space.

Nicole Stott
NASA Astronaut
Shuttle Missions STS-128, STS-129 and STS-133
International Space Station Expeditions 20 and 21

Introduction

The International Space Station is the most ambitious habitat contrived by mankind to support its existence beyond Earth. Day in, day out, it orbits the globe, half in blinding sunlight and half in blackest darkness. If the Station were a building on the ground it would need foundations to support its weight. In weightlessness on orbit, the Station is in a state of perpetual free fall around the planet, in which inward gravitational pull is equalled by outward centrifugal force. It is a perfectly balanced form of motion. Inside the Station, cocooned and safely sealed off from the hostile space environment, are six highly trained men and women working on experiments at the cutting edge of science for the benefit of the planet and mankind. They live in a cluster of bus-sized modules that are bolted on to a long structural truss that is almost the length of Centre Georges Pompidou in Paris. The vantage point of the Station's orbital path as it cycles back and forth across the equator provides it with stunning views of nine-tenths of the planet's habitable surface. From an altitude of 400km, features such as mountains, glaciers, forests, fields, rivers, lakes and cities stand out clearly. Also visible are hurricanes, eruptions, fires, floods, landslides and from time to time the ravages of war.

As the International Space Station is in a state of continual motion around the Earth, its orientation to the Sun and the heavens is always

Goodbye from the Space Shuttle This photograph of the International Space Station was taken by the crew of the Orbiter Atlantis on Shuttle Mission STS-135 as it departed and carried out a fly-around of the Station on 19 July 2011. STS-135 was the final Space Shuttle mission to the Station before the Shuttle programme was terminated by NASA.

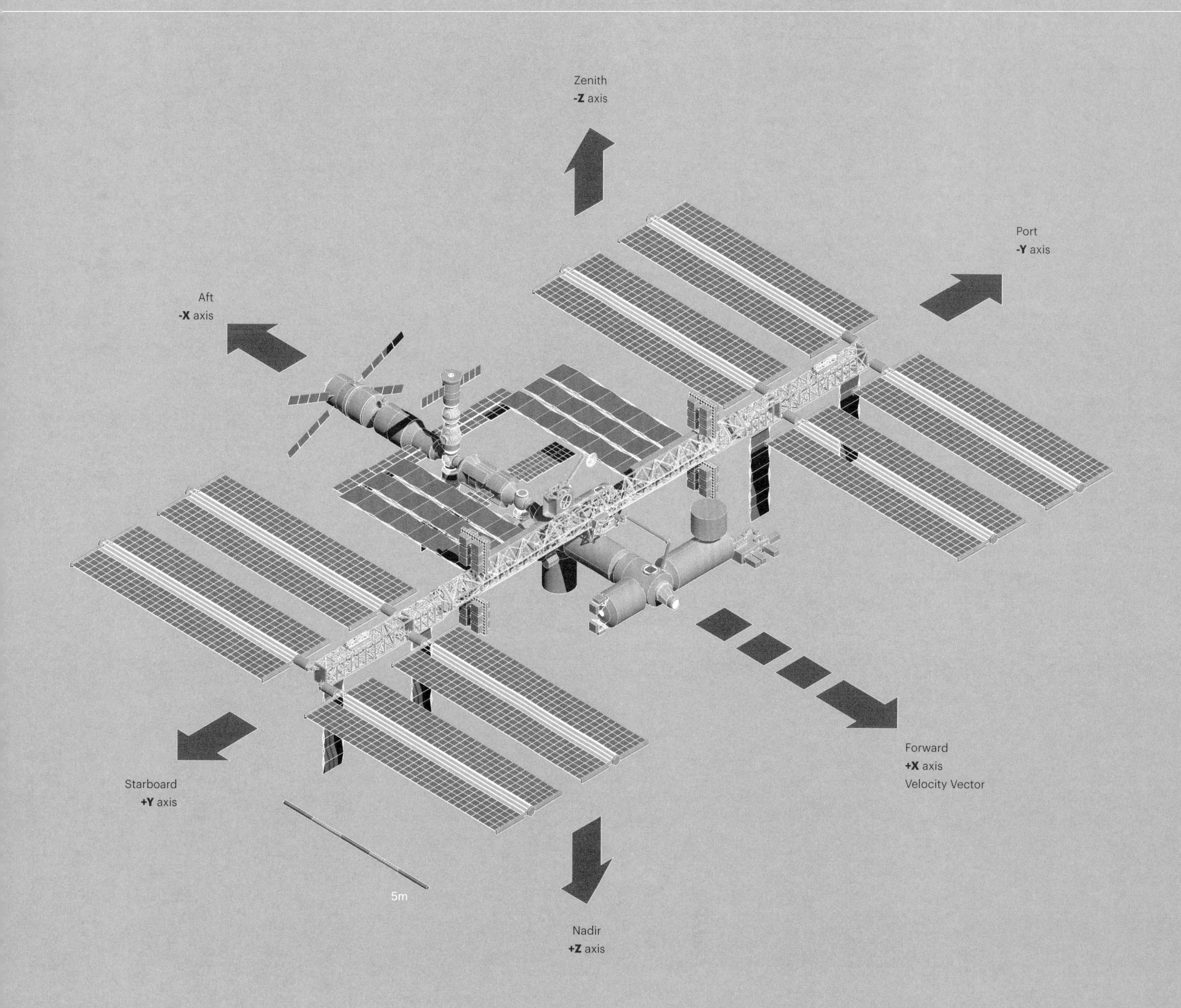
Zenith
-Z axis
Port
-Y axis
Aft
-X axis
Starboard
+Y axis
5m
Forward
+X axis
Velocity Vector
Nadir
+Z axis

changing. Of particular importance is the Station's need to point its solar arrays at the Sun to draw electrical power, requiring their constant adjustment to track the Sun's ever changing position relative to the Station. To help define its orientation on orbit, the Station has its own X, Y and Z axes based on a right-handed Cartesian system. The X axis is the Station's 'roll' axis and it runs through the line of interconnected pressurised modules from the front to the back of the Station like the centreline from stern to the bow of a ship. The forward X axis is often called the velocity vector because the Station is travelling at an average speed of 7.66km per second relative to the Earth's surface in that direction. The Y axis is the 'pitch' axis and it is perpendicular to the X axis looking down from above and extends outwards at right angles from the centreline, like port and starboard directions on a ship. The Station's long structural truss is aligned on the Y axis. The Z axis is the 'yaw' axis and is a line from the Earth's centre out into space that passes through the point where the X and Y axes intersect. The Z axis direction up towards the heavens is often called the zenith and down towards the Earth it is called the nadir. The X axis is +X forward into velocity vector and -X in the aft direction, the Y axis is +Y to starboard and -Y to port and the Z axis is +Z down to nadir and -Z up to zenith. This book will make much use of these various references and it will be useful to compose a picture of them in one's mind's eye.

The Station revolves around the Earth on its own orbital plane that is inclined at 51.65° to the equator. In the course of one orbit, which takes about ninety minutes to complete, the Station cycles across the equator twice, from the northern to the southern hemisphere and back again. In that time, the Earth has rotated about 22.5° beneath the Station's path so the Station passes over a different swathe of surface each time it goes around the Earth. This is what makes the International Space Station such an extraordinarily good place from which to view and study the planet.

The first recorded idea for a Space Station dates back to 1870. That year, a Unitarian minister and author from Boston, named Edward Everett Hale, wrote a story called 'The Brick Moon' about a satellite made of brick that would be launched to orbit as a visible navigational aid for ships at sea. It would become a space lighthouse. The Brick Moon would be about 65 metres in diameter and built of bricks to survive the heat generated by orbiting the Earth at great speed.[1] He envisaged several Brick Moons, launched to orbit by gigantic counter-rotating flywheels like stones from a catapult. The idea was not entirely fanciful, as the Space Shuttle Orbiters were sheathed across their undersides with brick-like tiles made from carbon to survive the heat of re-entry.

To most people alive in the 1940s, 1950s and 1960s, the idea of a Space Station orbiting the Earth would have been fantastic and inconceivable. Space was an alien and mysterious place seen in science fiction films, through telescopes on lawns at night or read about in paperbacks or

International Space Station The Station has its own reference coordinate system that defines its orientation as it orbits the Earth. Its flight direction is Forward on the +X axis, also referred to as the velocity vector. In the opposite direction to its rear is Aft on the -X axis. To its left, looking down at it from above, is Port on the -Y axis and to its right is Starboard on the +Y axis. Up towards the sky is Zenith on the -Z axis and down towards the Earth is Nadir on the +Z axis. Image: author.

Below: The greatest legacy of the Station may not be its scientific advances and discoveries but its role as an international project that has brought together many nations – former bitter enemies among them – in a spirit of mutual trust and shared commitment. It is a project that has helped to keep the peace on a global scale. Image: author.

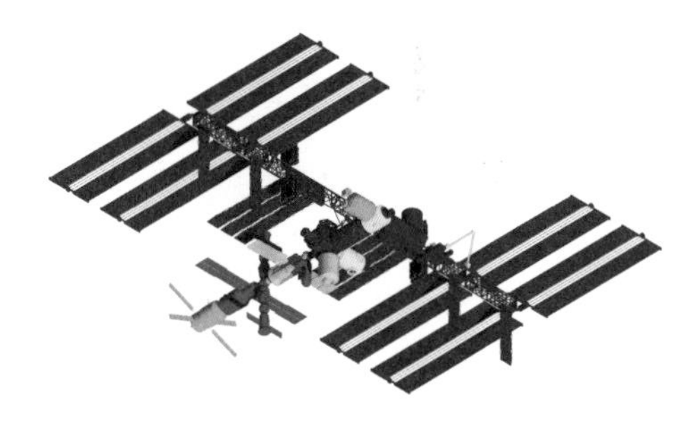

- USA-Europe
- Italy
- Russia
- USA-Russia
- Japan
- USA
- Europe
- Canada
- Brazil

children's comics. Occasionally people saw or read something on it that was more thoughtful. In a magazine article in 1952, the rocket engineering genius Wernher von Braun, who was the driving force behind the early development of America's rocket programme, envisioned a Space Station on Earth orbit shaped like a giant bicycle wheel that slowly rotated to provide artificial gravity.[2] A long forgotten film called *On the Threshold of Space*, made in 1956 by Twentieth Century Fox, realistically depicted the risky tests needed to prepare astronauts for spaceflight.[3] Then, with the pioneering orbital flights of Yuri Gagarin on Vostok 1 in 1961 and John Glenn on Mercury Friendship 7 in 1962, sending humans into space evolved from fiction to fact, though newspaper photographs of Gagarin and Glenn in their spacesuits crammed into their tiny capsules made the short flights seem like dangerous joyrides.

Apollo 11 Astronaut Buzz Aldrin unpacking scientific experiments from an equipment locker on the side of the Lunar Module a few hours after the historic landing of the Apollo 11 mission on the Moon, 20 July 1969. Built by Grumman Aircraft Engineering, the Lunar Module had a gross mass of 4,700kg and a 100 per cent oxygen atmosphere at a low pressure of 33kPa. The internal habitable volume for the crew of two astronauts was just 4.5m³.

A few years later, two extraordinary events changed common perceptions about going into space. The first was Metro-Goldwyn-Mayer's stunning 1968 science fiction movie *2001: A Space Odyssey* inspired by a short story by Arthur C Clarke. Written, directed and produced by Stanley Kubrick, the film's superb special effects and its convincing portrayal of a giant rotating space station, lunar colony and interplanetary spacecraft conveyed the imagined reality and predicted detail of future habitats in space to the public for the first time.[4] The second event was the spectacular Apollo 11 mission to the Moon in July 1969 and the successful lunar landing and return to Earth of American astronauts Neil Armstrong and Buzz Aldrin. The blurred and jerky video footage of Armstrong and Aldrin ambling across the Moon's surface banished the suspicions and superstitions that the astronauts and their Apollo lunar module would sink into a sea of lunar dust.

With these two events just a year apart, fact and fiction intermingled to portray human spaceflight as a complex, exciting and fascinating field where the dangers, if not avoidable, could be brought under control by sophisticated design and engineering. Armstrong and Aldrin showed that the Moon could be visited safely. In fact, after five subsequent Apollo missions landed there, it began to feel routine. The Apollo missions that had caught the world's attention ended in 1972. Following them were two missions to Earth orbit that seemed humdrum by comparison but would prove to be vitally important for any future Space Station. The first was Skylab, America's first space laboratory and outpost.[5] Three crews spent twenty-eight, fifty-nine and eighty-four days consecutively on Skylab in 1973 and showed that humans could live for long periods in weightlessness with the aid of physical exercise to combat adverse side effects. Up to that point, there was doubt that astronauts could survive in weightlessness for long periods without some kind of artificially induced gravity. The second mission was the Apollo-Soyuz project of 1975, the first demonstration of the orbital docking of two spacecraft launched by different nations –

in this case America and the Soviet Union.[6] Apollo-Soyuz showed that two small spacecraft could join together to make a larger one, that two spaceflight rivals could work together on a common project and that international rescue of a stranded spacecraft was potentially feasible. By the mid-1970s, the successful Skylab and Apollo-Soyuz missions and the go-ahead for the Space Shuttle had set the scene for the Space Station that is the subject of this book. It was to become an epic engineering development saga beginning in 1979 with some outline ideas and ending thirty-two years later when the final building blocks were plugged into the International Space Station as it orbited the Earth.

What is a Space Station anyway? A leading English dictionary defines it as a 'large artificial satellite used as a base for operations in space'.[7] This definition leaves much to be desired. Today, a satellite is universally understood in the aerospace world to be an object that operates remotely on orbit for purposes of telecommunications, navigation, observation, sensing, surveillance and suchlike. Calling a Space Station a satellite leads to potential misunderstanding of its function in a field where terminology and comprehension require clarity and precision. These days, only astronomers might refer to a satellite launched from Earth as artificial because satellites to them are natural orbiting objects such as moons, moonlets and various lumps of rock. The dictionary gets it right when it calls it a base for space operations.

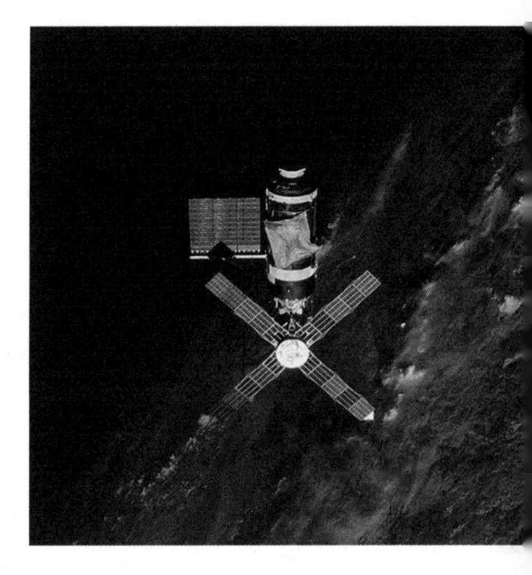

Skylab Skylab as seen from the Apollo command module after undocking of the Skylab-2 mission on 22 June 1973. Skylab was America's first crewed outpost on orbit. Skylab originated in an idea for converting an empty upper stage of a Saturn V rocket into a habitable laboratory and workshop. Three crews of astronauts occupied it on three consecutive missions of twenty-eight, fifty-nine and eighty-four days respectively. Skylab gave American astronauts their first experiences of long-duration spaceflight under conditions of weightlessness.

In its present form the International Space Station is a base dedicated to scientific research operations. This is made clear in a recent guidebook produced by NASA. The agency describes it as a national laboratory concentrating on research in biology, human physiology and psychology, physical science and materials, Earth and space science and technology for exploration beyond low Earth orbit.[8] The Station's job is to return the resulting discoveries and new knowledge to Earth for the benefit of mankind and the planet. This emphasis on research is quite narrow compared with the much broader role envisioned for the Station when President Reagan gave it the go-ahead in 1984. Then, it was intended to perform eight functions: space laboratory, astronomical observatory, transportation node, servicing facility, assembly platform, manufacturing plant, storage depot and staging base.[9]

The gradual erosion of expectations for the Station before its completion in 2011 is marked by multiple design changes to the Station's configuration, forced on the project by financial and political realities. These changes are chronicled in this book and they make a fascinating story. However, it is important to stress that if the Station's functions have diminished over time, its scale, size and presence have not. The International Space Station as a completed piece of engineering has, if anything, gained in stature for only recently can we stand back, take it in and appreciate the true significance of what it is.

Why call it a space station, rather than a base, port or outpost? There are twenty definitions of the noun station in the dictionary, excluding those that are ecclesiastical. For example, in early nautical use from the seventeenth century a station was a port, harbour or roadstead for ships. In aviation it became an aerodrome where personnel were employed or garrisoned. In ground transportation it was a place where trains regularly stopped for taking up and setting down passengers or for receiving goods for transport and including the buildings erected for these purposes. In industry and science it was a place where people were stationed and apparatus set up for some particular kind of industrial work or scientific research.[10] None of these definitions embraces the unique characteristics of the International Space Station. First and foremost, it is a habitat for astronauts and cosmonauts without whom it would be impossible to carry out the scientific research in the laboratories that constitute the greater part of the Station's anatomy. It is also a strongly international project, has been since the early 1980s and is today dependent on international operation and support for its survival. So, if we consider the stated research aims and international perspective, we might adopt the following working definition: the International Space Station is a port, habitat and laboratory on Earth orbit that is internationally operated for purposes of research in the physical, medical, biological, terrestrial and astronomical sciences and for the return to Earth of the discoveries and knowledge that result.

NASA has described the Station as one of the greatest technological, geopolitical and engineering achievements to benefit humanity. Few people familiar with the Station's development story and the challenges the project faced would disagree with that statement. Many would argue that the International Space Station is the single greatest engineering enterprise of the twenty-first century so far. Some would go further and claim it is the single greatest achievement in the history of engineering. For that title, however, it would encounter some stiff competition. There are bridges, dams, towers, canals and tunnels that are outstanding examples of engineering.[11] A list of contenders might include five key projects: the Panama Canal, the Eiffel Tower, the Golden Gate Bridge, the Aswan Dam and the Channel Tunnel. They were extraordinary engineering wonders at the time they were built and remain so today.

Each of these five projects overcame great challenges, became outstanding engineering achievements at the time they were built and have remained so. They have all proved financially successful over time, are all in continuous use today and have all become indispensable as cultural, environmental or technological objects in the service of civilisation. Their places in the hall of fame of the greatest structures built on Earth are firmly assured. They are complete life-cycle achievements. Though it is a sensational engineering achievement, the International Space Station is

Apollo-Soyuz Test Project View of the Soviet Soyuz spacecraft from the window in the American Apollo command module that docked with it on 17 July 1975. The Apollo-Soyuz Test Project of the two nations proved to be a vital stepping stone in the quest for a permanent Space Station on Earth orbit. It showed that two spacecraft could dock together on orbit, that a crew could potentially be rescued from a defective spacecraft and that two nations who were adversaries on Earth – America and the Soviet Union – could team up successfully in space.

not yet a complete life-cycle achievement because it is too new and has yet to prove its worth as a success measured by the terrestrial benefits flowing from its research. Not until the Station is close to the end of its life and a judgement can be made on the accumulated impact of its scientific advances and discoveries, or on other future roles and responsibilities the Station assumes, can such a claim be advanced or considered.

There is, however, one project to which the Station can be compared right now. It is the Large Hadron Collider. The International Space Station and the Large Hadron Collider have much in common. Both are dedicated to cutting-edge scientific research, were very ambitious in scale, had very large budgets, needed extreme engineering precision and involved a group of international partners. Both required construction in extreme environments and these are so hazardous that access to both projects is severely limited. In the case of the Large Hadron Collider, which is deep underground, only a small number of authorised technical staff can enter the tunnel. It has weak radioactivity, which requires a radiation protection technician to measure the tunnel's dose rate and determine how long the visitors can spend there.[12]

Large Hadron Collider In February 2007 at CERN, the Compact Muon Solenoid detector containing a magnet with a weight of 1,920t (equivalent to five Boeing 747 aircraft) was lowered into position 100m below the surface by a huge gantry crane. The detector is part of the Large Hadron Collider at CERN. Just 200mm separated the detector from the walls of the vertical shaft. During its descent the detector was suspended by four, fifty-five strand cables with sensors to monitor its sway and tilt. The detector helped in the search and discovery of the Higgs boson that CERN confirmed in 2013. With its superconducting magnet ring that is 27km in circumference, the Large Hadron Collider is the world's largest particle accelerator. Image: CERN.

The Large Hadron Collider is part of an organisation called CERN that in French originally stood for the *Conseil Européen pour la Recherche Nucléaire*, known in English as the European Council for Nuclear Research. CERN originated in 1949 in a proposal by French physicist Louis de Broglie to create a European laboratory to study atomic physics. The Synchrocyclotron that started up in 1957 was CERN's first nuclear accelerator, followed by the Proton Synchrotron in 1959. Early discoveries led to the commissioning of the first giant underground ring, the Super Proton Synchrotron, in 1971. The torus-shaped tunnel was 7km in circumference and ran beneath the Franco-Swiss border. Meanwhile, an even bigger ring called the Large Electron-Positron Collider was under construction. Completed in 1988, it was 27km in circumference and the largest European civil engineering project before the Channel Tunnel. The new Collider began operations in 1989 and hosted eleven years of valuable research. It closed in 2000 to make way for the construction of the Large Hadron Collider in the same tunnel.[13] CERN has made a succession of extraordinary discoveries in its sixty-five year history. Most recently, in 2011 and 2012, two experiments in the Large Hadron Collider confirmed the existence of the elusive Higgs boson, one of the most important and sensational discoveries in physics in modern times.[14], [15]

The evolution of CERN from its beginnings in 1949 up to the launch of the Large Hadron Collider followed a set of carefully paced milestones in which many discoveries eventually led to the Collider as a kind of 'grand-slam' experiment. It was a different story with the International Space Station, which had a roller-coaster ride during its development over several years before its final design emerged. This book follows the Station's development adventure in some detail but it is useful here to provide a brief introduction to it.

NASA began studies of both an orbiting space station and a reusable space vehicle system in 1970. Forced for budgetary reasons to choose between them, it opted for what became the Space Shuttle. Meanwhile in 1971, the Soviet Union began to launch its series of Salyut space outposts. America followed up in 1973 with Skylab, which housed three consecutive crews that gave America its first experience of long duration spaceflight. Nevertheless, the Soviet Union had taken a clear lead in human spaceflight endurance and as the Space Shuttle neared operational readiness at the end of the 1970s, NASA turned its full attention to a Space Station and began serious studies of design concepts. NASA also made overtures to foreign space agencies to consider joining the project and it soon attracted the interest of space agencies and their governments in Europe, Japan and Canada.

After President Reagan approved the project in 1984, NASA carried out a successful marketing campaign to sell it overseas and the foreign space agencies and their governments signed up as partners. It was simply called the 'Space Station'. NASA developed two ambitious proposals for the Station between 1984 and 1987 called the Power Tower and Dual Keel designs, both of which had major technical problems and projected costs that were climbing well above NASA's original budget. The agency produced a scaled-down design called the Revised Baseline Configuration and President Reagan renamed it Space Station 'Freedom' in 1987. NASA authorised its aerospace contractor teams to begin full development in 1988 but the Station's costs continued to escalate. NASA, struggling to bring the project's budget under control, produced another slimmer design called Restructured Space Station Freedom in 1991, but it was not enough and the financial crisis surrounding it grew. The Station ran into bitter political opposition in Washington and was very nearly cancelled. It was saved after Bill Clinton's election as President in 1992 when he invited Russia to join the project as a senior partner. The project was reincarnated as the International Space Station.

The International Space Station's purpose was manifold when President Reagan announced its predecessor in 1984. Behind this was NASA's need to sell the project to politicians at home and overseas, to participating national space agencies and to the public. It was going to cost a lot of money and the more jobs it would be able to do, the better. Concept studies carried out for NASA by aerospace companies at the beginning of the 1980s examined all manner of objectives the Station might achieve. As mentioned earlier, it was intended to be a space research laboratory, an astronomical observatory, a transportation node, a servicing facility, an assembly platform, a manufacturing plant, a storage depot and a staging base all rolled into one. Any evaluation of the International Space Station's long-term

Large Hadron Collider This photograph shows two of the Large Hadron Collider magnets before their interconnection. The blue cylinders contain the magnetic yoke and coil of the dipole magnets together with the liquid helium system that is necessary to cool the magnet so that it becomes superconducting. These connections were later welded together so that the beams are contained within the beam pipes. Image: CERN.

value as a project must recognise the chasm that exists between early expectations and final reality, for today the Station performs just one of these original roles extensively – that of a space research laboratory. Its growing use today as a platform for Earth observation was not one of its stated original purposes. At the time of completing this book, in early 2015, it is debatable whether any of the scientific research carried out on the Station as a leading-edge laboratory could be called a major scientific discovery or breakthrough. The Station has produced a vast quantity of valuable results in many fields of science but nothing that ranks in the same class as the discovery of the Higgs boson. The Station's scientists and researchers will be quick to point out that the pursuit of a single major discovery was never the Station's purpose. They are correct. Nevertheless, it would be a welcome justification for the Station's continued existence if such an event came to pass.

Mir Space Station Mir was the first Space Station. It was an outstanding achievement and forerunner of the International Space Station. Consisting of a cluster of seven modules and dating from its first assembly mission in 1986, Mir was begun by the Soviet Union and completed by Russia. It was deorbited in 2001 after the International Space Station had begun assembly on orbit. This photograph of Mir was taken from the departing Orbiter Endeavour on Space Shuttle Mission STS-89 on 9 February 1998. The dark grey docked spacecraft on the left is a Progress cargo vehicle. Another dark grey spacecraft docked on the right is a Soyuz crewed vehicle.

The International Space Station's supreme achievement is its construction. This places it in a class of its own. No other celebrated edifice in the last few hundred years comes close – not even the five projects mentioned earlier. The Station's engineering, transportation and construction were carried out so successfully over many years in the face of grave dangers and great difficulties. Every habitable module, every piece of metal framework, every nut, every bolt, every spacesuit, every computer, every apple, every orange, every chocolate bar, every toothbrush, every family photograph, every molecule of air, every drop of water, absolutely everything had to be launched using powerful rocket engines to supply the brute force needed to fight the Earth's gravitational pull and make a difficult and dangerous delivery flight to a remote place about 350km overhead. The place itself was empty, vast and so hostile that it would bring a quick end to any human exposed to it. The return flight to the ground was just as daunting with the risk of incineration during atmospheric re-entry. Performing just one of these voyages safely was a major challenge but the Station's design called for thirty of them just to deliver the Station's basic building blocks. Against the odds, all arrived on orbit safely and flawlessly where they fitted together correctly and precisely. Beyond these, the Station's operation required many flights to exchange crews and deliver supplies. So many things could have gone so wrong with the Station's assembly. None did. It was an astonishing triumph.

The International Space Station is a supreme engineering achievement but it is also a great architectural one. This book makes a claim. It is that the International Space Station is the first example of great architecture beyond Earth – of space architecture. A definition of space architecture is 'the theory and practice of designing and building inhabited environments in outer space'.[16] Architecture deals with everything built on the surface of our planet. Space architecture deals with everything beyond it. The boundary between the two is where something travelling beyond Earth's gravity begins to float freely in the weightlessness of space, bearing in mind that gravitational pull is always there, however slight, depending on the distance from the massive body or bodies in question. As far as Earth is concerned, that boundary is a stable orbit a few hundred kilometres overhead.

Preceded by the Salyut and Skylab orbital outposts, space architecture has emerged as a new field of architecture with the Soviet-Russian Space Station Mir, which orbited the Earth from 1986 to 2001. Mir was assembled on orbit from a series of building blocks, beginning with a core module in 1986. The Soviets added a second module in 1987 with a docking port for arriving Soviet spacecraft and a third module in 1989 with an airlock to enable spacewalks. Fourth and fifth modules brought up respectively in 1990 and 1995 were devoted to scientific research. The Space Shuttle delivered a sixth module for improved docking in 1995, for by then the Soviets and Americans were cooperating on Mir. A seventh and final module arrived in 1996. Mir vastly improved the pool of knowledge on the effects – physiological and psychological – of long-duration spaceflight on humans and it presently holds the record for the maximum length of a continuous tour of duty by humans in space. Space architecture continues today with the International Space Station. This is space architecture delivered to space and assembled and occupied there. It is the space architecture of reality. There is also the space architecture of near-reality and the space architecture of the imagination. They are three separate realms. Space architecture is a trinity.

The space architecture of the imagination is a vital realm because it abounds with ideas about the future that help, perhaps just as subliminal stimuli, to inspire plans for near-future missions to other parts of the Solar System. The space architecture of science fiction is full of dreams and sometimes nightmares. Notable among cinematic visions were the vast underground machine complex for nuclear-generated power watched over by Dr Morbius in the 1956 movie *Forbidden Planet*;[17] the rough and tough titanium ore mining colony on the Jovian moon Io in the 1981 movie *Outland*;[18] and the huge black and brooding edifice with a face built by advanced humanoids in the 2012 movie *Prometheus*.[19] In a class by itself was *2001: A Space Odyssey*, which marked a turning point in the popular perception of space.[20] It stands apart because it transitioned from the realm of imagination to the realm of near-reality. It featured a space station on Earth orbit, based on potentially feasible technology. Shaped like a giant bicycle wheel, the station rotated to provide its occupants with artificial gravity. It followed in the footsteps of the bicycle wheel design envisioned by Wernher von Braun in the 1950s.

Perhaps stimulated by the visions of von Braun and Arthur C Clarke, scientists and engineers began to study giant space colonies in earnest – several years before NASA began serious work on a Space Station. In the mid-1970s, Princeton University physicist Gerard O'Neill led the development of concepts for huge rotating space colonies at precisely balanced places in space called Lagrangian points, where the gravitational pull of massive bodies such as the Earth and the Moon cancel each other out. Objects placed there exist in a state of gravitational equilibrium. O'Neill called his space colonies 'islands.'[21] A 1975 summer study organised by NASA Ames Research Center at Stanford University in California resulted in a concept for a colony of 10,000 (or more) people. It became known as the Stanford Torus. It was another bicycle wheel design but this time of enormous proportions. Standing apart from it at a safe distance was a circular mirror of equal size. Like von Braun's and Clarke's earlier concepts but colossal in scale, the Stanford Torus wheel had an overall diameter of 1.8km. It rotated at one revolution per minute to provide artificial gravity around its rim that was equivalent to the gravity of Earth.[22] It was protected from cosmic and solar radiation by shielding wrapped around it like a bicycle tyre. The mirror reflected sunlight into the rim's tubular interior, which had a diameter of 130m. The Stanford Torus had agricultural fields amounting to 650,000m^2 to feed its population and another 980,000m^2 of verdant parkland and villages.

Stanford Torus The Stanford Torus was a design concept for a huge space colony on Earth orbit. Named after a NASA-sponsored study of space settlements held at Stanford University in 1975, the torus was 1.8km in diameter and would be home to a population of 10,000. Like von Braun's, Clarke's and Kubrick's earlier visions, the torus rotated to provide artificial gravity inside the circular ring that housed the colony's accommodation levels. The colony would be completely self-contained and independent of Earth. It would grow its own food, replenish its own atmosphere and recycle its own wastes. A giant mirror would reflect sunlight into its interior. Image: Don Davis.

The Stanford Torus and giant space colonies like it may be far in the future but bases on the Moon and Mars are firmly in the realm of near-reality. The Clavius lunar base in *2001: A Space Odyssey* was a massive circular structure sunk into the crater of the same name, with a protective shielding dome that separated into retractable tapered panels, enabling spacecraft to arrive and depart. Though the Clavius base was a highly ambitious concept, it embodied advanced construction technology used on Earth.

Its tapered protective panels were not much different from the huge sliding roofs above sports stadiums. At a more modest scale, NASA and others have studied and proposed outposts and bases on the surfaces of the Moon and Mars for years. The longest time that humans have spent on another body beyond Earth occurred on the 1992 Apollo 17 mission, when the crew spent seventy-five hours on the Moon's surface. Today, plans for bases on the Moon or Mars envision much longer periods of residence. Present space technology is fully capable of delivering and constructing an outpost on the Moon, but Mars will be more challenging because of its greater distance from Earth, its thin atmosphere and its stronger gravitational pull. Mars, however, is a much more interesting place to explore from a scientific viewpoint for it was once more like Earth and offers the tantalising prospect of evidence of life – past or present – deep below the surface.

Outposts or bases built on either the Moon or Mars must be able to withstand the hostile environmental conditions and protect the occupants safely to ensure that they will survive. This is by far the most important task for space architecture. With safety and survivability in mind, there are three major ways to approach the design of surface habitats.[23] To begin with they can be derived from the pressurised modules developed for the International Space Station. These are fully prefabricated and outfitted metal structures generally in the shape of long or short cylinders with conical ends that are sized to make the most of the available volume in a rocket's payload compartment or a spaceplane's payload bay. Once they touch down, they need enough mobility to move across the surface to an assembly point as they may land far apart. There the modules can be plugged together – probably by precursor robots instead of humans – like pieces of a giant toy construction kit. Many early concepts assumed that the Space Shuttle would deliver pressurised metal module building blocks for a Moon or Mars base to Earth orbit first, from where they would continue on the long voyage to their destination propelled by bolted-on engines known as transfer and landing stages. The limiting factor was the cross-sectional size of the Shuttle's payload bay, which restricted the diameter of the modules to 4.2m. With the end of the Space Shuttle era, this scenario no longer applies except perhaps with the potential salvage of some modules from the International Space Station at the end of its life to refit for space exploration missions. New generations of rockets with larger payload compartments will be able to deliver large diameter metal modules to Earth orbit and beyond. The big advantage of prefabricated-metal-module technology is that the engineering of it is fully understood and it has proved itself to be safe in space over several decades.

Next, surface habitats can be made from partly prefabricated materials and structures that are tailored for efficient transportation in spacecraft but require more assembly work at their destination. Leading technology candidates for these are inflatable structures made of advanced laminated fabrics that are flexible enough to fold up into compact cargo form to fit into a rocket or spacecraft payload compartment and then unfold, inflate and deploy on a remote surface. Different shapes and sizes are possible for these structures ranging from spheres to domes, tubes and torus-shaped rings. They can be independent structures or hybrid combinations with pressurised metal modules that might act as entry-exit airlocks or storage chambers.

Automated space missions have successfully incorporated lightweight inflatable structures, for example, to enable descending Mars landers and rovers to survive the shock of surface impact and bounce across the terrain before coming to rest. The problem with them is that, as of the time of writing, there has been no thorough testing of their technology in space. They are not declared to be safe for use in human habitats there. Until inflatable structures prove to be durable and reliable for space habitat applications over the long term – and there are plans to add an experimental inflatable module to the International Space Station for this purpose[24] – engineers cannot be confident of their safe use for space architecture

Mars Base This recent NASA concept for a Mars Base envisages the major surface exploration elements that are likely to form part of the mission. To the far left is an ascent and return stage with a cylindrical habitat module and conical capsule supported on legs above the surface. To the far right is a non-returning habitat also on legs with an inflated module mounted above an entry/exit airlock, a vertical shaft connecting the two and resource and equipment compartments on each side. In the foreground to the right is a surface exploration rover with a chassis and wheel arrangement derived from the Mars rover Curiosity.

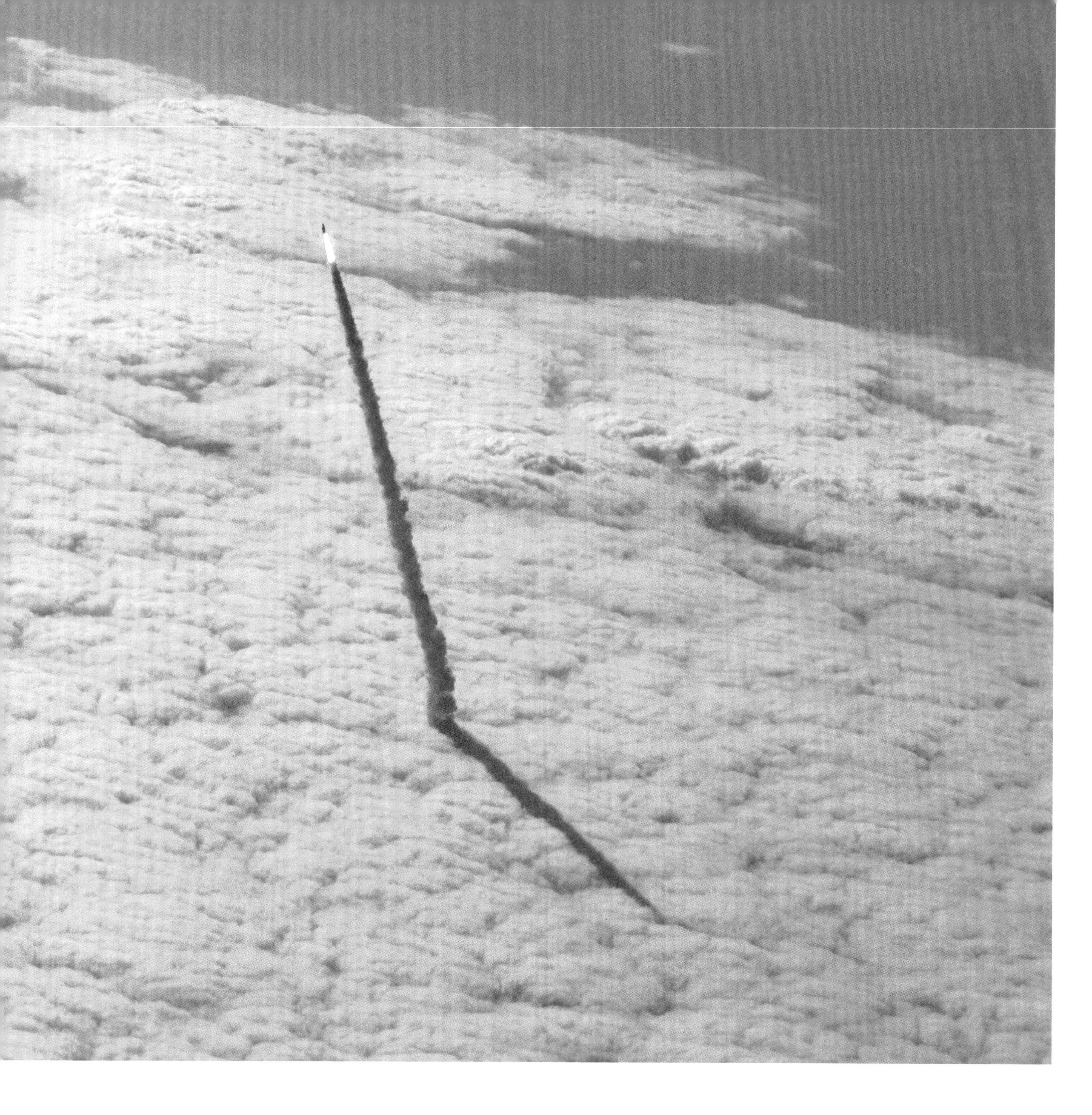

on the surfaces of the Moon or Mars or anywhere else in space where humans might go. It is a technology that is not yet ready for human exploration missions.

Lastly, surface habitats can make use of raw materials that are locally available such as small rocks, pebbles and loose soil of the consistency of sand that are found distributed across the surfaces of the Moon and Mars. This is described in the space field as in-situ resource utilisation. At this point, this approach is purely speculative as no experiments of any kind have been carried out to determine whether it is possible. Testing under local site conditions will be essential to determine the durability, reliability and safety of this technique. The simplest application of local surface materials is for loose mass shielding to protect against solar and cosmic radiation and the impact of micrometeoroids. Habitats can be buried, using the material; it can be directly deposited on overhead canopies or used to fill bag-like containers to make blankets or walls. Wherever possible it should, like the Clavius base, take advantage of local geographical features that already provide some protection such as craters, escarpments, bluffs, ledges, channels or rilles.

We do not know whether local surface materials on the Moon and Mars could be processed into a monolithic form resembling concrete with the ability to withstand the internal pressures of a habitat. It seems highly unlikely. It may be possible, though, to cast or form shells or domes to protect inflatable bladders that line their inner surfaces to maintain internal atmospheric pressure. Yet, the rapidly emerging technologies of 3D printing and additive manufacturing may usher in wholly new space construction technologies using local materials that we cannot foresee at the moment.

Over 2,000 years ago in the reign of the Roman Emperor Caesar Augustus, the architect and engineer Vitruvius wrote a treatise of ten books entitled *De Architectura*. He intended them to be 'how to do it' guidebooks for the design and construction of buildings throughout the Roman Empire. His approach rested on three cardinal design principles that in Latin were *firmitas, utilitas* and *venustas*. Andrea Palladio, the celebrated sixteenth-century Venetian architect, translated them into Italian as *comodità*, *perpetuità* and *bellezza*. In the early seventeenth century, Sir Henry Wotton, an English diplomat and connoisseur of the arts, took Palladio's words in turn and in English they became *commodity*, *firmness* and *delight*.[25] A later scholar described them as *convenience*, *durability* and *beauty*.[26] In today's world of commercial English, *strength*, *value* and *style* may be more appropriate. These three cardinal principles have come down to us through the centuries and continue to be revered today by many architects. The skilled combination and manipulation of the three usually leads to architecture that is good and sometimes great. How do these cardinal principles fare in the space field?

The Station's workhorse – the Space Shuttle
Space Shuttle Mission STS-134 with the Orbiter Endeavour rocketing skywards to the International Space Station after its launch from Kennedy Space Center on 16 May 2011. Commanded by astronaut Mark Kelly, this was the last flight of Endeavour. STS-134 was the penultimate Shuttle mission before the retirement of the Shuttle fleet. It took thirty American and Russian missions to deliver the Station's major building blocks to orbit, most of which were Shuttle missions like this one. This photograph was taken from a Space Shuttle training aircraft.

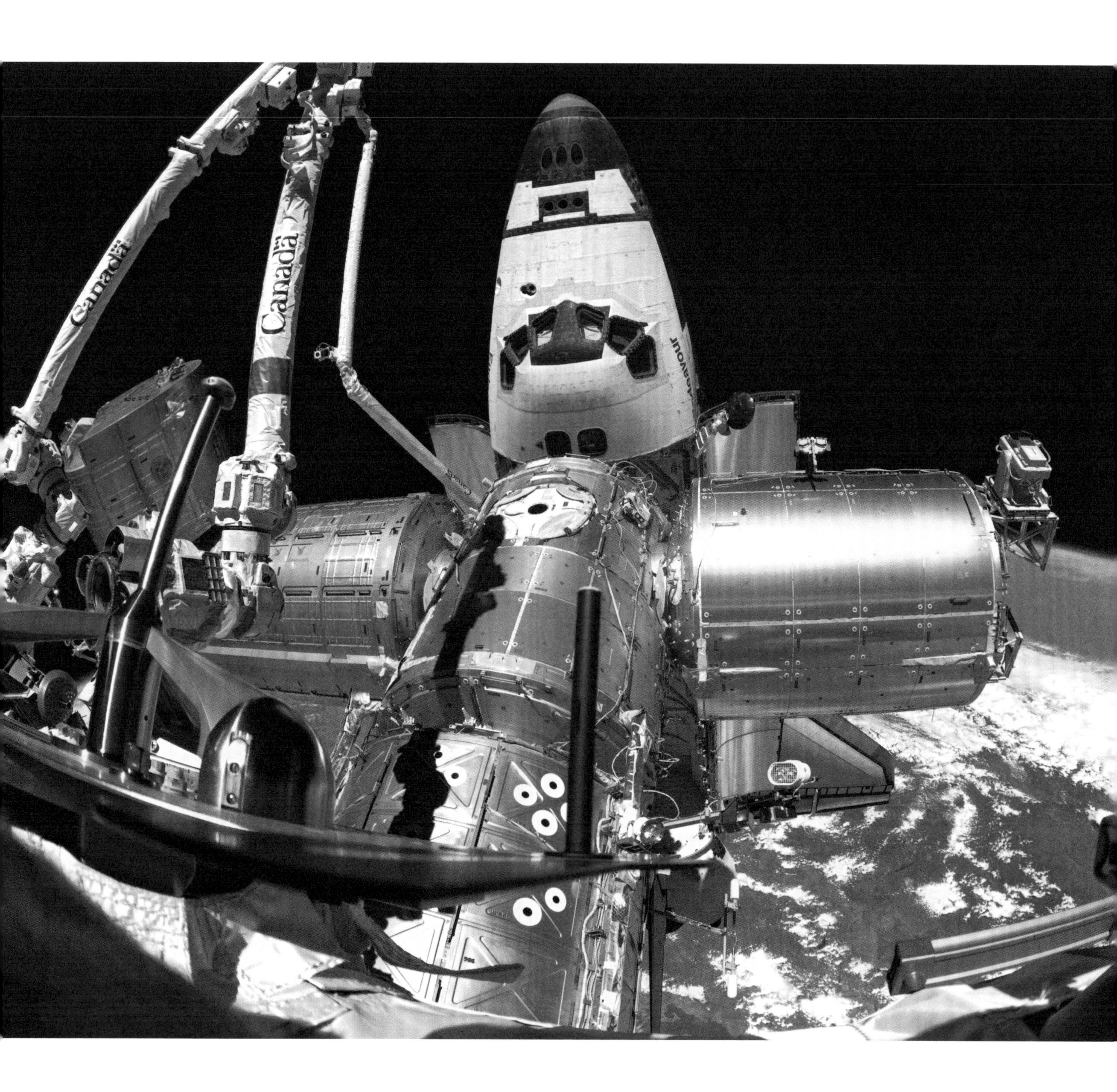
Canada
Canada

In the challenging and hazardous context of space, the role of *strength* is by far the most critical. It ensures human safety and survivability using robust space vehicles and habitats. Second comes *value*, also essential because the purpose of human spaceflight is to extend human knowledge through 'hands-on' exploration and scientific research and, where possible, to derive terrestrial benefits. The third, *style*, though present among the inspired and stimulating designs of illustrators in the space architecture realms of imagination and near-reality, has not existed much in reality. That is to say, it has not existed up until now. Of the two space projects in that realm – the Mir and International Space Stations – the International Space Station is, without doubt, a very stylish design. Astronauts and cosmonauts on arriving and departing spacecraft have taken stunning photographs of it orbiting the Earth. The way it sparkles in the sunlight from a distance gives it the appearance of an enormous jewel. Up close, its intricate composition of precisely fabricated metal parts in a wide array of shapes and sizes takes the machine aesthetic of the Modern Movement in architecture to new heights. If we could visit it and touch it, we would be awed and transfixed. The International Space Station exhibits all the characteristics of *firmitas*, *utilitas*, and *venustas*.

At the beginning, this Introduction stated that the International Space Station is the most ambitious habitat contrived by mankind to support its existence beyond Earth. Later it explained why the Station is the greatest engineering and construction masterpiece of modern times. Now it claims an architectural lineage for the Station that goes all the way back to the design principles of Vitruvius. These arguments are evidence of the International Space Station's greatness. The Station's legacy of peaceful cooperation among nations confirms it. Its border-free spirit of goodwill is perhaps its greatest achievement. Its strong internationalism is an asset that has been used effectively by its advocates when it has flown into political headwinds or budgetary storms. It has become an instrument of foreign policy for countries wishing to bury old hostilities, strengthen friendships and forge new alliances. They have come together on the Station in a close relationship based on mutual trust, for that is the only way to survive and succeed in the hazardous and hostile environment of space. Trust over many difficult years of development and assembly has been the hallmark of the International Space Station programme. It is above the quarrels that surface from time to time in the diplomatic relations of some of its partners. Up on orbit, where astronauts and cosmonauts have intensive and challenging daily work schedules planned down to five-minute slots, there is no time to worry about the squabbling of their governments on the planet below.

Shuttle docked at the International Space Station View of the International Space Station with the docked Shuttle Orbiter Endeavour on Shuttle Mission STS-134 in May 2011. At the centre just in front of Endeavour is the American laboratory module named Destiny. Berthed to it on the port side (left in this view) is Japan's laboratory module Kibo. Berthed to it on the starboard side (right) is the European laboratory module Columbus. Canada's robotic arm named Canadarm2 is at the top left with the Orbiter's robotic arm just behind it.

Solar arrays The International Space Station with all eight solar array pairs in position photographed from the Orbiter Discovery after its departure from the Station on STS-119 in March 2009.

International Space Station with clouds
A crew member on Space Shuttle Mission STS-133 took this photograph from the Orbiter Discovery after undocking with the Station on 7 March 2011. The alternating grey and white panels are thermal radiators. Four pairs of solar arrays are pointing towards the camera viewpoint and appear foreshortened.

Russian and American solar arrays and a radiator This photograph by a Station Expedition 38 crew member shows partly folded-up Russian solar arrays on the left, American solar arrays in the centre and a radiator panel at the top.

Diversity and vision 1979–1983

Precursor laboratories

The Skylab and Spacelab programmes of the 1970s–1980s

President Dwight Eisenhower and the United States Congress established the National Aeronautics and Space Administration (NASA) under the National Aeronautics and Space Act of 1958. It was an act 'to provide for research into the problems of flight within and outside the Earth's atmosphere, and for other purposes'. The formation of NASA was partly a product of the Cold War between America and the Soviet Union in the 1950s and partly the result of the Soviet success with the launch of Sputnik 1 on 4 October 1957. Sputnik was the world's first satellite and its success caught America unawares. It had a sensational effect on national opinion and created a nagging feeling in the public's mind that Soviet technology was forging ahead of its American counterpart. The surprise and reaction forced the White House to introduce measures to boost spending on aerospace research and development and launch a new space agency to manage new spaceflight initiatives and programmes.[1]

NASA began to think about Space Stations soon after its creation as an agency. By the early 1960s, NASA research centres and the aerospace industry had come up with several ideas for them, beginning with a modified version of the Apollo Command and Service Module spacecraft with an attached laboratory module as a minimal option. In 1962, while development

The Skylab space outpost This view of the Skylab space outpost was taken from the departing Skylab Command/Service Module during the Skylab 2's final fly-around inspection. The single solar panel is quite evident as well as the parasol solar shield, rigged to replace the missing micrometeoroid shield. Both the second solar panel and the micrometeoroid shield were torn away during a mishap in the original Skylab 1 lift-off and orbital insertion.

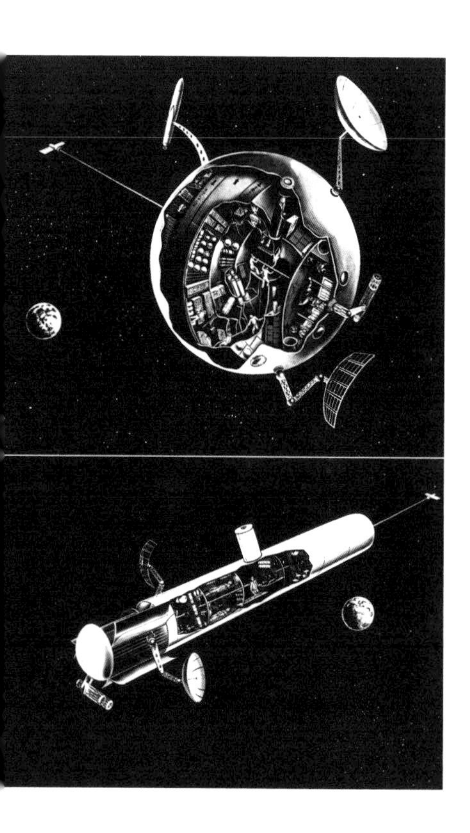

Early Space Station ideas Top: A NASA concept for a spherical Space Station dating from 1959, shortly after NASA was established. The pressurised sphere contained a mixture of different compartments outfitted as living quarters or laboratories for a large crew. Bottom: Also dating from 1959 was this NASA concept for a small space outpost shaped like a cigar. Inside it had stacked sleeping compartments for a small crew and a wardroom. A docking hatch and airlock projected out from the cylindrical side.

of the Apollo programme was picking up speed, following President John F Kennedy's announcement that the Moon was to be America's next space goal, Langley Research Center in Virginia – one of NASA's leading field centres – began to work on ideas for large Space Stations. One Langley concept was a hexagonal wheel that rotated to induce artificial gravity with spokes that radiated outwards from a central command module. It had about 1,400m^3 of living and working volume and sufficient accommodation for a crew of thirty-six. Another Langley idea was called a Manned Orbiting Research Laboratory comprising a cylindrical module, 4 metres in diameter and 7 metres long, with its own life support systems for a crew of four. Beyond NASA, the Douglas Aircraft Company and Lockheed Corporation came up with designs for large rotating and non-rotating Space Stations for crews of twenty-four persons.

By 1963, the main purpose of a Space Station had become clear. It was to find out whether humans could live and work efficiently and, from the biomedical viewpoint, safely in weightlessness for long periods and then return to readjust to the gravitational environment of Earth. At that time, the long-term effects of weightlessness in space were unknown as there had been no way to examine and evaluate them. There had never been any long-term missions to generate the desired medical knowledge. If there was ever going to be a Space Station on Earth orbit to provide a base for 'hands-on' research lasting months or years, knowledge of the reaction of the human body to weightlessness was vital. An early Space Station would be a first generation space laboratory that would study this critical biomedical issue as well as carry out research in areas of physics, astronomy and biology.

In 1964, attention shifted to an intriguing Douglas Aircraft Company idea of salvaging a spent third-stage shell of a Saturn V rocket (the Saturn S-IVB stage, normally empty of propellant at the end of the Saturn V's ascent) and fitting it out in space as a laboratory for a crew of two with appropriate additions and modifications for habitability. It became known as the 'wet' workshop design to signify that it had once contained liquid propellant. A 'dry' workshop design, with the rocket stage fully outfitted on the ground before launch, soon joined it as an alternative. Both versions would require an airlock and a docking module called a Multiple Docking Adapter to enable arriving Apollo spacecraft to dock and their crews to enter the pressurised cylindrical workshop. The Multiple Docking Adapter would also provide mountings for external experiments and telescopes as well as docking for resupply spacecraft. NASA thought that a first mission of twenty-eight days would require just two launches – the Apollo spacecraft with its crew and then the cylindrical workshop, airlock and Multiple Docking Adapter, all pre-attached to each other. The second mission would last fifty-six days and require two more launches – another Apollo spacecraft with the second crew combined with a resupply module and then an external solar telescope. Prospects for further missions were unknown. In 1966, President Lyndon Johnson approved the project and NASA began to receive budget allocations to initiate development.[2]

The internal debate about whether the workshop should be 'wet' or 'dry' intensified at NASA while America's growing military commitments in Vietnam and their escalating costs began to undermine Congressional support for the project. In 1968 Congress slashed funding but the development pace, though much slower, continued and in July 1969 NASA chose the 'dry' workshop as the preferred version, just two days after the historic lunar landing of Apollo 11. NASA gave the project the name 'Skylab' in February 1970.

The agency had developed a formal project review procedure during the Apollo programme to track its progress from the drawing board to the launch site. It was an important management tool and NASA also applied it to Skylab. The procedure involved seven review milestones of which the first three occurred in the design and engineering phase. In order, these were a Preliminary Requirements Review to assess different concepts and the concept chosen to meet mission objectives, a Preliminary Design Review to evaluate the basic design early in its detailed engineering phase and a Critical Design Review to examine technical specifications and drawings near the end of the detailed engineering phase. After the project moved off the drawing board into fabrication, the next two reviews were a Configuration Inspection to compare manufactured hardware with specifications and drawings followed by testing, and a Certification of Flightworthiness to confirm the qualification and documentation of the completed hardware. The last two reviews dealt with the entire assembled spacecraft and its operation. These were a Design Certification Review held a few months before launch to verify the spacecraft's flightworthiness, safety and pre-launch operations at the launch site, and a Flight Readiness Review held a few weeks before launch to check that the mission was ready to go.

Appendix 1 shows a typical programme flow sequence for a major space project and mission with an evolved and expanded set of review milestones drawn from later NASA project management documents. The reviews are placed throughout the project phases to ensure that completed work from one phase is reviewed and approved before moving on to the next.

Skylab's design phase reached a climax in mid-1970 with a Critical Design Review of all the major hardware items that lasted several weeks.[3] Fabrication work proceeded among the contractors, and testing of all the hardware elements took place between 1971 and 1972. In September 1972, NASA officially accepted the completed Skylab laboratory and it was transported by ship to prepare for launch at Cape Kennedy in Florida.[4]

By the time Skylab was ready for launch in 1973, it had four major research objectives. The first, and by far the most important, was to investigate the effects of weightlessness on the human body. A NASA

Early Space Station ideas Top: This inflatable design by NASA shaped like a torus made it as far as the prototype stage, around 1961. The prototype was about 10m in diameter. Deflated and stowed into a compact payload for launch, it would inflate and deploy itself once in space. Bottom: NASA called this 1977 Space Station design the 'spider' because its solar arrays would unwind from the exterior of a spent Space Shuttle propellant tank like a spider's web. The tank's interior would become a habitat and laboratory.

guidebook published before the completion of Skylab's missions pondered what the various biomedical results would show. 'Will the human body, which has been accommodated to the gravity field of the Earth ever since it has existed, readily adapt to life under weightlessness? ... It is believed that gravity may possibly be of influence in a number of biological and medical processes, such as the germination of seeds, the cleavage of cells, the growth of certain tissues, the regulation of metabolic processes, the adaptation of acceleration sensors, the control of cardiovascular functions, and perhaps the functioning of time rhythms.' Skylab would enable observation of the Earth's resources, covering agricultural productivity, timber harvesting, oil and mineral exploitation, water supplies, and urban and rural growth. It would also advance scientific knowledge of the Sun and the stars, particularly the effects on the upper atmosphere and Earth's weather of high-energy particles flowing outwards from solar flares. Lastly, it would study the processing of materials under weightless conditions. Promising materials included improved semiconductor crystals, superconductors and new types of metal alloy.[5]

Skylab Skylab seen from the Apollo command module after undocking of the first mission on 22 June 1973. Skylab was America's first crewed outpost on Earth orbit. Skylab began as an idea for converting an empty upper stage of a Saturn V rocket into a habitable laboratory and workshop. Crews of three astronauts each occupied it on three consecutive missions of twenty-eight, fifty-nine and eighty-four days respectively. Skylab gave American astronauts their first experiences of long-duration spaceflight in weightlessness.

Skylab's final design comprised a cluster of elements centred on the large Orbital Workshop that would provide the main living and working accommodation for the crews. Its pressurised shell was the third stage of the Saturn V rocket that the Douglas Aircraft Company had proposed back in 1964. Its roomy interior was 14.6 metres long and 6.7 metres in diameter and it had a pressurised volume of 275m^3, about the same as a two-bedroom apartment. In designing Skylab's interior, NASA and its consultants and contractors made a conscious effort to improve its habitability conditions beyond the spartan interiors of the earlier Apollo spacecraft. The Orbital Workshop had two accommodation decks separated by a perforated floor. The aft deck housed a wardroom with a table for preparing and eating food, a window, a dormitory area with three bedroom cubicles that provided full privacy for their occupants, a lavatory and personal hygiene cubicle, and an experiment workstation. The forward deck was a laboratory for large experiments with scientific airlocks for external viewing and exposure. Food, water and clothing storage containers and utility systems occupied part of the volume on each deck. Outside, two solar arrays projected outwards to port (-Y) and starboard (+Y). At the aft (-X) end of the Workshop were a waste disposal chamber, attitude control thrusters and a thermal radiator.

Around the Workshop's forward (+X) end was the Instrument Unit, a hollow ring of the same diameter as the cylinder, which contained the Saturn V rocket's guidance and navigation systems to control its launch and ascent. Attached to the Workshop's forward end was the Airlock Module, which stepped the Workshop's large diameter cylinder down to the small diameter cylinder of the Multiple Docking Adapter. The Airlock Module had three functions. It was the main structural link between the forward and aft parts of Skylab: it contained a port through which a crew member could carry out a spacewalk; and it housed the utility systems control centre and pressurised gases for Skylab's internal atmosphere. Next forward was the Multiple Docking Adapter, a small pressurised cylinder equipped with an in-line docking port for arriving Apollo spacecraft that would approach and rendezvous with Skylab from forward of its flight path. There was also a back-up docking port to the side for emergencies. Pointing upwards from Skylab and mounted directly on top of the Multiple Docking Adapter was the Apollo Telescope Mount for solar observation. Four solar arrays projected from the telescope in a cruciform arrangement. The Apollo Telescope Mount stowed along Skylab's centreline during launch and ascent and then rotated 90° into position once safely on orbit.[6] Overall, it was a highly efficient and compact design.

NASA successfully launched Skylab on 14 May 1973. It reached its operating altitude of 435km safely, but not all had gone smoothly. A micrometeoroid shield had torn away from the Orbital Workshop during launch and ascent. It took with it one of the two main solar arrays and jammed the other one, preventing its proper deployment. Skylab lost its electrical power and protection from the heat of the Sun. NASA launched the first Skylab crew on 25 May 1973. After a difficult docking, the crew's first job was to force a hastily made parasol through the scientific airlock and then unfold it to shade the exposed portion of the cylinder where the micrometeoroid shield had torn away. The crew followed this with a hazardous but successful spacewalk to free the jammed solar array. The first crew completed their twenty-eight day mission safely and returned to the ground. NASA launched the second crew on its fifty-nine day mission (increased from fifty-six days) on 28 July 1973 and the third crew on its eighty-four day mission on 16 November 1973 (a third mission had been added during Skylab's development).[7] The second and third missions were as successful as the first.

NASA had originally designed Skylab for a nine-year lifespan. After the third mission ended in February 1974 its future was uncertain. There were no plans for further missions. It was left on an orbit that NASA thought would be durable for several years with its atmosphere vented, its flight attitude stabilised and its systems shut down. However it had major problems. A gyroscope had failed, the cooling system was erratic and the power supply was running low. In the late 1970s, NASA's planners hoped that the Space Shuttle – then well into advanced development – would be able to deliver a new propulsion module on an early mission to keep Skylab safely aloft while they decided its future. The Space Shuttle comprised four elements – a spaceplane called an Orbiter with a crew and a spacious payload bay, a large liquid propellant External Tank to power the Orbiter to orbit and a pair of Solid Rocket Boosters for extra punch during early ascent. The Space Shuttle would deliver the new propulsion module to

Skylab A view of the Skylab Workshop interior taken by astronaut William Pogue on the third Skylab mission in January 1974. He took the photograph looking through a perforated aluminium partition towards the crew's living quarters. The partition separated the Workshop into two zones - one for living and one for working. Constructed from triangular isogrid panels, the partition provided attachment and storage for the spacesuits, equipment and experiments visible in this view. It also functioned as a foot restraint surface for the crew as it floated around in weightlessness.

Skylab in the Orbiter's payload bay. Despite NASA's preservation efforts, delays in the start of Shuttle missions and worries about Skylab's degrading orbit led to NASA's decision in December 1978 to terminate it.[8] America's first Space Station disintegrated in a fiery re-entry on 11 July 1979 and ended up as a shower of debris over the Indian Ocean and Australia.

The three crews had occupied Skylab for a total of 171 days. NASA had established a long-term human presence on Earth orbit for the first time. Skylab was the site of nearly 300 biomedical and technical experiments. The biomedical experiments encompassed a wide range of specialised fields including mineral balance, bioassay of body fluids, specimen mass measurement, bone mineral measurement, lower body negative pressure, vectorcardiology, haematology/immunology, human vestibular function, sleep monitoring, time and motion study, metabolic activity and body mass measurement. NASA added further tests during the course of the three missions.[9] The biomedical results from the Skylab missions were quite positive about human adaptability to, and tolerance of, weightlessness. Humans were evidently able to adapt and function effectively in a weightless environment for extended periods (for the durations of the Skylab missions). Daily in-flight personal exercise, a good diet and well-planned sleep, work and recreation periods were essential.

Skylab An overhead view of Skylab photographed from the Command Module during its final fly-around in February 1974 before returning home on the last of the three Skylab missions. The large module in the foreground is the Workshop. Its projecting solar array on the left is missing because it had torn away during launch and ascent. A gold-coloured metal parasol covers the exposed portion of the Workshop's exterior that lost its protective shield during launch. Skylab later remained unoccupied on orbit until its final destruction during atmospheric re-entry in July 1979.

There appeared to be no adverse physiological changes that would preclude longer duration missions, though the mechanisms responsible for such changes needed further investigation. Remedial or preventive measures would be considered for mission durations longer than nine to twelve months, particularly with bone demineralisation. Finally, there was a need for further biomedical observations for uninterrupted six-month periods in weightlessness.[10] The biomedical results came as good news for NASA's early planning for a future Space Station. They meant that the agency could shift to simpler Station designs that did not need artificial gravity to maintain the health of crews.

While the third Skylab crew was living and working on Earth orbit in January 1974, another space laboratory initiative called Spacelab was deep in the middle of engineering reviews in Europe. It was Europe's contribution to the American Space Shuttle programme. In January 1972, President Richard Nixon had given the go-ahead to the Space Shuttle, the world's first reusable spaceplane system. NASA had initiated Phase B design studies for both a Space Shuttle and a Space Station in 1970 (Phase B is the final design study phase that precedes project approval and contracts for hardware development. See Appendix 1.) Financial belt-tightening forced the agency to choose between them and it had opted for the Space Shuttle.[11]

In his statement announcing the Space Shuttle, President Nixon noted that it would provide routine access to space, reduce launch costs and increase opportunities for international participation in space missions.

NASA had been considering the idea of a flying laboratory for a while. The agency had called it the 'sortie can'. Impressed with the conversion of a Convair 990 aircraft into an airborne laboratory by Ames Research Center – another NASA field centre – and its success as a flying astronomical and Earth observation platform, NASA thought there was a promising role for something similar on the forthcoming Shuttle, then due to enter service in around 1979. At that stage in its development, the Shuttle spaceplane design already had a large payload bay running the length of the fuselage with doors that opened up to expose the payloads to space. There was plenty of room to accommodate a laboratory.

NASA and its counterpart in Europe, the European Space Research Organisation (later to become the European Space Agency), had been holding talks on the prospects of Europe participating in a new American space initiative. In June 1972, a NASA technical team went to Europe and proposed that the Europeans should develop a laboratory for the Shuttle. They made it clear that it would need a firm decision by August 1973. Were that not forthcoming, NASA would press on with the laboratory on its own.[12] European space ministers were aware of the long-term technological benefits of teaming up with NASA on the Space Shuttle. They were keen to proceed and in February 1973 they agreed to go ahead with Spacelab. They followed this with a formal intergovernmental agreement in September. Participating nations included Belgium, Denmark, the Federal Republic of Germany, France, Italy, the Netherlands, Spain, Switzerland, the United Kingdom and later Austria. Spacelab was a truly international endeavour. Europe would design, develop, manufacture and deliver to NASA both engineering prototype and flight hardware versions of Spacelab together with ground support equipment and spare parts. NASA would provide management and technical support and details of the tunnel link between the Shuttle Orbiter and forward end of the Spacelab module.[13]

Phase B studies of Spacelab by competing contractors in Europe throughout 1973 led in stages to the selection of a German-led team. This was because Germany was the main supporter of the project and would provide most of its financing; and also because a German team had come up with a highly flexible design that was perfectly tailored to the Shuttle Orbiter payload bays. In July, the German aerospace company ERNO proposed a Spacelab system that comprised a cylindrical laboratory for 'hands-on' experiments in a shirtsleeve environment with unpressurised U-shaped pallets for experiments that required exposure to space. It was a modular system. The laboratory came with two cylindrical segments and two conical endcaps. It would be able to fly in a short version with one segment or a long version with two segments. The forward endcap had a hatch that led to a crew access tunnel that, in turn, led to the Shuttle Orbiter's decks. The pressurised laboratory would therefore fly towards the forward end of the Shuttle Orbiter's payload bay. Behind the pressurised

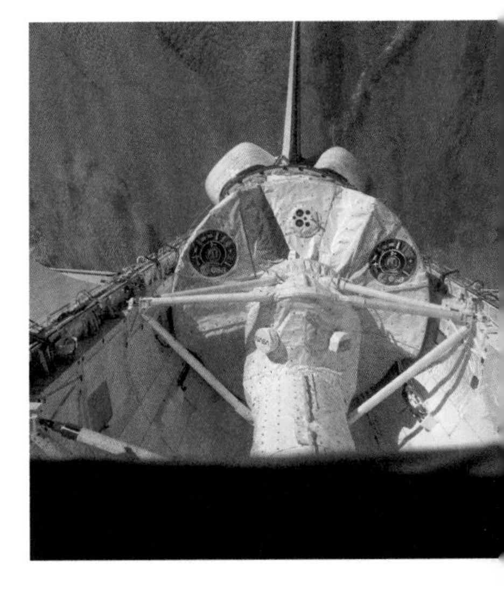

Spacelab View of the European Spacelab Module in the payload bay of the Shuttle Orbiter Columbia during Shuttle Mission STS-9 in November 1983. It was the first Spacelab mission on the Shuttle. The photograph was taken through the aft window of Columbia's flight deck. In the foreground is a tunnel through which the crew moved between Columbia and Spacelab during the mission. The tunnel is firmly attached to the payload bay sills on each side by three struts. The mission patches are those of the European Space Agency.

laboratory were the pallets. There would be up to five pallet segments attached to each other, end to end.[14]

The Spacelab system was a kit of parts designed for assembly into a specific combination before a mission, then disassembly after it, and then reassembly into a different combination for the next mission. It would have considerable versatility to fly a wide range of internal and external experiments. In June 1974, a team led by VFW-Fokker/ERNO won the development contract. The team included ten co-contractors and thirty-six subcontractors. About 2,000 people were involved in the project at the height of its development phase. McDonnell Douglas (which by then had incorporated the Douglas Aircraft Company, designer of Skylab) brought forty-two consultants over from America to help with Spacelab's project management. McDonnell Douglas had much experience in the design of pressurised space modules and the company's involvement was to prove very valuable to the Europeans.[15]

Spacelab's development cycle continued for the remainder of the 1970s and it made its maiden flight on the Space Shuttle in November 1983. Spacelab evolved to become an extraordinarily successful scientific research programme, making a total of twenty-four flights on the Space Shuttle between 1983 and 1998,[16] the year that the International Space Station eventually began construction on orbit. For Europe, Spacelab was a major technological leap into the world of human spaceflight. It was also the most complex project management challenge that European space companies had dealt with up to that time. Spacelab would give Europe the experience it would need to build its own pressurised laboratory module for a future Space Station.

Spacelab European astronaut Ulrich Walter demonstrates weightlessness in the Spacelab Module during the D-2 Mission on the Shuttle Orbiter Columbia in April 1993. The Spacelab Module contained a set of racks running down each side of a central corridor. At the base of the racks was a floor with life support system installations beneath. At the top, the racks angled inwards to follow the hull's curve. Mounted on the ceiling were two rows of light fittings. Grabrails projected into the corridor from the rack faces to assist astronaut mobility. Photo: DLR.

First ideas
Study for a Space Operations Center, 1979

By any measure, the earlier Skylab missions had been a huge success story. They had produced a body of research information that kept space scientists, doctors and engineers busy for years and amply demonstrated America's ability to maintain crews on Earth orbit for reasonably long durations. But it was time to move on. As Skylab fell to Earth in 1979 a study group at NASA's Johnson Space Center was at work on ideas for Skylab's successor. The Space Shuttle system was approaching its maiden flight and NASA needed a new objective for the forthcoming fleet's activities and a new vision for a human presence in space. Skylab had arrived on orbit on a single Saturn V flight with its three main elements – the large diameter Workshop, the Airlock Module and the Multiple Docking Adapter – firmly locked together as a single entity and, damaged shielding and solar arrays notwithstanding, ready to switch on and occupy. Skylab was a turnkey space laboratory. It would be a very different story with its successor. Expectations were that the Orbiters would deliver a series of prefabricated elements to a space destination of some kind. There, they would be put together in a carefully planned sequence, like large components arriving at a construction site on flatbed trucks, ready for bolting together to form a building. The high reusability of the Space Shuttle system would result in lower transportation costs compared with those for expendable rockets. NASA would develop a kit of space-ready parts with their shape, size and mass tailored to the delivery ability of the Shuttle Orbiter payload bays. Whatever form it was destined to take, Skylab's successor would herald a new era of expanded human activity in space.

NASA's 1970 decision to develop the Space Shuttle prompted the beginning of Space Station design studies that revolved around the Orbiters as delivery vehicles. A concept that year for an early Space Shuttle station showed a dense cluster of about fifteen modules pointing outwards on four axes from a long connecting passageway spine, with solar arrays mounted at one end and an Orbiter docking port at the other. NASA was also sensitive to the success of the Soviet manned space programme of six Salyut Space Station missions that had begun in 1971. NASA saw how crewed and uncrewed spacecraft periodically visited the Salyut outposts on orbit, how a Soyuz was always docked at a Salyut for emergency crew return, how the crew went on spacewalks to carry out exterior maintenance and how they conducted experiments inside the modules. NASA noted Soviet Premier Leonid Brezhnev's comment in 1978 that: 'permanently manned Space Stations with interchangeable crews will be mankind's pathway into the universe'.[17]

In November 1979 as the Space Shuttle programme neared operational readiness, work on a new Space Station began to pick up pace. The Johnson Space Center study group conceived a concept for a Shuttle-serviced, permanently manned facility in low Earth orbit. They called it a Space Operations Center and made clear that it was a tentative goal only. At that time, NASA had no plans to implement anything and the government had allocated no funds to do so. There were four different mission options for the new vision: a Shuttle-serviced, permanently crewed facility on low Earth orbit; visiting crewed missions to geosynchronous orbit; a permanent crewed facility on geosynchronous orbit; and visiting crewed missions to the Moon with or without a base there. It was ambitious. The study report noted that the last two options would bypass important incremental steps with technical and operational challenges that might exceed available funding. The first two options seemed more realistic. Together with the options were four main objectives: the construction, commissioning and operation of a large, complex Space Station; the assembly, launch, rendezvous and servicing of visiting spacecraft; the servicing of satellites in locally accessible orbits; and more development to reduce dependence on Earth for control and resupply.[18]

Spacelab View of the Spacelab Module and pallets after loading into the Shuttle Orbiter payload bay. The module and pallet system was extremely versatile. It enabled different module and pallet combinations to fly on different missions. In this view, the Spacelab Module made up from two cylindrical segments is to the right towards the front of the payload bay. To the left, the U-shaped pallet contains a cluster of experiments and equipment. Several early Space Station studies followed the Spacelab approach to pressurised module design.

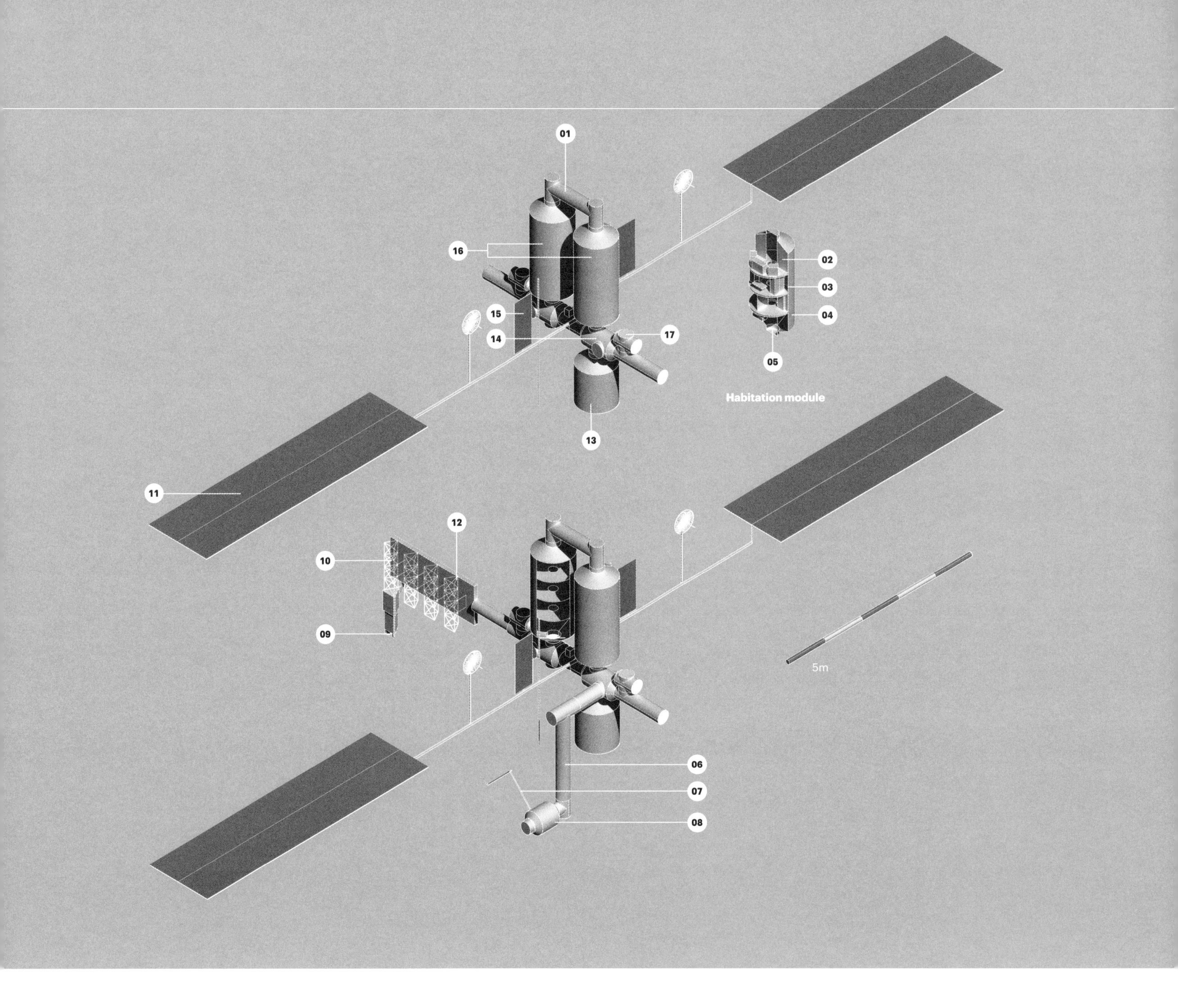

01	Connecting tunnel	**06**	Access tunnel	**12**	Work platform
02	Control centre deck	**07**	Robotic arm	**13**	Logistics module
03	Health maintenance & personal hygiene deck	**08**	Assembly & construction control module	**14**	Service module & tunnel (2)
04	Wardroom & exercise deck	**09**	Beam builder machine	**15**	Radiator
05	Private crew compartments	**10**	Fabricated beam sections	**16**	Habitation modules (2)
		11	Solar array	**17**	Airlock (2)

The study group proposed two versions of the Space Operations Center design – an initial phase with a crew of four and an extended phase with more elements attached to increase the Station's capabilities. The initial phase comprised five separate building blocks launched as payloads in the following order: twin service nodes with a passageway; two habitation modules; a supplies module; and an interconnecting tunnel. The service nodes and passageway together formed a long service tube aligned parallel to the circular disc of the orbital plane. The Space Operations Center would fly Earth-oriented, which is to say that its zenith-nadir (-Z/+Z) axis would always point to Earth's centre throughout its orbit. Orbiters would approach and dock at an airlock at the forward end. The aft end had a back-up airlock. The tube linked the modules and incorporated an entry/exit airlock, eight module berthing ports, two spacecraft docking ports and a service systems artery. There were two long booms extending outwards on the port and starboard sides that supported solar arrays, radiators, antennae and attitude control thrusters. As the first launch element alone on orbit, the service tube would have to function as an independent spacecraft controlled from the ground before the modules arrived. Berthed side by side to the tube's nodes, the two habitation modules contained the crew's living and working quarters. They pointed to zenith and a tunnel joined them at the top to provide the crew with an exit route from one to another in an emergency. The supplies module was a store that housed consumables, spare parts and other equipment. It pointed to nadir. Visiting Orbiters would periodically exchange it for another with fresh supplies and return the used one to the ground for replenishment. After docking, an Orbiter would use its own robotic arm for payload removal and berthing.

Each habitation module had conical endcaps and contained four accommodation compartments separated by three intermediate bulkheads perpendicular to the module longitudinal axes. The vertical alignment of the modules on the Space Operations Center on orbit would perceptually transform these compartments into floor levels and the bulkheads into floor decks if the crew required a familiar architectural reference in weightlessness. The first module comprised a command and control centre, a medical check-up laboratory and personal hygiene facility (bathroom), personal quarters (bedrooms), a communal wardroom (lounge/dining/kitchen) and a gymnasium. Circular holes through the centre of the floors provided circulation passages through the modules. The second module was identical to the first except for a scientific laboratory, which replaced the gymnasium. The two modules could together provide enough room for an eight-person crew though evidently with no attempt to combine communal facilities such as the twin wardrooms for better volumetric use. The total internal pressurised volume was 623m^3 or 78m^3 per crew member for an eight-person crew, but internal facilities and equipment occupied much of it.

1979 NASA Space Operations Center study
As the Space Shuttle neared its inaugural flight, NASA began to focus on Space Station designs tailored to the use of the Space Shuttle Orbiters as delivery vehicles. In 1979, Johnson Space Center carried out this exploratory study of a Space Operations Center with an emphasis on space construction, spacecraft assembly and satellite servicing. Above, an initial phase has twin habitation modules for a crew of four. Below, an extended phase adds a construction platform and operations control module. Image: author.

NASA noted that the interior volume was 'very small relatively to current US living standards' but hoped that the arrangement of internal volumes and equipment would provide visual interest so that 'the space will appear to be much larger than it actually is'. Interior design would include 'expert attention to all details affecting habitability, including surface materials, upholstery, colours, textures, lighting, reflectivity and transmission of sound as well as the basic architectural arrangement and subsystems'. Despite this concern about interior comfort there was no evidence of a window in the early illustrations, a feature that the Skylab crews had found so valuable. In any case, the modules pointed upwards and offered no Earth observation potential. Each module had its own environmental control and life support system that lasted ninety days before the need for replenishment by a visiting Orbiter.

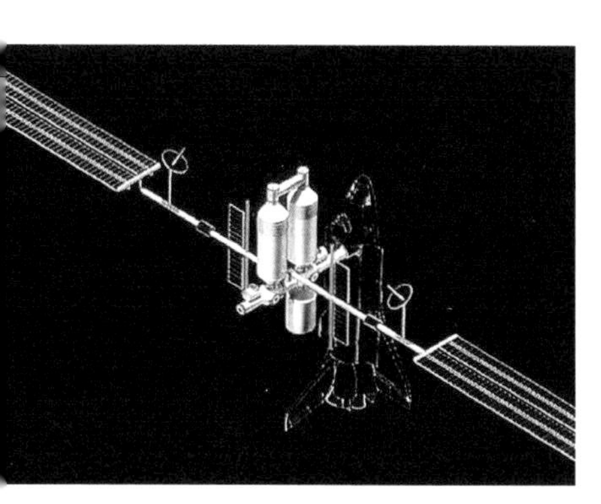

1979 NASA Space Operations Center Study Artist impression of the initial phase of NASA's 1979 Space Operations Center design. A Shuttle Orbiter is docked at one end of the long service tunnel that linked the modules together. Solar arrays extended outwards on long booms to port and starboard of the Center's centreline.

The system used recycling in order to reduce the need for resupplied consumables. It regenerated all drinking and washing water and atmospheric oxygen and treated its atmospheric carbon dioxide by the Sabatier method (carbon dioxide is reacted with hydrogen to produce water and methane – the water is recycled and the methane is vented into space). The major essential consumable was food. Each module had its own thermal control system using coolant loops to collect internal heat and feed it via a heat exchanger to external radiators that discarded it into space. This approach avoided the need to run coolant loops across module interconnections. The long booms supported radiators that served the service modules that were too slender for external mounting. The two large solar arrays on the long booms would together produce gross electrical power of nearly 100kW feeding a net power level of 35kW to the Center's busbars.

The extended design for the Space Operations Center added more features to expand its capabilities. These were a construction platform and a flight support facility. The purpose of the construction platform was to fabricate large space systems. This involved the fabrication and assembly of large structures, the installation and checkout of subsystems, and the launch of the completed system from the Center to its operational orbit. The main platform parts were a construction deck, a beam builder and holding jigs. The construction deck was a long rectilinear frame extending out from the aft end of the service tunnel and aligned parallel with the orbital plane. Attached to the deck's face were jigs to hold the fabricated pieces and their fabrication tool. The tool was a mechanical beam builder that could form lightweight triangular truss beams out of flat stock to any reasonable length. NASA would use the triangular beams to build different orbiting structures such as large solar arrays, communications platforms and antennae. The study was vague about how this would be done. It did not explain the workings of the production arrangement despite the recent demonstration of a full-scale beam builder prototype by a NASA contractor.

The construction platform was at the opposite end to the Orbiter's main docking port and there was no indication of how bulky construction payloads would be moved along the Center's length from the delivery point to the deck. Complementing the construction platform was a flight support facility. Its purpose was threefold: to assemble, check out and launch multi-stage spacecraft; to do the same for a space tug (orbital transfer vehicle) with fuelling; and to recover and service orbiting satellites. The space tug was a reusable, crewed vehicle that would carry out sortie missions to geosynchronous orbit and stay docked at the Center when not in use. The main parts of the flight support facility were a 'strongback' pressurised passageway, a 'cab' module and a robotic arm. A crew member would enter the cab module from the passageway to work the robotic arm, like a crane operator on a building site. The study suggested in outline how the robotic arm could manipulate the rockets and tug but there was little explanation.

The study group estimated that the Space Operations Center would cost $2.1 billion, excluding the costs of fabrication materials and six Shuttle launches. The construction platform and flight support facility would cost a further $600 million and need four more Shuttle launches. This was perhaps optimistic given the significant size of these elements illustrated in the study. Neither the initial nor extended designs appeared to include a pressurised laboratory module for research in a shirtsleeve environment. This was odd considering that the European Spacelab laboratory was shortly to become an important and regular payload on Shuttle missions. The first two Shuttle launches would deliver the twin nodes and the passageway. These amounted to pressurised service tunnels only that offered no habitation or laboratory space whatsoever and would be useless until the modules arrived. There was no attempt to achieve an operationally useful facility with the first missions.

The Space Operations Center would take ten years to complete. The Johnson Space Center group concluded by proposing the Center as a likely candidate for the next goal after the Space Shuttle because of its 'desirable combination of need, usefulness, scope, evolutionary development and funding requirements'.

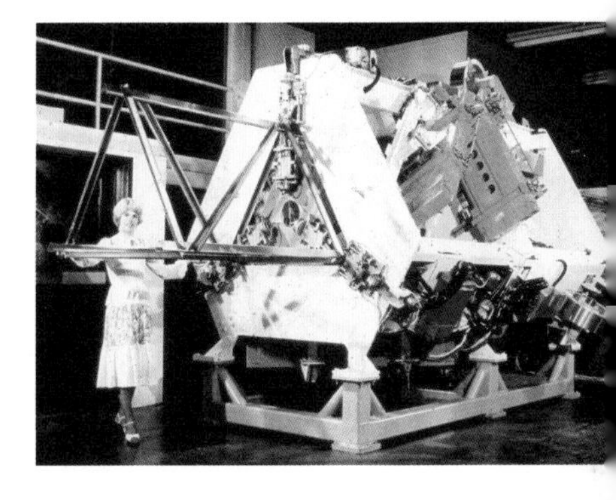

Beam Building Machine NASA's 1979 Space Operations Center design called for a mechanical beam builder that would fabricate long triangular trusses at the Center for use in the assembly of larger structures. NASA got as far as producing this full-scale prototype. The machine roll-formed coiled aluminium from spools (bottom and top right of white frame) into long V-shaped struts and then automatically welded on short braces stored in canisters (yellow).

Opposite viewpoints
Two parallel studies for a Space Operations Center and an Evolutionary Space Platform, 1981-1982

The 1979 study by Johnson Space Center began to focus attention on mission models, systems and technology requirements driven by the Space Shuttle fleet that was shortly to enter service. It was now time to begin to broaden the enquiry and engage the aerospace industry in some early studies. (Known as pre-Phase A studies, the feasibility stage of an aerospace

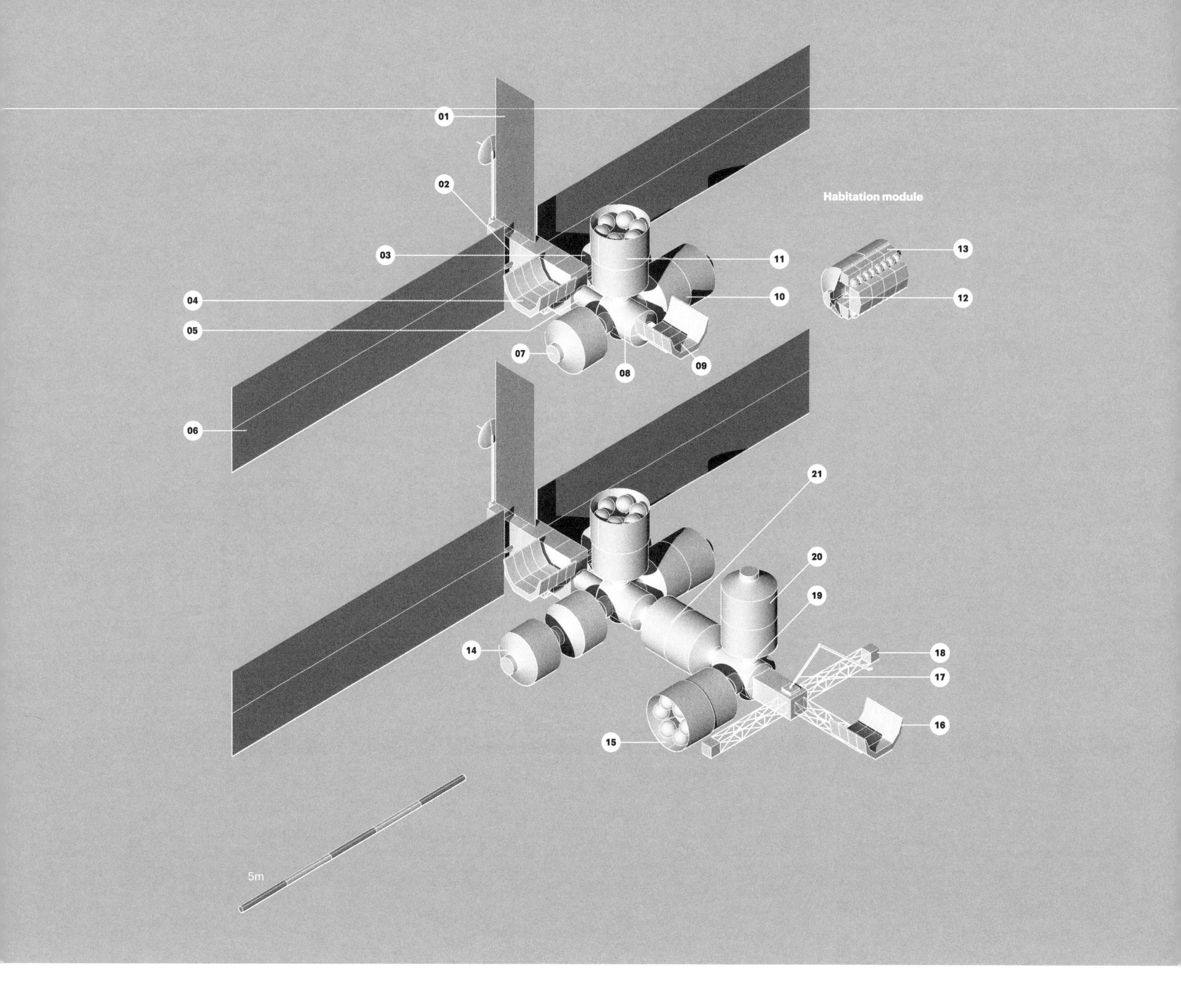

01 Radiator
02 Spine
03 Electrophoresis research module (behind spine)
04 1st science pallet
05 Access tunnel
06 Solar array
07 2nd habitation module (single segment)
08 1st core module
09 2nd science pallet
10 1st habitation module (double segment)
11 1st logistics module
12 Corridor
13 Storage racks
14 Life science module
15 2nd logistics module
16 2nd science pallet (relocated)
17 Robotic arm
18 Experiment truss
19 2nd core module
20 Command & control module
21 3rd habitation module (double segment)

development programme. See Appendix I.) In early 1980, NASA awarded two parallel contracts for these to Boeing and McDonnell Douglas. Johnson Space Center managed the Boeing study and Marshall Space Flight Center managed the McDonnell Douglas study. Johnson Space Center was NASA's leading field centre for human spaceflight while Marshall Space Flight Center was NASA's leading field centre for launch vehicle development. Work lasted from the middle of 1980 to the end of 1981 and the contractors published their reports in early 1982.

Boeing described its Space Station as a Space Operations Center, the same name that Johnson Space Center had used for its 1979 study. McDonnell Douglas described its Space Station as an Evolutionary Space Platform. The two concepts were quite different in their approach and reflected fundamentally different attitudes at each of the two NASA managing centres about what a Space Station ought to be. Boeing proposed a crewed Station from the outset, evolving into an operations centre that would build large space structures, refuel spacecraft and prepare deep space missions. In contrast, McDonnell Douglas proposed a modest automated platform visited by the Orbiters that would evolve gradually into a permanently crewed Station. The two field centres were traditional rivals and each was pushing NASA Headquarters in Washington to adopt its individual approach to the Station's design.[19]

The provenance of the McDonnell Douglas design of the Evolutionary Space Platform was an uninhabited laboratory called the Science and Applications Space Platform that the company had been working on since 1980 for Marshall Space Flight Center. This science platform would operate automatically on orbit with occasional Orbiter visits to tend experiments and carry out maintenance. The aim was to provide plenty of accommodation for science and technology in a low cost, low risk design with simple maintenance and long-term growth potential. The laboratory comprised twin solar arrays, a radiator, three attachment ports for science pallets, a propulsion package, batteries, attitude control and communications and data systems, all mounted on a rectilinear structural spine. Illustrations showed solar arrays extending out to port and starboard with only one axis of rotation and their inner edges almost touching the spine faces. McDonnell Douglas designed its space platform to fly in an attitude that made Sun tracking easier but the proximity of the arrays to the rest of the structure suggested shading and interference problems. The laboratory used existing hardware wherever possible, including parts of satellites, solar arrays, radiators, propellant tanks, thermal pumps, communications equipment and sensors, and control-moment gyros for attitude control.[20]

By early 1981, Marshall Space Flight Center and McDonnell Douglas were ready to begin the full development of their automated space platform but the project was soon overtaken by events. The Space Shuttle was flying and momentum was building to give it a major job: build a permanently

1981-82 McDonnell Douglas Evolutionary Space Platform study In 1981, NASA began to involve aerospace companies in Space Station design studies with a focus on the use of the Space Shuttle as the delivery vehicle. NASA gave parallel contracts to Boeing and McDonnell Douglas. McDonnell Douglas proposed a cautious build-up approach that it called the Evolutionary Space Platform. Its emphasis was on scientific research. Wherever possible, the design used existing space hardware. Above, an initial phase shows habitation, research and logistics modules and science pallets. Below, a growth phase adds more modules and a truss for attaching external experiments. Image: author.

1981-82 Boeing Space Operations Center study In their parallel 1981-82 Space Station studies for NASA. Unlike McDonnell Douglas, Boeing's emphasis was on space operations such as spacecraft assembly and satellite servicing. Boeing followed NASA's earlier 1979 studies and used the same name. The result was a more ambitious design with some large hardware elements. To the left, an initial phase has just a habitation module and a logistics module. To the right, a fully operational phase adds a second habitation module and two hangars but no laboratory modules. Image: author.

inhabited Space Station on Earth orbit. The Science and Applications Space Platform approach was no longer relevant, so McDonnell Douglas added pressurised modules to it and called it the Evolutionary Space Platform. Its design was heavily influenced by Spacelab, Europe's contribution to the Space Shuttle programme and the first science laboratory and pallet system designed to fly in the Shuttle Orbiters. McDonnell Douglas had been deeply involved in Spacelab's development in Europe up until the delivery of the Spacelab's engineering prototype to NASA in 1980 for crew training. Spacelab's influence was evident in McDonnell Douglas's proposed use of the segmented cylindrical modules that formed the pressurised modules and the U-shaped open pallets used for external equipment and experiments. Spacelab was about to become the first module system to fly on the Space Shuttle. It was ready and waiting for its maiden flight due in 1983. In a situation where the cost of a future Space Station was going to be a challenge from the outset, achieving economies through technology twice used was an attractive idea. Spacelab's versatile design made different module lengths possible, enabling them to fit into the payload bay together with the pallets and fly on a single Space Shuttle mission.

In their recommended initial concept, McDonnell Douglas first attached a specialised research (electrophoresis) module and a Spacelab pallet on the port and starboard sides of the laboratory spine respectively, very close to the inner edges of the rotating solar arrays. Then, at the spine's forward end they attached a central core module that comprised several ports for module additions, an Orbiter docking port, a pressurised passageway, a mini-control centre and a safe haven. To this they added double and single segment habitation modules on the port and starboard sides respectively and a supplies module above, which pointed to zenith. Forward of the new central core module they added another science pallet.[21] The minimal amount of pressurised volume in the two habitation modules was considered adequate for a two- to four-person crew for ninety days but it compared unfavourably with the size of the Russian Salyut 6 Space Station shown at the same scale. The two modules contained private crew compartments, a control centre, a workbench, equipment racks, and food and hygiene facilities. All this filled up their interiors except for narrow access corridors down the middle, but there was a window in the larger module. The supplies module contained the ninety-day supplies and consumables. The environmental control and life support system also used technology borrowed from the Spacelab laboratory, with supplied oxygen and nitrogen and carbon dioxide removed by a solid amine system and vented overboard.

Just three launches were needed to deliver it to orbit. The Shuttle would deliver the laboratory platform and a pallet on the first launch, the central module and larger habitation module on the second launch and the supplies module, small habitation module and another pallet on the third launch.

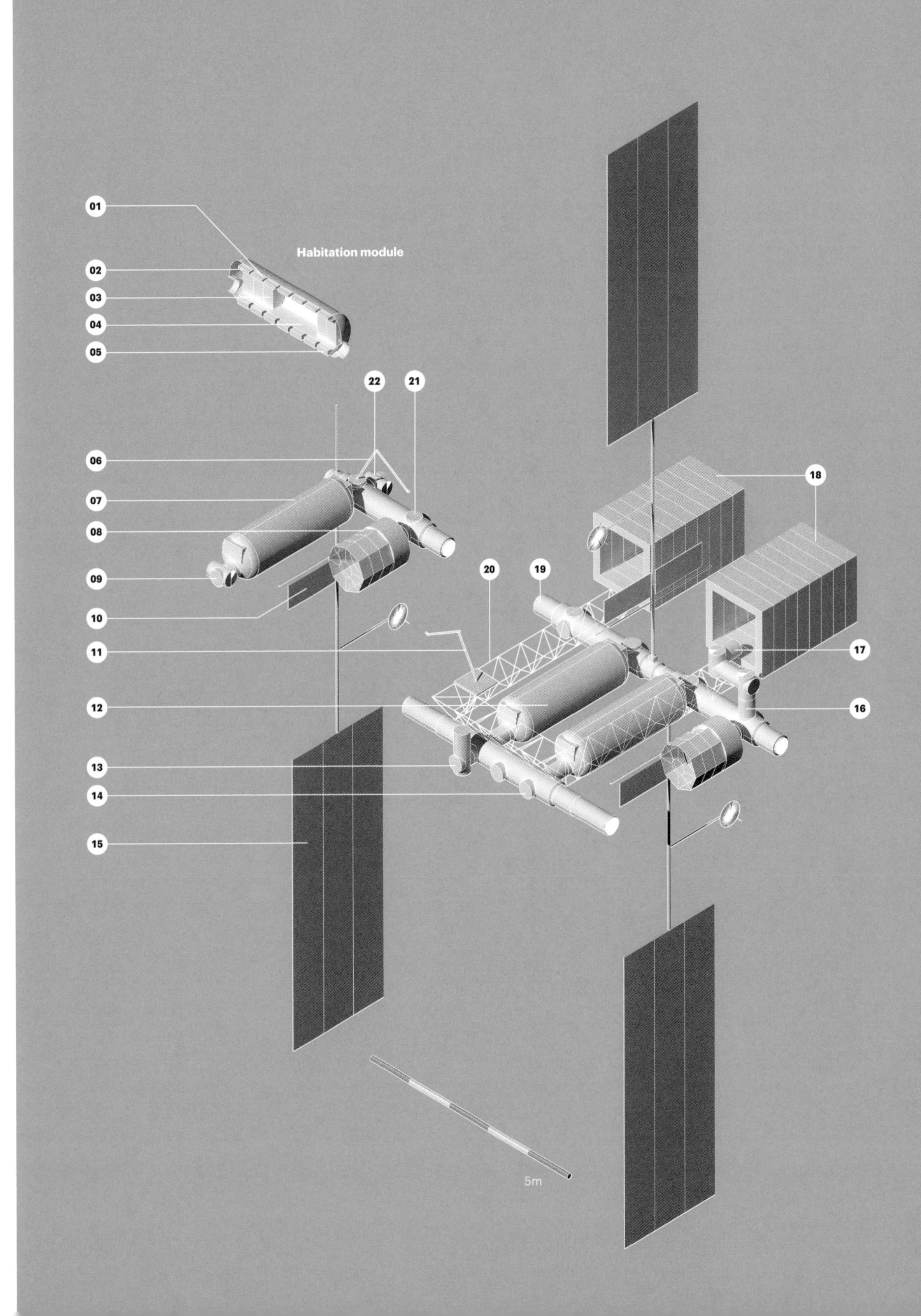

01 Ceiling life support systems
02 Ceiling deck
03 Corridor
04 Floor deck
05 Floor life support system
06 Robotic arm
07 1st habitation module
08 Logistics module (partially loaded)
09 Airlock (alternative orbiter docking port)
10 Radiator
11 Mobile robotic arm
12 2nd habitation module
13 2nd airlock module
14 Docking port and tunnel
15 Solar array
16 1st airlock module
17 Crew access tunnel
18 Hangars
19 2nd service module
20 Rails (along truss edge)
21 1st service module
22 Airlock

The concept would 'recognise the realities of budget constraints and payload availability, both of which combine to prescribe a vehicle of modest beginnings and yet flexible for growth into service for those major orbital operations that are emerging'. However, it was debatable whether this modest concept that was based on European-built hardware would provide the American aerospace industry with the work it expected out of the new Space Station initiative. It was also questionable whether the small crew and lack of a general-purpose pressurised laboratory module would be acceptable to the scientific community.

McDonnell Douglas's growth concept for a Space Station resembled an expensive toy construction kit in which a set of elaborately made pieces plugged together in different ways. Buying extra pieces would expand the range of possibilities. McDonnell Douglas suggested several growth directions that offered many scientific capabilities. These encompassed pharmaceutical production, materials production, life science, solar science, Earth science, oceanography, spacecraft recovery and servicing, and deep space mission preparation and launch. Typical of these was a near-term growth design involving the temporary relocation of the Spacelab pallet at the forward end of the initial design. This would make way for the forward attachment of another habitat module, another central core module, a robotic arm and the reattached pallet, in that order. Added overhead would be a command and control module and on the starboard side another supplies module and a life sciences module. Exterior experiment support trusses would extend out to port and starboard of the second core module.[22] The initial design and evolved options for the Evolutionary Space Platform assumed the possibility of switching between Earth and inertial flight orientations depending on the pointing requirements of the on-board experiments. This suggested that Orbiters would need to perform rendezvous and docking from different directions depending on the Station's attitude at the time, an added navigational and operational challenge for the forthcoming Space Shuttle fleet.

1981-82 Boeing Space Operations Center study Three artist impressions of Boeing's 1981-82 Space Operations Center design. Opposite above, the initial operational phase has two habitation modules, side by side, linked at their ends by long slender pressurised tunnels to form a loop with a box-like hangar projecting from one corner. Above, the Orbiter has docked and the hangar doors are open (left side of image) to reveal a spacecraft inside. Opposite below, an evolved phase adds another pair of modules, a second corner hangar and more truss structure (in the foreground).

Boeing's design was much closer to Johnson Space Center's design of the 1979 Space Operations Center and shared the same name. Its focus was on the assembly and construction of complex spacecraft, the servicing and basing of orbital ferry vehicles and the servicing of satellites. It concluded that a Space Shuttle delivered Space Station could perform all anticipated missions up to the year 2000, that operations facilities should be designed for workplace ease and flexibility, that a Station should be Earth-oriented and fly at 370km altitude, and that there should be plenty of duplicated systems and safety features to avoid accidents. Boeing asked itself the question 'Why Man?' and answered that space operations could be automated in principle, although some would be very difficult and expensive and that human involvement remained a practical and cheap way of getting the job done.[23] Boeing pointed out that space operations are not routine or repetitive and no two servicing events are alike, implying that human decision-making ability and manual dexterity were indispensable.

Boeing explored initial, operational and growth Space Station designs on low Earth orbit and a design on geosynchronous orbit in its study. Three Orbiter flights would deliver enough building blocks for a crew of four in the initial design. There would be a habitation module for crew living and working, a thinner service module for systems distribution and emergency habitation, a pair of airlocks, a ninety-day supplies module and a robotic arm. The habitation, service and supplies modules were aligned horizontally in a U-shaped arrangement. Pointing to nadir from one end of the service module was a long boom that supported twin radiators, an antenna in the middle and a single large solar array at the bottom. The habitation module had a longitudinal deck and ceiling that ran the full length of the module. Beneath the floor deck was a service storage zone that housed water tanks, waste tanks and pressurised gases. Above it was the life support system with carbon dioxide removal, ventilation, thermal control, dehumidification and ducting. The open volume between the floor and ceiling contained private crew compartments, shower, washbasin and toilet compartments, food preparation workstations, spacesuit storage, a communal table and an exercise area. At one end of the module at mezzanine level was a command and control deck and at the other end an observation deck. Though both these decks had windows, they appeared to face away from Earth. Boeing evaluated closed and open loop life support systems for the internal atmosphere and proposed an open loop that converted to a closed loop in increments. Initially a solid amine system would absorb carbon dioxide that was occasionally steamed out and combined with hydrogen in a Sabatier reactor to recover the oxygen.

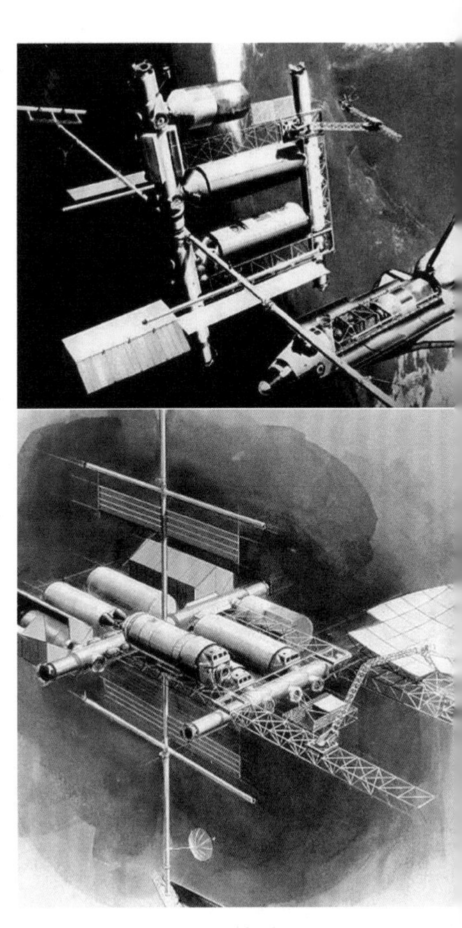

Boeing's operational design could be permanently manned by a crew of eight after a further three Orbiter deliveries to expand the initial design. A second habitation module was aligned alongside the first with both sandwiched between the first and second service modules. This provided two module exit paths in an emergency. Each service module had an airlock. Above the module cluster was an open truss framework in a U-shape with movement tracks for a mobile crane or robotic arm. The two open ends of the U-shaped framework each supported a large box-shaped hangar. These were unpressurised but there were no details about how they would be delivered to orbit or used for space operations. The addition of a second long boom with radiators, antenna and a solar array pointing to zenith in the opposite direction to the first one completed the operational design. The growth design took the operational design further with the extension of the trussed framework to support a work platform for the assembly and launch of orbital space tugs and very

USA

large satellites using the crane or robotic arm as the principal assembly tool. It would need a total of nine Orbiter flights. The illustrations of the growth and operational designs omitted any logistics modules but these essential building blocks would continue to need allocated berthing ports throughout the Space Station's evolution. Finally, Boeing's Space Station design in geosynchronous orbit comprised habitat, supplies and solar storm shelter modules, two airlocks, a crane, an external storage platform and twin solar arrays on long booms with two axes of rotation. Revell, a manufacturer of plastic model kits, produced a kit around this time based on Boeing's design. It was called the Space Operations Center, its scale was 1/144 and it came with a Space Shuttle docked at one end of the service module.

As the parallel Boeing and McDonnell Douglas studies were in progress, the Space Shuttle era began. NASA launched the Orbiter Columbia on 12 April 1981. Ten years after its conception the Space Shuttle system was now operational. Columbia, the first winged and wheeled spacecraft, spent just over two days on orbit and returned to the ground safely despite some technical problems. These included the first evidence of ice breaking away from the External Tank and striking the Orbiter as well as the loss of some thermal protection tiles during the flight.[25]

The Space Shuttle's maiden flight coincided with the publication of a report by a space lobbying group that urged the government to refocus and revitalise its national space agenda. The Citizens Advisory Council on National Space Policy had been set up jointly by the American Astronautical Society and the L5 Society to propose a realistic space policy to the advantage of the nation. It published a report of its first meeting in spring 1981 entitled *Space: The Crucial Frontier*. The Council had a membership list that included many eminent individuals in fields of astronautics, aeronautics, science, engineering, literature and medicine.[26] The report noted that: 'the United States has a world mission' and that it needs 'visible goals: a reason for the nation to exist'. It came to conclusions that spanned national security, technology development, free enterprise, commercial exploitation, geostationary strategy, space solar power, space heritage, environmental protection and space access. Taking its cue from the 1979 Johnson Space Center study, it recommended that 'the United States must immediately establish a permanent manned presence in space' and that a Space Operations Center 'should be given highest priority and be so recognised in funding decisions'. The emergence of the Citizens Advisory Council on National Space Policy (no longer in existence) at the time signalled that an audience well beyond government and the aerospace industry had begun to sit up and take notice of the nation's space programme as it emerged from a period of passivity after the Apollo years.

1981 inaugural launch of the Space Shuttle NASA launched the first Space Shuttle Mission STS-1 with the Orbiter Columbia from Cape Kennedy on 12 April 1981. Columbia's commander was John W Young and its pilot was Robert L Crippen. Columbia returned and landed safely just over two days later. The successful beginning of the Space Shuttle era was a critically important event for the future Space Station. It enabled NASA to proceed with the Station's design and planning in the knowledge that the Shuttle fleet would be available to deliver the building blocks to orbit.

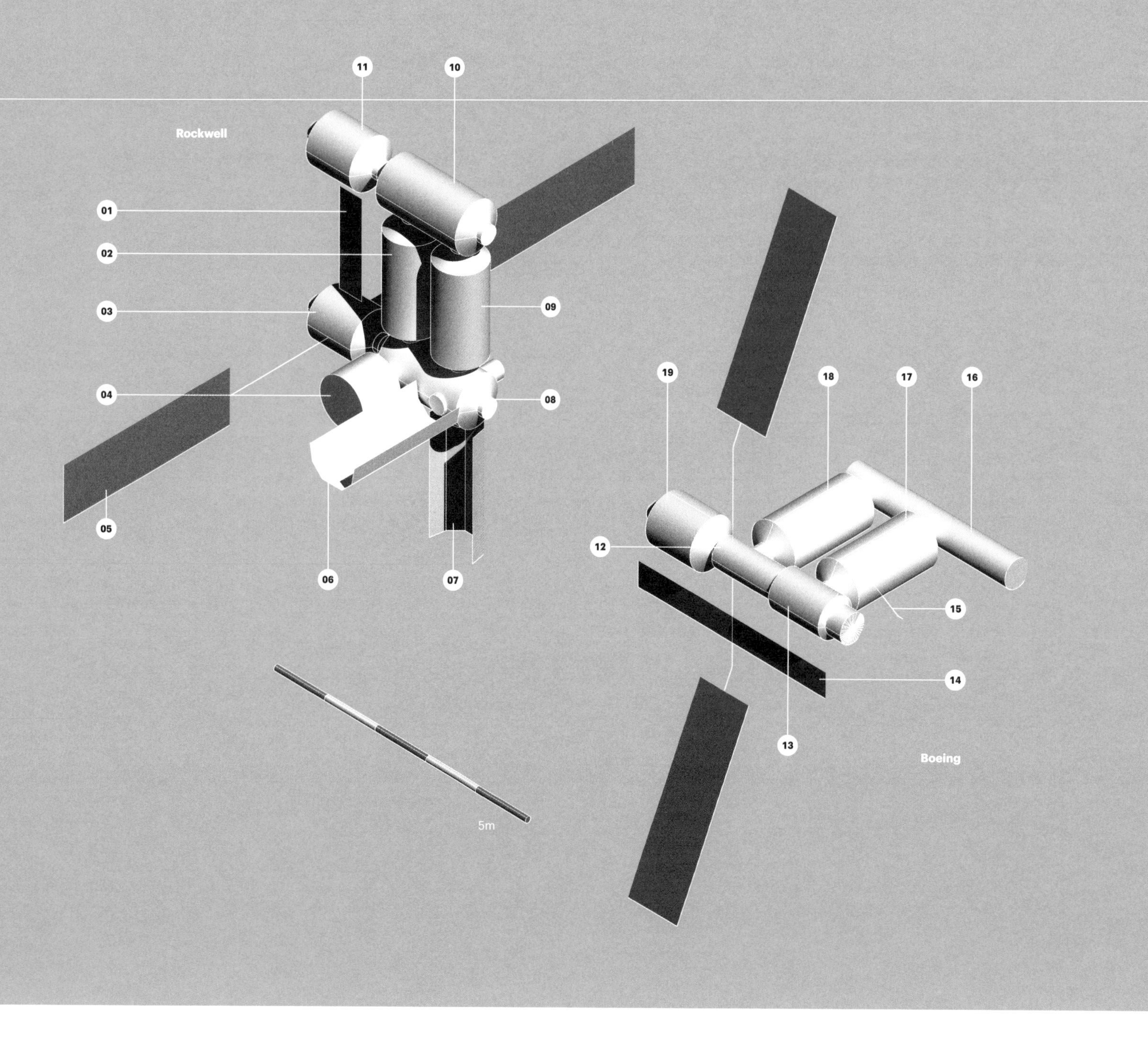

01 Radiator
02 1st habitation module
03 Energy module
04 Logistics module
05 Solar array
06 Astronomy services pallets
07 Payload services pallets
08 Command & control module
09 2nd habitation module
10 Life sciences module
11 Life sciences resources module
12 Service tunnel
13 Service module
14 Radiator
15 Robotic arm
16 Tunnel & supplies module
17 2nd habitation module
18 1st habitation module
19 Command & control module

Diverse approaches
Eight parallel studies on Space Station needs, attributes and architectural options, 1982-1983

Following Ronald Reagan's election as President of the United States in 1980, his administration appointed James Beggs as the new NASA Administrator. Beggs' first major task was to ensure that the Space Shuttle system went operational. After the Orbiter Columbia's successful first flight in April 1981, he and NASA Deputy Administrator Hans Mark went public at Senate hearings with their shared enthusiasm for a new Space Station. It was to be 'the next logical step' in space but it would take time to get off the ground and the continuing feud between Johnson Space Center and Marshall Space Flight Center was an unhelpful start. From now on NASA Headquarters in Washington would take charge and the agency formed and formalised a Space Station Task Force in May to direct the project.[27]

Casting its net as wide as possible, the task force awarded eight contracts in August to aerospace companies to carry out parallel investigations under the umbrella title of the Space Station Needs, Attributes and Architectural Options Study. The contractors would examine a broad spectrum of issues from the viewpoint of future users of a Space Station. The objectives were to identify suitable missions, conduct market research, establish requirements, describe the benefits to society as a whole and compare the project's value to its likely cost. NASA did not want detailed designs, and architectural concepts were used to interpret and illustrate the requirements visually. NASA wanted to justify having humans in space in broad terms that would appeal to politicians and the public alike. It wanted to use the results to help strengthen the national case for a new Space Station. The contracts went to Boeing, Rockwell, Lockheed, McDonnell Douglas, General Dynamics, Martin Marietta, TRW and Hughes Aircraft.

As the contractors began work, NASA formed a Concept Development Group at its Headquarters in Washington DC to lead the design of the new Space Station. In September 1982, NASA invited prospective international partners to a meeting, which the agency called an 'International Orientation Briefing'. NASA had begun discussions with other countries earlier in 1982 with the suggestion that those interested might explore Space Station missions and capabilities from their own perspectives. Europe, Japan and Canada responded positively and began their own studies. At the September meeting, NASA laid out its thoughts about international cooperation and information exchange during the requirements definition stage. The agency invited foreign observers to attend the final reviews of the eight parallel studies in the following spring.[28]

In March 1983, as all eight aerospace contractor teams neared the end of their studies, NASA held a Space Station technology workshop in Williamsburg, Virginia. There was no doubt that a new Space Station was going to present the aerospace world with a multitude of engineering

1982-83 Space Station Needs, Attributes and Architectural Options Study – Boeing and Rockwell designs In 1982, NASA formed a Space Station Task Force to lead the project. One of its first actions was to give contracts to a group of major American aerospace companies to study a spectrum of Space Station issues covering missions, requirements, market research, terrestrial benefits and long-term value. The contractors produced outline architectural designs to show what their Stations might look like. In the right image opposite, Boeing's incremental Station design had two parallel habitation modules that formed a loop with service tunnels at each end. In the left image opposite, Rockwell's initial Station design had two parallel habitation modules that formed a loop with two life sciences modules at one end, and command and energy modules at the other. Image: author.

challenges and it was vital to find out what types of new technologies were needed to meet them. Workshop participants were expert engineers and managers drawn from the aerospace industry at large. NASA divided them into groups with each group tackling a specific engineering field. The list of fields covered habitability, life support, spacewalks, flight control, propulsion, fluids, communications, structures, mechanisms, data, power, thermal control and a new field called systems/operations technology. There were some important conclusions and recommendations. The habitability group wanted a proper set of design guidelines, more attention to acoustics and a versatile, portable workstation. The life support group urged the development of regenerative 'closed loop' systems for water supply and internal atmosphere. The structures and materials group identified many action areas including improved thermal control coatings, space debris impact resistance and radiation shielding as well as studies comparing deployable to erectable structures, robotic manipulation techniques and solar array designs. The power group was concerned about the life-cycle reliability of solar arrays and the effect of their orbital drag on a Space Station. The systems/operations group stated that the objectives for its new field were 'to make new things possible' and 'to make old things cost less or work better'. It recommended research on the relationship between missions and technologies, automation and autonomy, simulation and emulation, engineering standards and systems analysis tools. NASA noted that information generated by the workshop would 'establish a stepping stone' for users of Space Station technology to plan for and develop what was needed.[29]

Using the Space Shuttle as a temporary Space Station President Ronald Reagan and his wife Nancy greet astronauts Thomas Mattingly and Henry Hartsfield after their return from a mission on the Space Shuttle Columbia in 1982. Rockwell International built the Shuttle Orbiters and proposed fitting them out for extended duration visits to orbit as the first step towards a Space Station in its 1983 Space Station Needs, Attributes and Architectural Options Study for NASA.

The eight contractors presented NASA with their results of the Space Station Needs, Attributes and Architectural Options Study in April 1983.

In carrying out its market research, Boeing Aerospace Company had found it effective to hold telephone interviews with potential users to obtain opinions and envision missions. The civilian user community divided into three categories of scientific research and commercial and technology development. The science missions focused on Earth and solar observation, plasma physics, life sciences, materials and astrophysics. The commercial missions focused on semiconductors, pharmaceuticals and optical glasses. The technology missions focused on space structures, flight control, large optics, fluids, robotics and energy. The payoff would be fourfold. First, there would be a better understanding of the Earth, the Solar System and the universe. Second, there would be advances in materials and life sciences. Third, there would be cutting-edge electronic, pharmaceutical, optical and robotic products. Fourth, there would be higher productivity for space transportation. Boeing compared the architecture of a Space Station to that of a speculative office block. Both needed feasibility studies and pre-leasing and marketing at the planning stage. Both involved structural mechanical, electrical, thermal, hygiene, data and crew circulation functions. Both encompassed an architectural envelope, specific dimensions, codes and regulations, construction budgets, life-cycle costs, storage and economies of scale. There were, however, some major unstated differences. A speculative office block could be built fairly quickly at little risk and reliable cost and leased immediately. A Space Station might take a decade just to design and engineer and its fabrication and assembly on orbit would involve great risk, great cost and much caution. Suggesting that a Space Station could be pre-leased like an office block was somewhat simplistic.

Boeing recommended three architectural approaches, which it described as incremental, unified and derivative. The incremental Space Station architecture provided maximum adaptability and flexibility through a choice of different types of module tailored to different purposes. Two habitation modules side by side were linked at one end by a tunnel and supplies module and at the other end by a service module, command and control module and an Orbiter docking port. This looped cluster flew horizontally relative to the flight path with two solar arrays and radiators on booms that extended to zenith and nadir from the service module. The unified architecture emphasised maximum commonality between the modules and their outfitting for particular activities. Two habitation modules, a command and control module, a supplies module and a laboratory module were squeezed together side by side. This cluster also flew horizontally with solar array and radiator booms pointing to zenith and nadir as before. The derivative architecture was an unmanned platform that evolved into a crewed Space Station. It began life as a large rectilinear structural frame that supported solar array and radiator booms, to which Orbiter missions gradually added modules. Boeing concluded its study by underlining the dependency of benefits on life-cycle costs and highlighting the advantages of humans over robotics in space.[30, 31]

Rockwell International was the developer of the Shuttle Orbiters and it was natural for the company to stress the vital role that the Space Shuttle system would play in delivering anything to orbit for a Space Station and operating it there. In the opening sentences of its study summary, Rockwell recommended Extended Duration Orbiters as the first step in a Space Station development plan. They would be able to stay on orbit for a few weeks at a time to carry out Space Station advanced work, test operational techniques and develop technologies. Next would come teleoperated manoeuvring and multi-mission spacecraft. A permanent four-crew Space Station would follow, comprising a command module, an energy module, a supplies module, two airlocks and a payload service structure. This would expand to accommodate a crew of eight with increased operational capabilities including the attachment of various types of science platform. Alternatively, a pair of small four-crew Stations could replace a single eight-crew Station. Rockwell saw the need for a division between space and shirtsleeve environment operations. In space were the deployment

1982-83 Space Station Needs, Attributes and Architectural Options Study – Lockheed and McDonnell Douglas designs
Eight American aerospace companies took part in NASA's 1982-83 Space Station Needs, Attributes and Architectural Options Study. To the left, Lockheed's initial Station design had two loops of modules on either side of a central spine. One loop had two parallel habitation modules and the other had two laboratory modules. Both loops had connecting tunnels at their outer ends. The spine comprised a service module with truss structures mounted at each end. To the right, McDonnell Douglas's incremental Station design was modest with a habitation module, a laboratory module, a core module, a service module, a spine at one end and a truss structure at the other. Image: author.

of low Earth orbit satellites, the launch of satellites to geostationary orbits, the servicing of orbital space tugs and the storage and construction of large assemblies. The crew would carry out research and development in a shirtsleeve environment inside the Station. Rockwell based its proposals on low, medium and high mission locations on orbit with each location tailored to specific user requirements. It envisaged 981 missions to all locations between 1991 and 2000, a staggeringly high number. These missions would lift a total cargo mass of over 6,000 tonnes into space, supported by over 150,000 astronaut hours of effort. Rockwell's fleet of Orbiters and NASA's astronaut corps would be very busy for years to come. This rich bounty of missions represented a very ambitious vision of the future.

Rockwell used the term 'command module' instead of 'habitation module' to describe a single module that, to begin with, contained living, sleeping, cooking, hygiene, experiment, medical, exercise and Station operations control facilities. The command module's long axis orientation was parallel to the orbital plane. Berthed in-line at its forward end was the energy module, to which Orbiters would dock from the flight path ahead. Berthed near the front of the command module were the supplies module and payload service structure. The supplies module pointed out on the port side. The payload structure pointed down to nadir. It consisted of a long train of European Spacelab pallets and was another example of the use of these pallets in early Space Station studies. Both airlocks faced nadir with one mounted on the energy module and the other on the command module. Mounted on each side of the energy module and pointing horizontally out to port and starboard were solar arrays on long booms with a radiator above pointing to zenith. The growth design for a crew of eight added several more modules. There were two habitation modules side by side berthed to the command module and relieving it of habitation functions. These pointed upwards and connected to a life science module to form a loop. At the forward end of the life science module was a separate biological research chamber. There was a processing laboratory module berthed to the rear of the command module that pointed out on the starboard side. Completing the growth design were propellant storage tanks added alongside the laboratory module.[32, 33]

The Lockheed Missiles and Space Company began its summary by saying that from a technical point of view a Space Station could be up on low Earth orbit by the end of the decade. The main problem was the question of need. Lockheed had contacted scientific users who had shown a modest level of interest and commercial users who had shown a lot of interest but in the company's view, the role of a Space Station as a space operations hub had the most potential. However, there was not an immediate and pressing need for it as a civilian facility. A Space Station's role in the space operations field would cover space radar and space telescope maintenance, large structures assembly, astronomy platform support, satellite replacement and resupply

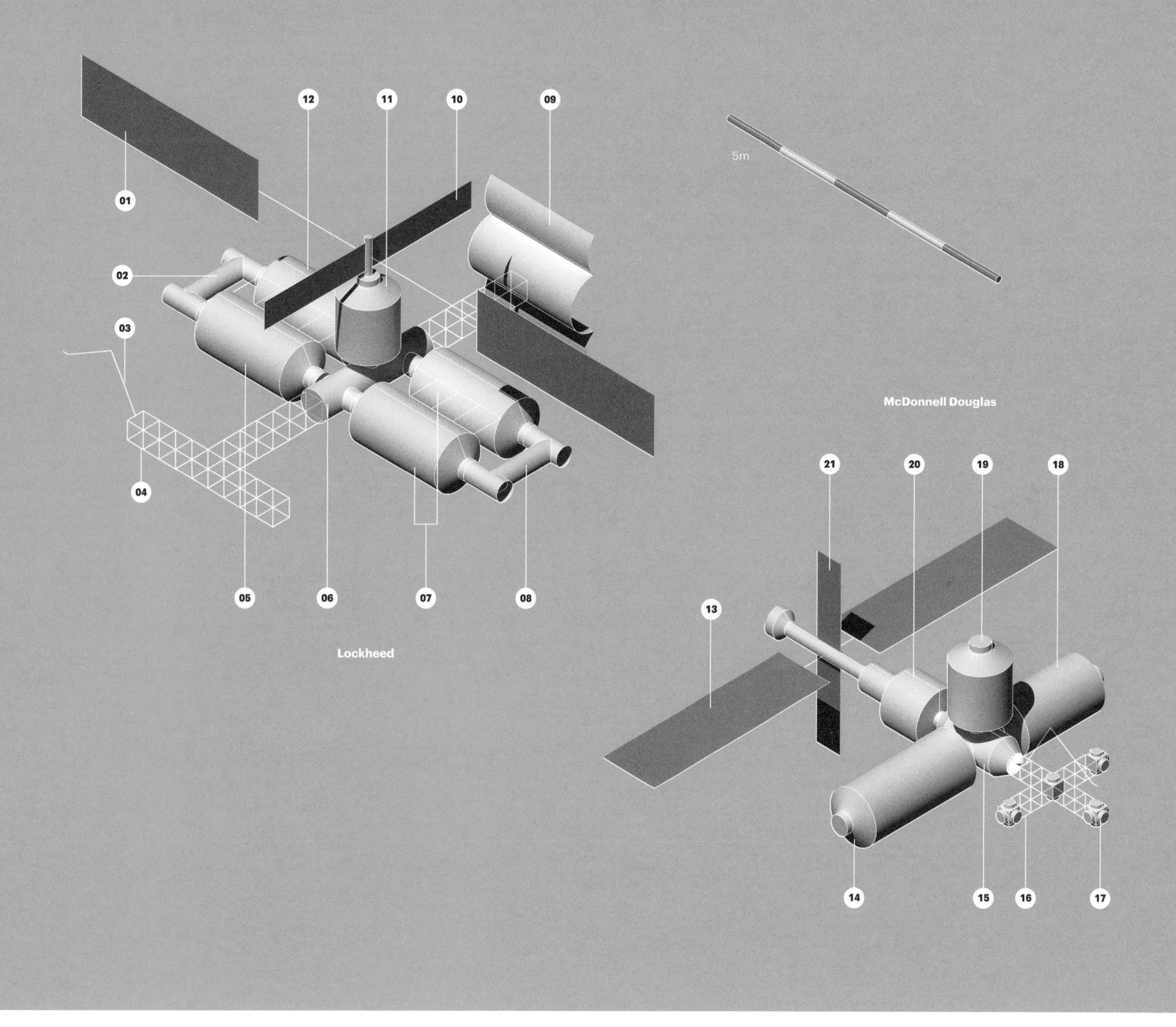

01 Solar array
02 Connecting end tunnel
03 Robotic arm
04 External truss structure
05 Laboratory module
06 Service module
07 Habitation modules
08 Connecting end tunnel
09 Hangar hinged doors
10 Radiator
11 Logistics module
12 Laboratory module
13 Solar array
14 Laboratory module
15 Core module
16 External truss structure
17 Docking ports
18 Habitation module
19 Logistics module
20 Service module
21 Radiator

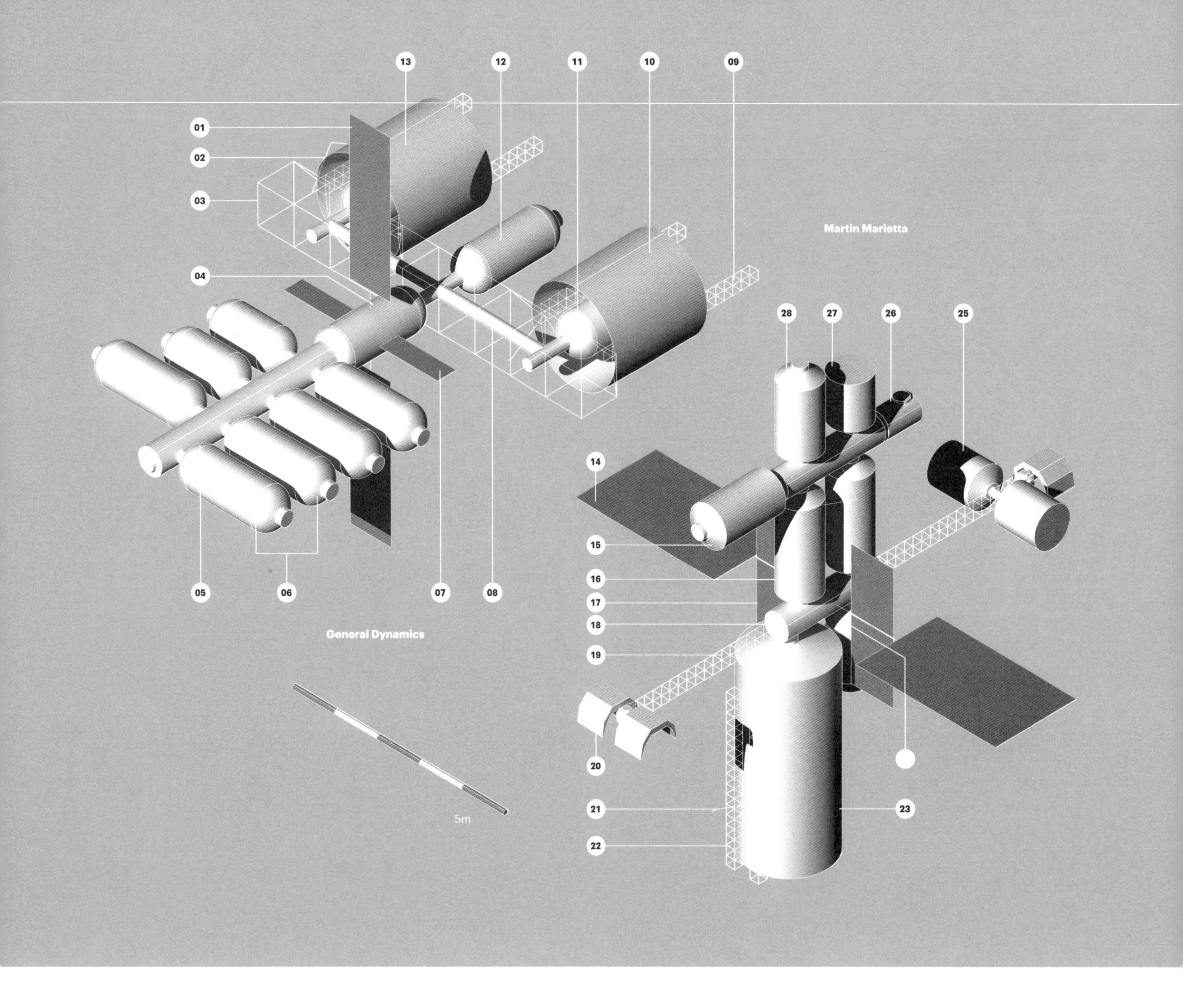

01	Solar array	**08**	Access tunnel	**15**	Materials processing laboratory	**22**	Construction area
02	Robotic arm	**09**	Rail beam	**16**	Habitation modules (2)	**23**	Sliding service hangar
03	Structural support truss	**10**	Sliding services hangar	**17**	Radiator		Propellant storage (behind radiator)
04	Core module	**11**	Maintenance module	**18**	Energy module	**25**	Materials payload
05	Service tunnel	**12**	Propellant storage	**19**	Trussed beam	**26**	Service tunnel
06	Mixed habitation and lab modules (TYP)	**13**	Sliding service hangar	**20**	Science pallets	**27**	Logistics module
07	Radiator	**14**	Solar array	**21**	Robotic arm	**28**	Life sciences module

and Space Shuttle rescue. Lockheed extracted four system concepts from civilian user requirements and mission scenarios, all built around a basic Space Station. The first concept's focus was on Earth and space observation and remote sensing. That of the second was on life sciences and materials processing. The third dealt with celestial and meteorological observations and automated materials production and the fourth dealt with satellite servicing and assembly of large satellite structures.

Lockheed's first assembly phase would begin in 1990 with the delivery of a habitation module for a crew of three, an attached electrical power module and a sensor experiment testbed, all on a single Orbiter mission. This basic facility would take a year to prove its reliability, leading to incremental build-up over six years at an average delivery rate of two missions annually. By 1996 it would become a fully capable operations and research facility for a crew of six, with all its building blocks delivered by the Space Shuttle system. The fully operational Space Station comprised a core of four full-size modules connected in two parallel pairs by end tunnels and a central narrow module to form a figure of eight. This flew horizontally, perpendicular to the orbital plane with the Orbiters docking at one of the corners. The service module at the centre extended outwards on each side in the form of T-shaped external structures. On one side the T was made from rectilinear trussed frames. On the other side the T comprised a cylindrical hangar attached to a section of trussed frame. The hangar had hinged curved doors that opened outwards. Above was a short module pointing to zenith that supported twin solar arrays on booms and twin radiators. Lockheed did not specify what the individual modules were for but a combination of habitation, laboratory, supplies and service modules would be needed. The exterior structure and hangar would support different space operations functions. The evolutionary build-up could be halted at any point and the facility would remain useful. Lockheed concluded that a capability for human operations in space would be of national benefit and that Space Station development should begin immediately.[34, 35]

McDonnell Douglas placed great emphasis in its study on sets of missions and their requirements as the drivers of a Space Station. Each mission set was analysed in depth and generated its own conceptual design. There was no single ideal design solution. There were six science and applications mission sets covering solar science and astrophysics, communications, Earth environment, planetary exploration, life science and materials processing. There were four mission sets concerned with space operations. These encompassed satellite launch and servicing, low-cost space transportation, orbital assembly and checkout, robotics and remote operations and large structural assembly. There were five mission sets concerned with technology that dealt with the evaluation of the human role, fluid storage and management, satellite servicing technology, crew manipulated robotics and orbital space tug servicing. Finally, there were

1982-1983 Space Station Needs, Attributes and Architectural Options Study – General Dynamics and Martin Marietta designs
The eight aerospace contractors taking part in NASA's 1982-83 Space Station studies produced very different designs. To the left, the evolved Station of General Dynamics was a T-shaped design with a cluster of several habitation and laboratory modules attached to the T's mast that functioned as a service tunnel. The T's crossbar was a narrow access tunnel with a large hangar and maintenance module at each end and a propellant storage tank in the middle. To the right, Martin Marietta's evolved Station design stacked a pair of parallel habitation modules, a life sciences module and a logistics module on top of a large cylindrical hangar with experiment facilities at the end of trussed beams on each side. Image: author.

six commercial mission sets covering reusable orbital transportation, communications, platform leasing, remote sensing, orbital services, and materials processing and manufacturing. It was an extensive mission menu.

Using this menu, McDonnell Douglas conceived a range of design options for a Space Station ranging from a '100 per cent capture architecture' that aimed to provide optimal accommodation for all the missions to several less ambitious single or dual facility approaches aimed at mission groups. It would be a question of budget and McDonnell Douglas considered that a modular approach would enable a Space Station to pass through three phases of initial capability, expanded capability and evolutionary growth. A modular approach was already obvious with the Orbiter payload bays determining the maximum shape and size of the modules. The continuous delivery of building blocks by the Orbiter over a period of ten years would enable a Space Station to embrace more and more missions gradually. The choice of Space Station design was governed by the choice of orbit it would fly on. As well as high 57° and low 28.5° orbital inclinations, McDonnell Douglas unusually added a 90° Sun synchronous orbit as a possibility. What emerged was a series of three 'emphasis' architectures with a Space Station tailored to support the triad of science and applications, operations and technology, and commerce. A fourth architecture would be tuned to the needs of orbital space tugs. The study concluded by saying that 100 per cent of the identified missions could be met by four separate facilities on orbit – two habitable Space Stations and two platforms visited by crews – with various smaller combinations of mission groups met by single or dual facilities. McDonnell Douglas made it clear that the missions drove the requirements and the requirements drove the architecture. A Space Station design would not begin to coalesce until NASA defined the missions and requirements properly.[36, 37]

After carrying out a mission requirements analysis that came up with nearly 150 options, General Dynamics, Convair Division settled on a single Space Station on a 28.5° inclination orbit as its preferred solution. It would be able to support human-conducted research and development, free-flying satellite servicing, staging and servicing of an orbital space tug, and construction of large structures and payloads. There could be separate automated platforms on other orbits and a future second Space Station on a polar orbit. General Dynamics focused its attention on those areas it felt would produce the highest payoff from the presence of a crewed Space Station. In the company's view, an orbital space tug based at a Space Station would produce the greatest economic benefits and its development should begin as soon as technically possible as it would take ten years to develop. The main functions would comprise crew habitation and life support, an electrical power system, attitude control, communications and data, supplies and storage, and Orbiter docking ports. Other features would include pressurised laboratories and especially an orbital space tug hangar, servicing and payload facilities and propellant storage. There were two space tug hangars in General Dynamics' design. These were substantial structures shaped like large cylinders mounted on the external structure. The cylinders were hollow with thin walls. They were open to space at their outer ends with pressurised maintenance modules at their centres that resembled wheel hubs. The hangars would slide in and out on a pair of trussed beams that projected out beyond the hangar ends. The beams would provide docking points for incoming spacecraft and enable the cylinders to slide over them during operations. Each hangar was equipped with a robotic crane to assist maintenance and assembly work. Propellant tanks were mounted behind the hangars to refuel the docked spacecraft through feed lines.

In General Dynamics' scenario, a Space Station would become a space garage. However, the diameter of the hangars substantially exceeded the diameter of the Orbiter payload bays. They would have to be delivered to orbit in parts, requiring assembly into their final form together with the trussed beams, central hubs, propellant tanks and other equipment. All this would amount to a major constructional undertaking before the Station could be put into service and used as an orbital space tug base. As part of its study General Dynamics also produced a 'Space Station prospectus' that outlined what the company thought were realistic opportunities for private financing.[38, 39]

Most of the aerospace companies taking part in the study based their Space Station concepts on building blocks delivered by the Orbiters. Though not a stated NASA requirement, it was a logical and sensible approach as the Space Shuttle system had now become operational, its ability to lift payloads to orbit was proven and it was ready for a big job. Martin Marietta Aerospace was the fabricator of the External Tanks, the large propellant tanks that provided the Space Shuttle system with liquid oxygen and hydrogen propellants during launch and ascent. These normally separated from Orbiters just before reaching orbit and the tanks fell back into the Earth's atmosphere and burnt up. In its study, the company used the Space Shuttle as just one of three possible delivery approaches. In the other two it proposed different launch vehicles to loft much larger payloads to orbit.

Its first proposal featured building blocks that would fly in Orbiter payload bays in a carefully orchestrated Space Station build-up sequence, comprising modules added gradually on four axes to a long pressurised core service tunnel. The result was a cluster that flew parallel to the orbital plane with the core service tunnel aligned in the direction of flight. The tunnel continued as a trussed beam fore and aft with Spacelab pallets mounted at both ends for external experiment and equipment attachment. Solar arrays extended out on the port and starboard sides. As some other contractors had done, Martin Marietta proposed adding large hangars to enable a Space Station to service spacecraft.

Space Shuttle External Tank Martin Marietta built the giant External Tanks that provided the Shuttle Orbiters with cryogenic propellant during launch and ascent to orbit. In its 1983 Space Station Needs, Attributes and Architectural Options Study for NASA, the company proposed to add a cylindrical extension at the bottom of the tank called the aft cargo carrier to deliver some of the Station building blocks.

1982-83 Space Station Needs, Attributes and Architectural Options Study – TRW and Hughes designs To the far right in the image opposite, TRW's evolved Station design in NASA's 1982-83 Space Station studies had a loop comprising two habitation modules (one long and one short), a junction module and two tunnel modules. Projecting out from one tunnel module were laboratory and logistics modules. To the left, the design of Hughes Aircraft Company was utterly unlike any of the others. It utilised four empty Space Shuttle propellant tanks as spokes of a rotating wheel Station, outfitting them with living quarters on the inside. The circular rim comprised four ribbons of solar arrays with each forming a 90° curve. At the wheel's hub was a laboratory module. At each of the four intersections of the spokes with the rim were propellant storage tanks. Image: author.

Martin Marietta's second proposal did not involve the Shuttle Orbiters. The company proposed to add a cylindrical extension called the aft cargo carrier to the back end of its External Tank. The company pointed out that Space Shuttle transportation costs were going to be significant and that its payloads were limited in volume. The aft cargo carrier could be more cost effective by delivering larger building blocks for a Space Station. Martin Marietta's third proposal involved the creation of an entirely new large payload carrier called a Shuttle Derived Vehicle. It was a large cylindrical fairing that incorporated a large diameter pressurised volume for delivery to orbit on a single launch, like the earlier Skylab laboratory. It had a nose cone at the front and the Orbiter engines at the back and it replaced the fuselage, cockpit and wings with a fully outfitted habitable module. It was perhaps natural for the company as the builder of the large External Tanks to promote Space Station approaches that could use its large-scale fabrication skills. Martin Marietta concluded that a Space Station would be affordable within predicted budgets and beneficial to its user community by providing continuous research time in space. It would 'pay for itself' by adding value to the Space Shuttle transportation system.[40, 41]

TRW Space & Technology Group 'elected to come down on the side of minimum programme cost and risk' in its study. It believed that a modest initial Space Station that had a good chance of timely deployment for a stated budget was preferable to a programme that could easily experience substantial cost overruns and schedule slips. It proposed choosing early Space Station capabilities to generate the greatest benefits and accommodate a wide range of users. TRW canvassed views about a Space Station from individuals and organisations in the fields of science, satellite servicing, communications, materials processing, assembly and testing, space tourism, foreign interests and remote sensing. TRW came up with a single design for a Space Station on a 28.5° inclination orbit supplemented by free-flyer platforms on the same orbit and on a polar orbit. Four Orbiter flights would deliver an initial set of building blocks to support a crew of five comprising a habitation module, a laboratory module, a supplies module, a node module, a service spine with attached solar arrays and radiators, and an external assembly platform. The node module connected the laboratory and habitation modules on each side of it in a line. Pointing to zenith from the node module was the supplies module. Pointing out on one side was the services spine with twin solar arrays and radiators. On the opposite side was the external assembly platform aligned parallel to the habitation and laboratory modules. The modules had different lengths made up of standard cylindrical segments and again followed the design approach of Europe's Spacelab module. This initially compact design would evolve over a decade to a fully capable Space Station for a ten-person crew. There would be more modules and airlocks, more solar arrays, a hangar, propellant storage, robotic arms and a service station for an orbital space tug.

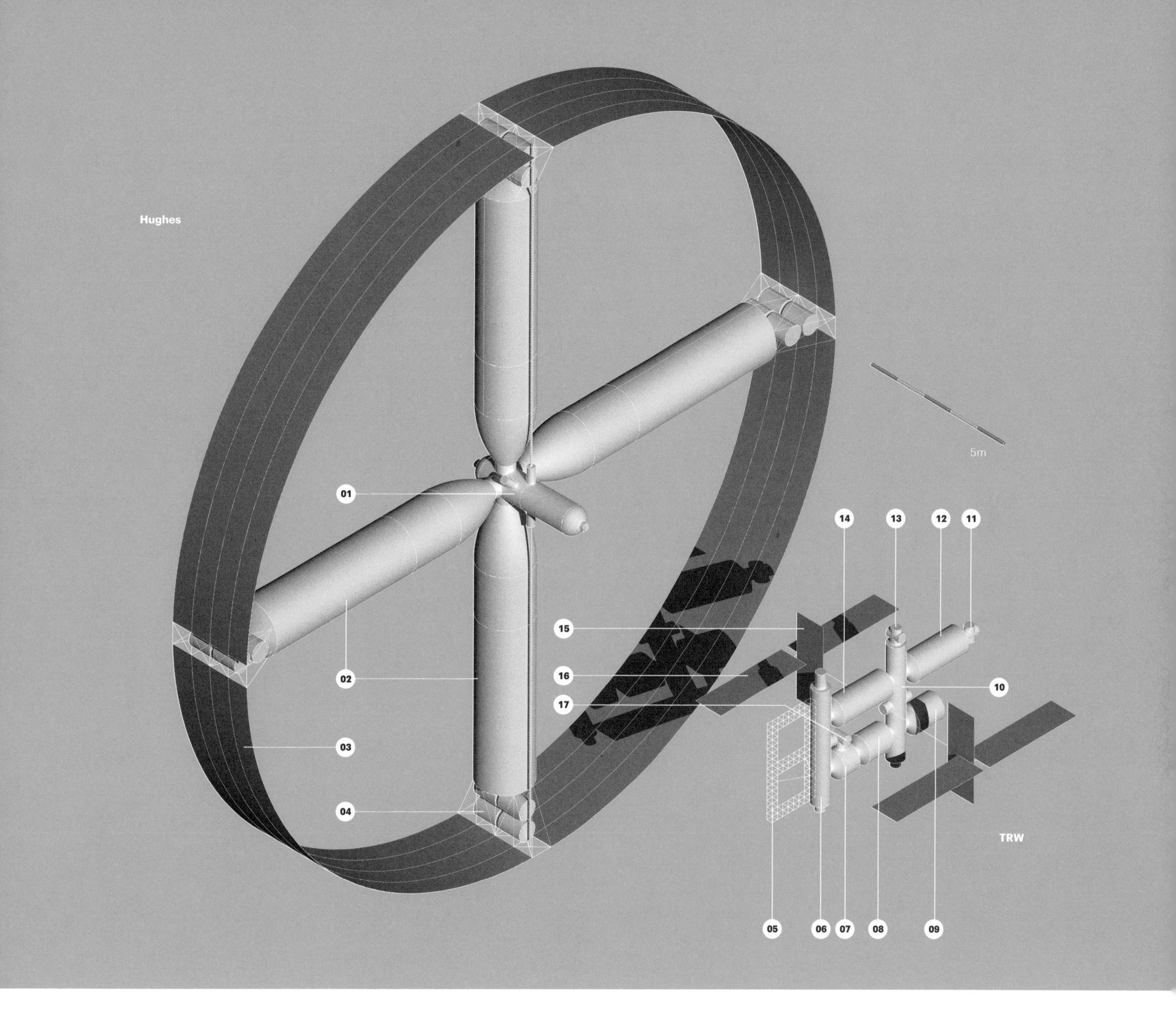

01 Laboratory module
02 Shuttle propellant tanks outfitted as habitation modules (4)
03 Solar arrays (4 × 90º segments)
04 Propellant storage

05 Assembly platform
06 Tunnel module
07 Junction module
08 Short habitation module
09 Logistics module

10 Tunnel module
11 Airlock
12 Laboratory module
13 Airlock
14 Long habitation module

15 Radiator
16 Solar array
17 Airlock

TRW described its design as an evolving modular system based on the use of the Orbiters for all deliveries. All the building blocks fitted into the payload bays. There would be no 'evolutionary dead-ends' and all modules would stay in continuous use. Nothing was discarded or wasted. TRW commented that a Space Station would be a substantial but profitable investment. Its transportation savings would lead its benefits. This was because the average 64 per cent load factor of Orbiter flights would increase to about 82 per cent by delivering extra equipment, space parts and propellant when spare payload bay volume permitted. There would be benefits from satellite delivery and servicing on orbit, a permanent laboratory, a base for space tugs and the ability to assemble large structures. In the long term, the high number of Space Shuttle missions to a Space Station would decrease. TRW concluded by anticipating that a Space Station would become a portal to future space exploration. Like the Wright Brothers' first aircraft and the first Sputnik spacecraft, a permanently crewed Space Station on low Earth orbit would open a door to the future.[42]

Of the eight contractors that NASA commissioned on the Space Station Needs, Attributes, and Architectural Options Study, the joker in the pack was Hughes Aircraft Company. Hughes came up with the most intriguing and radical design that ignored an incremental assembly approach using the Orbiters. Instead, Hughes proposed the use of four empty Space Shuttle External Tanks refitted on orbit for habitation. It arranged them like spokes in a giant wheel with a circular ribbon of solar arrays forming the wheel's rim and providing a massive 150kW of power. The wheel would rotate to generate centrifugal force to create artificial gravity inside the habitat tanks. At the wheel's centre was a non-rotating laboratory that provided weightlessness. Hughes was an expert at spin-stabilised satellites and saw an opportunity to use the same approach at a much larger scale.[43] Hughes' innovation was a refreshing change but it is questionable whether the idea would ever have worked as envisioned because the level of artificial gravity in a rotating Space Station is determined by its rate of spin and the radius distance of its accommodation rim from the hub. The shorter the radius, the higher the rate of spin needed to achieve artificial gravity approaching that of the Earth and a high spin rate induces sickness and nausea. In the Hughes concept, the lengths of the tanks resulted in a short radius. Rotating Space Station concepts need very large radii to keep the spin at a tolerable rate.

It is important to give recognition to the influence of Europe's Spacelab as a role model on several of the early Space Station design studies between 1979 and 1983. While the studies were in progress, module and pallet elements had arrived at Kennedy Space Center and were undergoing processing and integration for Spacelab's maiden flight. The pressurised laboratory fitted snugly into the Orbiter's payload bay and made maximum use of its cross-sectional volume. Inside the laboratory were a floor, experiment racks and overhead storage. Racks at the forward end and in subfloor volume contained Spacelab's utility systems. The laboratory's cylindrical segments, manufactured in Italy by Aeritalia, were made of roll-formed waffle pattern aluminium alloy sheets welded at the seams with end flanges of forged aluminium to join the segments and the end cones together using bolts. The U-shaped pallets, manufactured by Hawker Siddeley Dynamics (later part of British Aerospace) in the United Kingdom, were also made of aluminium alloy. Their design allowed for mounting lightweight experiments on the inner surfaces of the U-shape or heavy experiments on hardpoints that transferred the loads to the main pallet structure and thence to the Orbiter through attachment fittings.[44, 45] Spacelab was real module hardware and it was highly versatile. Both the pressurised module and the pallets amply demonstrated their potential for use on the Space Station, whether directly or in some evolved form.

NASA had tasked the eight contractors of the 1982 Space Station Needs, Attributes and Architectural Options Study with describing the requirements, concepts, missions and benefits for a new Space Station. The agency certainly got what it wanted. The studies offered a wealth of advice on the requirements, concepts, missions and benefits for a new Space Station, providing NASA with plenty of information to promote and argue the case for a Space Station at both national and international levels. Less clear was the design direction a Space Station was taking. It was not the priority of the studies and the contractors had given it varying amounts of attention. Seven out of eight of them had envisioned designs based on the incremental assembly of modular building blocks sized for flight in the Orbiter payload bays. Many had proposed evolutionary growth stages that started small and ended large. Some had included complicated and extensive structures that would require assembly on orbit. Modules were all shapes and sizes with some pressurised and some not. There were narrow connecting tunnels of different lengths, airlocks that poked out here and there, and bits of structure tacked on as external platforms. Little attempt had been made to optimise the building blocks for delivery in the Orbiter payload bays. Orbiter approach and docking and Space Station flight attitudes were not always clear. Neither was the robotic choreography needed for step-by-step assembly. Internal accommodation arrangements were generally ignored.

Despite these shortcomings, NASA for the first time had engaged a group of aerospace companies in helping to define the direction for a Space Station. Something else was becoming clear. The modular and incremental assembly approach favoured in the studies could work well for NASA's prospective international partners. With common berthing and interface techniques, the partners would be able to develop and label their own building blocks, thus preserving continental (in the case of Europe) and national identity, which was so essential for domestic political and public consumption.

Space Station genesis President Ronald Reagan shares a joke on the telephone with the crew of Space Shuttle Mission STS-2 orbiting the Earth in November 1981. Standing from left to right are Terry Hart, Hans Mark (NASA Deputy Administrator), James Beggs (NASA Administrator) and Christopher Kraft (Director of Johnson Space Center). Seated from left to right are astronaut Dan Brandenstein (CAPCOM) and the President. With the Space Shuttle flying in 1981, NASA was beginning to focus on the Space Station. The agency had commissioned some early contractor studies at the time this photograph was taken, but it would be three more years before President Reagan would formally announce the project. Beggs and Mark became the driving force behind the Station in the early 1980s. They campaigned in Washington, often against institutionalised opposition, and took the project overseas on a marketing tour.

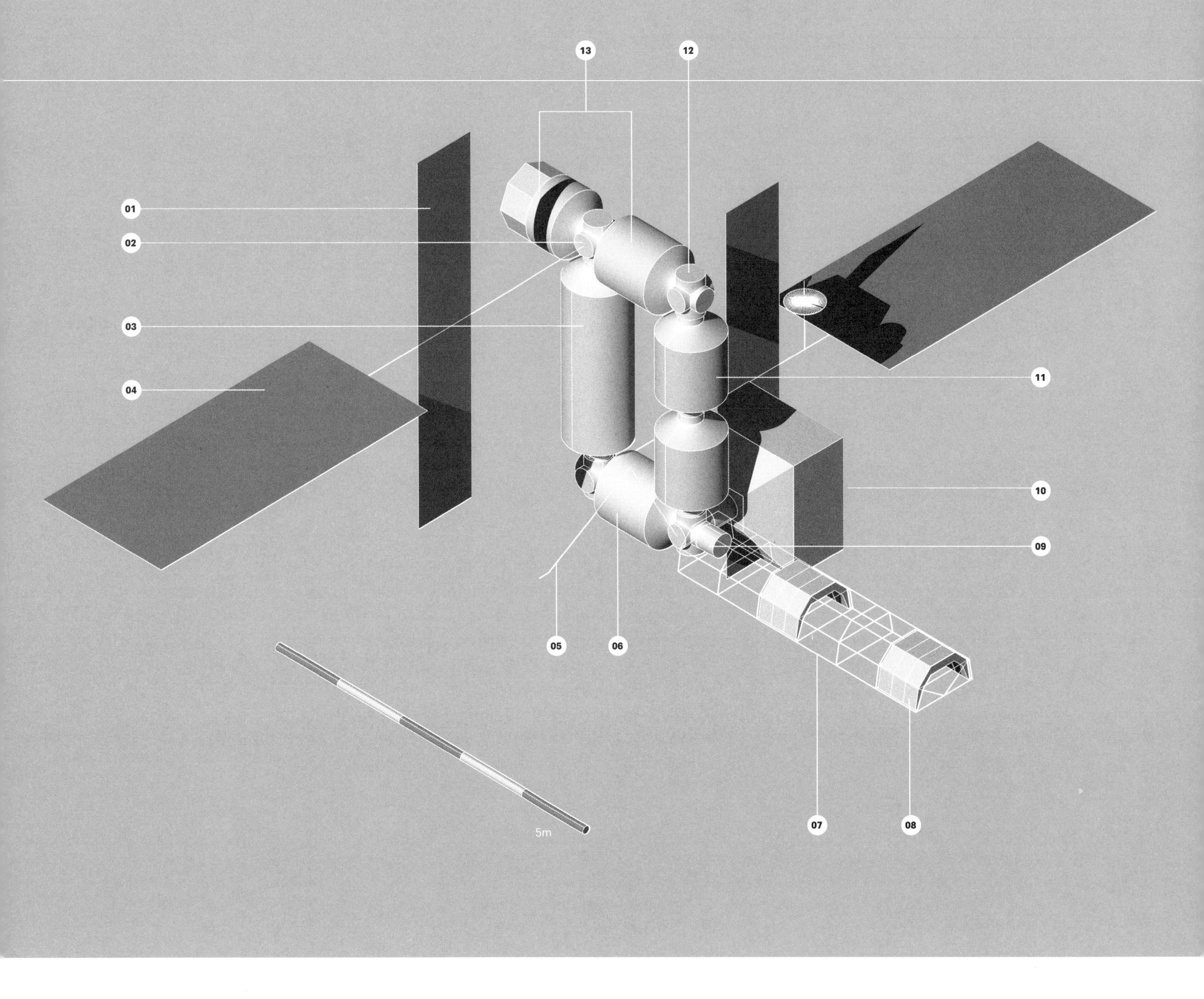

01	Radiator	**05**	Robotic arm	**09**	Airlock and node	**13**	Logistics module
02	Node	**06**	Command module	**10**	Hangar		
03	Habitation module	**07**	Exterior payload support structure	**11**	Electrical power module		
04	Solar array	**08**	Science pallets	**12**	Nodes (4)		

Many issues
Space Station Symposium, 1983

The time was approaching when the Reagan Administration would have to make a decision on whether to proceed with a Space Station. In July 1983, NASA held its first public forum on the project with the aim of raising public awareness and interest and airing the multitude of issues and challenges involved. The American Institute of Aeronautics and Astronautics organised it as a wide-ranging symposium and it took place at Arlington across the Potomac river from Washington DC.

NASA Administrator James Beggs gave the keynote address and described a Space Station as: 'the node or point in space in which we would do our long-duration work'. It would give America the means to venture beyond Earth orbit. He was concerned about growing competition from the Soviet Union, Europe and Japan and hoped that Space Station planning that had been under way for over a year would result in an approved programme in the near future. He concluded with a quote from the Bible that 'where there is no vision, the people perish'.[46]

The chairman of the symposium was Philip Culbertson, NASA Associate Deputy Administrator, who stated that the symposium's objective was: 'to present in a concise but fully documented manner to all interested sectors and individuals the current thinking on the designs, technology, and applications of an Earth-orbiting Space Station'.[47] He believed that public debate on the Station's 'pros and cons' was important to the decision-making process. The symposium covered a great deal of ground with papers on how the Station would host activities in life sciences, Earth sciences, commercial telecommunications, pharmaceutical manufacturing, materials processing, satellite servicing, space transportation and on-orbit assembly. There were papers on architectural options and issues, national security implications, user community opinions, automation, artificial intelligence, communications, information and life support.

Hans Mark, NASA Deputy Administrator, recalled a yacht race with Wernher von Braun off the California coast in the 1970s. On a long downwind run back to port when the crew had time to relax, von Braun began a discussion about the future of space exploration. Von Braun compared space exploration to the early exploration and settlement of Antarctica following the introduction of aircraft services. Thinking far ahead, he thought that the Apollo missions would eventually lead to a lunar settlement and the introduction of lunar spacecraft based at a Space Station to service that settlement.[48]

Luther Powell, Concept Development Manager on NASA's Space Station Task Force, gave an account of the agency's thinking on the architectural anatomy of a Space Station. Its functional capabilities would include power, thermal control, data processing and transmission, stability and attitude control, a liveable environment, scientific research, pilot production and

1983 NASA-JSC Design and Evaluation of Selected Space Station Concepts – The Building Block design In 1983, Johnson Space Center in Houston carried out a study of three Space Station concepts that the Space Station Task Force at NASA Headquarters thought were worth investigating. The first of these was called the Building Block with many similarities to designs proposed by the aerospace contractors in their 1982 studies. It utilised a loop of modules that NASA called a 'quadrangle'. The loop comprised a habitation module, two electrical power modules, two logistics modules and a command module, all connected by nodes. A payload support structure with science pallets projected out from an airlock at one corner. Image: author.

1983 NASA-JSC Design and Evaluation of Selected Space Station Concepts – The Delta design In the second of its concept studies for NASA's Space Station Task Force, Johnson Space Center explored a megastructure design for the Station, called the Delta. Much larger in scale than anything that NASA had contemplated before, it comprised a prism-shaped structure made up of three trussed panels. The triangular volume inside was open and intended for future space operations such as satellite servicing or construction. A solar array covered the whole outer face of one panel. Attached along the bottom lateral edge was NASA's 'quadrangle' of pressurised modules. Image: author.

manufacturing, all supported by humans and machines working together. Its architectural options could begin with extended Orbiter missions and an automated platform, followed by a habitable platform occupied by crews for limited durations, followed in turn by a permanently staffed Station. Its principal architectural features would comprise a habitat, laboratory and production facilities, science and applications experiments, servicing equipment, logistics storage and a transportation docking capability. Supplementing a Space Station on orbit would be free-flyer platforms, reusable space tugs, and regular visiting Orbiters. There would be many development challenges to tackle including safety, maintainability, modularity, autonomy, controllability, stability and habitability.[49]

Bold visions
Studies of Building Block, Big-T and Delta Concepts, 1983

As 1983 drew to a close, Johnson Space Center in Houston carried out a study of three different designs for a Space Station. The study was completely independent of the contractor studies that had preceded it. It was called the Conceptual Design and Evaluation of Selected Space Station Concepts and Johnson Space Center published the results in December. As NASA's leading field centre for human spaceflight, Johnson Space Center had spearheaded studies of a Space Station for many years, on and off, both internally and through contractors. The Space Station Task Force and Concept Development Group at NASA Headquarters had identified a large number of Space Station trade studies and Johnson Space Center had carried out several of these based on an internal statement of work. NASA Headquarters had now pinpointed three existing design concepts that merited further examination. In November, an interdisciplinary team of designers, engineers and scientists in Houston tackled them, with visionary results in two of the three cases. The first concept was named the Building Block. It explored a cautious, modular approach that Johnson Space Center and some of NASA's contractors had followed in earlier studies. The second and third concepts, however, seemed to throw caution to the wind with their proposals for a giant triangular trussed megastructure called the Delta and another giant trussed structure called the Big-T.[50]

The Building Block design was relatively modest. Its aim was to minimise structure and subsystems hardware by using pressurised modules as the structural foundation to which other elements would be attached. The initial design had a cluster of six modules linked together in a loop that the team described as a 'quadrangle'. Like some but not all of the earlier Space Station designs, this was a safety feature that provided two exit paths in an emergency as well as efficient crew circulation. The quadrangle consisted of a long habitation module, a short command and control module, two short logistics modules and two electrical power units built in modular form.

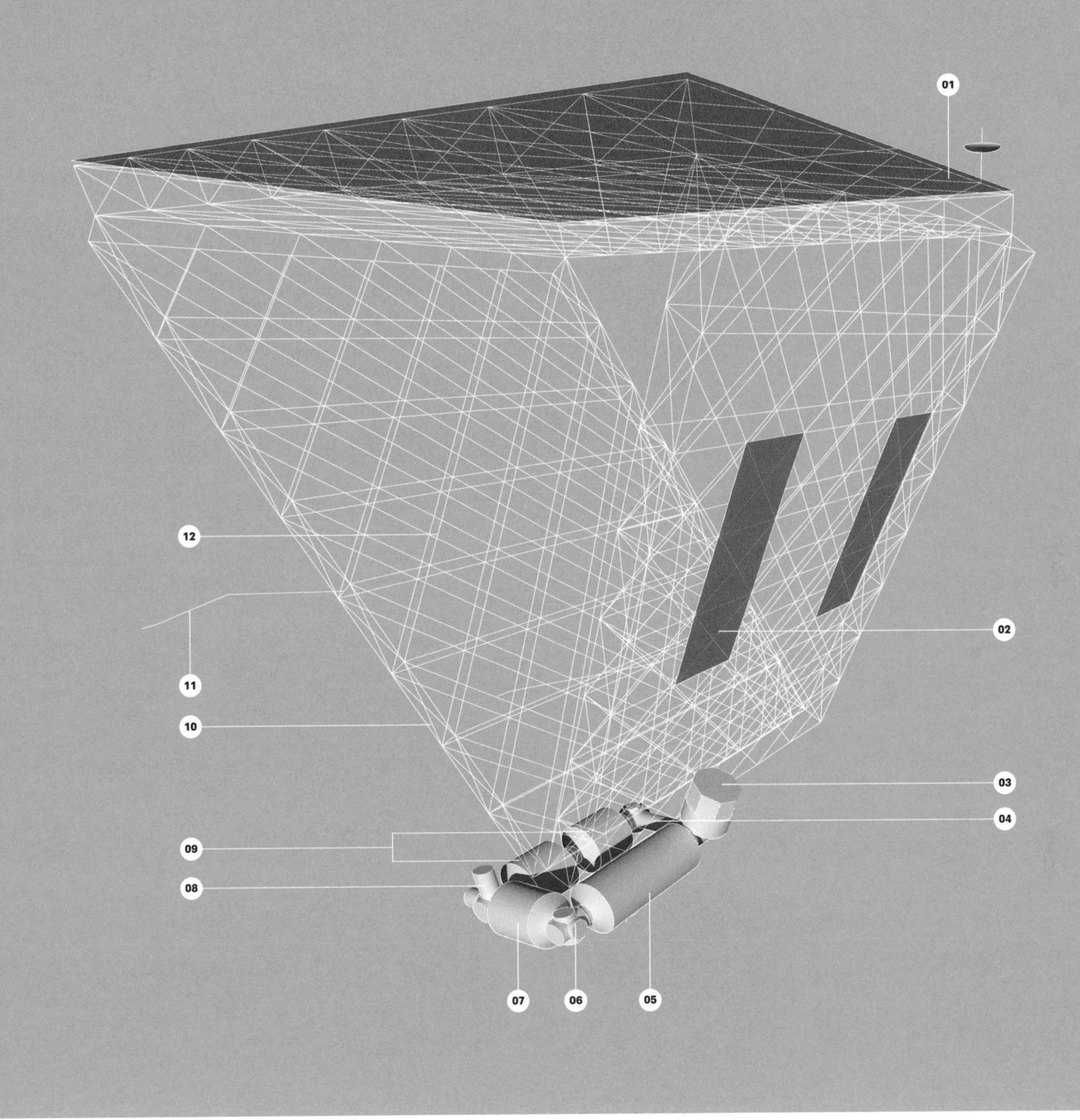

01 Solar array deck
02 Radiator
03 Logistics module
04 Connecting tunnel
05 Habitation module
06 Nodes (4)
07 Command & control module
08 Airlock
09 Laboratory modules
10 Delta truss structure sides (3)
11 Robotic arm
12 Internal hangar volume for future equipment additions and space operations

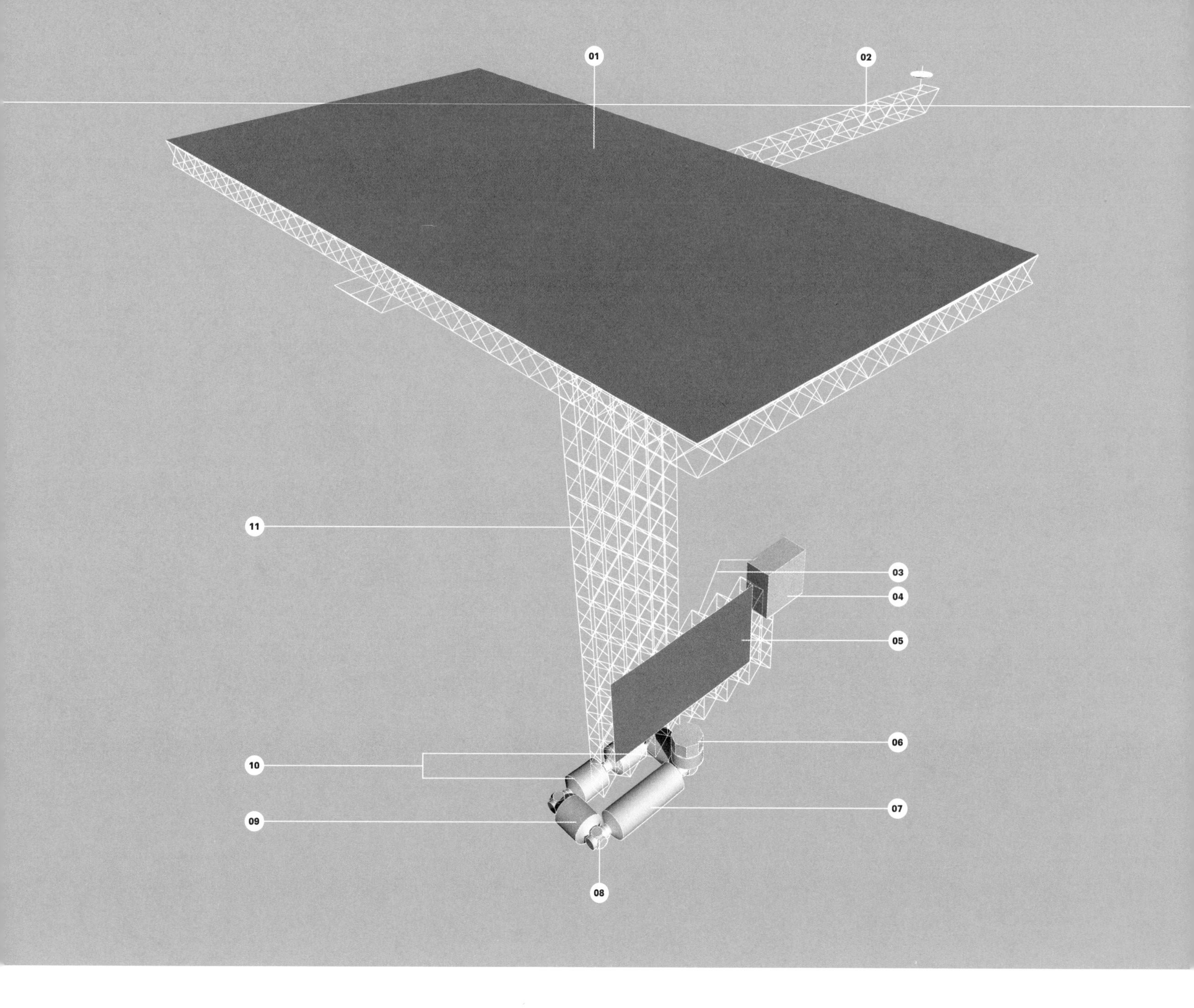

01 Solar array deck
02 Truss structure boom
03 Robotic arm
04 Hangar
05 Radiator
06 Logistics module
07 Habitation module
08 Nodes (4)
09 Command & control module
10 Laboratory modules
11 Truss structure mast

There were four nodes to connect the modules, each with six berthing ports. Two booms perpendicular to the plane of the quadrangle provided mountings for solar arrays, radiators and antennae. Extending out from the quadrangle corners were a box-like hangar, a device that resembled a robotic arm, and an external experiment structure. The Space Station flew with the quadrangle parallel to the orbital plane and the long side of the quadrangle vertical. This provided gravity gradient stability, clean forward approach paths for Orbiters and space tugs and, good Earth and celestial viewing. There were two Orbiter docking and berthing ports for attaching temporary modules, manipulator arms and hangars. The Station would need to rotate on the port-starboard axis to reboost its orbit because of the thruster locations. The provenance of the Building Block design is evident in the Johnson Space Center and Boeing designs for a Space Operations Center from 1979 and 1982. The Building Block replaced the earlier paired module and narrow service passageway loop with the module quadrangle, making all modules useful for habitability functions. The Orbiters docked with either a short module or a node. The design borrowed the Spacelab pallets from earlier McDonnell Douglas designs but inserted them into a long, framed cage to form an experiment and equipment beam that pointed in the aft direction, adding a tail to the quadrangle.

Deeply offset solar array and radiator booms replaced the earlier in-line or nearly in-line boom arrangements. Gone from the initial design were extraneous bits and pieces. It focused on the essentials that a small-crewed Space Station needed to become operational. Accommodation layouts inside the modules followed the approach shown by Boeing in its 1982 Space Operations Center study. Longitudinal decks and ceilings ran the full length of the modules. Under the floor deck was a service zone with water tanks, waste tanks and chiller equipment. Above the ceiling was the life support system with carbon dioxide removal, oxygen revitalisation and ventilation equipment. In the habitation module, the volume between floor and ceiling contained private crew compartments, a shower, a washbasin and toilet facilities, food preparation workstations, a communal table and an exercise area. In the command and control module, the volume contained three crew workstations, a food preparation area, a toilet and equipment racks.

After a trade-off study, Johnson Space Center chose a life support system with a partially closed cycle for all three Space Station designs. A water electrolysis process drawing on a supply of hygiene water provided the oxygen. A reduction system collected the carbon dioxide to convert it to water via hydrogenation. Water produced in this way was combined with humidity condensate to produce water for drinking and food preparation. Shower waste water and urine were processed through multiple treatments to produce clean shower water. Nitrogen gas came from high pressure or cryogenic tanks. In the thermal control system, coolant fluid from the boom radiators circulated through a looped heat pipe system to heat exchangers

1982 NASA-JSC Design and Evaluation of Selected Space Station Concepts – the Big-T design Equally impressive in scale as Johnson Space Center's giant Delta Space Station concept was this third concept called the Big-T. In 1983, NASA's design ideas for the Space Station were becoming more ambitious than those of its aerospace contractors. The T's mast and crossbar both used open trussed panels. At the base of the mast was NASA's 'quadrangle' of modules with a hangar mounted on a panel piece that extended outwards. Solar arrays covered the outer surface of the crossbar panel. The idea of big structure on the Space Station had caught on and would recur in the first detailed design for the Station the following year. Image: author.

Artist impressions of 1983 NASA-JSC Space Station concepts Top: The Building Block Space Station had a 'quadrangle' of seven modules with a docked Shuttle Orbiter at one corner and a projecting U-shaped external experiment structure at another. Bottom: An expanded version of the Building Block design that included three more modules above and spacecraft and satellite servicing hangars below. In this view, the Orbiter is docked at the Station's underside.

in the modules where it collected module heat, was vaporised and then circulated back to the radiators. The Building Block growth design added another module quadrangle and experiment structure mounted in a mirrored orientation above the first quadrangle and pointing to zenith. Below the first quadrangle and pointing to nadir was another module cluster, propellant storage tanks and spacecraft servicing hangars. Added to the two booms were more radiators and more solar arrays extending out at right angles from the base of the existing arrays.[51]

It is not easy to fathom what NASA's motive was in pursuing the second Space Station concept named the Delta design, as there was little explanatory information in the study's reports beyond a reference to the Delta's structural stiffness. Johnson Space Center described the design's principal feature as 'a large deployable triangular truss structure'. This was an understatement. The Delta design was a megastructure of such a size that, if placed on the ground, it would have made a good triumphal arch. Its geometry was that of a hollow extruded equilateral triangle. The triangle's sides were trussed in profile, each 44.5 metres long, 2.2 metres thick and joined at the corners. Extruding these three sides for a distance of 38 metres turned them into three-dimensional tetrahedral truss structures. Their combined surface area amounted to 5,085m², equivalent to 71 per cent of a football pitch in size. Solar arrays covered one complete face of the structure and produced 60kW of power. The other two faces supported radiators, power conditioning equipment, experiments, payloads and pressurised modules. In the initial design the modules formed a quadrangle that repeated the Building Block design but with a tunnel added to close the quadrangle's loop. The quadrangle's long axis was mounted along one of the Delta's leading edges and parallel to it. The quadrangle's modular arrangement was tweaked slightly to make it fit with the 60° angle. The growth design swelled the size of the megastructure by doubling the area of the Delta's side with the solar arrays, boosting them to 120kW and increasing the length of the other two sides for more radiators and equipment and another two quadrangles of modules. Hangars occupied the hollow prismatic volume inside.

The Delta Space Station would fly in a solar orientation perpendicular to the orbital plane. The solar array side always faced the Sun regardless of where the Station was on orbit.[52] The orientation rotated slightly for seasonal sunpath variations in the northern and southern hemispheres.

What was the purpose of the Delta's vast expanse of structure? The documents stated briefly that it would provide maximum rigidity, enhanced controllability and mission versatility. Would the Shuttle Orbiter fleet be able to deliver such a huge quantity of components to orbit to build the Delta's structure? There were basically three ways to build large structures on orbit. The first way used on-site assembly. An Orbiter would deliver a payload bay full of structural components such as struts and nodes and astronauts would go on spacewalks to plug them together manually, perhaps with robotic support. The second way used prefabricated elements. Three-dimensional structural segments would be prefabricated before launch in shapes and sizes that made the most of the Orbiter's payload bay volume. After arrival on orbit, a mechanical arm would remove them from the payload bay and attach them to each other one at a time, mission after mission. This is how the Space Station's pressurised modules would be linked together. The third way used deployable structures. Large structural elements would be built on the ground with hinged joints enabling them to fold up and stow inside a payload bay, tailored to fit the payload bay's shape and size. Each structural bay would have some of its edges hinged at their midpoints enabling them to fold inwards like spokes on an umbrella. Once on orbit they would deploy outwards in a similar manner to a pantograph. Johnson Space Center proposed this third way for the Delta design as it was the most efficient of the three for a large, lightweight structure.

Also difficult to comprehend was NASA's choice of the third concept, named the Big-T design, as again there was little explanation beyond another reference to its structural stiffness. Like the Delta but even bigger, the Big-T was a megastructure of heroic proportions, shaped like a giant letter T with its crossbar perpendicular to the orbital plane and its post orientated from nadir to zenith. Like the Delta, the mast and crossbar were trussed in profile and extruded to turn the T into a three-dimensional structure. The post widened out at its base to support modules and equipment. It would also offer rigidity (though less than the Delta's triangle, as a T is not a self-rigidising shape), versatility and plenty of surface area for solar arrays on the crossbar's topside and radiators on its underside. The module cluster repeated the quadrangle of the other designs with the loop completed by a tunnel at one end. It was placed at the bottom of the Big-T's mast, the lowest point of the design. Above the modules, the mast's sides supported hangars and other external equipment. The initial design had half of the crossbar structure in place with the other half added in the growth design, both supported by twin girders that attach them to the post. The Big-T would fly on orbit in an Earth-oriented attitude with the mast and crossbar flying end-on into the flight path to minimise drag. The structural approach followed that of the Delta design with structural segments stowed for delivery in Orbiter payload bays and then deployed and unfolded on orbit.[53]

Johnson Space Center stated that its three design concepts were neither recommended nor optimised designs but merely tools for criteria evaluation. It carried out an in-house evaluation of them and discussed the benefits and penalties of each. The study took place quickly without the level of attention to requirements, mission objectives or user benefits paid by the eight contractors in the 1982 Space Station Needs, Attributes and Architectural Options study. The argument given for the introduction of

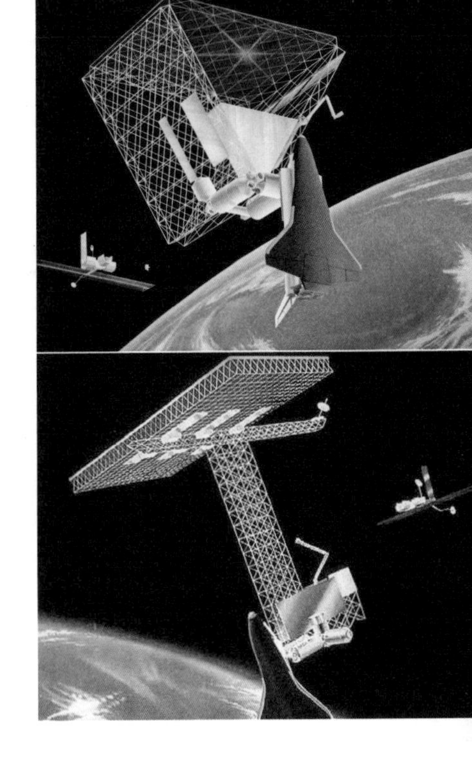

Artist impressions of 1983 NASA-JSC Space Station concepts Top: The Delta Space Station with a docked Orbiter, a triangular hangar inside the truss structure and a separate free-flying platform for science research. Bottom: The Big-T Space Station shown with a docked Orbiter and a free-flying platform. In these two images, the free-flying platforms would not fly close to the Stations as depicted.

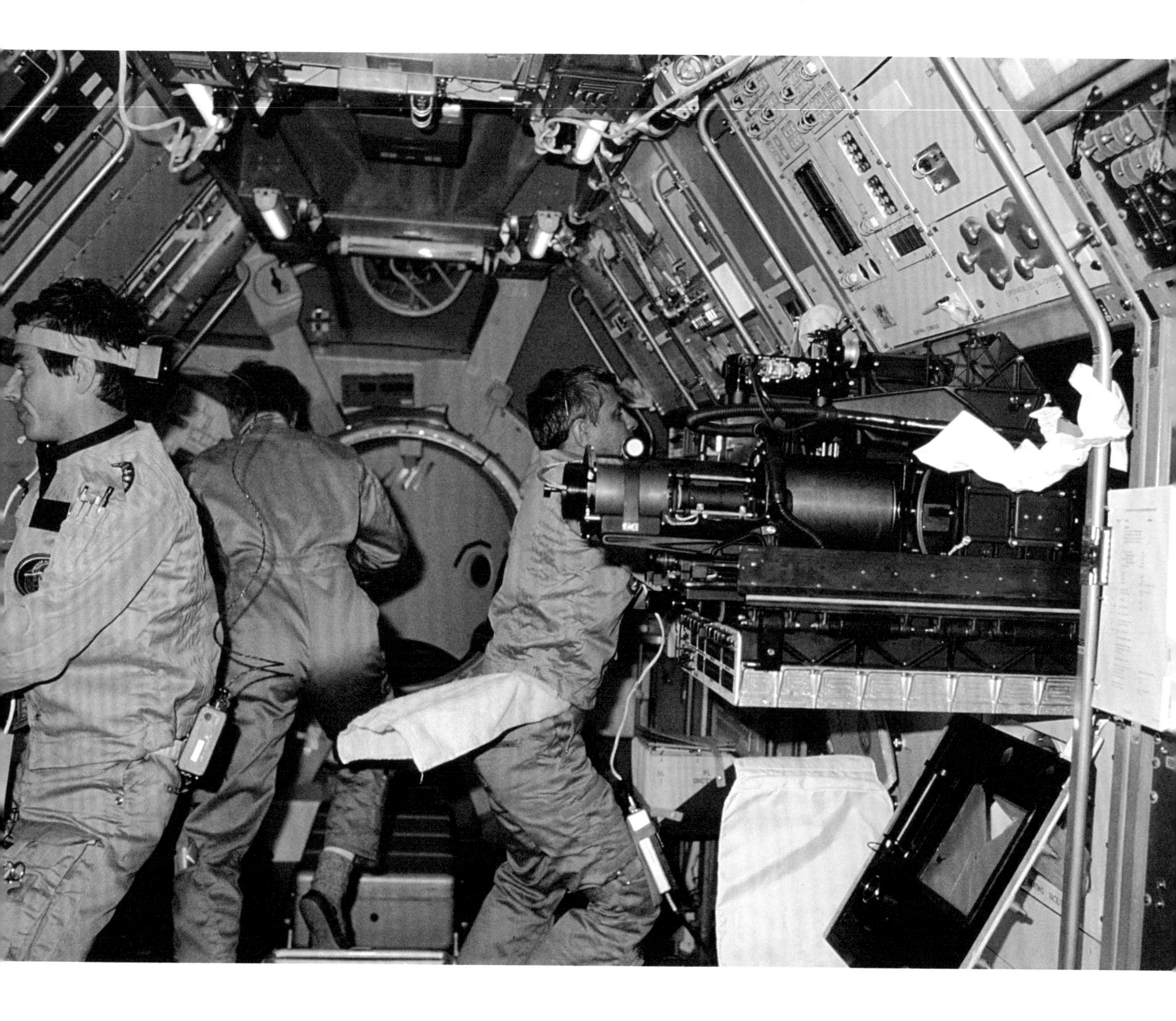

a lot of structure was the need for stiffness to avoid the dynamic control challenges of flexible designs that had solar arrays and radiators extending outwards from clustered modules on long flimsy booms. There were some assumptions about the Delta and Big-T designs that the study did not evaluate. One was that the huge deployed structural frames of which NASA had no prior experience would fit efficiently and economically into the Orbiter payload bays in stowed form and deploy reliably and safely on orbit. Testing such huge structures under weightless conditions on the ground before launch was impossible. Another assumption was the outline cost estimates, which claimed, incredibly, that the Delta design with its vast additional structure would be cheaper to build than the Building Block design. The dynamic controllability of large flexible structures later proved to be manageable as subsequent Station structural designs showed. The International Space Station's solar arrays bear witness to the feasibility of using flimsy structures in space that can turn to face the Sun as they orbit the Earth. Perhaps further engineering analysis and evaluation of the Delta and Big-T could have disposed of this assumption and more efficient and slimmer structures would have emerged.

The Delta and Big-T concepts were unprecedented and marked the first time that NASA had proposed large structural frameworks as the principal ingredients in a Space Station recipe. They were bold and visionary, but also flawed and unrealistic. Perhaps the Space Station Task Force wanted to widen and enliven the debate about what a Space Station should be after a series of more modest contractor concepts. Possibly Johnson Space Center hoped that one of its own ideas would prove to be a seed that would germinate and blossom into the preferred solution at NASA Headquarters. What is certain is that it had been a valuable design exercise. The introduction of such physically adventurous structures had introduced the idea that some sort of large exterior structure had an important role to play. This new element would affect the course of the Space Station's design from that point on.

In November 1983, an important event took place that had much significance for the Space Station programme. Launched on Space Shuttle Mission STS-9 inside the Orbiter Columbia was Spacelab-1. It was the first flight of Europe's space laboratory and the first time that a pressurised module tailored to the shape and size of the Orbiter payload bays had flown to space. Working in a shirtsleeve environment between two banks of experiment racks, the crew showed what life inside the future Space Station would look like. It was also the first time that six humans had travelled together into space on a single spacecraft.[54]

Spacelab inaugural mission View inside the Spacelab Module on its first flight, Spacelab-1, on the Space Shuttle in November 1983. The module's diameter and length were sized to enable it to fit snugly into the Orbiter's cargo bay. The future Space Station modules would be similar. The central corridor bordered on each side by two banks of experiment racks gave an idea of what the Station's future working environment would look like.

ПРОГРЕСС

Super Typhoon Neoguri A photograph of Super Typhoon Neoguri taken from the International Space Station on 5 July 2014. Neoguri was headed towards Japan at the time. At the top centre is a docked Russian Progress resupply vehicle with the Russian word 'Progress' on its side. At the bottom left is a docked Russian Soyuz crewed spacecraft with the Russian word for 'Soyuz' on its side.

Deployment of NanoRack CubeSat experiments from the Station NanoRack CubeSats are experiments developed as miniature satellites for space exposure. The experiments cover fields such as Earth and space observation, space electronics testing and educational research. This photograph taken in February 2014 shows two sets of them just after their ejection from their launcher box. The launcher box was mounted on the end of the Japanese Kibo robotic arm.

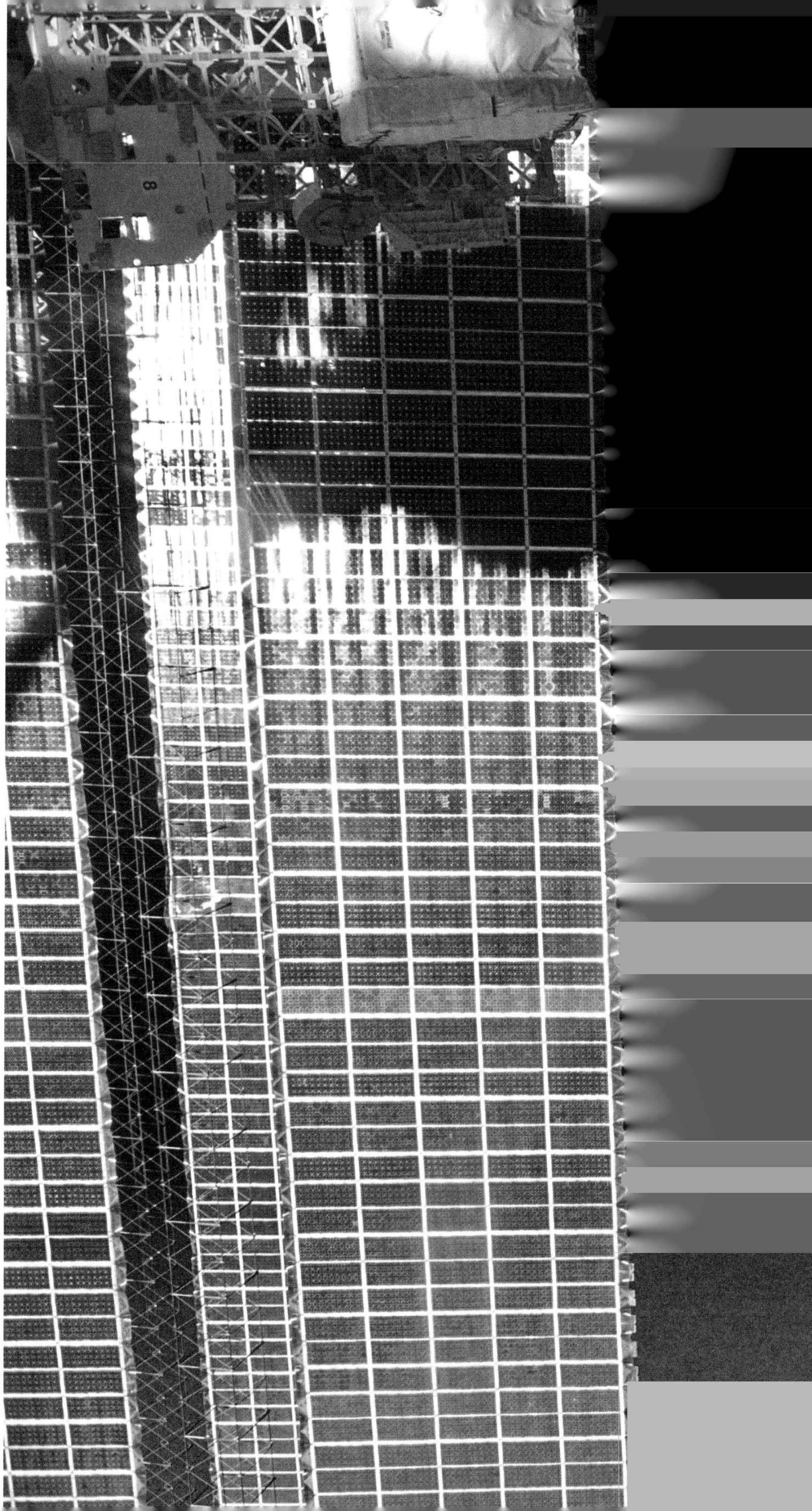

The Station's robotic arm in action over South Africa The Station's Canadarm2 articulated robotic arm is about to release the Cygnus commercial cargo spacecraft after its visit to the Station in August 2014. Cygnus spacecraft are built by Orbital Sciences Corporation. To the lower right is the Cape of Good Hope in South Africa.

MCDONNELL DOUGLAS
MEXICO
HUGHES
MORELOS
ACCESS

Ambition and grandeur 1984–1988

Ambitious proposition

Presidential announcement and Power Tower configuration, 1984-1985

In December 1983, President Reagan made the decision to proceed with a Space Station. It followed a successful White House briefing that featured a large-scale model of the Station supplied by NASA. The agency had first attempted to persuade Reagan in 1982 and the path to Presidential approval had not been easy. In Washington political circles there had been scepticism about the project's international nature and opposition to allocating any budget to it. Presidential Science Advisor George Keyworth had been against it and Department of Defense Secretary Caspar Weinberger had been strongly opposed. There had been concern at the Department of Defense and among the national security community about the transfer of American aerospace technology to other nations. Neither had the Department of State been enthusiastic about the foreign policy potential of the project. The Office of Technology Assessment, however, had pointed out that thousands of aerospace engineers and technical support staff who had worked on the Space Shuttle were available to work on a new project as the Shuttle's development had ended. NASA overcame the Washington objections by appealing to President Reagan and his advisers. NASA Administrator James Beggs was 'absolutely convinced that a space station is the next bold step in space'.[1] NASA had recommended to the Reagan

Structure assembly tests on a Shuttle mission On Space Shuttle Mission 61-B in November 1985, astronauts Jerry Ross and Sherwood Spring test-assembled two structures to explore the feasibility of astronauts constructing the Space Station's large truss structures by hand. In the ACCESS (Assembly Concept for Construction of Erectable Space Structure) test shown in this photograph, the astronauts spent thirty minutes building a triangular section truss from ninety-three struts and thirty-three nodal joints. The test was successful.

administration that the Station should have a crew of six to eight persons, 75kW of electrical power, a laboratory for life sciences, another laboratory for materials processing and two automated free-flyer platforms – one that co-orbited with the Station and the other that was on a polar orbit.[2] These were the Station's requirements at their most basic.

The right moment had arrived. The White House inserted an announcement on the Station into Reagan's State of the Union Address in January 1984 and used the opportunity to signal that the Reagan Administration would welcome international participation. Prospective international partners had been aware of the project for some time. The first international briefing on it had taken place at NASA Headquarters in September 1983. Representatives from the space agencies of Canada, France, Germany, Japan and Italy had been present and interest had been growing. The section of Reagan's State of the Union speech dealing with the Space Station was crafted for foreign as well as domestic consumption. It would be a bold political move aimed at securing and maintaining national space leadership while offering an appealing international venture that could be played to foreign policy advantage. In his speech before a joint session of Congress on 25 January, Reagan began by describing space as the next frontier and the best place to demonstrate the nation's technological leadership and ability to make life better on Earth. He then said the following: 'Tonight, I am directing NASA to develop a permanently manned Space Station and to do it within a decade. A Space Station will permit quantum leaps in our research in science, communications, in metals, and in lifesaving medicines, which could be manufactured only in space. We want our friends to help us meet these challenges and share in their benefits. NASA will invite other countries to participate so we can strengthen peace, build prosperity, and expand freedom for all who share our goals.'[3]

Shuttle Orbiter Enterprise The Shuttle Orbiter Enterprise is seen mounted on top of NASA's Shuttle Carrier Aircraft, a modified Boeing 747, in April 2012 at Washington Dulles International Airport. Enterprise was the first Orbiter that never flew to orbit but was vitally important for aeronautical flight tests during the Space Shuttle development years. In the early 1980s, NASA Administrator James Beggs sent Enterprise mounted on its Boeing 747 on a tour of Europe to boost interest among Europe's spacefaring nations in joining the Space Station project as partners. Image: NASA/Bill Ingalls.

James Beggs and his team at NASA had been busy drumming up international support for the Space Station in the run-up to the Presidential announcement. Beggs had sent the prototype of the Shuttle Orbiter spaceplane, mounted on top of a specially adapted Boeing 747, to show it off at various airports in Europe to boost interest and it had the desired effect. By 1982, the European Space Agency's Council had adopted a resolution to work with NASA on the Space Station. Some countries were more supportive than others. The British were keen at first, the French were cautious and the Germans were somewhere in between. With their experiences and know-how acquired during the development of the Spacelab module and pallet system, Germany and Italy soon took the technical lead in the new initiative. The first formal meeting at which NASA briefed prospective partners on a Space Station had taken place at Johnson Space Center in 1982, two years before the Presidential approval. By the end of 1983, German and Italian interest had come into focus with a proposal for a pressurised laboratory module contribution to the Station. After President Reagan announced the Space Station in January 1984 though, there were some odd reactions in Europe. President François Mitterrand of France stated that Europe should launch its own Space Station for military purposes and his comment rang alarm bells at the European Space Agency. Mitterrand later dropped this position.[4]

It had taken five years since the start of serious technical studies and much time and effort to bring a Space Station served by the Space Shuttle from a concept in Johnson Space Center's mind's eye in 1979 to a Presidential announcement for the world to note in early 1984. NASA and its contractor teams had worked on ideas and developed concepts for the Station, identified its requirements, scoped out its missions, delved into its technologies, defined its benefits and guessed at its costs. Rooted in the success story of Skylab, and with Spacelab as a role model, a generic Space Station concept had emerged as a cluster of pressurised modules of different shapes and sizes, garnished with a variety of bolted-on pieces of equipment. The idea of adding big structure to the module cluster to give the bolted-on equipment room to breathe and the Space Station a stable foundation for growth had been mixed in as another key ingredient by Johnson Space Center while the Space Station at large had been politically seasoned to give it international flavour and appeal. But there was still no firm design to drive all the technical work that would now have to begin. With Reagan's go-ahead, preparing the first comprehensive proposal for the Space Station's overall architecture would become NASA's big task for 1984.

Ronald Reagan announcing the Space Station to Congress President Reagan announcing the Space Station to Congress in his State of the Union Address in January 1984 and signalling that he would welcome international involvement. Behind him are George HW Bush, Vice President, on the left and Thomas O'Neill, Speaker of the United States House of Representatives, on the right.

The President's announcement on the Station set the clock ticking for NASA and its prospective international partners. Two months after Reagan's speech, NASA Administrator James Beggs testified to the House Appropriations Committee. He told them the Space Station would cost America about $8 billion to build in 1984 dollars. The figure was little more than an educated guess at that early design stage. With no firm design on which to base the cost, it could be nothing else. It did not include Shuttle launch costs, operational costs, ground facilities costs or international partner costs. It covered just the Station's hardware construction costs. Beggs highlighted eight important functions that the Station would serve. It would be a space laboratory for scientific research and new technology development, a permanent Earth and astronomical observatory on orbit, a transportation node and operations base for payloads and vehicles, and a servicing facility for the maintenance and repair of these payloads and vehicles. It would also be an assembly site to put large structures together and check them out, a manufacturing plant where human intelligence and Station services combined to enhance commercial opportunities, a storage depot to house space payloads and parts for later use, and finally a staging base for more ambitious future missions.[5] The overall design comprised

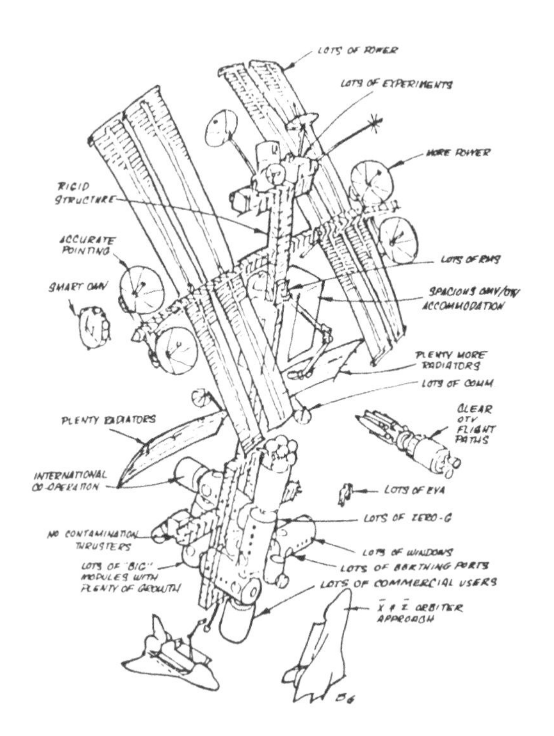

1984 Space Station Power Tower Requirements This sketch of the 1984 Power Tower design by architect Brand Griffin summarizes the major requirements for the Space Station as seen at the time. The Station's requirements covered many different disciplines, were challenging to incorporate and often conflicted with each other. Image: Brand Griffin.

three independent orbiting elements, the Space Station itself for a crew of eight on a 28.5° inclination orbit, an automated platform nearby on the same orbit and another automated platform on a polar orbit.[6] This was the first time that the Space Station's purpose, design and cost had been articulated clearly and concisely and in language that those within and beyond the space community could understand.

To stimulate the international interest that had already been aroused, Beggs went on an overseas tour to meet with the governments and agencies concerned. He was tasked by the White House to seek a framework for collaboration that Reagan could announce at a seven nation economic summit (the G7) in London in June 1984. Around that time, Beggs went to London, Bonn, Rome, Paris, Tokyo, and Ottawa.[7] He emphasised that the Space Station would be entirely civilian in nature. The European Space Agency gave him a warm reception. Beggs asked for an expression of interest from Europe within six months and a commitment to go ahead with Phase B studies six months after that. He wanted the Europeans 'to take a close look at our plans and concepts and then, based on your long-term interests and goals, share with us your ideas for cooperation that would expand the capabilities of the Space Station.' By the end of 1984, Europe's envisioned contributions to the Station had become elaborate and ambitious. They centred on a pressurised laboratory module they had named Columbus. It would be permanently attached to the Station. Another pressurised laboratory module named the Man-Tended Free Flyer would co-orbit independently and then dock with the Station for astronauts to tend its experiments and carry out maintenance. There would be one or more additional modules to house systems such as power, communications, data management and attitude control. There would also be an automated co-orbiting space platform and another automated platform placed on a polar orbit.[8] These were Europe's proposed contributions as NASA began work on the first detailed design for the Space Station following the President's go-ahead.

Despite the positive reactions, the issue on the prospective international partners' minds was how much it was all going to cost and how high the membership fee would be to join the exclusive Space Station club. In London, Reagan had meetings with the national leaders present for the G7 summit. A photograph shows the President explaining a scale model of the Space Station to Prime Minister Margaret Thatcher at Lancaster House. The resulting summit communiqué was cautiously worded while the prospective international partners considered their own interests and positions in response to Reagan's invitation. The outcome was positive and by the end of 1984 the space agencies of Canada, Europe and Japan had all signed agreements with NASA to provide Station hardware elements.[9] Like Europe, Japan would build its own pressurised modules and it also proposed to develop a Station resupply space vehicle.

Meanwhile, it was time for NASA to produce a design for the Station that would focus the minds of everyone involved on the project as it began development. NASA's Space Station Program Office set up a 'skunk works' team and tasked it to produce a comprehensive Space Station proposal over a four-month period. (The expression 'skunk works' had apparently originated on a Lockheed advanced aircraft development project in Burbank, California in 1943. Strong smells from a neighbouring plastics factory wafted through the development building and reminded workers of the name in Al Capp's cartoon strip.) NASA emphasised that the design and accompanying information should be 'considered as a potential point of departure for the definition phase, not a set of approved design solutions'. The effort would 'provide a focal point for the definition and assessment of program requirements, establish a basis for estimating program cost, and define a reference configuration in sufficient detail to allow its inclusion in the definition phase Request for Proposal'. The design would be a distillation of the previous studies that NASA had carried out or commissioned since 1979. It would extract their best features and remix them with new ones in a design founded on programmatic, technical and economic feasibility.

As a starting point the team chose five previous concepts for evaluation. They were the Delta and Big-T designs from Johnson Space Center's 1983 study, a 'Planar' design based on a module cluster with solar arrays in the same plane that had broadly originated in several of the eight 1983 contractor studies, the 'Spinner' rotating Station conceived by Hughes Aircraft Company that was one of the contractor studies and a new design called the 'Power Tower' that resembled a huge cross. The team soon narrowed the five concepts to three – the Delta, Planar and Power Tower designs – noting that they incorporated features taken from all five. The team then converted the Station's mission requirements to design requirements. Out of this the Power Tower emerged as the preferred design. NASA said it resolved all the user requirements in the short and long term and enabled effective orbital operations. NASA unveiled the design and published the results of the Power Tower effort in August 1984.[10]

The genesis of the Power Tower lay in a 1983 idea explored by Boeing that did not form part of their contract work for NASA at the time. It preceded Johnson Space Center's 1983 studies of the Delta, Big-T and Building Block Station concepts. Boeing's idea was simple. It comprised a single long backbone truss with a module cluster at one end and solar arrays and radiators at the other. The need to give the Space Station design room to breathe by locating major equipment away from the modules on some kind of large external structure was not then recognised and Boeing's idea was innovative. Boeing's design received little notice at the time but was later credited as the ancestor of the Power Tower design.[11]

The Power Tower was shaped like a giant cross with its mast and crossbar made of long trussed beams of square section and a 2.7-metre

1984 Space Station Power Tower design
Artist's impression of the 1984 Space Station Power Tower design. Shuttle Orbiters and other visiting spacecraft would dock at the cluster of modules at the base of the tall mast. The short crossbar at the mast's top would provide support for astronomy and astrophysics instruments.

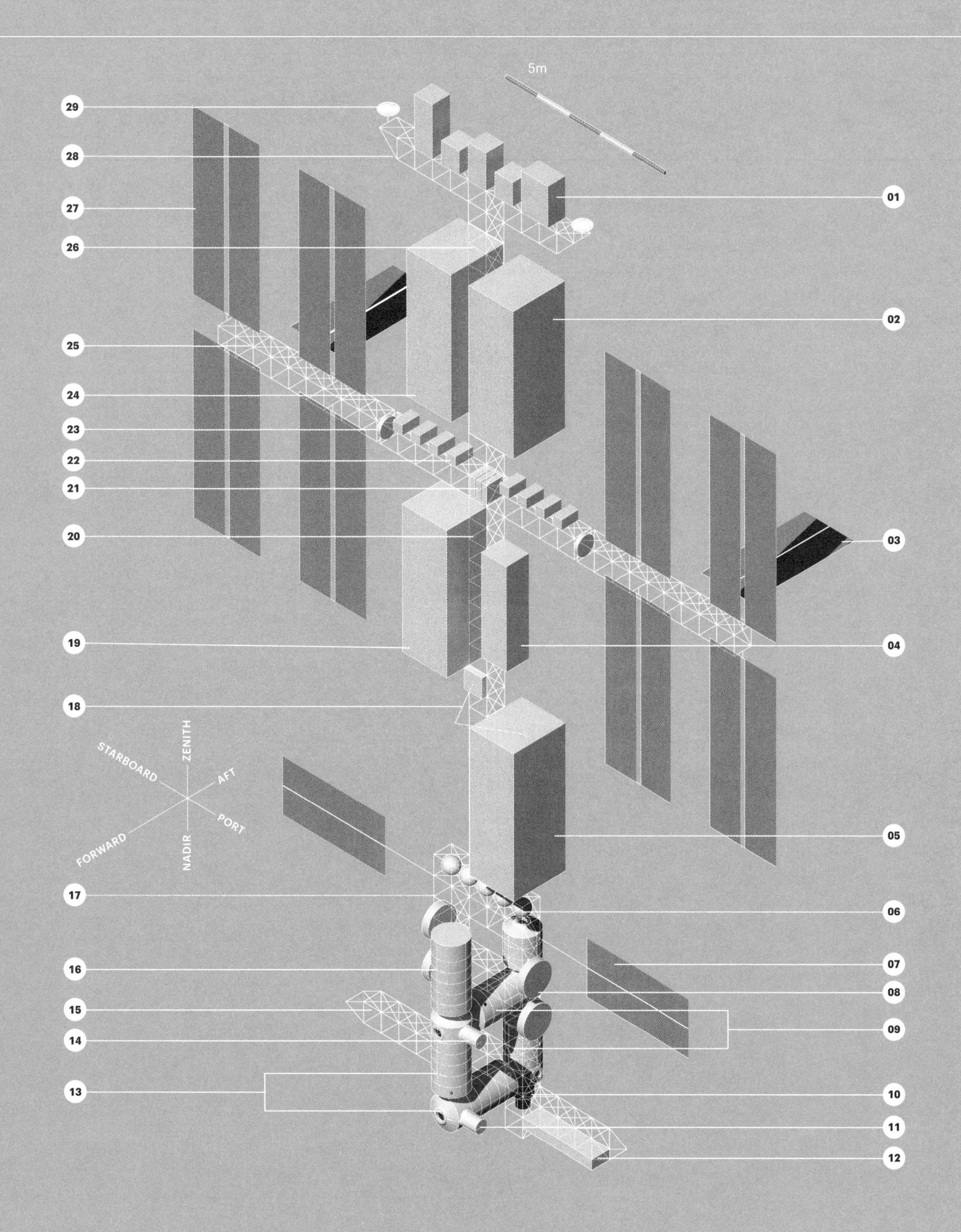

- **01** Astrophysics and space science experiments
- **02** Satellite servicing hangar
- **03** Radiator
- **04** Satellite instrument storage bay
- **05** Technology demonstration hangar
- **06** Propellant storage tanks
- **07** Radiator
- **08** Space vehicle equipment storage
- **09** Laboratory modules (2)
- **10** Earth observation instruments
- **11** Airlock
- **12** Technology demonstration experiments
- **13** Habitation modules (2)
- **14** Airlock
- **15** Lower boom
- **16** Logistics module
- **17** Mast extension
- **18** Mobile robotic arm
- **19** Technology demonstration hangar
- **20** Lower mast 2.75m bay size
- **21** Attitude control
- **22** Tool boxes and spare parts storage
- **23** Rotating joint
- **24** Satellite storage hangar
- **25** Main crossbar 2.75m bay size
- **26** Upper mast 2.75m bay size
- **27** 8 solar arrays in 4 pairs (1,784m² total)
- **28** Top crossbar 2.75m bay size
- **29** Antennae

side dimension. Attached symmetrically to the mast's bottom were two back-to-back L-shaped extensions of the same truss structure. These were spaced apart to accommodate a cluster of modules in their midst. The mast was 121 metres long including the L-shaped extensions. On each side, rotating portions of the crossbar supported four independent solar arrays in two pairs. One half of each pair pointed up and the other half pointed down. The total solar array power output exceeded 75kW. The arrangement of solar arrays, long crossbar and tall mast called to mind the large sails of a square-rigged ship. The crossbar structure was 80.5 metres wide. A variety of elements and equipment adorned the mast. At the bottom just above the module cluster were propellant storage tanks, maintenance equipment for space tugs and radiators projecting out on each side. Above these was a box-shaped refuelling hangar, a large diameter antenna and some spherical propellant tanks. Further up were box-shaped satellite instrument and engineering stores, Station attitude control thrusters and some small antennae. Just above the junction of the mast with the crossbar were more radiators, spare parts stores, toolboxes, more propellant storage tanks and more box-shaped hangars for satellite storage and servicing. Right at the top of the mast was a short crossbar piece, made of the same truss structure, which supported a stack of science and technology experiments pointing to zenith. Front and back faces of the mast's truss structure were kept free of attachments and reserved for tracks for mobile robotic arms that would travel up and down the mast's height.

The Power Tower Station was Earth-oriented with the mast continually aligned to point to Earth's centre as it orbited the planet. The crossbar and solar arrays were perpendicular to the orbital plane and faced directly into the flight path. NASA noted that the long, slender shape of the Station made it sensitive to gravity gradient attitude control. The effects of aerodynamic and spacecraft docking forces and unbalanced masses on the structure would need to be minimised to reduce the attitude control system's station-keeping task. The design provided Station elements and payloads with good fields of view. At the bottom, one of the modules had an unobstructed view of Earth. At the top, science experiments and astronomy instruments had an unobstructed view of the heavens. There was plenty of clearance room for Shuttle Orbiters approaching and docking with the module cluster and for construction and servicing activities along the mast. NASA recognised that the assembly of the Station on orbit would pose many challenges. A large structure had never been built in space before. Yet, the number of assembly spacewalks carried out from Orbiters would always be quite limited – typically two excursions for a two-man team for each Orbiter flight. Spacewalks would be few before a permanent Station crew arrived to carry them out. The Station would need to be self-powered and controllable from the first building block onwards in case of delay in the flight sequence. This was due to its need to maintain a safe orbit and attitude on its own.

1984 Space Station Power Tower design
NASA published the first detailed study of the Space Station in August 1984 and formally entitled it the Space Station Reference Configuration. It became known as the Power Tower design and it resembled a giant cross. It had a tall mast with a cluster of modules at the bottom and a crossbar about two-thirds of the way up. The crossbar carried four pairs of solar arrays on each side. The mast and crossbar were open trusses based on a 2.7m × 2.7m × 2.7m structural bay. They were pre-assembled with hinged joints. Two Space Shuttle missions would deliver them to orbit as compact payloads. Described as single-fold structures, they would deploy and expand outwards on one axis like an accordion. Image: author.

1984 Space Station Power Tower design – alternative structures In its 1984 Space Station Reference Configuration design, NASA explored both deployable/pre-assembled and erectable/space-assembled approaches to the truss structure (opposite). To the left is the deployable, single-fold 2.7m structure used in the Power Tower Station design. At the centre is an erectable structure based on a 4.6m × 4.6m × 4.6m bay. Teams of astronauts would build it from stacks of struts delivered on Space Shuttle missions. To the right is a deployable structure based on a prism-shaped structural bay 3m on edge. It is a double-fold structure. It would deploy and expand outwards on one axis and then on another. Image: author.

The Power Tower Station would need to become habitable quickly to enable assembly operations, commissioning and checkout without the presence of a docked Orbiter. The proposed launch and assembly sequence showed one Orbiter flight for each assembly phase, one logistics module to support initial occupation, spacewalks from the Orbiter on the first five flights and use of the Orbiter's robotic arm on every flight. NASA identified a total of thirty-six major elements that made up the entire Station. Just seven Orbiter flights would be needed to deliver all these, partially assembled into building blocks that made the most of the Orbiter payload bay shape and size. The first flight would bring the central part of the crossbar and the second flight the lower part of the mast and both L-shaped extensions. The third flight would bring the first module and two airlocks. The fourth flight would bring the second module and the top part of the mast. The fifth flight would deliver the third module and the outer parts of the crossbar. The sixth flight would deliver the fourth module and the seventh flight the fifth module. NASA laid out the Orbiter's payload bay stowage schematically for each of these flights to show how it would be done.

The major innovation of the Power Tower design was the use of the mast and crossbar to provide a structural backbone, an idea that had been born in Boeing's 1983 sideline study. NASA explained that the Power Tower's structure would function as the Station's structural foundation, provide surfaces for the attachment of payloads and utility lines, give the Station structural stiffness to help its dynamic control, act as the rail track for the mobile robotic arms, and supply structural redundancy and repair capability in the event of damage. NASA described three different concepts for the structure aimed at maximising its stowage efficiency in the Orbiter payload bay and then constructing or deploying it in orbit.

A deployable single-fold truss was the first concept and the one used for the mast and crossbar shown in the drawings included in NASA's report on the design. The 2.7-metre side dimension was the maximum that could be stowed comfortably in an Orbiter payload bay when folded along one axis. The truss had an orthogonal tetrahedral design with four inward-folding longitudinal struts and telescopic diagonals enabling it to collapse into a compact payload. Each structural bay reduced to the equivalent of two longeron (long strut) diameters. The result was that a 66-metre long truss made of 50mm diameter tubes shrank to become a stowed package a little over 2.4 metres long. Once on orbit, it would unfold slowly, one structural bay at a time. The whole mast and crossbar structure would need at least two Orbiter flights for delivery.

The second concept was an erectable truss. It arrived on orbit as a packaged payload of individual structural members, like sections of steel bundled on to a flatbed truck for delivery to a building site. Its main advantages were that it eliminated the need for folding techniques and enabled the assembly of a truss with a side dimension unconstrained

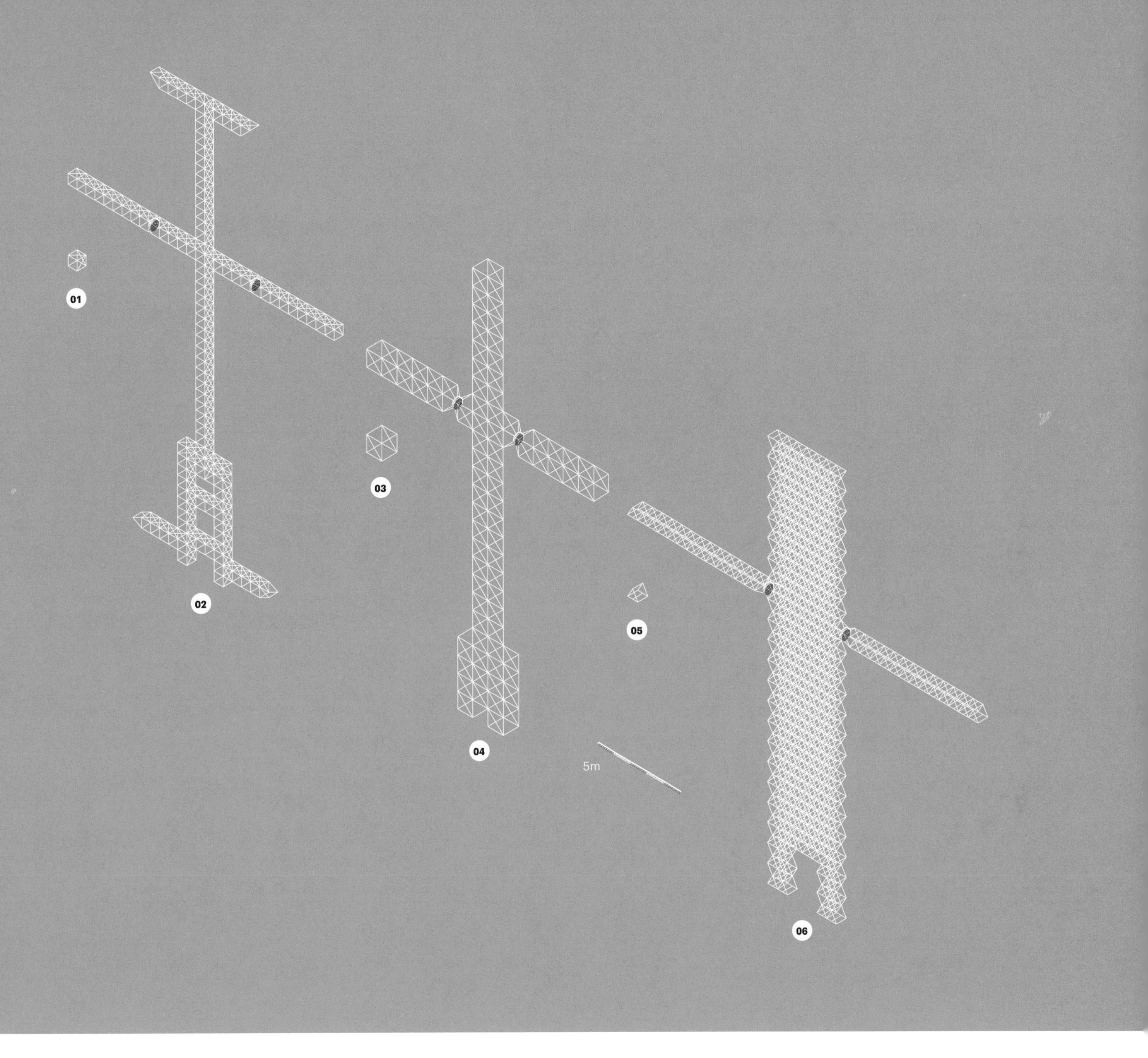

01 2.75m bay size
02 Deployable single-fold structure
03 4.6m bay size
04 Erectable structure
05 3m bay size
06 Deployable double-fold structure

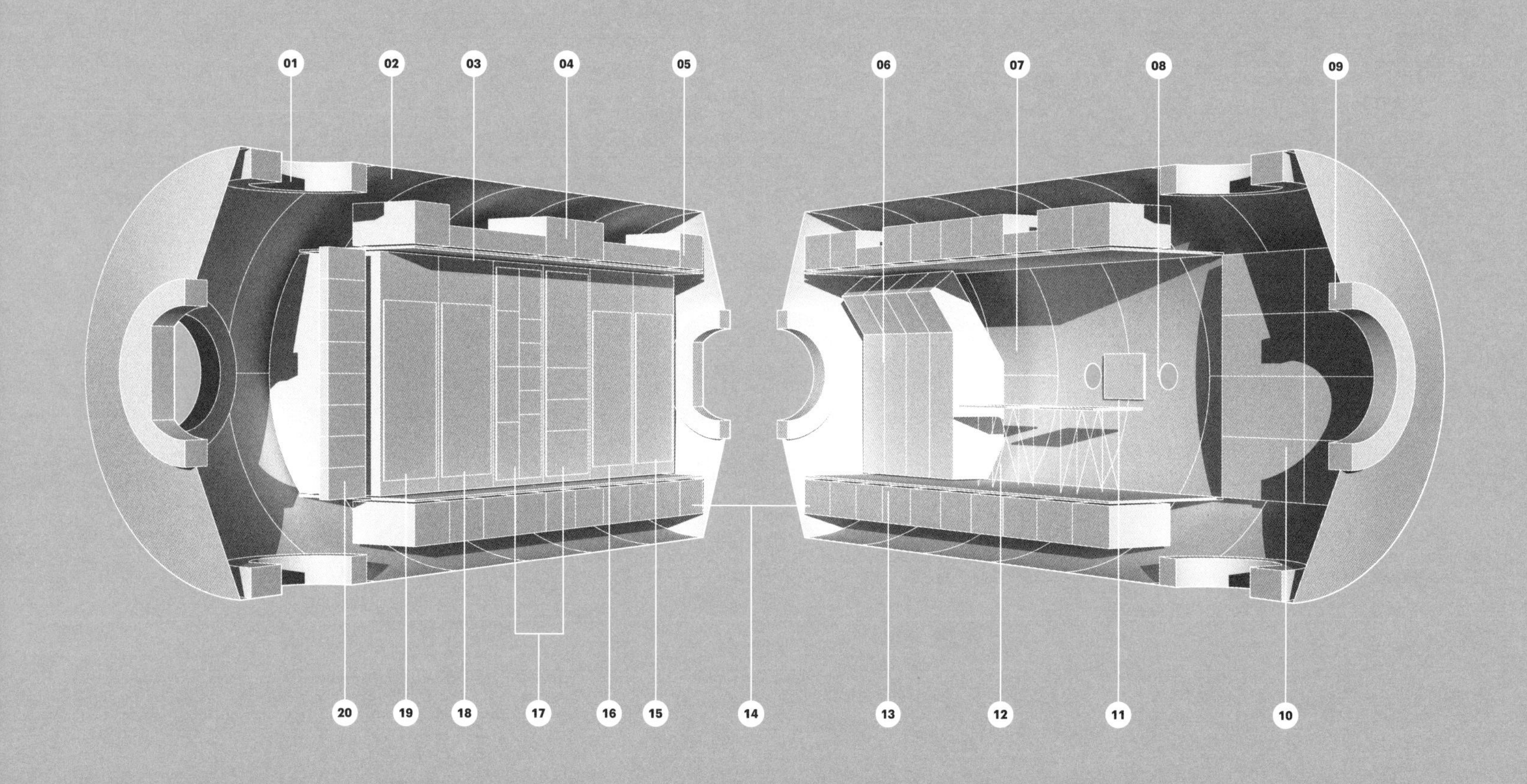

01 Side berthing port
02 Pressurised hull
03 Ceiling
04 Life support systems
05 Storage
06 Health maintenance, data management and communications
07 Operations area & wardroom
08 Viewing windows
09 End berthing port
10 Power equipment, safe haven & thermal control
11 TV screen
12 Wardroom tables
13 Floor
14 Life support systems
15 Toilet
16 Shower/urinal/handwash
17 Galley
18 Food storage
19 Vegetable chamber
20 Life support & oxgen masks

by the Orbiter payload bay. NASA chose a side dimension of 4.6 metres for the erectable truss, which added a lot of stiffness. The main problem with the erectable truss was that it would need many astronaut spacewalks to assemble its structure and then route utility lines through it. Just one Orbiter flight would be needed to deliver all the parts to orbit.

The third concept was a 3-metre deployable double-fold truss. Like the first concept, it was a structure that folded up for stowage in the Orbiter payload bay, but its ability to fold up on two axes made it even more compact. This, in turn, enabled it to evolve from a linear truss to a planar platform with the platform's length and width defined by the double-fold capability. It had a tetrahedral geometry based on a single member, 3 metres long. Upper and lower surface members hinged at their midpoints. The folded geometry was apparently going to be so compact that NASA claimed that the entire structure for the Station would fit into one Orbiter payload bay. The concept had a lot of structural space capacity in the event of damage while its single member size would reduce manufacturing costs. Its deployment on orbit, however, would need a lot of care and caution.

NASA's evaluation of the three structures showed that each had advantages and disadvantages. The larger strut length of the 4.6-metre erectable truss resulted in the fewest component parts, the lowest mass and a greater stiffness-to-mass ratio than the others, but it would need a great deal of assembly on orbit. The 2.7-metre deployable single-fold truss would involve the least amount of spacewalk time and its ground assembly and utilities integration would enable full checkout before launch, but NASA had no experience of fold-out structures in space.[12] NASA held back from making firm recommendations on one type of structure.

The module cluster utilised the looped pattern explored in several earlier Space Station studies. Instead of the earlier 'quadrangle', NASA now described it as a 'racetrack' that gave each module two exit paths to improve evacuation and safety in an emergency. The design requirements for the modules were clear: they would be made of materials with a service lifespan of ten years without upgrading; their size and weight would be within Orbiter payload bay dimensions and lift capability; they would support an internal atmospheric pressure of 101.4kPa (a normal atmospheric pressure on Earth); they would offer internal attachments for accommodation and outfitting; they would incorporate shields to micrometeoroid and debris penetration; they would permit docking of Orbiters and berthing of other modules to themselves; they would include crew circulation passages along their lengths and windows for observation; and they would be structurally designed to ample pressure, thermal and mechanical load safety factors. The cluster comprised two habitation modules and two laboratory modules that together formed the racetrack and the logistics module that pointed outwards on its own. The laboratory and habitation modules had a standardised design with a cylinder, conical endcaps that tapered to

1984 Space Station Power Tower design – pressurised modules The standard module design in NASA's 1984 Space Station Reference Configuration was a pressurised aluminium cylinder with conical endcaps. It was 10.4m long and 4.6m in diameter. It had one berthing port with hatch at the centre of each endcap and four berthing ports with hatches at 90° radial intervals at one end of the cylindrical hull. Plugging an end port of one module into a side port of another formed them into a shape like the letter L. Repeating this twice more resulted in a loop of four modules that NASA called a 'racetrack'. Inside the modules, floor and ceiling planes defined central living and working volumes for the crew. Lining these volumes on each side were a mixture of compartments and racks containing equipment, amenities and storage. One of the habitation modules had an alcove for a wardroom with tables and windows. Above the ceiling and below the floor were life support systems and more equipment. Image: author.

circular ports and an efficient module-to-module berthing technique that eliminated the need for interconnecting nodes. The modules were 10.4 metres long and 4.6 metres in diameter. The end of each cylinder had four circular berthing ports spaced at 90° intervals around the circumference. These ports were recessed into concave depressions in the cylinder to fit within the module's cross-section. This recessing was essential to enable the modules to fit into the Orbiter payload bays. The end port of one module plugged directly into one of the circumferential berthing side ports of another module, forming a right angle between them. Repeating this for all four modules resulted in the racetrack shape. The main disadvantage was the loss of useful volume at one end of each module due to the intrusion of the four berthing ports. NASA had conducted a comparative study of different module patterns with and without racetracks, with and without nodes, and with combinations of long and short modules and found this racetrack pattern to be the best.

The proposed method of module fabrication used all-welded, integrally machined skin-stringer panels of aluminium plate. This eliminated mechanical fastener technology that was more susceptible to pressure leaks. Inside the pressurised skin were ring frames to support internal accommodation, equipment and experiments. Outside were 'bumper' panels, spaced 50mm from the pressurised skin to provide micrometeoroid and debris impact protection. Multilayer thermal insulation was sandwiched between the two skins. The hatches to all the docking and berthing ports had clear diameters of about 1.3 metres. The module cylinder walls incorporated four observation windows 400mm in diameter with double full-pressure glass panes for safety.

The two habitation modules provided living accommodation for a crew of six. The first of these contained a wardroom, galley, shower and toilet, safe haven area, an operations workstation, a health maintenance workstation and storage. Also included were waste management, life support, thermal, data, communications and power systems. The second habitation module contained six private crew compartments, a workshop workstation, a laundry, another shower and toilet, another safe haven area and an operations centre. This module also included waste management, life support, thermal, data, communications and power systems. Internal architectural arrangements followed the designs of several earlier Space Station studies. Placing the wardroom in one module and the sleeping facilities in another module would provide some degree of acoustic isolation of the private compartments from a potentially noisy wardroom. Floor and ceiling planes defined a central open volume along the module and gave it a single level horizontal orientation. Compartments and equipment racks were fitted along each side and narrowed the free volume. The racks were left out in several places to create alcoves for the wardroom and workstations that reached to the module's curved inner surface and its windows. The one innovation was the crew sleeping area, formed radially as six private compartments around the cylinder's inner surface in 60° segments with a tubular passageway through the middle, a highly efficient and compact design that made the most of the weightless environment in an architecturally interesting manner. Each sleeping compartment had a relatively generous volume of 4.2m^3 with a length of 2.2 metres.

The laboratory modules generally followed the same internal layout as the habitation modules. One was mainly intended for life sciences and contained a dense row of experiment racks down each side of the volume. These included an animal research laboratory, a human medical research station, an astrophysics station, an operations workstation and a safe haven area. Set into the floor was an exercise treadmill. The tops of the racks angled inwards like the science racks inside Europe's Spacelab laboratory that flew on the Space Shuttle. The second laboratory was for microgravity and materials processing. More dense rows of racks contained banks of materials processing experiments and chambers, two workstations and another safe haven area. Each laboratory module had life support, thermal, data, communications and power systems, and the life sciences module had a separate life support system for the animal laboratory.

NASA proposed a partially closed system for the Power Tower's life support. This would involve regenerative processes for recycling atmospheric oxygen and water combined with the resupply of nitrogen to replace leaked air. NASA considered that a fully closed ecological life support cycle was not possible at that time because of the high cost and risk of developing on-board food growth and harvesting technology. A partially closed system was a good compromise. The functions of the life support system involved maintaining atmospheric pressure and temperature, controlling atmospheric composition and humidity, revitalising and recycling atmospheric oxygen, processing and recycling hygiene and drinking water, dealing with solid and liquid waste, and servicing spacesuits. It would be a combined centralised-distributed system. Each module would have its own ventilation ducting, temperature and humidity control, fluids plumbing and fire detection/suppression. All modules would share systems for carbon dioxide removal, carbon dioxide reduction, oxygen generation and waste water reclamation. NASA did not go as far as giving details of the technologies involved and left the life support system design at the schematic stage. System equipment and ductwork occupied most of the volume above the ceiling and below the floor of all the modules. The thermal control system was based on a heat pipe approach with thermal control fluids of water on the inside and ammonia on the outside operating through heat exchangers. The proposed system used projecting radiators mounted on the sides of the Power Tower mast as well as curved radiators on module exterior surfaces.[13]

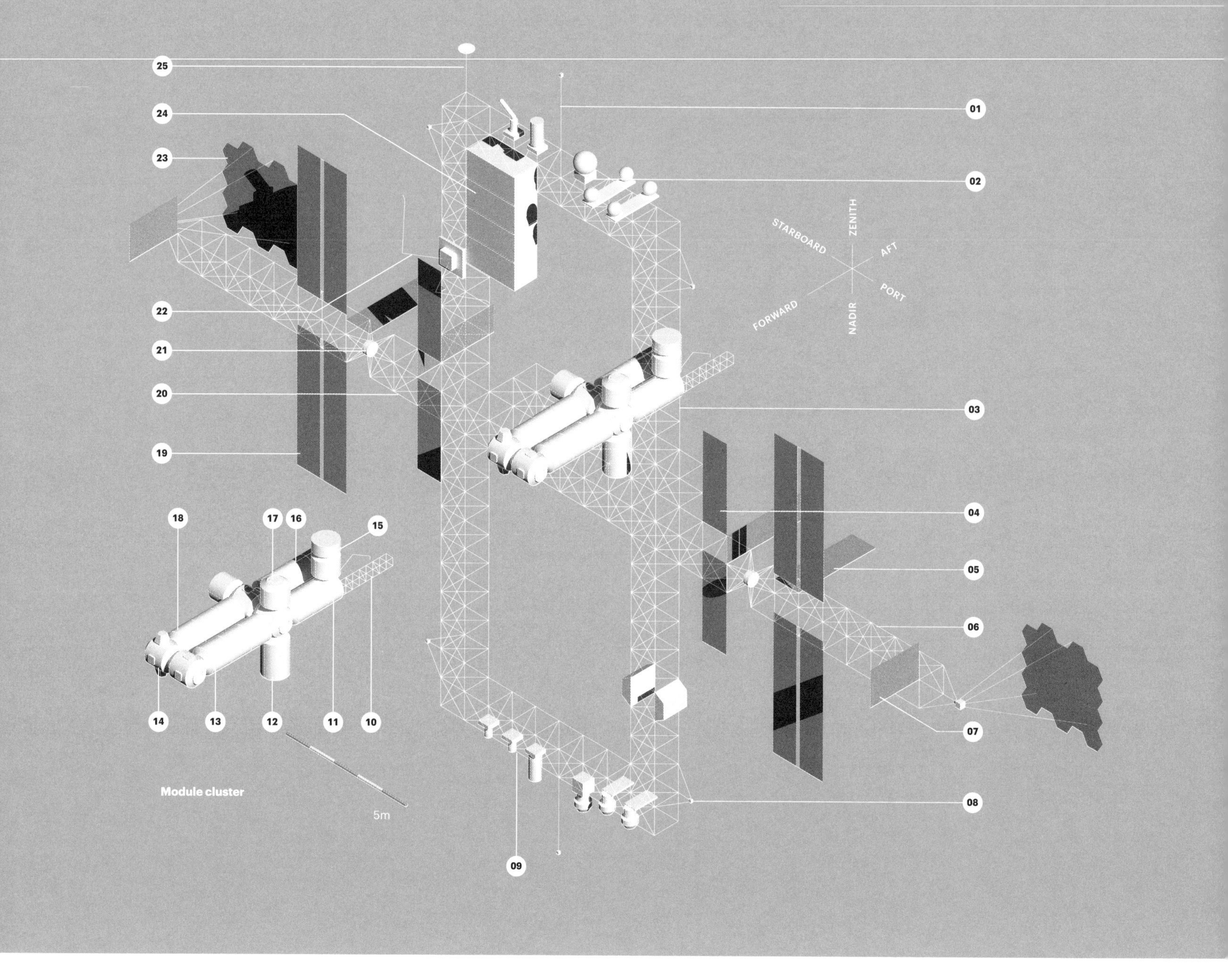

01 Tether
02 Zenith experiment & instrument deck
03 Top & bottom keels 5m bay size
04 Power radiator
05 Thermal radiator
06 End crossbar 5m bay size
07 Power radiator
08 Propulsion unit (4 places)
09 Nadir experiment & instrument deck
10 Japanese experiment truss
11 Japanese laboratory module
12 US logistics module
13 US habitation module I
14 Nodes, cupolas & tunnel
15 Japanese logistics module
16 European laboratory module
17 Airlock
18 US laboratory module
19 Photovoltaic solar array
20 Centre crossbar 5m bay size
21 Rotating joint (alpha joint)
22 Robotic arm on mobile platform
23 Solar dynamic collector
24 Satellite servicing facility & hangar
25 Tracking & Data Relay Satellite antenna

The Power Tower study was the equivalent of Phase A in NASA's multi-phase project development sequence. Appendix 1 shows what Phase A covers in a typical space project and programme. All the previous NASA and contractor Space Station studies had been at the pre-Phase A conceptual feasibility stage, aimed at producing a broad spectrum of ideas and alternatives from which a project would emerge. The purpose of Phase A was to determine the feasibility and desirability of a project and its compatibility with NASA's strategic objectives,[14] which the Power Tower had done. It was now time to move the project forward into more detailed development using the Power Tower as a reference tool. In September 1984, one month after NASA's release of its report on the Power Tower design, it issued a request for proposals from the aerospace industry across America for four Phase B contracts, each involving a different mix of contractors. A different NASA field centre would manage each of the four contracts.[15]

In November 1984, the Office of Technology Assessment of Congress in Washington published the results of a study on civilian Space Stations and the future of America in space.[16] The Senate Committee on Commerce, Science and Transportation and the House Committee on Science and Technology had requested the study. Its purpose was to examine the technical issues concerning the choice of a Space Station and to enable Congress to look at the larger context of space exploration. America's proven ability to develop top quality space hardware required that 'equally sophisticated and thoughtful decisions must be made about where the US space program is going, and for what purposes' as the Congressional report put it. There were numerous expert contributors and participants to the study. The report that resulted made frequent use of the word 'infrastructure'. Borrowing the word from terrestrial engineering fields such as transportation and utilities distribution where it was in common use, space infrastructure was to be a network of space facilities in low Earth orbit and beyond that included a Space Station, a space transportation node and a set of resources and utilities that may or may not be operated independently. The rationale was the need for an infrastructure to support sustained investigation into life and materials sciences, storage of propellants and supplies, staging of missions to destinations beyond Earth, instrumentation proof testing and satellite maintenance and repair. There were several approaches to delivering this infrastructure and the Space Station was just one of them.

The report was not particularly enthusiastic about NASA's Space Station plans. It acknowledged that, though 'the case to be made for acquiring some long term, inhabitable infrastructure in low-Earth-orbit is persuasive', there was no compelling case that it should be a Space Station as NASA had defined it. The report took the view that 'simply put, there is no such thing as the "Space Station". What is under discussion is a variety of sets of infrastructure elements'. Congress had enacted the creation of the

1985-86 Space Station Dual Keel design
Scientists predicted that placing the laboratory modules at the base of the tall mast in the 1984 Power Tower Space Station design would adversely effect the sensitive laboratory experiments due to the dynamic behaviour of the tall mast in flight. This caused NASA to replace the Power Tower with this design in 1985. NASA called it the Dual Keel because of the twin keel-shaped truss structures above and below the main horizontal crossbar. The Dual Keel design brought the module cluster close to the Station's centre of mass where there would be the least disturbance to experiments from the dynamic movement of the Station as it orbited the Earth.
Image: author.

National Commission on Space in July 1984 with the aim of investigating the future of the entire civilian space programme, including the future course of a Space Station. The report noted that 'over the next year and a half, the deliberations and eventual findings of the National Commission could offer NASA, and others seriously interested in the space future, the opportunity to develop new program options, and to compare these new options, new methods, and new attitudes with the civilian "Space Station" program as currently defined. A fresh, basic and uninhibited review of policy issues might well result in a fundamental change of NASA views on ... the appropriate character of the "Space Station" program.'

Though the study and its report had sensible objectives in furthering the broader interests of America in space, their timing was sluggish and out of step. NASA had spent five years exploring Space Station concepts, had been give the go-ahead by President Reagan, had just settled on a reference design and was about to hand out contracts to begin the Station's detailed development. Congress seemed to have just woken up to Reagan's national Space Station decision and had responded by muddying the waters around a project that had just achieved some clarity for the first time. The Congressional study and its report were several years too late to do much good.

Great expectations
Dual Keel configuration, 1985-1986

In 1985, NASA began the Phase B phase of the Station's progress that would last for two years. NASA gave parallel Phase B study contracts to several aerospace companies. Contract management was split between Johnson Space Center with a responsibility for studies by Rockwell and McDonnell Douglas on the one hand and Marshall Space Flight Center with a responsibility for studies by Boeing and Martin Marietta on the other.[17]

According to NASA, the purpose of Phase B in a project's development cycle is to establish its initial baseline. This transforms initial requirements into preliminary designs and system and subsystem specifications. In Phase B a project is fully defined for the first time.[18] Appendix 1 shows what Phase B typically covers.

As it entered Phase B, NASA's Power Tower Space Station design was running into trouble. It appeared to have some major design flaws, particularly with its gravity gradient and with crew-induced vibrations. A gravity gradient is basically a slight change in the gravitational forces acting on an object on orbit due to the shape and size of the object itself. Any part of an object on orbit that is not at its centre of mass will experience a slight gravity bias. The further away a part is from the centre of mass, the greater the bias, resulting in rotation or torque effects. The laboratory modules were down at the bottom of the Power Tower's mast and a long way from its centre of gravity and they would experience these rotation and torque effects, even if very slight. Additionally, dynamic models developed for the Station's structure showed that the crew living and working in the modules at the bottom of the Power Tower's mast would generate structural vibrations that the long mast would tend to amplify. The combined effects could disturb the microgravity environments inside the laboratory modules and spoil the scientific research and experimental materials processing taking place there. Typical of materials experiments that would be adversely affected would be crystal growth that required the least amount of microgravity disturbances possible. There was also a lack of room on the Power Tower's structure for large observation, monitoring and sensing instruments that required unobstructed pointing down to nadir or zenith.[19]

The scientific community was concerned with the Power Tower's design problems. NASA Headquarters hosted a Space Station briefing at the National Science Foundation's study centre at Woods Hole in Massachusetts to present the Power Tower design to scientists representing all the partners – Canada, Europe, Japan and America. It did not go down well. The scientists complained loudly about the microgravity problem. They wanted the laboratory facilities as close as possible to the Station's centre of mass. They were irritated that NASA had not consulted them earlier and that the project's design had got so far without their input. A structural shake, rattle and roll, however minuscule, would be a serious defect.

The Station's primary purpose was to conduct science and engineering investigations in the microgravity conditions of space and the quality of that microgravity was of paramount importance. To some extent, the Power Tower appeared to have inherited this design flaw from previous NASA studies. In 1983, Johnson Space Center had produced its three Space Station concepts – the Planar, Delta and Big-T designs – which had similar problems. The Planar design had linked the modules end to end and orientated them vertically, the Delta design had put the modules at one of the corners of the prism-shaped structure and the Big-T had put them at the bottom of the tall mast.[20] In each case, habitability and laboratory modules were placed away from the centre of mass. In the Delta and Big-T designs they had been perched at extremities of large structures – the worst possible places. The Power Tower's defects were serious and forced NASA and its contractors to set it aside and come up with something better.

By summer 1985 a revised design called the Dual Keel had emerged that aimed to correct the major defects of the Power Tower. It used most of the Power Tower's elements but put them together in a different way. A major feature of the Power Tower had been its resemblance to a giant construction toy made up of a kit of parts. To change it was a matter of taking the parts to pieces and reconfiguring them. By October, both designs appeared to be still under consideration at NASA meetings. McDonnell Douglas had switched over to the new Dual Keel but Rockwell was still working on the

1985-86 Space Station Dual Keel design
Top: Artist's impression of the 1985 Space Station Dual Keel design. The crossbar consisted of three truss segments with the two outer segments attached to the central segment by rotating joints called alpha rotary joints. Each solar array wing had a similar joint at its attachment point to the truss called a beta rotary joint. The alpha and beta rotary joints would work together to keep the solar arrays facing the Sun as the Station orbited the Earth. This double rotating joint feature remained an important part of the Station's design as it continued to change throughout the late 1980s and early 1990s. Bottom: The central module cluster in NASA's 1985 Space Station Dual Keel design. Two pairs of long modules were linked at their ends by nodes and tunnels to form a cluster that resembled the number 8. The truss structure supported the modules. To the right, a robotic arm extended upwards from a platform mounted on the truss.

Structure assembly tests in a water tank NASA carries out underwater tests to examine the unique challenges of assembling or servicing space hardware in the absence of gravity. Ballasting parts, equipment and spacesuited astronauts for neutral buoyancy means that they neither float nor sink in the water but have a total weight that is equivalent to the water they displace. This condition resembles the weightlessness of space. This test in a water tank at Marshall Space Flight Center dates from about 1986. Supported by divers, test crews including astronauts assembled a tetrahedral structure called EASE (Experimental Assembly of Structures in Extravehicular Activity) from struts and nodes. EASE was a prototype of a potential design for the Space Station's truss structure.

old Power Tower.[21] The Dual Keel featured two major changes. The first was the elimination of the tall vertical mast of the Power Tower's cruciform shape and its replacement by two U-shaped truss sections added to the crossbar. These were the dual keels that would provide a lot more room for mounting Earth and space observation experiments and instruments. Imagining the crossbar as the hull of a yacht, one keel extended down into the water on the fore-aft line in the normal fashion. The other keel mirrored the lower keel and pointed up in the opposite direction. This gave the new design the appearance of the Ancient Greek letter Phi turned onto its side. The second major change was the elevation of the module cluster from the bottom of the Power Tower's mast to the level of the crossbar. This placed the habitation and laboratory modules and their occupants close to the Station's centre of mass and disposed of the Power Tower's gravity gradient problem.

Like the Power Tower but perhaps more so, the dominating feature of the Dual Keel was the vast amount of truss structure. It dwarfed all the other elements in scale and size. There were now two favoured versions of the truss. One was a deployable single-fold beam based on a 2.7-metre module. The other was an erectable beam based on a 5-metre module – a slight increase over the previous 4.6-metre module of the Power Tower. At about 154 metres wide for the 2.7-metre truss and 170 metres wide for the 4.6-metre truss, the crossbar had now become a backbone of impressive size. At each end it had rotating lengths of truss that supported twin solar arrays on the inside and a solar dynamic collector perched at each extremity. The 5-metre truss structure had a lower keel, 50 metres deep, with a transom 75 metres wide and an upper keel 40 metres high with a transom 60 metres wide. Altogether it comprised ninety structural bays amounting to 450 metres in total cumulative length. There were some design variations in versions studied by the contractors. Sometimes the keel's transoms extended either side of their twin posts and sometimes the keels contained additional transoms close to the main crossbar.

At some point soon NASA and its contractors would have to choose between different versions of the truss structure for the Station. Was it going to be the 2.7-metre deployable single-fold version, the 5-metre erectable version or something else? NASA and its contractors had settled on the Dual Keel design but the type of structure it would have was still an open question. In July 1985, NASA Langley Research Center published the results of a study that compared different types of truss structure.[22] Although the Langley team had based its study on the now abandoned Power Tower, the results were relevant to other Station designs. The Langley team had explored four design options, the 2.7-metre deployable single-fold truss, the 5-metre erectable truss, the 3-metre deployable double-fold truss and a new 5-metre deployable truss called a Pactruss. Each had advantages and disadvantages. All the deployable trusses offered pre-launch installation of cables, joints, some fluid lines and compact stowage in the Orbiter payload bays. The 5-metre erectable truss offered excellent growth potential, plenty of payload attachment capacity, greater stiffness and it needed only a little more assembly time on orbit.

The study concluded that strong consideration should be given to the 5-metre truss because of its growth and payload benefits. However, as a warning, the study found that all the trusses required too many spacewalks per Shuttle mission. There had been very little effort to develop techniques to reduce the number of hours astronauts would be forced to spend building the structure on spacewalks. By October 1985, the choice of truss structure had become the focus of attention. There were differences of opinion between groups of project offices about which version was best. One group preferred the 2.7-metre truss, another preferred the 5-metre truss and a third had no preference.

It was a competition in which the larger truss emerged as the winner because its long-term benefits outweighed the short-term benefits of the smaller truss and the Space Station was a long-term project. The 5-metre truss also had superior physical properties. It offered nearly three times the stiffness of the smaller truss, lower free-play sensitivity, lower thermal deflections, lower mass and a lower number of component parts. An analysis also showed that the 5-metre truss with one longeron missing or removed would retain reduced structural stiffness and load carrying capability.[23] When it came to access for spacewalking astronauts, they would be able to float relatively easily through the open interior of the 5-metre truss from one end to the other but in their spacesuits they would barely squeeze past diagonal struts inside the 2.7-metre truss.[24] The smaller truss was doomed.

NASA had been carrying out tests of astronauts putting together structures in large water tanks for a while under conditions of simulated microgravity called neutral buoyancy. To achieve neutral buoyancy, the weight of an astronaut in a watertight spacesuit is calibrated to equal the weight of the water displaced by the suit and its occupant. The result is that the astronaut neither floats to the surface nor sinks to the bottom, a physical condition similar to that of a spacewalk in the weightlessness of orbit. At Marshall Space Flight Center in Huntsville, Alabama, NASA performed preliminary tests on assembling prototypes of the structures that were under consideration for the Station. Then, the agency carried out similar tests in space for the first time.

On 26 November 1985, Shuttle Mission 61-B launched into orbit with two experiments called ACCESS and EASE. ACCESS stood for Assembly Concept for Construction of Erectable Space Structure and EASE for Experimental Assembly of Structures in Extravehicular Activity. NASA gave these lengthy names to two experiments that achieved several objectives. They provided valuable astronaut hands-on experience of building structures in space; they gave directions for improving the performance of erectable structures;

they enabled a comparison of space assembly procedures with those from ground tests and they explored the feasibility of the 5-metre truss that was now favoured for the Space Station.[25]

NASA described ACCESS as a 'high-rise tower composed of many small struts and nodes'. It was a triangular cross-section truss made up from ninety-three tubular aluminium struts that were 25mm in diameter and thirty-three nodal joints to connect the strut ends. The struts formed the longitudinal struts, diagonals and battens of the structure with a face width of 1.4 metres and an overall length of 13.7 metres. EASE was 'a geometric structure that looks like an inverted pyramid and is composed of a few large beams and nodes'. It was made up from six struts 3.7 metres long and four nodal joints. Working in the Orbiter payload bay, two astronauts put the two structures together and then took them apart. They used no tools and simply snapped the parts together. Both experiments were successful. The two astronauts took just thirty minutes to assemble the ACCESS truss while the EASE pyramid was put together and taken apart eight times.

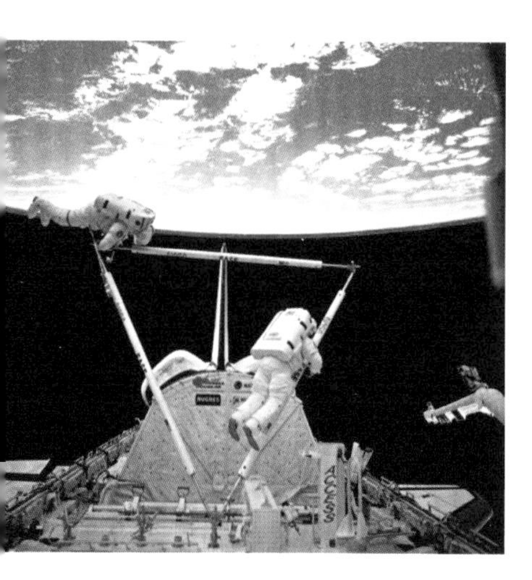

Structure assembly tests on a Shuttle mission In the second test on Space Shuttle Mission 61-B in November 1985, the crew built the larger EASE structure. It was made up of six struts and four nodal joints. Two astronauts assembled it by simply snapping the parts together. They did this successfully eight times. Based on extrapolations of the two tests, NASA concluded that the larger EASE truss would be the best structural solution for the Space Station.

Extrapolating the work performed and the time taken, suggested that seven Shuttle missions would be needed to build the complete Space Station truss structure. Astronaut assembly of the 2.7-metre truss would take ninety-six hours compared with 116 hours for the 5-metre truss.[26] The experiences of Shuttle Mission 61-B convinced NASA that the larger truss was the best solution for the Space Station. Repeating the reasons for the final choice near the end of 1986, NASA summarised its advantages as greater stiffness, lower mass, lower part count, lower cost, better growth potential and more payload accommodation.[27]

However, a question mark continued to hover over the amount of time astronauts would need to build the truss structure and then lay the utility systems through it. NASA had estimated the 116 hours of spacewalk time based on their experiences with the structures in the water tank tests and the results of the ACCESS and EASE experiments. The scale and complexity of the Dual Keel truss would be much greater than any tests could simulate and the challenges of spacewalking around or through such a vast structure would be immense. The basic Dual Keel skeleton contained eighty-three structural truss bays with each bay comprising four longitudinal struts, four lateral struts and five diagonals, making a minimum of well over 1,000 struts that spacewalking astronauts would assemble by hand. It was a daunting prospect. Noting that 'there is no precedent for an on-orbit structure as large and complex as that being considered for the Space Station', NASA proposed a Space Shuttle mission to carry out a large-scale flight test of sixteen bays of truss structure with internal utility lines to give astronauts more experience with tackling large structures ahead of the real thing.

Tragedy and shock
Loss of the Shuttle Orbiter Challenger, 1986

With the truss tests on Shuttle Mission 61-B, NASA had taken the first physical step towards building the Space Station on orbit. Like engineers arriving at an empty field with a drilling rig to take soil samples for the foundations of a future building, astronauts had arrived on orbit with structural rigs to carry out construction tests for a future Space Station. The Space Shuttle system had begun to show its versatility as a component delivery truck and construction work site for the Space Station. The new project was making steady progress and NASA was beginning to test the new technologies on a step-by-step basis. Then tragedy struck.

On the morning of 28 January 1986, Space Shuttle Mission 51-L with the Orbiter Challenger launched to orbit. Its flight ended seventy-three seconds later in an explosive burn of hydrogen and oxygen propellants that destroyed the Shuttle's big External Tank and exposed the Orbiter to severe aerodynamic loads that caused complete structural break-up. All seven astronauts on the mission perished. The two Solid Rocket Boosters flew out of the fireball and were destroyed by the range safety officer.[28] Later that day, President Reagan spoke to a shocked nation on television. Mourning the crew's loss and offering condolences to the bereaved families on behalf of the nation, Reagan commented that 'we've never lost an astronaut in flight; we've never had a tragedy like this ... we've grown used to the idea of space, and perhaps we forget that we've only just begun'. He reaffirmed that 'we'll continue our quest in space ... there will be more shuttle flights and more shuttle crews ... nothing ends here'.[29] NASA immediately grounded the entire Space Shuttle fleet.

Within a few days the President had appointed a Commission to investigate the disaster and it got to work quickly. Its mandate was to review the circumstances surrounding the disaster to pinpoint the causes and then make recommendations for corrective action. Attention soon focused on the Solid Rocket Boosters. The *New York Times* reported that the manufacturer of the twin boosters had said that they were not designed to operate below 40°F and that the air temperature was close to freezing at launch. The Shuttle had been standing on the launchpad all night in sub-freezing conditions.[30] The Commission found that the ambient air temperature at launch had been 36°F: 15°F colder than any previous Shuttle launch. Launchpad cameras had recorded smoke coming from a joint line in one of the boosters moments after the launch. Fifty-eight seconds later a flame had sprung up from the booster followed by a well-defined plume. Then a cloud had appeared and flowed down the side of the External Tank followed by a flash that rapidly brightened as the liquid oxygen and hydrogen tanks inside it apparently ruptured in a fiery conflagration. At seventy-three seconds the Shuttle began to break up and at seventy-six seconds fragments of the disintegrating Orbiter started to fall from the sky, enveloped in a reddish-

Module exterior outfitting tests in a water tank NASA also began the first underwater neutral buoyancy tests on Space Station modules in 1986. This view shows the neutral buoyancy facility at Marshall Space Flight Center in Huntsville, Alabama. Two test crew members – one of whom was architect Brand Griffin who proposed the tests – are wearing spacesuits adjusted for use in water. They are carrying out tests to simulate the attachment of debris shield and thermal radiator panels to a module's exterior. Additional tests explored procedures for repairing a hole in a hull after the penetration of a piece of orbital debris.

brown cloud. Over the next several months the Commission sifted through evidence, conducted interviews, held hearings and inspected debris. The cause was determined to be the field joint between two sections of a booster casing that had begun to emit smoke just after launch. The field joint was a circular tongue-and-groove joint enabling one casing section to slot into the other with pins securing them together. A pair of rubber rings of circular cross-section called O-rings and a band of putty provided a continuous seal inside the joint to prevent 'blow-by' of hot solid propellant gases. But how had the joint failed?

Shuttle Orbiter Challenger Disaster
The loss of the Shuttle Orbiter Challenger and its crew on 28 January 1986 was due to a faulty seal in one of the Solid Rocket Booster casings. The seal failed shortly after launch and allowed hot gases to leak through a casing joint and impinge on the large propellant tank. The tank ruptured and exploded, leading in turn to the destruction of the Orbiter Challenger and the loss of the lives of its crew of seven astronauts.

Among the thirteen-member Commission was distinguished physicist Richard Feynman. In a book he later wrote[31] he describes the events that had led up to the pinpointing of the cause. Air Force General Donald Kutyna, another Commission member, had asked Feynman over the telephone what happened to rubber when it got cold, a question that had come to Kutyna's mind while he was working on his car's carburettor. Feynman had replied that it became stiff. The failure's probable cause dawned on him. The rubber O-ring, compressed inside the tongue-and-groove joint under freezing temperatures, had lost its resiliency and become brittle allowing hot gases to blow past it as the joint flexed just after launch, a normal occurrence. Feynman realised that he might be able to demonstrate the failure at a recorded Commission hearing on the rubber seals that was about to take place. Equipped with a sample of the rubber O-ring and a small C-clamp that he bought from a hardware store, he arrived at the Commission hearing room, asked for a glass of iced water, clamped the rubber and dropped it into the glass. When it was his turn to speak he took the seal out of the water, unscrewed the clamp and said 'when you undo the clamp, the rubber doesn't spring back. In other words, for more than a few seconds there is no resilience in it at 32°F. I believe that has some significance for our problem'. Feynman's demonstration and revelation in front of cameras caused a sensation. On 6 June 1986, the Commission submitted its report to the President, explaining that the specific cause of the failure had been the destruction of the seals that were intended to prevent hot gases from leaking through the joint during the propellant burn of the solid rocket motor. With the cause of the tragedy identified, NASA acted to correct the defect with a new joint design and began the process of returning the Space Shuttle to operational status. Yet, the failure of a simple rubber seal and its impact on the Space Shuttle would have major repercussions over the next few years on the course of the Space Station programme.

Years later, NASA reflected on what had happened. 'The first sign of trouble had been a puff of grey smoke from the aft field joint of the right Solid Rocket Booster about half a second after Challenger's launch, the investigation revealed. It was followed by additional puffs, then flame and finally the explosion. The temperature on launch day at 11.38 was 36° Fahrenheit, about 15° colder than any previous launch. No one really understood how the seals on the boosters' field joints worked, or how potentially serious the partial burn-through of O-ring seals seen in those joints after some previous flights had been.'[32] NASA admitted that it had been 'a very public disaster, watched on live television by millions and millions more on subsequent newscasts. Thirty-two months later, after overhauls that brought more than 450 changes to each of the remaining three Orbiters, with redesigned Solid Rocket Boosters, a plethora of changes in processes and management procedures and more, NASA was ready to fly again.'

Second thoughts
Dual Keel configuration, 1986-1988

While the Presidential Commission had been carrying out its six-month investigation into the Space Shuttle Challenger disaster, NASA had pressed on with progress on the Dual Keel Space Station, but in July 1986 astronaut criticism of the design surfaced and was made public. Press agencies and newspapers reported that some astronauts had raised strong objections in a leaked internal NASA memorandum. Written by Gordon Fullerton, an experienced NASA astronaut, it listed several major concerns. The Station had no 'lifeboat' to enable the crew to escape in an emergency; the amount of habitable room was too limited; there was poor separation of work, exercise and eating areas from sleeping areas; there was no room for a second shower and toilet; there was no safe haven and crew shelter in an emergency; there was doubt that the eighteen-plus Shuttle missions needed to build the present design were feasible in the Challenger disaster aftermath.[33] A separate but equally serious concern was NASA's plan to move a significant portion of Space Station work from Johnson Space Center in Houston to Marshall Space Flight Center in Alabama. It was a move opposed by Texas's Congressional delegation because of the potential local job losses. Faced with these serious criticisms, at the end of July 1986 NASA Administrator James Fletcher put the Space Station programme on hold for ninety days and set up two teams to look into the design and the criticisms of it.[34]

As if the multiple burdens of assisting with the Challenger disaster inquiry, conserving the grounded Shuttle fleet, responding to astronaut criticisms and finessing proposed project management moves were not enough, NASA had to contend with the findings of the National Commission on Space that President Reagan had appointed in October 1984 to envision America's future for human space exploration. The Commission of fifteen prominent individuals drawn from the aerospace world, chaired by Thomas Paine, published its report in May 1986. It was called *Pioneering the Space Frontier*.[35] They had devised an ambitious and visionary plan for the human exploration and eventual settlement of the Solar System. Illustrated with colour renderings of spaceports, lunar colonies, Mars bases, sophisticated spaceships and orange groves inside huge greenhouses, it was a dream of America's future in space that appeared to border on science fiction. There was no cost plan, just a brief reference to a price tag and a graph with a wavy line labelled 'programme cost'. There was barely a mention of the roles that the Space Shuttle and Space Station would play in delivering the vision that suggested scepticism about both programmes among the Commission's members. The report recommended that President Reagan and Congress direct NASA to produce a long-range implementation plan and specific five-year agenda by the end of December 1986. As NASA struggled to recover in the aftermath of the Challenger disaster, the timing of the report's publication seemed awkward and unhelpful.

At the time that James Fletcher put the Space Station on hold in July 1986, the Dual Keel was evolving into an advanced design. It was moving on from preliminary design in Phase B to final design in Phases C/D. NASA and its partners had provisionally agreed who was going to carry out the fabrication work on the Space Station's various building blocks. NASA had published an illustration of the Space Station that for the first time clearly showed the roles and responsibilities of the NASA field centres and the Station's international partners.[36]

Appendix 1 shows where Phase C/D occurs in the flow sequence for a major space programme. NASA states that the purpose of Phase C in a project's cycle is to establish a complete design that is ready to fabricate, integrate and verify. Phase D then moves it forward into full development with hardware fabrication, integration and verification.[37]

Marshall Space Flight Center and its contractor Boeing were responsible for the development of one laboratory module, one habitation module, the logistics module, the pressurised and unpressurised module hulls and the life support and internal thermal control systems. This was given the title of Work Package 1. Johnson Space Center and its contractor McDonnell Douglas were responsible for the development of the truss structure, nodes, airlocks, a mobile platform for a robotic arm, the Station's external thermal, data, communications, propulsion, guidance and navigation systems as well as all the astronaut spacewalk operations. In a recent change NASA had reduced the number of American modules from two habitation and two laboratory modules to one of each but they were now 13.6 metres long, the maximum payload length that the Orbiters could carry, so the total internal volume was about the same. This was titled Work Package 2. Goddard Space Flight Center and its contractor GE Astro Space were responsible for the attached payloads, a servicing hangar, a free-flying, co-orbiting platform and a free-flying, polar platform. This was titled Work Package 3. Lewis Research Center and its contractor Rocketdyne (Rockwell) were responsible for the electrical power system and its distribution. This was titled Work Package 4. Then there were the international partners. Europe through the European

Space Agency would provide a laboratory module that it had named Columbus and its own free-flying polar platform. Japan would contribute a laboratory module, an experiment logistics module and an exterior experiment platform. Canada would supply a long robotic arm, using its expertise gained with the Space Shuttle's robotic arm. Fabricated in different places around the world, the building blocks would arrive on orbit where Canada's robotic arm would manipulate and plug them together.

The pressurised modules formed a cluster as they had done in the Power Tower design but they were now positioned in the middle of the crossbar and directly above it, as close as possible to the Station's centre of mass. This gave the module cluster the best possible conditions of microgravity that had been missing in the Power Tower. The American habitation and laboratory modules lay side by side with twin six-port nodes shaped like dumbbells plugged into their endcap ports to form a loop. Plugged into the outer ports of one node pair and facing aft were the Japanese and European laboratory modules. The parallel longitudinal axes of all four modules lay horizontally in the fore-aft flight direction. The means of their attachment to the Dual Keel crossbar truss was not yet clear. Plugged on to the top ports of the twin central nodes were two airlocks and the top port of another node contained a cupola with large window panes. Plugged underneath one central node and pointing to nadir was the American logistics module. Plugged into a port in the upper curved surface of the Japanese laboratory module was the Japanese experiment logistics module, while its exterior experiment platform cantilevered off its outer endcap.

Orbiters would approach from forward of the Space Station's flight path and dock with a forward-facing port on one of the loop nodes. An Orbiter would be able to exchange a depleted logistics module for a freshly supplied one in its payload bay using its robotic arm.[38] Like the Power Tower, the Dual Keel flew with the horizontal crossbar facing the flight path and the rotating trusses at the port and starboard ends supporting the solar power generation elements. The final Phase B design showed two types of these. On the inside close to each rotating joint were twin solar arrays pointing in opposite directions. At each end and pointing out were solar dynamic collectors. Each comprised hexagonal mirror tiles formed into a parabolic dish that reflected sunlight on to a heat engine and dynamo. Also mounted on each side of the crossbar were thermal radiator panels facing different directions. Studies of growth versions of the Dual Keel design showed each end of the crossbar extending out to port and starboard to support more solar generation equipment. One version showed six solar dynamic collectors on each side capable of generating an overwhelming 450kW.[39] Encrusted like barnacles on the faces of the trusses that formed both keels was an extensive inventory of equipment, facilities, experiments and instruments. Among the largest of these were an 8-tonne box-shaped hangar for space tug servicing, a 4-tonne space construction experiment,

Human space exploration ambitions In 1986, as NASA was hard at work returning the Space Shuttle to flight status after the Challenger disaster, the National Commission on Space published a report entitled *Pioneering the Space Frontier* that advocated a highly ambitious plan for future human exploration of the Solar System. Barely mentioning the roles that the Space Shuttle and Space Station would play, the Commission urged Washington to direct NASA to prepare a five-year agenda for the plan. Among other objectives, the Commission recommended establishing a human presence on the Moon. Proposals for a lunar base had been under study in the space community for some years. An example was this concept created during a 1984 NASA summer study at the California Space Institute, Scripps Institute of Oceanography, University of California, San Diego. Image: NASA/Dennis Davidson.

Canada

a 4.5-tonne production unit, a 5.7-tonne radiation experiment and a 3.6-tonne observation experiment. The impact of these large masses on the Station's inertia and controllability on orbit and on its centre of mass meant that they needed careful positioning on the truss structure.[40]

A vast amount of smaller equipment was needed to support astronaut spacewalks. There were eight astronaut movement aid and attachment storage compartments, six general toolboxes, eight customer equipment toolboxes, six stores for astronaut work platforms, two stores for crew emergency equipment, eight power outlet points for tools and equipment, fourteen spacewalk lighting fittings, two contamination monitors and a decontamination facility. Four propulsion modules projected out from the dual keels on arms to provide stationkeeping control. The robotic arm on a mobile trolley ran on rails along the trusses.[41] Then there were the scientific experiments that comprised a host of zenith-pointing and nadir-pointing instruments, telescopes and sensors arranged along the top and bottom transoms of the dual keels. Access corridors from the module cluster ran through the crossbar and dual keel post trusses to enable astronauts to reach all the facilities and equipment sites.

As the Space Station entered the detailed design phase, NASA and its contractors were tackling a growing number of detailed issues of critical importance. Prominent among these was the start of work on key technologies that needed long development and testing cycles. There was a list of new technology needed for the life support system. It included an advanced air-treatment test device called a molecular sieve to remove the carbon dioxide gas and a device for liquefying it, multifiltration and reverse osmosis test units, prototype equipment for recycling urine and testing water quality and system controls. The thermal control system needed new types of heat pipe radiators and heat exchangers. The propulsion system needed a disconnect mechanism that would produce minimum spill. There was an effort to develop a prototype module structure that could be applied to all the American-built modules using plasma arc welding. Work began on prototypes of the hatches and latches that were essential for module berthing on orbit. An umbilical was needed to enable transfer of hazardous gases and liquids outside the modules.[42] It was a large shopping list of new technologies that the Station would need.

In September 1986, NASA established two task forces to carry out critical reviews of the Space Station and NASA's management of it. The first of these was called the critical evaluation task force and based at Langley Research Center. Its purpose was to re-examine the validity and safety of the design in the aftermath of the Orbiter Challenger disaster. The review drew support from other NASA field centres, the Station's contractors and the international partners. The review was prompted by Congressional concerns over project management changes, astronaut grumbles about the number of spacewalks needed for the Dual Keel assembly and concerns about the Space Shuttle

Space Shuttle robotic arm Canada had developed the robotic arm for the Space Shuttle. With its ability to grapple and move large objects in the Orbiter cargo bays, ranging from experiments to satellites, it became an indispensable manipulation tool on Shuttle missions in the 1980s. As the international partners' roles on the Space Station consolidated with the division of the project's development into engineering work packages in 1986, an advanced and more capable version of the Shuttle's arm that would plug the Station's building blocks together in space emerged as a vital element. This photograph shows the robotic arm on Space Shuttle Mission STS-72 in January 1996.

fleet's ability – reduced since the Challenger loss – to deliver all the assembly elements. The task force confirmed that the Dual Keel design remained valid and made some recommendations. They said that it ought to be possible to begin with experiments early in the Station's construction, that there ought to be fewer spacewalks needed to assemble the structure, that there should be a 'safe haven' on board to provide a refuge for crews in an emergency, that there should be a 'lifeboat' for crew evacuation and return to Earth and there should be more use of automation and robotics.[43] The second task force was called the Space Station operations task force. Composed of government and private sector experts, its purpose was to examine NASA's past operational experience with space programmes and to consider any new approaches that could improve efficiency when the Station became operational.[44] The same month, NASA began a financial review of the Station to provide a thorough updated assessment of its anticipated costs. As 1986 drew to a close, NASA established a Space Station Program Office at Reston, Virginia to coordinate and manage the project.

Reality check

Revised Baseline Configuration and Space Station Freedom, 1987-1989

NASA had officially changed the Space Station design from the rejected Power Tower to the Dual Keel at an internal systems requirements review on the Station in March 1986. By the beginning of 1987 it was the Dual Keel's turn to receive a volley of criticism when NASA sent its revised cost and schedule estimates to the White House. In 1984, NASA Administrator James Beggs had given the House Appropriations Committee an $8 billion figure for the construction of the Power Tower Station with a permanent crew occupation date of 1994. That figure had been based on minimal information on the Station's design and engineering at that early stage. By the end of 1986 after the Dual Keel replaced the Power Tower, the Station's estimated cost had risen to $8.3 billion in 1984 dollars or $12.2 billion in 1987 dollars, accompanied by a first launch slip to January 1993 and a permanent crew capability by 1994. Then in 1987, The Dual Keel's projected cost rose again to $14.5 billion in 1984 dollars.[45] It was a big jump and, perhaps anticipating the consternation it would cause in Congress, NASA and the Reagan Administration carried out a parallel design study to explore ways of reducing the Dual Keel's cost. They called the result the Revised Baseline design. It was in NASA's words a 'descoped' Station design. NASA held the estimated cost of the Revised Baseline design to $12.2 billion in 1984 dollars or $17.7 billion in 1989 dollars, accompanied by a slip of the first mission to March 1994 and a permanent crew capability by April 1995. It was somewhat less than the Dual Keel's $14.5 billion but much higher than the Power Tower's original $8 billion of 1984. Congress would have to digest the hefty increase along with the descoped design. NASA responded to complaints over the first launch date slip by offering an earlier date of March 1994, but discord and discontent rumbled on in Washington over the revised cost and design, causing NASA to delay the proposal requests to industry for the next development phase.[46]

Under NASA's descoping plan the Dual Keel Station would be split into two phases with the Revised Baseline design forming Phase I. This comprised the long horizontal truss of the Dual Keel with the upper and lower keels removed at their roots where they joined the transverse truss. The solar dynamic power collectors were plucked out and the solar photovoltaic arrays doubled up to 75kW to compensate. Also 'deferred for future consideration' were the large servicing hangar and the co-orbiting platform. A Phase 2 would add back the two keels at some later unspecified date at an estimated additional cost of $3.8 billion in 1984 dollars over Phase 1. The Station's final cost would therefore come in at $16 billion in 1984 dollars, twice the original figure quoted three years earlier. NASA thought that the new phased approach was technically feasible, that it dealt with the cost problems at the front end of the programme and that it responded to safety and other concerns raised in the critical evaluation task force review the previous year. The 5-metre erectable truss now formed a clean horizontal spine made out of three spans without any perpendicular structural projections. The reduction of the truss structure was severe, down to twenty-nine structural bays from the previous ninety bays of the Dual Keel design. The central span was 75 metres long and contained fifteen structural bays. The end spans were each 30 metres long and connected to the central span by the rotating joints. With an overall length of 145 metres the spine was slightly shorter than that of the Dual Keel.

Doubling up the four solar arrays of the Dual Keel resulted in two pairs mounted on each rotating end truss that pointed up and down as before. The roles and responsibilities for developing and managing the Revised Baseline design were the same as the Dual Keel's, with Johnson Space Center, Marshall Space Flight Center, Goddard Space Flight Center and Lewis Research Center dividing NASA's share of the work between them into four packages. Japan, Europe and Canada would make their individual contributions as before. There were a few additions. The European Space Agency had added its Man-Tended Free-Flyer platform that would occasionally dock with the Space Station for crew access and experiment tending and then undock and co-orbit with the Station to give the experiments the best possible microgravity conditions. It would also provide a destination for the European Hermes spaceplane that was under consideration at the European Space Agency at the time. Canada had added a versatile robotic hand called Dextre to the end of its long robotic arm on the Station. The international partnership remained intact. The Revised Baseline design incorporated the same module cluster as the Dual Keel's.

1987 Space Station Revised Baseline design
Artist impression of the Revised Baseline design that emerged in 1987 as the result of the need to reduce the scale, scope and cost of the Dual Keel design. The principal change was the omission of the top and bottom keels of the earlier design with a long transverse truss acting as a spine to carry all the other elements.

1987 Space Station Revised Baseline design
Escalating costs, construction challenges and concerns over the number of Station assembly flights in the aftermath of the 1986 Space Shuttle Challenger disaster caused NASA to divide the Dual Keel's assembly into two separate phases in 1987. Phase I of the newly named Revised Baseline design omitted the upper and lower keels of the 1985-86 Dual Keel design. The crossbar truss with its two rotating joints and four pairs of solar arrays now became the dominant structural element. The module cluster of the Dual Keel design remained as before, close to the centre of the structure. The upper and lower keels would be added in a future Phase 2. Image: author.

The cluster was still at the centre of the horizontal spine but this time just below it. There were the twin American laboratory and habitation modules, the American logistics module, the European Columbus Module, the Japanese laboratory module, the Japanese experiment logistics module and experiment platform, four 'resource' node modules and two airlocks. Two of the nodes were still located at the forward end facing into the Station's flight path to provide the docking points for the Shuttle Orbiters.

With the two keels gone, the long trussed spine of the Revised Baseline design would have to provide room for the multiplicity of external equipment and experiments that needed a new home. The two outer 30-metre spans were unsuitable because they rotated to help their solar arrays track the Sun. The central 75-metre span would have to carry everything and it would have to squeeze in. Its upper, lower and rear faces provided structural mounting points while distributed utility systems along its length supplied power, thermal control, communications, data, video and space environment monitoring. The truss's front face was reserved for the rail-mounted Mobile Transporter and its Canadian teleoperated robotic arm that would service the equipment and experiments. NASA and its contractors conceived a standard mounting platform that would work for all types of equipment and experiment. It comprised a flat octagonal deck supported on four bifurcated legs attached to all four nodes of a single truss bay. The legs raised the deck clear of the truss face like legs raising a table top off a floor.[47] Each mounting platform was independent and could be added or removed without disturbing the truss or other mounting platforms. The octagonal deck was versatile enough to host many types of payload such as multiple small experiments, one large science experiment, a gimballed and cradled telescope capable of accurate pointing, or a liquid propellant storage tank.

Early in 1987, the Office of Management and Budget, the Office of Science and Technology Policy, the National Security Council and NASA had jointly asked the National Research Council to take a look at the Space Station and carry out an independent examination of its design, cost and ability to satisfy user requirements. In its report published in September the National Research Council endorsed the Revised Baseline design, finding that none of the alternatives it had reviewed was as satisfactory. It pointed out that a national commitment in the form of multi-year funding would help to stabilise the programme and it advised not to proceed with Phase 2 until the nation's long-term space goals had come into focus. The programme management of the Space Station had been consolidated at a new office based in Reston, Virginia following NASA Administrator James Fletcher's direction in April 1986. Its purpose was to streamline the project's management and eliminate its fragmentation across different NASA field centres, specifically Johnson Space Center and Marshall Space Flight Center that were traditional rivals. In the summer of 1987, the Space Station

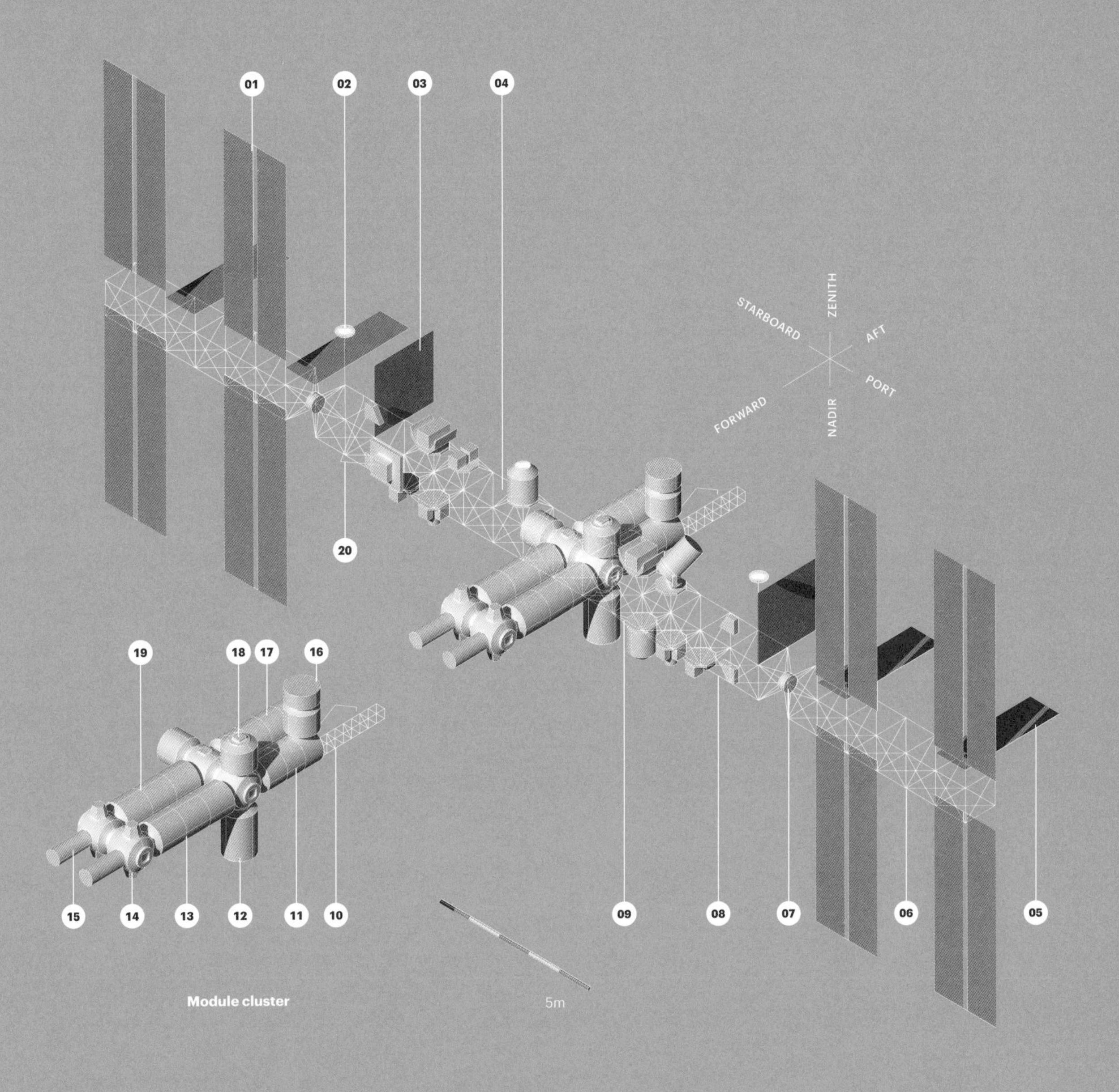

01 Photovoltaic solar array
02 Tracking & Data Relay Satellite antenna
03 Power radiator
04 Zenith experiment & instrument deck
05 Thermal radiator
06 End truss 5m bay size
07 Rotating joint (alpha joint)
08 Centre truss 5m bay size
09 Nadir experiment & instrument deck
10 Japanese experiment truss
11 Japanese laboratory module
12 US logistics module
13 US habitation module I
14 Nodes, cupolas & tunnel
15 Docking tunnel
16 Japanese logistics module
17 European laboratory module
18 Airlock module
19 US laboratory module
20 Robotic arm on mobile platform

Program Office at Reston in conjunction with the Office of Space Flight carried out a Space Station transportation study. It looked at ways of improving safety, space transportation planning, reducing the Station's assembly burden on the Space Shuttle fleet and using alternative launch vehicles to deliver the Station's building blocks. It concluded that the Space Shuttle should continue as the Station's delivery service but with an improved payload return capability, that the Shuttle should be limited to just five flights a year for Station purposes and that Station crews should be rotated four at a time with tour durations extending to 180 days. NASA incorporated these findings into the Station's programme in November. NASA also carried out a Station operations management study and a Station distributed systems study, concluding that both were basically sound in the Revised Baseline design.[48]

Sally Ride's report In 1987, Sally Ride, the first American woman in space, authored a report entitled *Leadership and America's Future in Space*. NASA's Administrator had commissioned her to produce it as a comment on the state of American human spaceflight. The report bluntly stated that, in the aftermath of the Challenger disaster, the Soviet Union was now the only nation that was a long-term inhabitant of Earth orbit. She pointed out that the latest Soviet Space Station, Mir, would be operational a decade before the American-led Space Station. In this earlier photograph taken in 1983, Sally Ride is communicating with ground controllers during a six-day mission on the Shuttle Orbiter Challenger.

In August 1987, NASA published a report to the agency's Administrator by Sally Ride, an astronaut and the first American woman in space. It was called *Leadership and America's Future in Space* and it dealt with the perceived loss of American prowess in space in the aftermath of the Orbiter Challenger tragedy. The report bluntly stated that the Soviet Union was now the only long-term inhabitant of Earth orbit. It had sent eight small Space Stations there since the mid-1970s of which the latest, Mir, would accommodate crews and experiments a decade before the first crew arrived on the Revised Baseline Station. By contrast, the first and only American Space Station, Skylab, had been left empty in 1974. When Sally Ride turned to the Space Station she queried its evolution and asked whether it would accommodate users with different requirements if those requirements were incompatible. She gave an example of a Station that included a laboratory that needed a long-term weightless environment without disturbances as well as a spacecraft assembly and checkout hangar that would generate significant disturbances with its workshop activities.[49] It was a valid and serious concern. The Revised Baseline design might be suitable for scientific research but future hardware additions could raise problems of the kind that led to the demise of the 1984 Power Tower design.

Late in 1987, the House Committee on Appropriations Subcommittee on HUD-Independent Agencies had asked NASA to carry out a study to explore an early crew-visited capability for the Station. Congress was concerned about delays and wanted the Station's assembly on orbit reshuffled to bring forward the date that the Station would start to do useful scientific work. In April, NASA Administrator James Fletcher outlined NASA's response in a letter to the Subcommittee. A version of the Revised Baseline design that could deliver early science was certainly feasible though bringing forward the date of its first launch was not possible. The ground rules were the provision of early pressurised volume and tended experiment capability, no interference with the Station's overall assembly plan, no new hardware beyond the Revised Baseline design, and use of the Space Shuttle only to deliver the building blocks and autonomous operation between Orbiter visits. Exterior experiments would, however, have to slip if the American laboratory module was brought forward. NASA proposed launching the laboratory on the fourth Shuttle assembly flight in the fourth quarter of 1995. The first three flights would deliver about half the truss structure, two resource nodes, one half of a solar array pair and initial propulsion, docking, guidance, station-keeping, thermal, communications and tracking systems. The laboratory module would contain a set of science and materials experiments ready to switch on. The next flight would deliver the second solar array. NASA anticipated no significant cost impacts to the re-phasing and that the date for a permanently crewed Station would be unaffected. In his letter James Fletcher underlined NASA's confidence, pointing out to Congress that 'the current configuration results from four years and over $600 million worth of definition analysis by government and industry. It has been reviewed by NASA and the National Research Council. It represents, in my view, the optimal balance between development costs, operations costs and satisfaction of user requirements in a safe Station.'[50]

Since the Space Station had formally begun as a project in 1984 the world had come to know it simply by that name. Now that it was moving into full development it needed something better. In June President Reagan announced that he had given it a proper name. From now on it would be called Space Station 'Freedom'. He picked the name himself from hundreds of suggestions by NASA staff and contractors working on the project. Neither the American public nor the international partners had been asked for ideas. It was certainly not a name taken from ancient Greek or Roman deities, such as Mercury, Saturn or Apollo, which NASA had favoured in the past. It was a name that perhaps captured the global political mood of the times from Reagan's own viewpoint. While the nations of the free world were happily teaming up on this grand vision, some other nations were plodding along wearily under the thumb of totalitarian regimes. Yet the person who had proposed the winning name may have meant something quite different by it. Equally 'freedom' signified the release from the gravitational confinement of the Earth's surface and the physical liberation offered by the weightlessness of space. After all, overcoming the force of gravity was the Station's chief challenge.

Watershed moment
Intergovernmental Agreement, Memoranda of Understanding and Contracts to Proceed, 1988

September 1988 was a pivotal month for Space Station Freedom's formal paperwork. NASA signed contracts to proceed to full development of the American segment with its main contractors Boeing, McDonnell Douglas, GE-Aerospace and Rockwell. That same month, the American government

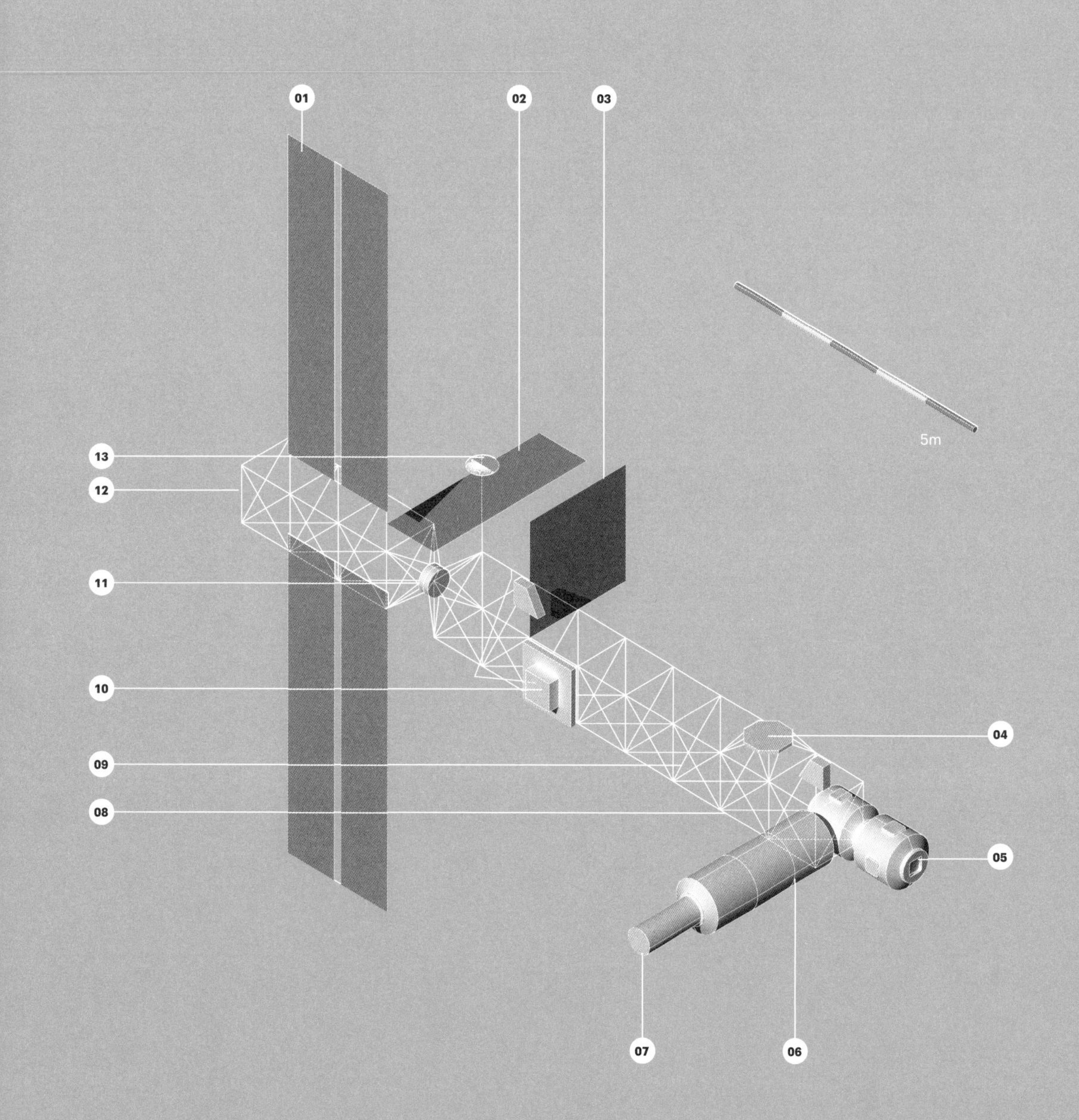

01 Photovoltaic solar array
02 Thermal radiator
03 Power radiator
04 Zenith experiment & instrument deck
05 Resource nodes (2)
06 US laboratory module
07 Docking tunnel
08 Nadir experiment & instrument deck
09 Centre truss 5m bay size
10 Robotic arm on mobile platform
11 Rotating joint (alpha joint)
12 End truss 5m bay size
13 Tracking & Data Relay Satellite antenna

signed a multilateral Intergovernmental Agreement with the governments of the nine member nations of the European Space Agency and the governments of Canada and Japan. The Agreement spelled out that all the partners would cooperate: 'in the detail design, development, operation and utilization of the permanently manned civil Space Station' and it would come into force in 1991. Parallel to the Agreement were Memoranda of Understanding signed between NASA and its counterpart space agencies in Europe, Japan and Canada. Specified in the various documents were the hardware and services that each nation or group of nations would provide. America would contribute Station infrastructure, a habitation module, a laboratory module, various resource modules, a polar orbiting platform and ground support elements. Europe would provide its Columbus Laboratory Module, its free-flyer experiment platform, a polar orbiting platform and ground elements. Japan would contribute a laboratory module, a support module and an experiment deck, and Canada would provide its robotic arm technology.

In November 1987, in the run-up to the Intergovernmental Agreement, the European Space Agency and Europe's space ministers had sought to craft some watertight clauses covering individual partner binding commitments, an equitable working partnership between them and the prohibition of military activities on the Space Station. The intent was to ratify the Agreement in the European political arena so that it would take precedence over national laws and legally bind all the nations involved. NASA was cautious in its response and urged moderation. In the first place, what the Europeans were proposing was virtually an international treaty and it would never be acceptable to the American government. NASA also had to damp down Europe's demand for a 'genuine partnership', whatever that meant. NASA worked with its European counterparts to identify what such a partnership covered, item by item. Europe was also emphatic about the peaceful role of the Space Station. There were concerns that the American Department of Defense wanted to use it for military research, in particular to support its Strategic Defense Initiative plans. This possibility was completely unacceptable to the International Partners. Secretary of Defense Caspar Weinberger tried to calm things down by giving the partners a list of possible military experiments on the Station that he thought would be acceptable but he made it worse by confirming Europe's suspicions. The final Agreements made it clear that the Space Station was for civil and peaceful purposes only but the American chief negotiator placed on record that his country had the right to use its Station elements and allocated resources for American national security purposes. Any Space Station partner, though, would have the right to deny the use of its module, if it had one, for military purposes. There was no satisfactory resolution of the different viewpoints and the disagreements that resulted almost brought the negotiations to a standstill.[51]

1987 Space Station Revised Baseline design – early crew access stage Late in 1987, Congress asked NASA to look at the possibility of accelerating the start of scientific research on the Station. NASA responded with this interim stage in the build-up of the Revised Baseline design. It comprised a laboratory module, two resource nodes, a docking tunnel, half of the overall truss structure, a pair of solar arrays, the robotic arm and its mobile platform. It would allow Shuttle Orbiters to dock on short visits so that crews could initiate and tend experiments. Image: author.

Part of NASA's difficulty in dealing with the Europeans was that so many partner nations were involved and they often disagreed among themselves over space issues and, despite the European Space Agency's best efforts, were sometimes unable to present a united front. Different governments had different agendas and developing a consensus had needed great care and patience. Capricious Britain had blown hot and cold on the Station. It had been very keen on the idea at first, in the early 1980s, but its interest had evaporated with the exception of the free-flying Earth observation platform that it wanted its aerospace industry to build. France, Europe's leading spacefaring nation, had wanted Europe to adopt its Hermes spaceplane arguing that European space independence meant Europe launching its own astronauts to orbit on its own Ariane rocket. Germany had been worried about a financial commitment to both Ariane and the Station and its scientists were sceptical of the need for an additional Station capability beyond that of the existing Spacelab system that flew on the Space Shuttle. Italy, on the other hand, had been very keen on the project from the start as its aerospace industry was the European expert on pressurised modules. The smaller spacefaring nations in Europe had wanted to be sure that a real opportunity existed for their industries and scientists to play a part.

Nevertheless, rejecting France's proposition to include Hermes in Europe's space plans but leaving the door open for some future Hermes development, the European ministers in Rome had earlier approved the European Space Agency's long range plan, subject to the negotiation of acceptable terms and conditions, that would lead to what their agency insisted was to be an International Space Station as opposed to an American Station with foreign participation. Europe had three strategic aims: the strengthening of its space transportation capabilities to stay competitive with systems elsewhere; the development of independent European facilities to support human activities in space; and cooperation on a Space Station with America in particular. Achieving these aims would result in significant long-term benefits for Europe. Attached as conditions to Europe's participation had been a resolution – at first kept confidential – on the objectives to be met. Among these had been a desire for European responsibility for the design and development of several Station building blocks, a wish for European access to all parts of the Station on terms equivalent to the most-favoured users, a need for fair operations cost-sharing, a need for a common approach to handling technology transfer, a need for the equitable provision of supplies and services, and a guarantee of maximum legal security and the mutual use of transportation and communications facilities.[52] Such was the wish list of things that Europe wanted to ensure in its negotiations on the Station. But past wishes were one thing and the new Intergovernmental Agreement was another. If Europe wanted to participate in the Station it would have to fall into line with America and NASA's way of doing business. Henceforth, Space Station Freedom's partners would have to march together in step towards their shared objective and it was time to bury the differences of opinion.

On 29 September 1988, the Space Shuttle system returned to flight status with the launch of STS-26 and the Orbiter Discovery, which delivered a NASA tracking and Data Relay Satellite into orbit.[53] Thirty months had passed since the Challenger disaster. There was much jubilation and emotional relief at NASA at the successful launch. The *Chicago Tribune* noted that the future of NASA and the nation's space programme was riding on the Orbiter Discovery. It pointed out that since the grounding of the Space Shuttle fleet, the Soviet Union had launched sixteen cosmonauts aboard six spacecraft to its Mir Space Station and that one of them had spent 326 days there.[54] America had some catching up to do. Much of the thirty months had been spent devising and implementing ways of improving the safety of the Space Shuttle system and its future missions. Improving the safety of the Space Station Freedom also continued to be a big issue. There were still no plans for it to have a permanently docked escape vehicle of some kind to enable a crew to leave in an emergency. It was rather like a small hotel with an entrance lobby but no fire exit. If fire blocked the lobby the guests would be trapped.

Following its review of the Revised Baseline design in 1987, the National Research Council continued with a workshop to examine the Station further. It found that the design, assembly and operations plan was too ambitious for the reduced Space Shuttle schedule, that crews would spend too much time on maintenance duties and not enough on experiments, that there was no provision for the unmanned resupply of American modules and that the project's management structure was complicating the project's development. It also found that the issue of evacuating the Station's crews in an emergency and returning them safely to Earth had not yet been addressed. The National Research Council thought that the Station ought to be equipped with an escape vehicle.[55] Such a vehicle would have three vitally important purposes. First, it would enable a crew to return to Earth in a non-emergency were the Space Shuttle fleet unavailable or unable to fly. A grounded fleet might result from the catastrophic loss of another Orbiter or a multitude of maintenance and repair issues or consistently bad weather or hurricane damage at the launch site. Grounding the fleet would interrupt the resupply of the Station with vital consumables such as water, food, atmospheric gases or propellants for stationkeeping. The crew would be forced to evacuate if existing supplies ran out.

Second, a crew escape vehicle would offer a means of escape and evacuation from the Station in the event of a life-threatening emergency such as the penetration of a module hull by a micrometeoroid or piece of orbital debris resulting in depressurisation or a fire such as the one that occurred on the Soviet Union's Mir Space Station.

The Hermes reusable spaceplane Hermes was proposed by Europe in the 1980s to give it an independent capability to launch European astronauts to orbit on Europe's Ariane rocket and to visit the Space Station. France, Europe's leading spacefaring nation, was the main advocate of the initiative. However, by 1988 with the need to reduce European space expenditure, interest in it had waned and it was excluded from Europe's space planning. This illustration shows the Hermes design. Image: ESA.

Third, it would offer a space 'ambulance' in a medical emergency that required the return of an ill or injured crew member to Earth. The Station would have a small medical facility capable of treating minor injuries or ailments and providing supportive care but it had no surgical facilities and would be unable to deal with a major trauma. An unexpected illness such as appendicitis or a kidney stone or a limb injury from a spacewalk accident could demand a rapid return to Earth for a crew member. This was impossible without some sort of ambulance vehicle. The medical issue was of particular concern as it was impossible to predict accurately the incidence of crew member illness or injury. One study suggested that such an emergency, which would require the urgent return of a sick or injured crew member to the ground, could occur every four to twelve years for a crew of eight.[56] As the Station moved into the hardware development phases there was no formal requirement for a crew escape vehicle as a program element and it did not feature in Space Station Freedom's design.

Close quarters
Module habitat development, 1984-1988

Survivability and habitability are the twin foundations of design for human life in space. Survivability is concerned with keeping humans alive inside a spacecraft, capsule or module, protecting them from the hostile environment outside and ensuring their safe passage into space at a mission's beginning and back to the ground at its end. It is rather like providing intensive care in a hospital except that the patient is not ill. In space, habitability is the deluxe version of survivability. It provides a modicum of comfort to crews through its limited replication of the living standards and amenities that they are accustomed to on Earth. Survivability and habitability are both tailored to the metabolism of the crew and the human body's need to stay alive through its intake of oxygen, water and food and its disposal of carbon dioxide, liquid waste, solid waste and heat.

At the birth of human spaceflight at the beginning of the 1960s, there was no such thing as space habitability. It was a challenge just to ensure the safety and survivability of the astronauts and cosmonauts during their relatively brief missions. This was the case in the early and mid-1960s with the American Mercury and Gemini programmes. Conditions inside the capsules were barely tolerable from a human factors standpoint. The capsules supplied just what was needed to keep the astronauts and cosmonauts alive. There was a need for atmospheric pressure to maintain human physiology and hold the human body together. It was possible to drop this down from Earth's standard atmospheric pressure but there was a lower limit. Coupled to this was the need for a breathable atmosphere with the right mixture of oxygen and nitrogen that the human body was used to on Earth. The lower the air pressure, the higher the amount of oxygen needed for respiration, but an oxygen-rich atmosphere was dangerous because of the risk of fire. Fire and smoke suppression was essential whatever the internal air mixture. Then there was a need to keep the spacecraft's internal atmosphere at a suitable temperature and humidity level. The temperature range for human physiological comfort was very narrow compared with the range a spacecraft experienced on the outside. Removing exhaled carbon dioxide from the atmosphere and replenishing it with oxygen was also essential. Astronauts and cosmonauts also needed nourishment and water to keep them alive during their missions and an acceptable means of dealing with bodily wastes. All these needs defined the basic life support systems inside the space capsule.

Outside the capsule, space was not empty but full of subatomic particles and tiny pieces of debris that were potentially hazardous to the capsule and those inside. The background radiation of space – cosmic radiation – penetrated spacecraft hulls and built up slowly in human tissues. Occasional solar flares produced bursts of radiation that were potentially lethal in a few hours of unprotected direct exposure. Micrometeoroids the size of dust particles or larger travelling at extremely high velocities could penetrate a spacecraft's metal alloy skin. Capsule exteriors needed shielding against hazards such as these. So, in the vacuum of space with its temperature extremes and particle hazards, the capsule needed an enclosing structural hull to contain an internal breathable atmosphere, impact shielding on the outside to prevent hull penetration and thermal insulation to keep the internal atmosphere at the right temperature. There were also the physical stresses and strains on the capsule of getting it into space and back. The launch and ascent regime subjected the capsule's structure to vibrations and strong g-forces that peaked at certain points and could approach the human threshold of tolerance. On the way back to Earth the capsule's exterior had to withstand massive heating during atmospheric re-entry and descent and then survive the shock of parachute-assisted splashdown or touchdown.

Habitability became essential for long-duration spaceflight lasting months or years. Most importantly, it introduced open volume inside a spacecraft for the crew to float about in freely and experience weightlessness without the confinements of a cramped cockpit where moving an arm or a leg required care to avoid accidentally activating controls or equipment. Open volume enabled the interconnection of adjacent modules at their ends with the crew moving between them through hatches. This physical freedom made it possible to design distinct accommodation zones for living, working and sleeping with the crew able to move easily from one to the other. Individual privacy became feasible. A division between day and night zones was possible with the latter containing individual sleeping compartments and personal hygiene facilities such as a lavatory cubicle and a proper shower. There was

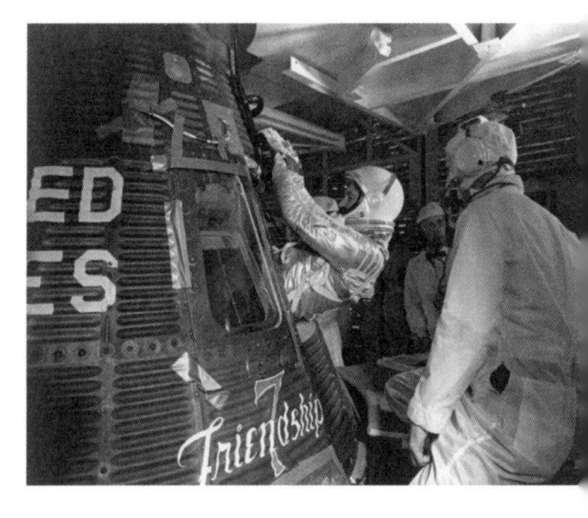

Human survival in space In the early days of human spaceflight, the main design and engineering challenge was to ensure the survival of astronauts. There was little or no concern over spacecraft habitability conditions on the missions as they were so short. This photograph shows astronaut John Glenn climbing into his cramped Mercury capsule 'Friendship 7' before launch on a Mercury-Atlas rocket in February 1962. His mission lasted for just under five hours.

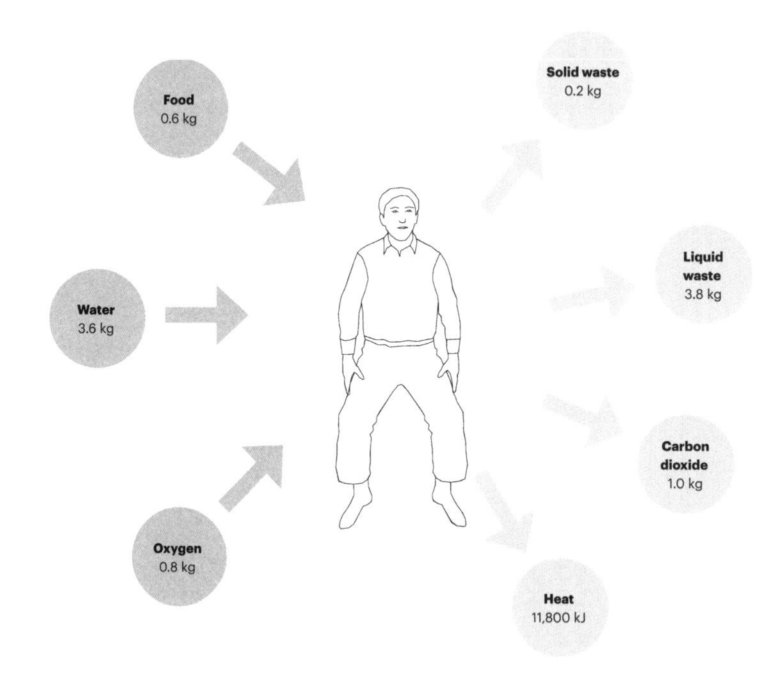

Daily human metabolism demands The fundamental requirement for enabling human life to exist in space is to provide for the metabolic needs of the human body through the entire mission from launch to landing. Every crewed spacecraft, capsule or module must incorporate the on-board life support systems necessary to meet these needs. This diagram summarises the typical daily metabolic input and output demands of one person. The metabolism demands become a considerable design and engineering challenge as crew sizes increase and missions lengthen. Image: author.

a clear separation of wardroom and laboratory accommodation in the day zone. The wardroom contained food preparation and eating facilities, where the crew could have a meal together, spend some leisure time and discuss their work plans. There was a window for looking out and taking photographs. The laboratory had various instruments and items of equipment housed in racks and workstations or mounted independently around the walls of the pressurised hull. There was a means of disposing of laboratory and other waste. On long-duration tours, physical exercise was essential for crew members to maintain their health and they regularly worked out on exercise equipment such as a bicycle machine or a treadmill. The inclusion of a medical station made health monitoring possible. Communications with the ground became vital with the exchange of technical data and work schedules between the crew and ground control as well as private conversations with family members. Skylab, America's first orbital outpost in the 1970s, embodied all these habitability features. They were a huge improvement over what had gone before.

The Space Station would attempt to raise Skylab's space habitability standards to new levels of comfort and efficiency, but it would be necessary to divide the crew accommodation into several smaller volumes. The spacious volume of Skylab's single workshop module was not possible. The Station's nucleus would be the cluster of pressurised modules in which the crews would live and work during tours of duty that NASA anticipated would last up to six months. Crews of six or perhaps eight astronauts would spend all their time inside the modules except for occasional spacewalks when, sealed in spacesuits, some would venture out of an airlock to carry out assembly tasks, maintenance duties or repairs. Early Space Station studies by NASA and its contractors from 1979 onwards had envisioned the pressurised modules in a multitude of shapes and sizes. They were long or short or somewhere in between. They had hemispherical, elliptical, conical and occasionally flat endcaps. They had berthing ports and hatches at their ends or around their sides or both. Some had porthole windows, some had cockpit windows and others had no windows at all. Some were quite narrow and more like tunnels. What governed them all was the payload bay of the Shuttle Orbiters, which limited a module's maximum length to about 13.6 metres and its diameter to about 4.5 metres. The resulting volume inside was precious and had to be put to the best use. It was possible to put a value on every cubic centimetre of it. Module internal design and outfitting would be a compromise between cramming in all the equipment, experiments and supplies so vital to the Station's operation and reserving enough free volume in which the crew could exist and breathe without constantly colliding with equipment or each other as they floated around inside. Satisfying the requirements of each was always going to be a matter of balancing conflicting priorities.

Though it was possible to quantify the exact amount of volume needed for the equipment, experiments and supplies, it was quite difficult to quantify the amount of free volume needed to make the Station's interior acceptable to the crews as a habitable environment. For some time the question of what constituted a habitable space environment had occupied the minds of sociologists, psychologists, physicians and others concerned with the physical and mental wellbeing of astronauts. Habitability revolved around human factors and the need to design space environments and equipment to fit the human body, engage the human mind and ensure the health of both. The quality of sleep and the repetitive work-rest cycles and the maintenance of crew skills and performance were major areas of concern. The physical proportions of humans should drive the physical features of a spacecraft with its interior designed for optimal habitability, life support, safety and operational efficiency. The human body was the starting point. Human anatomy was an essential reference as it changed in weightlessness when released from gravity or the confinement of cramped capsules. The evolution of humans on Earth owes much to the effects of terrestrial gravity on the human body. From the time that humanoids began to walk upright, their head, neck, spine, pelvis and legs assumed a vertical orientation that transferred their body weight through their bones to the ground as directly and efficiently as possible. Muscles developed to help to hold the bones in this upright posture, whether running, walking or at rest.

Overleaf: Survivability – main design and engineering issues In the early years of human spaceflight, ensuring the survivability and safety of astronauts and cosmonauts throughout the short missions was the major challenge for the capsule designers and engineers. These diagrams summarise some of the main issues they had to solve. The life support systems that ensured astronaut survival on the Mercury missions of the early 1960s had to be fitted, together with the astronauts, into capsule volumes of just 2.8m³. Image: author.

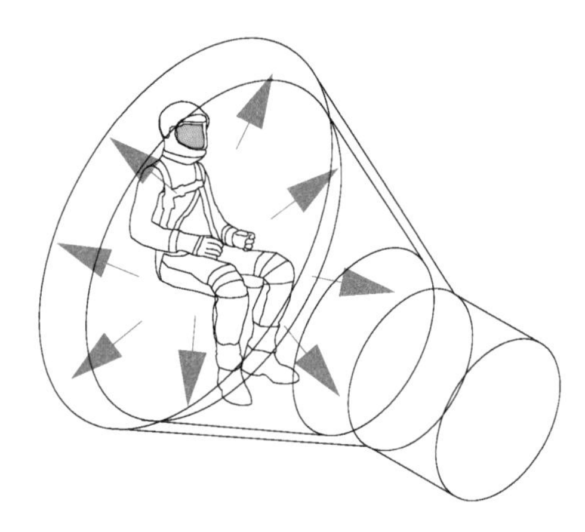

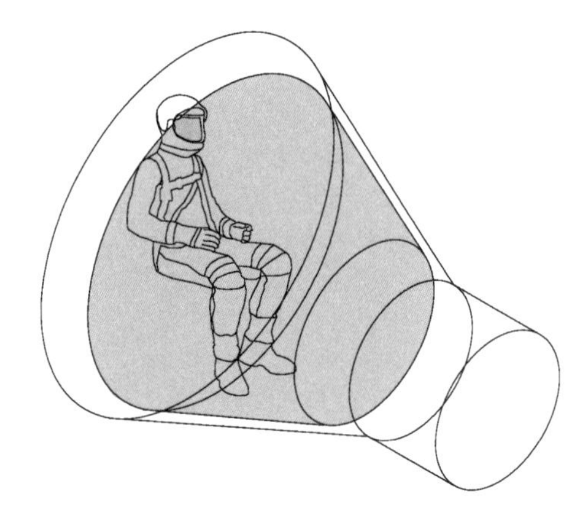

Pressure
Maintain internal ambient pressure

Respiration
Provide correct breathable air mixture

Temperature
Maintain physiological comfort range

Water + food
Supply sufficient water and solid nourishment

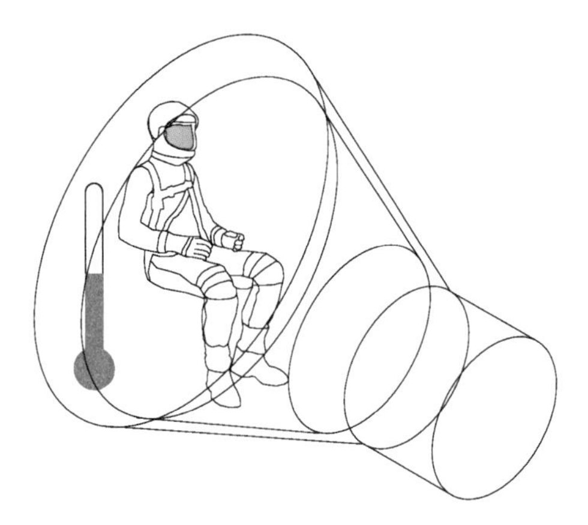

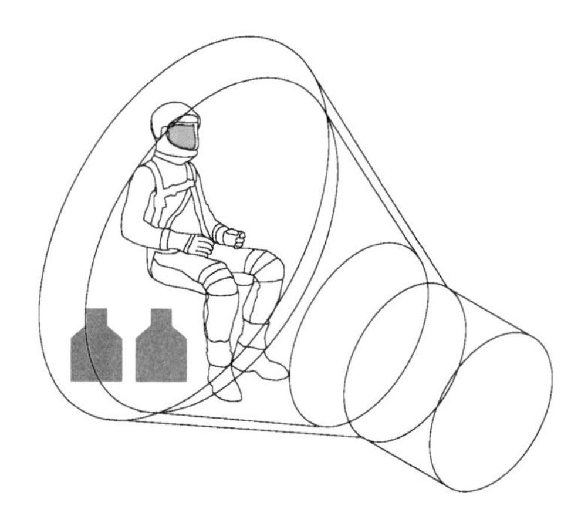

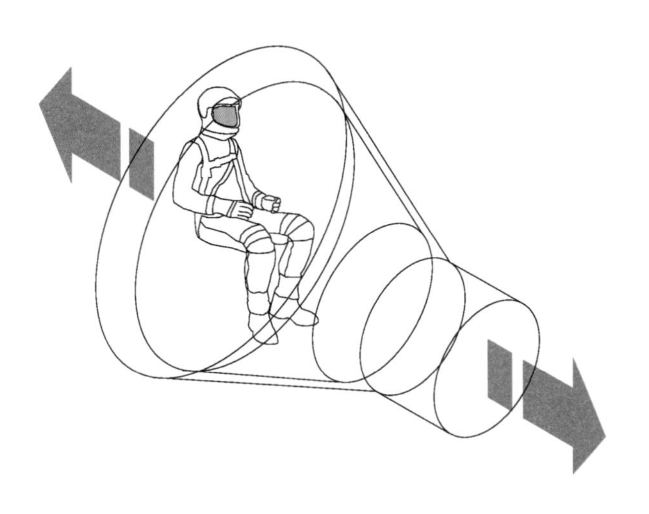

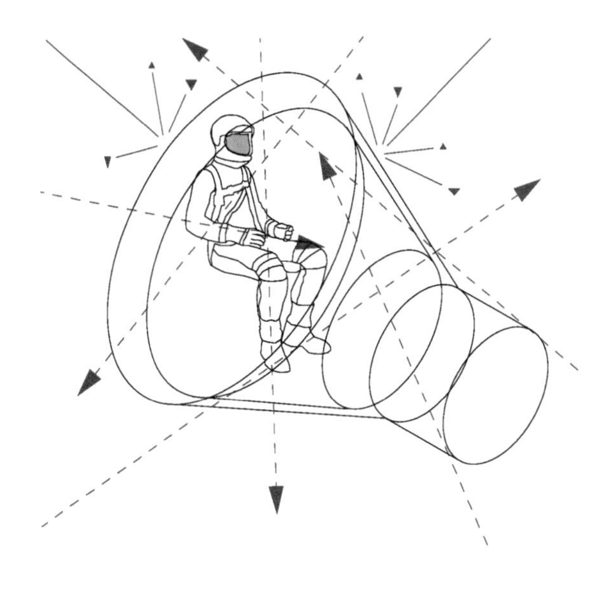

G-forces
Keep g-forces below human threshold

Debris + radiation
Protect from micrometeoroids, cosmic and solar radiation

Fire
Ensure interior fire and smoke suppression

Re-entry
Protect from atmospheric re-entry heating

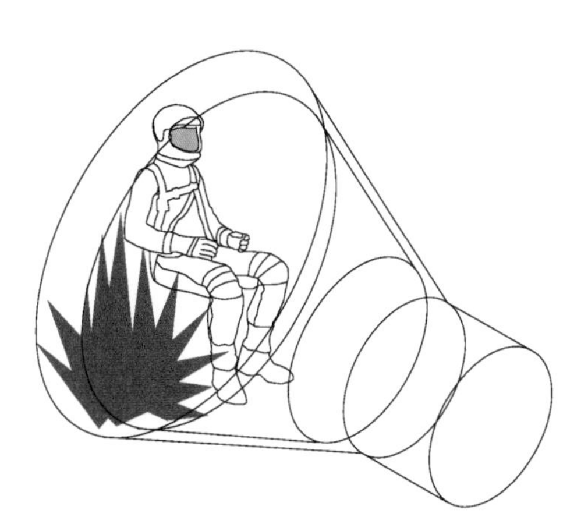

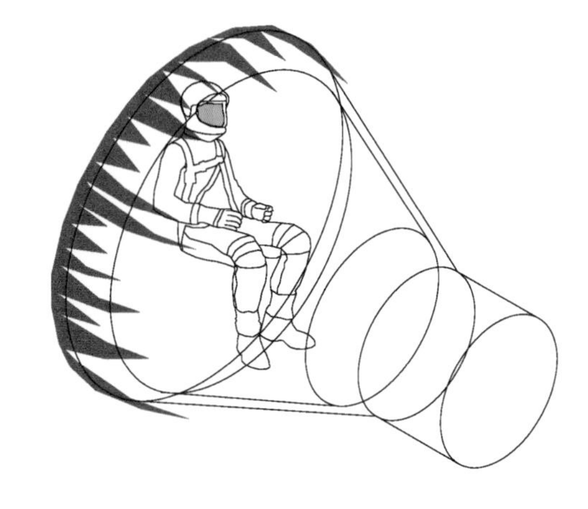

Habitability – main design and engineering issues Habitability is essential for long-duration human space missions of all kinds. It first appeared in significant form in the Skylab Orbital Workshop of the early 1970s. It introduces proper crew accommodation to space module interiors and, to the extent possible, some of the amenities and features of living and working environments on Earth. The key to habitability is the free internal volume through which the crew can move and this ought to increase with the mission length. The illustrations opposite depict some of the major habitability design issues. Image: author.

Removing the gravity causes the body's posture to change. It assumes a more comfortable position where the muscles can relax, released from their former burden. A pioneering design guide produced by architect Brand Griffin for Johnson Space Center in 1978 illustrated the anthropometric changes to the human body when this happens.[57] The upper torso and head bend forward, the elbows bend and the arms splay out to the front and side. The upper legs angle forward and outward from the pelvis, the knee joints bend and the feet point slightly. As a result, overall body height diminishes by about 15 per cent and the posture becomes slightly simian in appearance. These changes have important consequences for the ergonomic design of the accommodation and equipment inside the pressurised modules intended for operation in weightlessness.

A human factors study carried out by the National Academy of Sciences described a crewed spacecraft as a 'microsociety' – a social system comprising the crew, the spacecraft, the environments in which the spacecraft travelled, its relation to the organisation and society on which it depended and all other factors that affected its mission. Organisational arrangements in a spacecraft, including crew size, structure and dynamic organisation, would have a direct bearing on performance and the viability of the microsociety.[58] Writing from the viewpoint of the early 1970s, the study predicted that life on a long-duration mission (such as a Space Station) would be very different from the 'primitive' conditions of spacecraft of the 1960s and that future long-duration missions would result in many technological changes and advances to the benefit of the crews. It concluded that 'if man is to participate in long-duration spaceflight, his requirements – physical, psychological, behavioural and interpersonal – must be given far more attention than has heretofore been the case in the design of the spacecraft and the mission. Requirements for life support and safety, optimal habitability and comfort, and operational efficiency should be incorporated into spacecraft engineering from the beginning. Habitability in particular must be improved, especially with regard to capsule volume, configuration and noise as these relate to work, mobility, exercise, recreation and sleep.' Discussing habitability, another human factors study originating from NASA in 1985 noted: 'the need for a shift in emphasis from merely sustaining human life in space to maintaining a high quality of life in space'. It advocated measures that included the fuzzy phrase 'environmental richness' and noted that the use of crew leisure time and the need for privacy were key habitability issues.[59]

An important independent research study undertaken for NASA in 1986 drew up a list of recommendations for the Station's habitability based on a review and comparison of confined habitats in extreme environments on Earth.[60] It listed critical behavioural factors with design lessons for the Station and then examined terrestrial counterparts that could generate useful design guidelines. The research slant was on biological, psychological

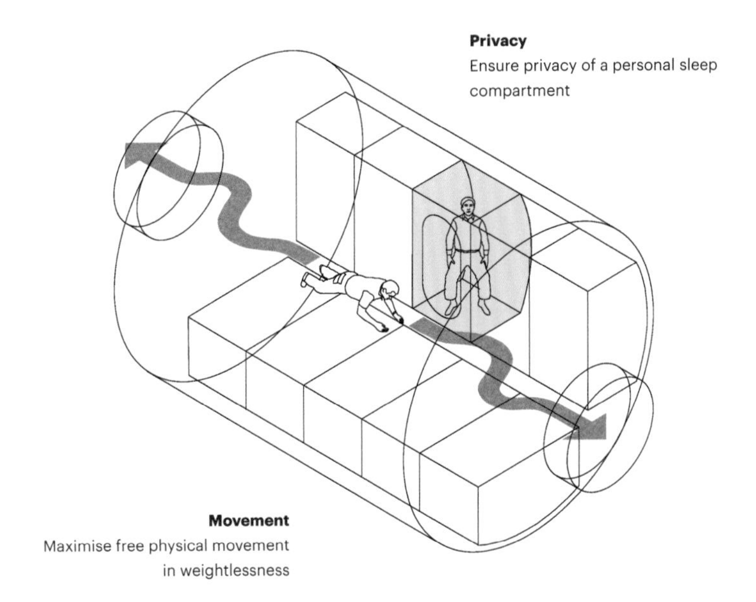
Privacy
Ensure privacy of a personal sleep compartment
Movement
Maximise free physical movement in weightlessness

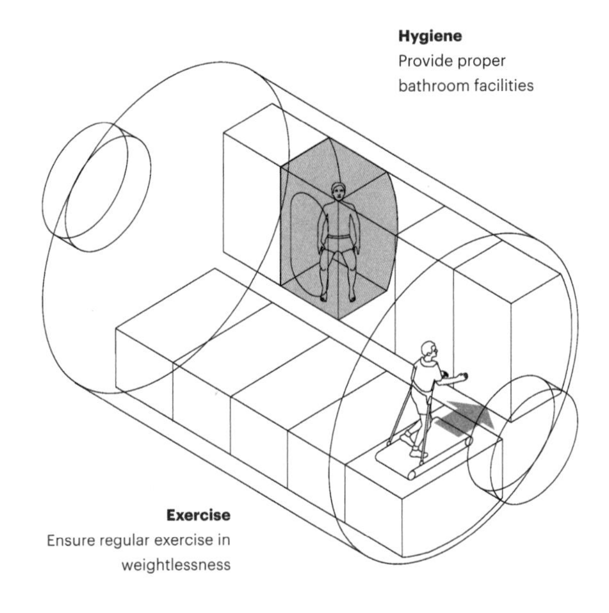
Hygiene
Provide proper bathroom facilities
Exercise
Ensure regular exercise in weightlessness

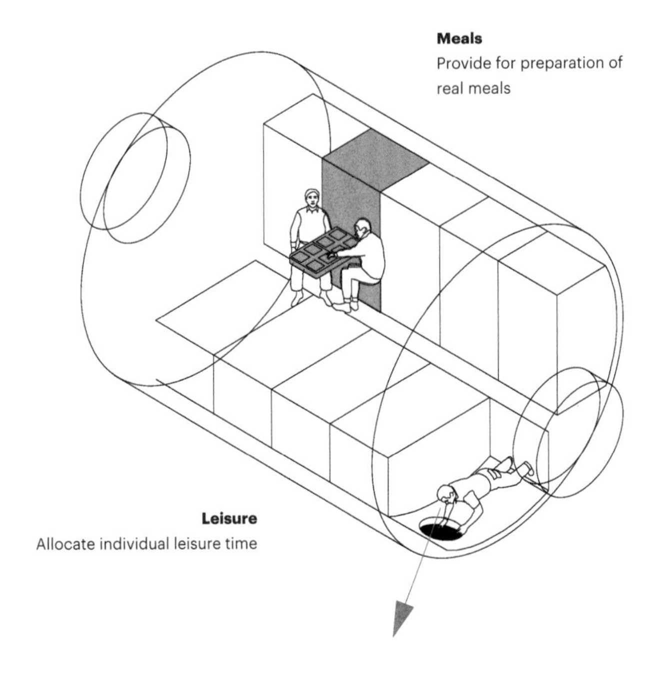
Meals
Provide for preparation of real meals
Leisure
Allocate individual leisure time

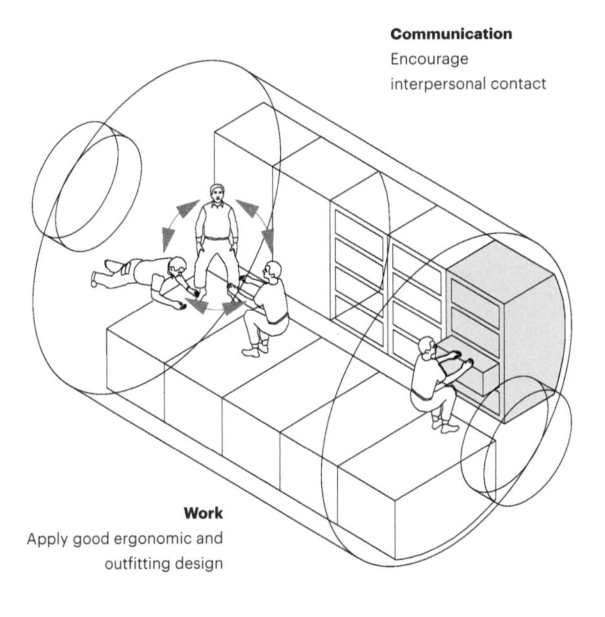
Communication
Encourage interpersonal contact
Work
Apply good ergonomic and outfitting design

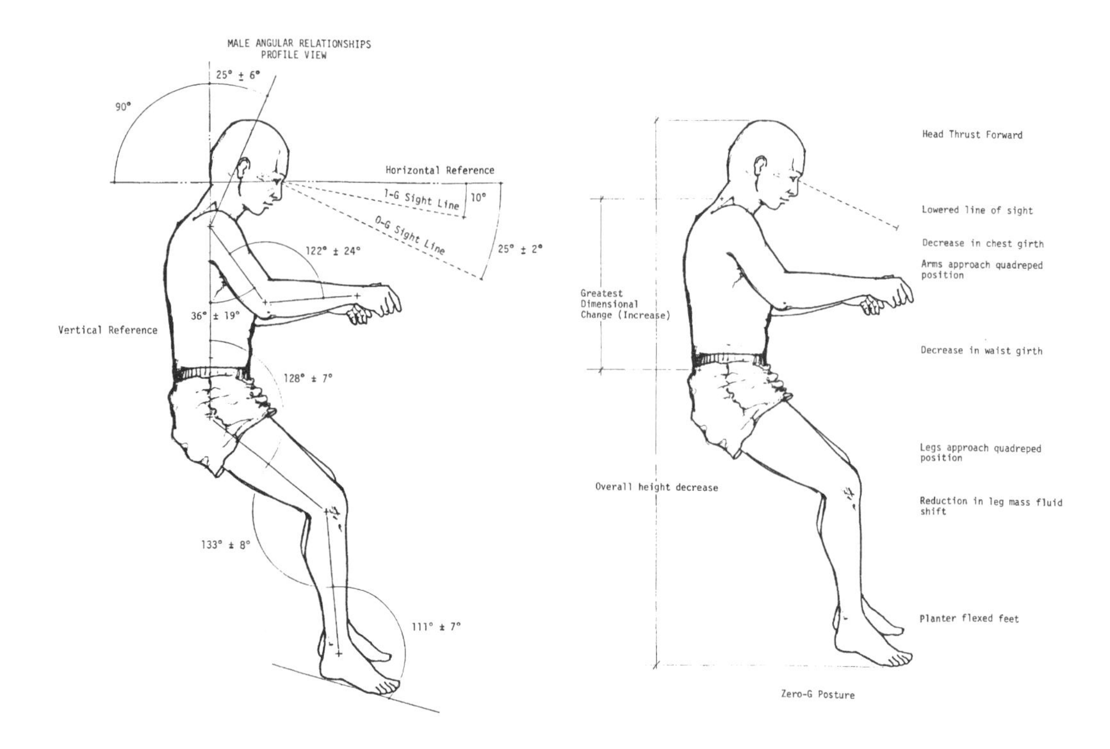

Anthropometric adaptation to weightlessness These diagrams are taken from a 1979 pioneering design guide produced by architect Brand Griffin for NASA that shows the effects of different gravitational conditions on human body posture. In the confined environment of a spacecraft or space module, the interaction of the human body with its surroundings takes on great significance. Changes to body posture caused by the removal of gravity have a major impact on the ergonomic design of equipment such as control consoles and laboratory workstations. Comfortable arm reach distances and eyesight lines are important to assist astronaut productivity. Replacing chairs, which are useless in space, are body restraints to keep floating astronauts in place while they carry out tasks and these also must be designed for comfortable use in weightlessness. Illustrations: Brand Griffin.

and sociological issues that together formed the 'intangibles of habitability.' The study began by identifying all the major behavioural issues. These were sleep, clothing, exercise, medical support, personal hygiene, food preparation, group interaction, habitat aesthetics, outside communications, recreational opportunities, privacy and personal space, waste disposal and management, on-board training and simulation, and the behavioural and physiological requirements of weightlessness. It then examined Antarctic research stations, remote military outposts, nuclear submarines, undersea habitats, long-distance yacht racing, fishing boats, offshore oil platforms, supertankers, the Skylab Station and even the Ra papyrus boat adventures of Thor Heyerdahl. The research methodology centred on parametric evaluations of these different extreme environments by over fifty participants who were experts in the fields. The correlation and synthesis of their responses resulted in the research conclusions. These added up to nearly 100 specific design guidelines spanning the spectrum of behavioural issues. Many of these directly affected the crew accommodation design in the modules. There should be a 'mini-gym' for the crew to take vital physical exercise separated from other areas; there should be a proper full-body shower for everyone to use at least once a week; there should be a wardroom for communal meals at a table and a galley to prepare the food; the size of a private sleep compartment should be at least 2.4m^3 and larger if there were no communal areas; husband and wife astronauts would need a larger private compartment; there should be as many windows as possible for viewing and relaxation; the interiors should incorporate familiar Earth-like features.

While human factors studies largely dealt with the qualities of habitability inside a module, another set of studies tackled the vexing question of how much volume an astronaut needed to live and work inside the same module. It was vexing because it depended not only on variables such as the length of the mission and the size of the crew, but also on the 'environmental richness' and those 'intangibles of habitability' that the human factors studies had alluded to. It was just not possible to pin it down precisely and precision was the hallmark of the space field. As anyone who has lived in a small city apartment knows, the layout of the accommodation and its design theme can make the difference between a comfortable refuge on the one hand and a claustrophobic cell on the other. It is the same in space where a confined, pressurised environment can easily become claustrophobic and unfriendly. Dating from 1963, research began to explore the relationship between liveable volume and mission duration in an attempt to discover an index or 'golden mean' of habitability.

Some early research had resulted in a simple graph that plotted the free volume per individual crew member against the time spent in space. The graph contained three lines (known as Celentano curves named after the lead researcher at the time) that showed levels of habitability volumetric quality designated as 'tolerable', 'performance' and 'optimal'.[61] The 'tolerable' curve rose steeply to begin with, then gradually fell off to three months duration after which it flattened out at $5.7m^3$. The 'performance' curve also rose steeply initially and then gradually fell off to seven months duration after which it flattened out at $11.0m^3$. The 'optimal' curve rose very steeply initially and then gradually fell off to seven months duration after which it flattened out at $19.6m^3$. No further volumetric increases were indicated after the curves flattened out. The implication was that future human missions to other planets taking many months or years would apparently need no more room. Neither would a Space Station for long-term crew tours on Earth orbit. For purposes of comparison, a small studio apartment of average floor-to-ceiling height has a typical volume of $140m^3$, which is about seven times the 'optimal' figure. From the viewpoint of early 1963, just after the beginning of human spaceflight when enough room to squeeze astronauts and cosmonauts into capsules may have seemed sufficient, it was perhaps impossible to grasp the human factors and habitability implications of future missions that might last many months or years. Despite its obvious defects, this early research many years later found its way into NASA's guidebook of standards for space vehicle accommodation design, known as the 'Man-Systems Integration Standards.'[62] Later researchers cast doubt on the value and validity of this early work and came up with volumes that

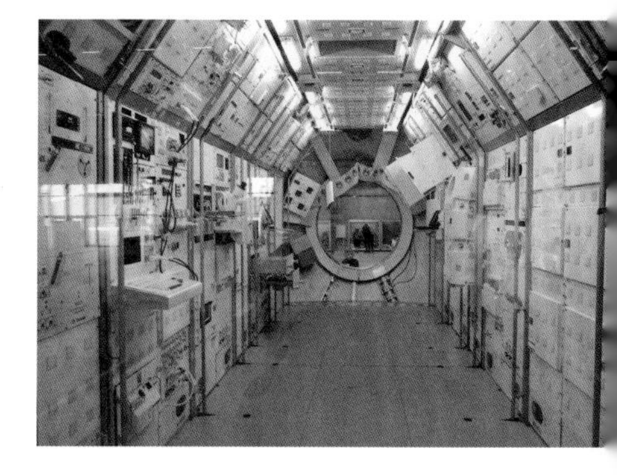

Spacelab module interior The Spacelab Module that flew on the Spacelab D1 and D2 missions on display at Technikmuseum Speyer in Germany. The Spacelab module was the first pressurised module system designed to fly on the Space Shuttle. Its rows of modular experiment and equipment racks ran along each side of a central corridor and faced inwards. It was the forerunner of the rack system design adopted for the Space Station. Photo: Kozuch.

Overleaf: architectural options for module interiors Eight diagrams that depict different architectural options for the Space Station module interiors. These options were studied by NASA and its contractors in the mid-1980s. Each option had advantages and disadvantages and none was perfect. NASA finally settled on the Four Stand-Off design (right-hand page, bottom right corner). Image: author.

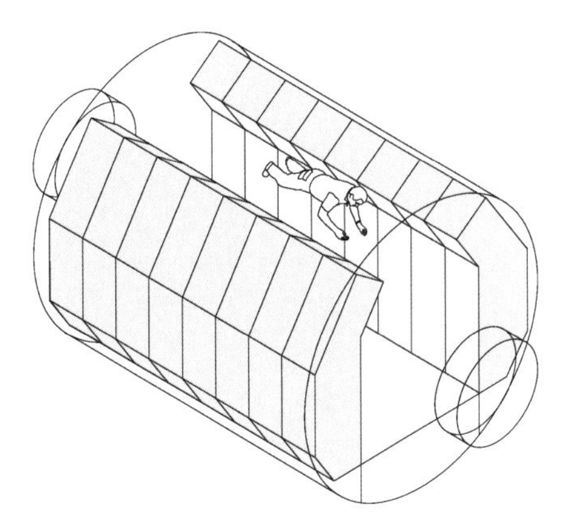

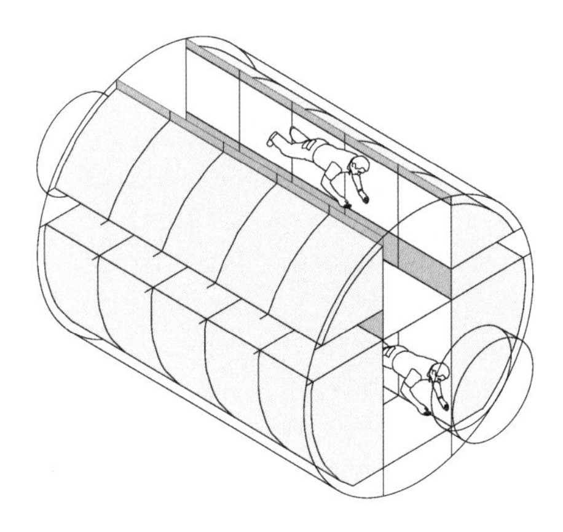

Spacelab Reference
The Spacelab module that flew on the Space Shuttle was an important reference for the studies. It contained two lines of experiment racks mounted on a floor with the rack tops angled inwards and life support systems under the floor. This left a central corridor for crew access and experiment operation.

Loft Corridor
This design divided a central corridor into lower and upper levels with a horizontal screen. A life support zone separated the racks and compartments on the screen line into upper and lower banks. The quality of the corridor volume was poor as there was no large communal area for the crew to gather.

Offset Core Wall
This design divided the module volume longitudinally with two banks of racks and compartments on one side, separated by a life support systems zone. The racks and compartments faced a generous volume for the crew, but the quadrant cross-sections of the racks and compartments produced inefficient shapes.

Double Side Wall
This was similar to the Loft Corridor design but it moved the corridor screen downwards to become a floor plane with life support systems below and to the side. Though this produced large rack and compartment volumes, their tapered tops were inefficient and the quality of the corridor volume was poor.

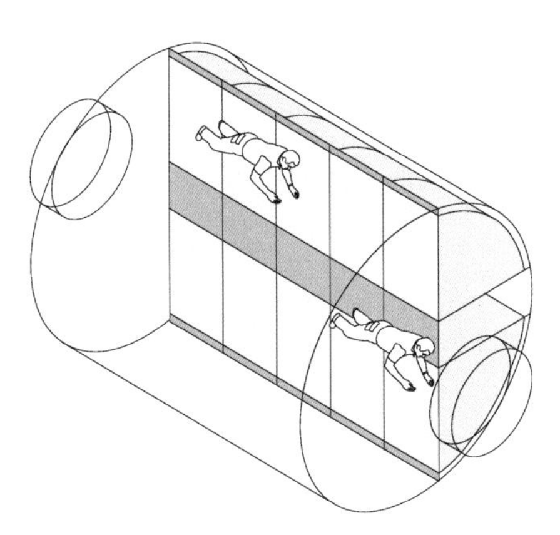

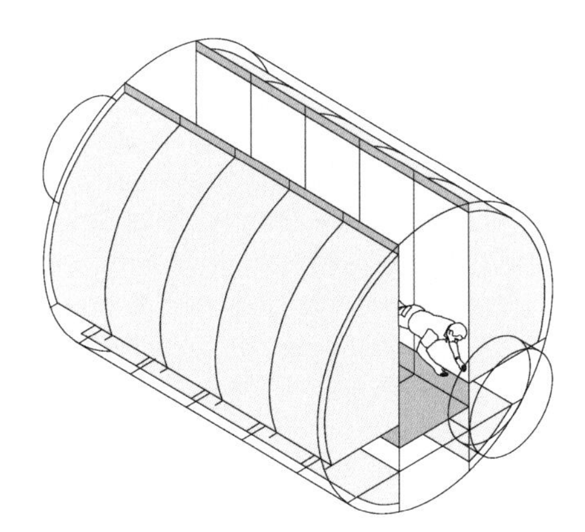

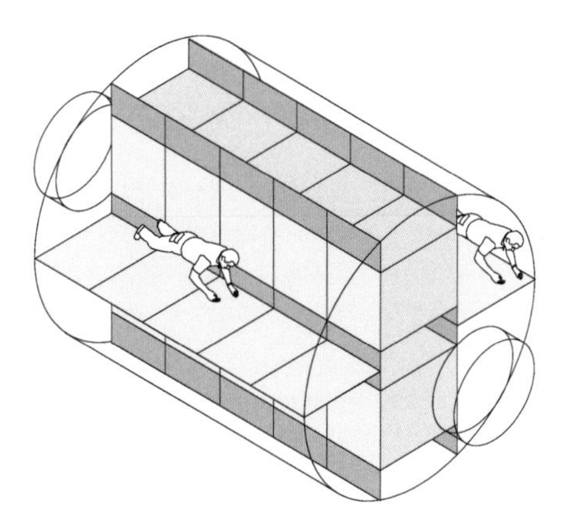

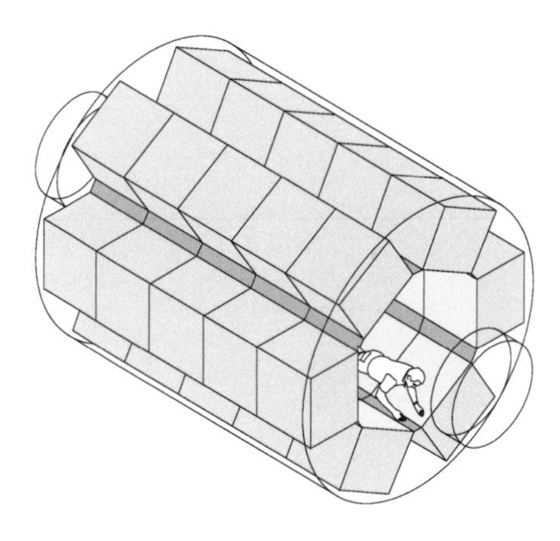

Central Core
All the racks and compartments were combined into one bank that ran longitudinally down the module in this design. Free volume on each side was divided into upper and lower levels by horizontal screens. Rack and compartment access was good but the central bank blocked crew access to the module's hatches.

Hexagonal Radial
This design had a central corridor with a constant hexagonal cross-section. Racks and compartments lined each face of the hexagon. The fragmentation of the racks and compartments resulted in six triangular-section volumes between them that wasted volume. The hexagonal corridor was oppressive.

Curved Cabin
This design had a loft corridor above open volumes for crew activities. Racks and compartments had different shapes according to their functions. They had curved surfaces that resembled the interiors of aircraft cabins. Life support systems routes ran in stand-off structure on opposite sides.

Four Stand-Off
The Four Stand-Off design had a central corridor with a square cross-section that was austere. Racks and compartments lined it on all four sides. Separating them at the corridor corners were four quadrants containing the rack and compartment stand-off structure and life support systems routes.

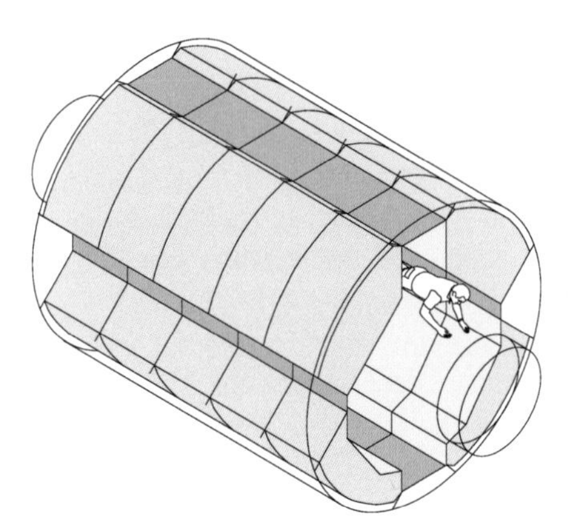

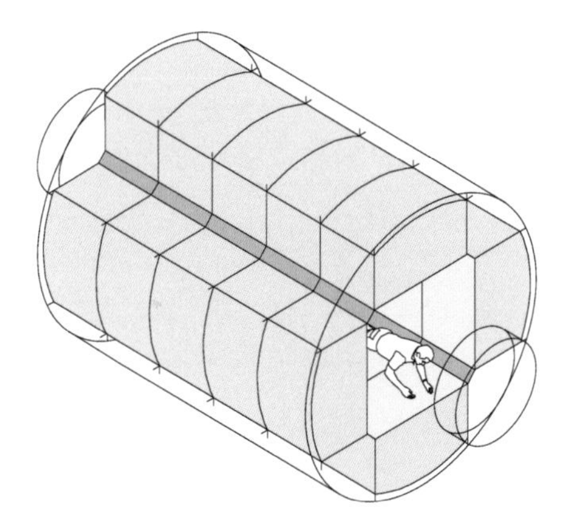

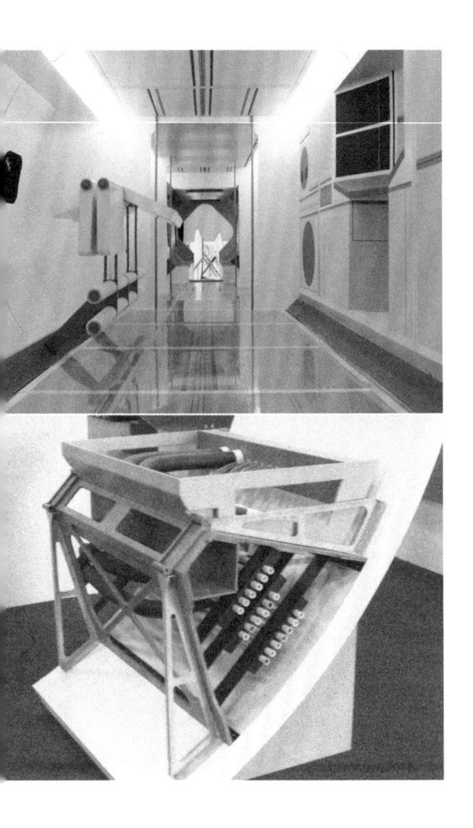

Module interior mock-ups and prototypes
Top: Four Stand-Off design – McDonnell Douglas Astronautics Company. This mock-up dates from 1984 and shows an early proposal for Space Station module interiors. On the left is an alcove and window in front of which is a folded table. On the right are angled computer screens. Bottom: A full-size model of a piece of the Stand-Off structure shows how it works. Two pairs of yellow brackets project inwards from the curved hull and attach at their apexes to a narrow grey faceplate that spans between the pairs. The faceplate supports a hollow rectangular tray that holds the base of a rack or compartment. The front edge of the tray has orange hinge joints with the faceplate, enabling the rack or compartment to swing inwards. Visible in the void between the brackets are air duct, fluid line and cable routes. Images: author.

increased steadily with mission length. Long-duration missions from the 1970s onwards that included Skylab, the Salyut series and the Mir Space Station demonstrated that adequate pressurised volume was vital to provide acceptable living and working conditions. On Mir there was about 90m^3 of pressurised volume for each crew member.[63]

Appendix 2 shows the evolution of space vehicle pressurised volumes according to mission length from the beginning of human spaceflight up to the International Space Station. Derived from a 1998 NASA study of a proposed human mission to Mars,[64] it shows the relationship between increased volumes and longer missions.

The module cluster of the Dual Keel and Revised Baseline designs was a collection of modules of different lengths with a total internal volume in the region of 948m^3. This worked out at 158m^3 and 119m^3 per person for crews of six and eight respectively, greater than on the Mir Space Station but not that much greater. Though this was pressurised volume to sustain a liveable, breathable environment, it was not the free volume for the crew to float through and occupy. It was far from it. There were three essential requirements for the module interiors to fulfil. First, they had to house large quantities of experiments, equipment and supplies, the structures needed to support them and the life support systems needed to maintain the habitable environment. There was going to be a lot of hardware to fit in. Second, where practicable, they had to accommodate volumetric subdivisions into different activity areas such as laboratories, a wardroom, a galley, bathrooms, sleeping compartments and supplies storage and, if possible, separate these into day and night zones. Third, they had to include a corridor for normal crew circulation and movement as well as emergency escape through the end hatches. There were other important requirements such as the need to secure all hardware firmly in place for launch and ascent in the Orbiter payload bays, to size all hardware to pass through the end hatches for pre-launch installation and to move it out of the way easily on orbit for cleaning and maintenance of the module's hull or even the repair of an air leak.

The result of all these requirements was that the total free volume of a module's interior shrank after subtracting the volume occupied by the essential hardware and systems. It shrank to a leftover volume for the crew to circulate through and it was a fraction of the total pressurised volume. Early Space Station studies up to and including the 1984 Power Tower design had explored two fundamental ways of designing the module interiors derived from two architectural forms familiar on Earth: towers and tunnels. In a tower such as a lighthouse, small circular floors are stacked on top of each other. Applying this approach to the interior of a module results in a series of compartments separated by partitions. Treating the partitions as 'floors', 'walls' and 'ceilings' and arranging facilities and equipment accordingly would give the interior a strong vertical feeling or orientation.

The crew would sense that there was an 'up' and a 'down' in the absence of normal gravitational cues experienced on Earth. In a tunnel such as an underground train station, there is a continuous floor or platform running along the cylindrical tube. Applying this approach to a module interior results in a compartment with an end-to-end 'floor.' Adding a 'ceiling' and again arranging the facilities and equipment accordingly gives the interior a strong horizontal feeling of orientation and the crew senses there is an 'along'. The Po wer Tower module cluster had two modules orientated horizontally and four modules orientated vertically with respect to Earth, though drawings of the module interiors at the time showed horizontal orientations and clearly defined 'floor' and 'ceiling' planes. By the time of the Dual Keel design, the module cluster had a dominant horizontal bias. The American habitation and laboratory modules, the Japanese and European laboratory modules and the four node modules connecting them were all in the same horizontal plane. Only the logistics modules were perpendicular and pointed up or down. This meant that all the living and working volumes could have a common 'floor' and 'ceiling' reference, if that was what was desired.

By the time of the Dual Keel Space Station in 1985, NASA and its contractors were studying ways of arranging and outfitting the accommodation inside the American modules. Within the module cylinders many architectural approaches were possible. The starting point was the interior layout of the Spacelab module that Europe had developed for the Space Shuttle system. By mid-1985, Spacelab had flown successfully twice. The main body of the module consisted of cylindrical segments, each 4.1 metres in diameter and 2.7 metres in length that housed the experiments. Shallow conical endcaps closed the module at each end. Assembly of the segments, installation of the experiments and closure by the endcaps took place on the ground before each mission. Spacelab could fly in two sizes known as the 'short module' with an overall length of 4.3 metres and a 'long module' with an overall length of 7 metres. Inside, the module had a full-length floor beneath which were the life support and other systems needed to maintain a habitable and functional environment for the crew. Entry to the module from the Orbiter's cabin was through a tunnel at the forward end. Down the centre of the module was a corridor with experiment racks on each side, sized to contain standard 483mm laboratory equipment trays.[65] The racks and their experiments were generally flush-faced with few projecting experiments or items of equipment other than grabrails to assist the crew as they floated around. The racks were tall, with the upper third of their height angled inwards to clear the module's cylindrical inside surface. The overall effect was similar to a locker room at a skiing resort. It was efficient and clinical and worked well for Space Shuttle missions of about ten days. However, it was hardly suitable for a tolerable long-term Space Station environment.

From the early-1980s onwards, NASA and its contractors set about exploring equally efficient but more tolerable designs for the module interiors. They used a 'mock-up' technique to test out ideas at life size. One of the advantages of designing space vehicles and modules is that their scale is sufficiently small to enable the fabrication of full-size models using lightweight materials. The level of fidelity of these mock-ups varies from rough to detailed, depending on how far the design has progressed.

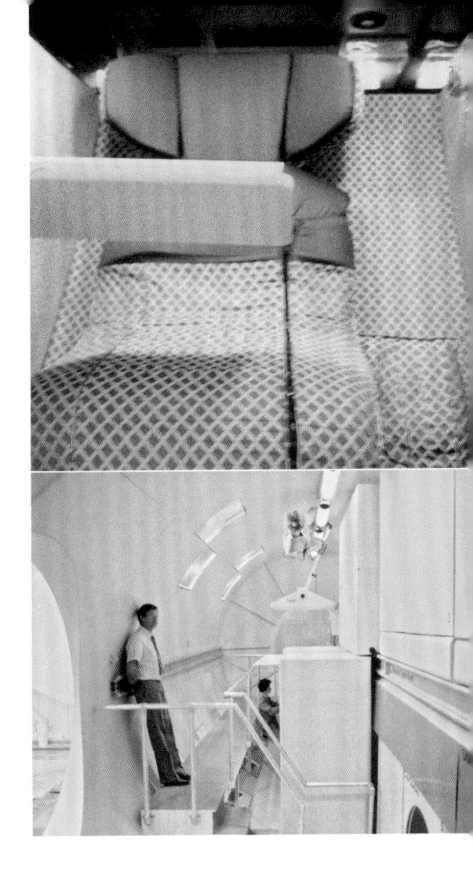

Module interior mock-ups and prototypes
Top: Private Sleep Compartment – NASA Johnson Space Center, 1984. This Space Station module mock-up contains a sleep restraint (bed), a workstation, storage for clothing and personal effects, and enough room for donning and doffing clothes. The sleep restraint is angled to follow the contours of the human body as its shape changes in weightlessness. Soft, adjustable lighting is an important feature in helping to induce relaxation and sleep. Bottom: Offset Core-Wall design – NASA Johnson Space Center, also dating from 1984. This mock-up shows another design option for Space Station module accommodation. On the right are two banks of racks and compartments. Separating them and providing structural support and air, fluid and cable routing is a beam running the full length of the module. Placing all the racks and compartments on one side opens out the other side to the curved inner hull and creates a large volume for crew use. Images: author.

1984 early concept for sleeping quarters
Coloured pencil drawing by the author of an early concept for Space Station sleeping quarters. This was similar to a NASA design at the time for one of the Station habitation modules, but here the compartments are separated by sliding partitions. Sliding the partitions on tracks around the circumference would make their volumes larger or smaller according to the number of crew members. They would fold together during the day to provide more communal living volume. Image: author.

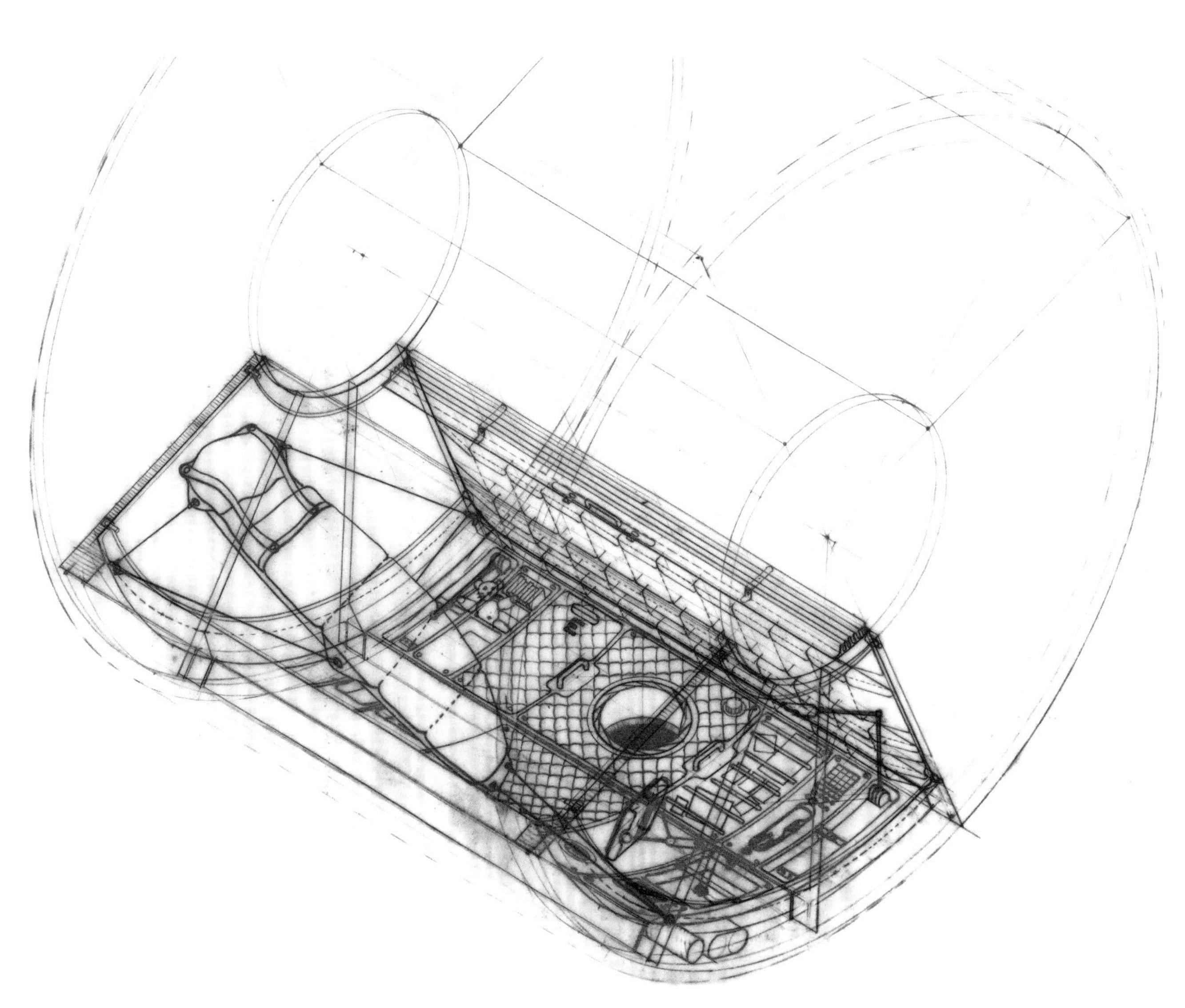

Mock-ups are early prototypes. They are used to test many issues, such as whether there is adequate room for crew movement and rack access or the best routing for mechanical ducts and electrical cabling. The use and value of mock-ups in the early stages of design development is a lesson that has yet to be learned by the terrestrial construction industry when it builds homes and apartments. A lot of poor residential design could be avoided if developers and architects used such techniques to evaluate their schemes before going ahead.

Several candidates emerged from the early module outfitting concept work by NASA and its contractors.[66] There was a Central Core design in which a bank of back-to-back racks and compartments ran full-height down the module's centre. On each side was an open volume extending out to the curved hull. Access to the racks and compartments was excellent. There was plenty of room for free crew movement with the possibility of separate day and night accommodation areas, but the curved volumes tapered to acute angles at the top and bottom and the central bank blocked the circulation path to the end hatches. There was an Offset Core design in which the bank of racks and compartments was moved to one side of the module, offering a generous free volume on the other side equivalent to about half the module's cross-section with a lot of curved hull visible. This produced some awkward quadrant shapes for the racks and compartments. They were deep with access from one side only and strongly curved where they met the hull. There was a Hexagonal Radial design in which long rectangular racks and compartments lined each edge of a hexagonally shaped central corridor. This provided plenty of equipment volume and a direct path to the end hatches, but racks were horizontal and difficult to remove and the austere and unfriendly hexagonal corridor restricted the crew's headroom and movement. There was a Loft Corridor design with two levels of racks and compartments separated by a horizontal screen. This provided two separate activity corridors inside the module with the possibility of day and night zones, but the upper corridor was shallow like a crawl space, the lower corridor was very cramped and the problem of the quadrant shapes returned. There was a Four Stand-Off design with a central corridor of square cross-section and racks and compartments lining each of the four sides.[67] This had a reasonably large circulation volume down the middle but was rather austere. There was excellent access to the racks and compartments but they all had to be moved out of the way to access the module's hull.

By the end of 1985, one of the Space Station contractor designs had taken a clear lead. It was the Four Stand-Off design proposed by McDonnell Douglas.[68] It drew the name of Stand-Off from the structural method of mounting the racks and compartments around the module hull, positioning them slightly apart from the hull's inner surface, in other words standing them off it. There were four Stand-Off zones spaced at 90º radials around

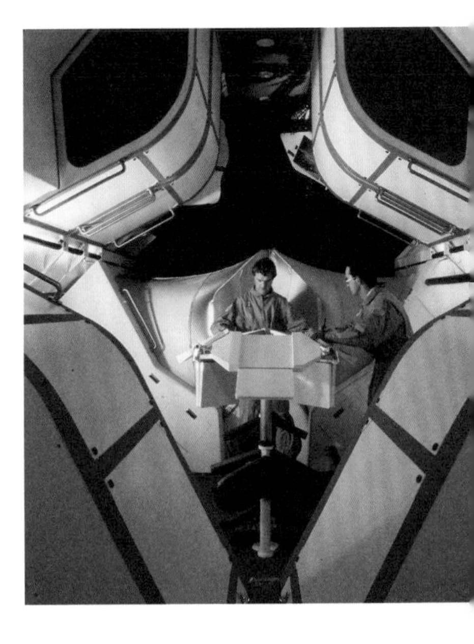

Curved Cabin design – Southern California Institute of Architecture Dating from 1987, this mock-up of a Space Station habitation module shows the Curved Cabin design that was an evolution of the Four Stand-Off design. It rotated the Stand-Off structures by 45°, departed from the rigid rack geometry and showed how it was possible to give module interiors the more comfortable look and feel of passenger aircraft cabins with curved and soft surfaces and less harsh lighting. Photo: author.

the module's circumference. An observer of the module's cross-section imagining it as a compass face would visualise the four Stand-Off zones at the north-east, south-east, south-west and north-west points. Each zone incorporated a lightweight structural truss of triangular cross-section with the triangle's hypotenuse bounded by a portion of the hull's surface and its apex pointing inwards to the module's centre. Each truss was open and ran the full length of the hull. It was large enough to accommodate life support ducts, fluid lines, gas lines, power and communications cables and other distributed systems. Deep racks and compartments of equal height slotted into the volumes between the Stand-Offs to form the square-section central corridor. Attached along the apex line of each structure were hinges and latches that matched counterparts at the four corners of each rack and secured them firmly in place for launch on the Shuttle Orbiters. Once in space, crew release of the latches at two corners would enable each rack to swing out into the corridor on hinges at the other two corners, allowing someone to move behind it to carry out maintenance, repair or pressurised hull inspection. Blunting the apex separated the hinges and latches of adjoining racks and ensured that the swing-out motion of a rack would clear its neighbour. The blunting produced a long narrow slot for a lighting strip at each corner of the corridor. McDonnell Douglas's design was a universal solution to the interior design of Space Station accommodation modules, applicable to habitation and laboratory modules alike and even to the shorter node modules. It was a functional and efficient solution that standardised the rack and compartment shapes and sizes. It ensured that racks and compartments were interchangeable and, effectively weightless, could pass through module end hatches pushed around gently by the crew. If installing them all in a module before launch was not possible due to launch weight restrictions, other Orbiter missions could deliver them to the Station at later dates.

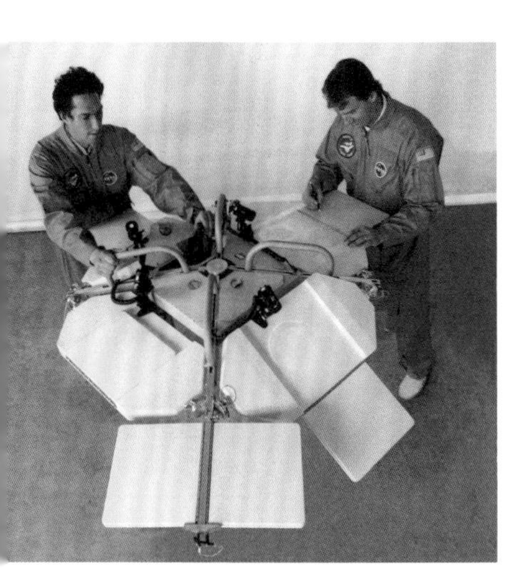

Space Station wardroom table – Future Systems, architects Above: Photograph of a 1988 prototype of a wardroom table intended for the Space Station habitation module. It was designed for four to eight crew members for use for meals, meetings, work or leisure. Four inner worktops could tilt to different angles with fold-out extension panels. Four corner worktops had fold-out butterfly panels that retracted out of the way when not in use. The table had individual task lighting and storage compartments. It would occupy an alcove formed in the habitation module by omitting a few racks or compartments. Opposite: coloured pencil development drawing of the table by the author. Image: author.

However, the Four Stand-Off design had some major disadvantages from the human factors and habitability standpoint. It replicated the ski-resort locker room effect of the short-mission Spacelab laboratory. The central free volume seemed clinical, bounded on each side with a regular geometrical grid of rack and compartment faces that concealed the curvature of the pressurised hull that could otherwise be used to beneficial architectural effect. It resulted in a potentially unfriendly living and working environment. The effect was not helped by the colours that NASA initially chose for the interior. Canadian astronaut Chris Hadfield later gave his impressions: 'Being in the American segment feels different. When the first piece of it – Node 1 (Unity) – was launched in 1998, the psychiatrists who were consulted thought that soothing colours were the key to mental health, so they chose ... salmon. Either they changed their minds or stopped dabbling in interior design, because the rest of the USOS is, mercifully, white. Even though the cylindrical segment is 15 feet in diameter, the racks that have been installed

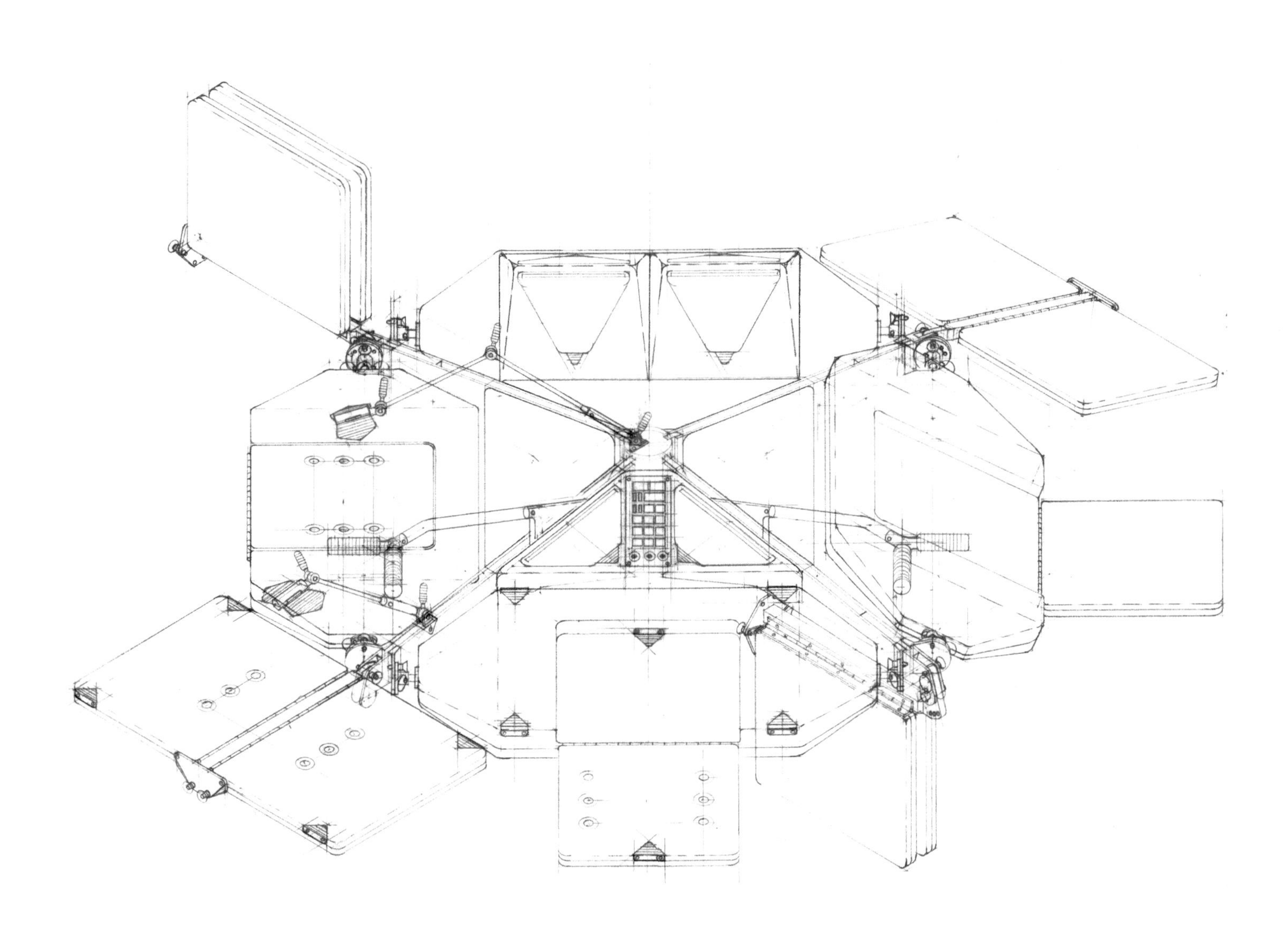

to hold experiments and create storage space reduce the interior to a square cross-section where, arms outstretched, it's not quite possible to touch both sides. The combination of bright lighting, no windows and white walls creates an atmosphere similar to that of a hospital corridor.'[69] In its basic geometrical form, the Four Stand-Off design ignored the need for a comfortable anthropometric and ergonomic fit between the human body and its surroundings. To some extent McDonnell Douglas recognised this and suggested ways of mitigating it. Racks could be left out to form alcoves extending to the curved hull for a communal area such as a wardroom that could then be given windows. Sleeping compartments could project slightly into the central corridor with 'bump-outs' to provide more dressing and undressing room for an occupant. Workstations could be recessed and hollowed out into rack faces to provide desk surfaces and control consoles.

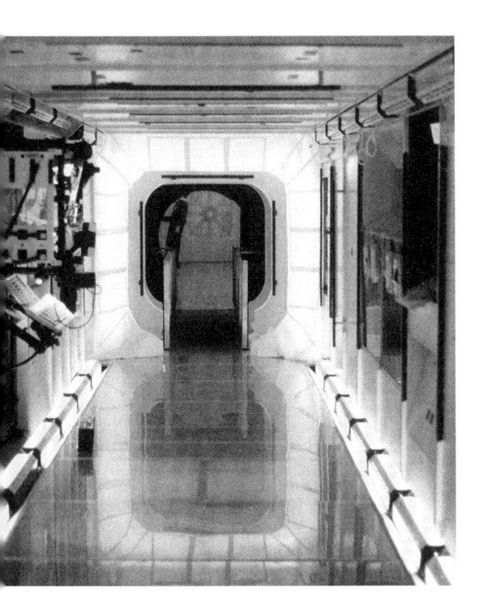

Four Stand-Off design – NASA Marshall Space Flight Center This mock-up dates from about 1988 and shows a more evolved version of the Four Stand-Off design proposed for Space Station laboratory module. By the late 1980s, the rack and compartment design had resolved into a standardised rack shape with a height of 2m, a width of 1.05m, a depth of 0.86m and a volume of 1.57m³. It had a curved back to enable it to fit snugly against the module's hull. This photograph shows the repetitive joint lines between the racks on the corridor sides and ceiling. Image: author.

A few years later, another mock-up study on habitation module accommodation explored an evolution of the Four Stand-Off design. Carried out at Southern California Institute of Architecture in the late-1980s and described here as the Curved Cabin design, it rotated the Four Stand-Offs by 45° so that they aligned with the cardinal points of the module (north, south, east and west) when viewed in cross-section. They were transformed into a Ceiling Stand-Off (north), a Floor Stand-Off (south) and two Wall Stand-Offs (east and west). Slimming down the depths of the Floor and Ceiling Stand-Offs introduced the potential for double-height open volumes. The design departed from the rigid rack geometries in favour of curved and contoured racks that were reminiscent of the curved luggage compartments and curved fuselage linings in passenger aircraft cabins. There was softer lighting. The module was for habitation and therefore did not necessarily need standard racks to house scientific experiments and harsh task lighting. Omitting some of the racks created an alcove for a wardroom lined with soft stowage bags that doubled-up as a decor lining. The double-height volume enabled the inclusion of a crew 'bypass' corridor above, something that was impossible with the Four Stand-Off design.[70]

While NASA and its contractors, after studying many variations, had homed in on the Four Stand-Off design for the American module interiors, Japan appeared to be following a design for its laboratory module that was almost identical to that of Europe's Spacelab module of the 1970s. It had a deep floor zone for life support systems and equipment. Above the floor was a line of racks either side of a central aisle with the upper portions of the racks angled inwards like the Spacelab racks.[71] Japan later adopted the Four Stand-Off design for its long pressurised laboratory module.

The 1986 Dual Keel and 1987 Revised Baseline designs for the Station both showed a module cluster with one American laboratory module and one American habitation module. It was obvious that the laboratory module would contain rack-mounted experiments densely packed from one end to the other to make the most of the available volume. In contrast, the habitation module would accommodate a range of equipment and facilities to provide a liveable environment for the crews. By the end of 1986 in a document redefining the Space Station and its requirements, NASA clarified what the accommodation for the habitation module would cover.[72]

The focal point of the habitation module was going to be a wardroom large enough for eight crew members to occupy at the same time. It would be a multipurpose area able to function at different times of day as a dining room, a meeting room, a lounge, a window viewing room and a recreation area. It would contain a variety of equipment, features and supplies for leisure activities. Close to the wardroom would be a galley or kitchen to store, preserve, prepare, heat and serve food and drink to crew members. It would be able to store frozen and perishable food as well as food at normal ambient temperature in sufficient variety to offer a choice of menus each day. The galley would be easy to clean, with dishwasher and trash collection and compaction equipment. Then there would be a medical facility of modular design to allow for its future expansion and upgrading. It would permit routine and emergency medical examinations and diagnostics, on-board treatment for the cure or stabilisation of illness or injury and monitoring of a critically ill crew member. It would have its own medical database as well as a communications link to NASA's medical team on the ground. Associated with it would be a hyperbaric airlock to provide raised air pressure to treat a crew member suffering from decompression sickness. The medical facility would also be able to deal with a deceased crew member, though NASA did not specify how.

Alongside the medical facility would be an exercise facility with equipment such as a treadmill and a bicycle ergometer for crew members to maintain their physical fitness in weightlessness. It would be able to measure body mass and dimensions and have its own ventilation system. Workstations were an important requirement for both the habitation and node modules. They were needed for Space Station operations control, experiment control, equipment maintenance and repair, crew health monitoring and window observation. There were two kinds that NASA described as a crew station and a window station. A crew station would service dedicated tasks and activities that excluded off-duty, housekeeping or leisure activities. It should be of good ergonomic design and conform to various safety standards. It would have electrical power and connections to the Station's command, control, data and communications systems. It might have tools and instruments for repairs and it would need a separate ventilation extract system with airflow filtration. A window station would be used for Earth observation, equipped with cameras, recorders, timers, maps, lighting, optical instruments and writing implements. Two crew members should be able to look out of one window for scientific observation or recreation at the same time. The window glass would filter out infra-red and ultra-violet radiation and control glare and it should be replaceable on orbit if damaged.

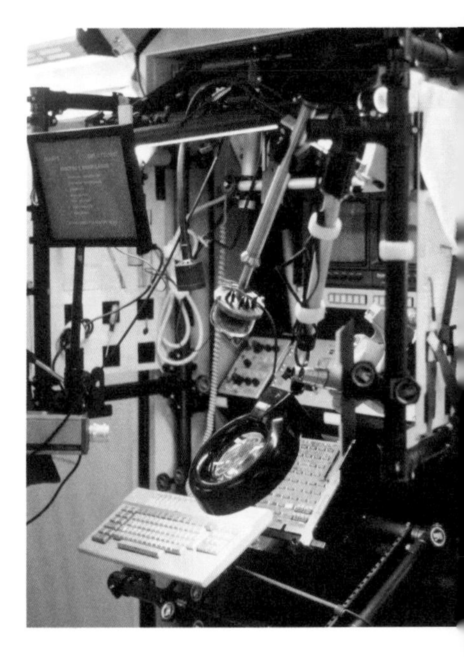

Multipurpose Workstation – NASA Marshall Space Flight Center This was a 1988 prototype of a multipurpose workstation for repair and maintenance tasks on the Space Station. It fitted inside a rack and slid out for use like a deep drawer in a filing cabinet. The workstation arrangement in the photograph shows a set of instruments and tools for electronics work. In a weightless environment, all work pieces and tools must be held down or restrained in some way to stop them floating away. Image: author.

Also essential was a private sleeping compartment for each crew member. There would be at least six of them. They would be big enough for sleeping, donning and doffing clothing, personal grooming, storage of clothes and personal effects and their enclosing partitions could be moved to make them bigger. Each sleep compartment would have acoustic isolation to exclude noise and enable its occupant to get a peaceful night's rest. Despite this, there was no mention of the desirability of creating separate day and night areas inside the module, which would certainly help with privacy and the need for peace and quiet. Each sleep compartment would also have its own ventilation and temperature controls and its airflow would be continuous, in one direction and filtered to remove hair, nail trimmings and lint. There would be a minimum of two bathrooms or 'personal hygiene areas' in NASA's parlance. They would have 'facilities for body waste collection and disposal'. This was NASA's functional description of a space toilet. This would be no ordinary toilet for there would be separate collection and disposal of liquid and solid bodily waste, anticipating the recycling of the first. Each bathroom would have a washbasin or 'personal hygiene station' with a mirror and facilities for bathing, though NASA did not specify if this was a full body shower with running water. The plumbing systems involved 'shall not contaminate the cabin atmosphere with waste material, bacteria, toxicants or noxious odors.'[73]

The first crew wardroom: Skylab By the end of 1986, NASA had defined exactly what the Space Station habitation module would contain in the way of crew accommodation. The focal point was to be a wardroom large enough for an eight-person crew and able to function as a dining room, meeting room, lounge, observation room and leisure facility. By contrast, this photograph shows the wardroom area for three-person crews on Skylab in the early 1970s.

There would be a laundry 'to clean and dry intravehicular apparel', towels, washcloths and sheets though again NASA did not specify if the laundry would involve running water. Finally, hatches and passageways would have an internal diameter or side-to-side dimension of at least 1.27 metres. Hatches would be capable of closing by hand in thirty seconds or less and have viewing portholes. Missing from NASA's description of the accommodation inside the habitation module was any reference to comfort and the desirability of incorporating it in some form into the crew's living environment. Perhaps comfort belonged to that fuzzy human factors world of 'environmental richness' that a NASA study had already identified. Comfort had yet to appear as a requirement in the precise designs and descriptions of the modules. There was no hint anywhere that it would ever do so.

As the American-built habitation and laboratory modules had the same shape and size, it made sense to develop a detailed pressurised module structure that would work for both functions, now that their internal accommodation arrangements had solidified around the Four Stand-Off design. By 1987, Boeing was proposing a structural solution that it called a Common Module and had already built a full-size prototype of it at Marshall Space Flight Center to test the manufacturing and assembly processes and structural performance.[74] Boeing's aim was to reduce the cost of the module's design, development, testing and launch, and to minimise the risk of structural failure on orbit through high load factor testing of the actual flight modules rather than a test prototype. The Common Module would have a thirty-year life, a maximum launch weight of 18,595kg and be of all-welded construction to eliminate the risk of air leaks in space. Its pressurised hull was, as before, a long cylinder with a conical endcap at both ends, sized to fit snugly into the Shuttle Orbiter payload bays to make the most of the available volume.

Panels machined in a 45° waffle grid pattern out of aluminium alloy formed the Common Module's hull wall. The hull skin that contained the internal air pressure was 3.2mm thick and stiffened by the waffle ribs that were 22mm in depth. Rolling the panels into 90° curved segments and then welding four segments together produced a cylindrical hull section. Two welded circular rings of roll-forged aluminium connected three hull sections together to form the full-length cylinder. Circular rings at each end provided mountings for the conical endcaps that were each formed from four machined and curved aluminium panels. All the forged rings had external flanges for attaching multilayer thermal insulation, 1mm thick micrometeoroid protection panels and support structure for external equipment and experiments. The gap between the pressurised hull and the protection panels was just 114mm. Trunnion and keel pins of stout design would safely secure the module in place in the Orbiter payload bays for launch and ascent. The module had a berthing port at the centre of each conical endcap, comprising a berthing ring and an inward-opening hatch with an aperture size that was 1.3-metre square with rounded corners. There was a window made of three panes of glass, sealed into the circular hull with attachment rings and seals.[75] Boeing's Common Module was a giant tin can but an elegantly designed and efficiently engineered one. Its simplicity and uniformity and elimination of the need for a test prototype offered the prospect of a reliable and economical fabrication phase further down the line.

Despite all the effort that had gone into the design of the Space Station's modules and their interiors, they would never be able to compete with Skylab for the quality of spaciousness and the freedom of movement that the orbital outpost of the early 1970s had given to its crews. The Space Station's standardised Common Module would have about 222m^3 of total pressurised volume. This was a gross volume, hull wall to hull wall and endcap to endcap. This would shrink to about 66m^3 of usable free volume for the crew after the subtraction of the rack, compartments and stand-off volumes. This was a net volume, rack faces to rack faces and running full length inside. By contrast, Skylab's workshop – the main accommodation element for the crew – had about 514m^3 of total pressurised volume.[76] This reduced to a working volume of about 275m^3. The Skylab workshop had had about four times the free volume of the proposed Space Station module. Cramming as many racks as possible of experiments and equipment into it so that, standing at its centre with outstretched arms, a crew member would almost be able to touch opposite rack faces, would never result in anything other than the hospital corridor effect that astronaut Chris Hadfield had described.

Space Station Common Module The prototype of the Boeing Common Module in 1987. Boeing had proposed the Common Module as the standard American module structure for laboratory and habitation use on the Space Station. Each module would have a thirty-year life, a maximum launch weight of 18,595kg and be of all-welded construction. The photograph shows the module's curved panels fabricated in a 45° waffle grid pattern. The pressurised hull thickness was just 3.2mm.

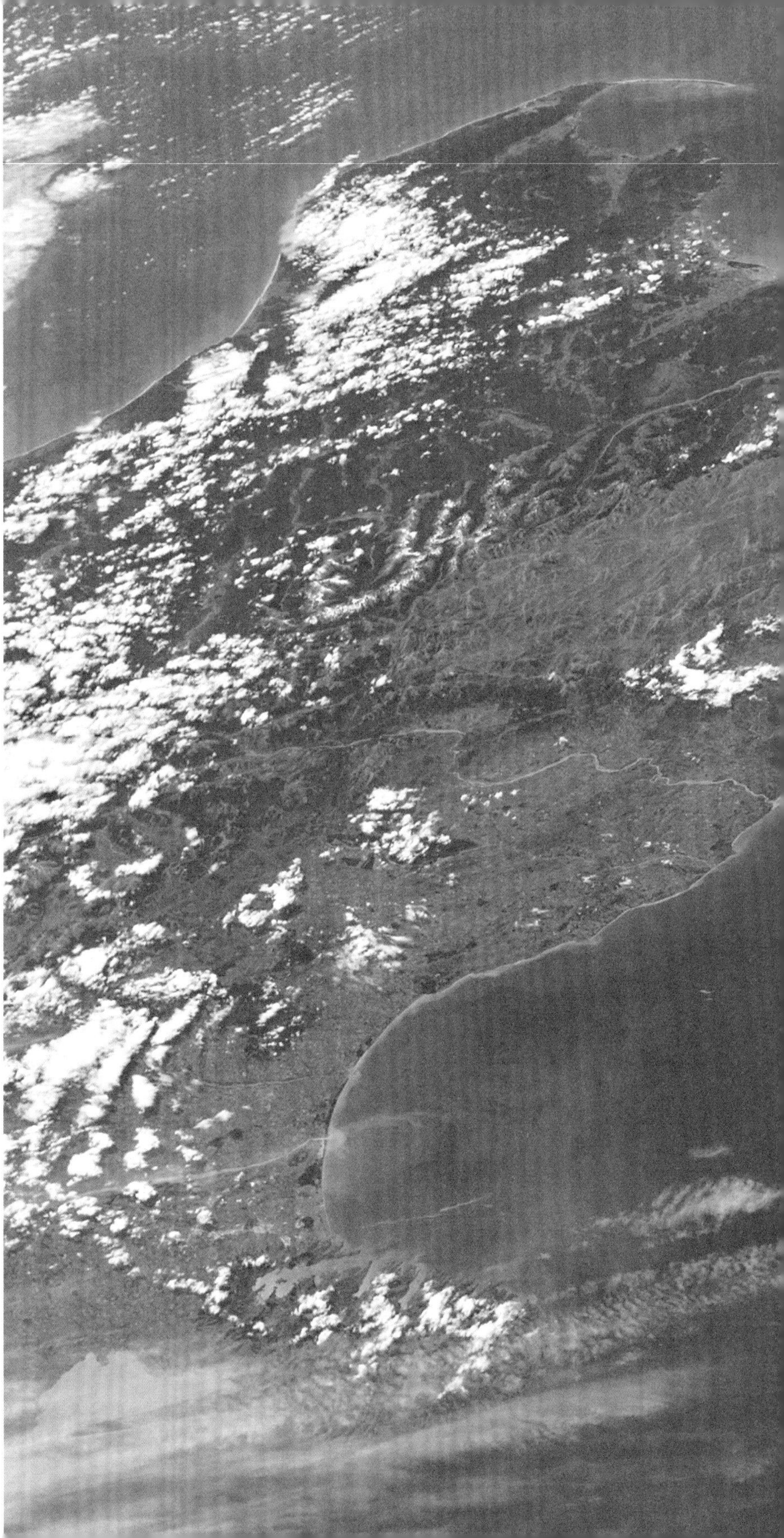

Spacewalk Astronauts Robert Curbeam (left) and Christer Fuglesang (right) photographed on a spacewalk at the end of the Station's S1 truss piece during Shuttle Mission STS-116 in December 2006. Below them is part of New Zealand. The two astronauts are tethered to the Station by safety cables as they work on the exterior of the truss.

Spacewalk Astronaut Stephen Robinson is firmly anchored to a spacesuit boot restraint mounted on the end of the Canadarm2 robotic arm as it slowly moves him around while he participates in a Station spacewalk on Shuttle Mission STS-114 in August 2005.

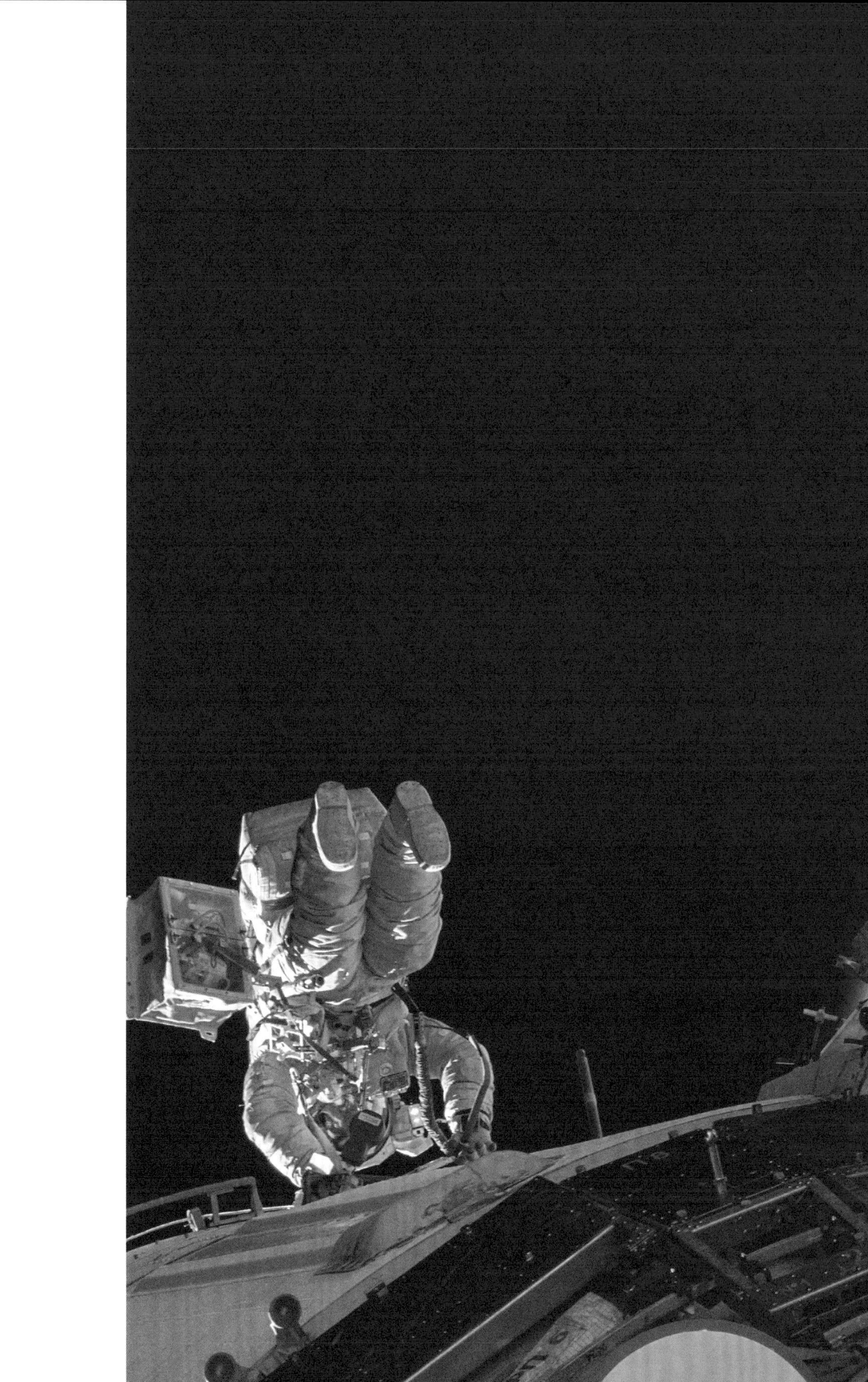

Spacewalk Russian cosmonauts Oleg Kotov and Sergey Ryazanskiy wearing Russian spacesuits carry out an eight-hour spacewalk on the Station's exterior.

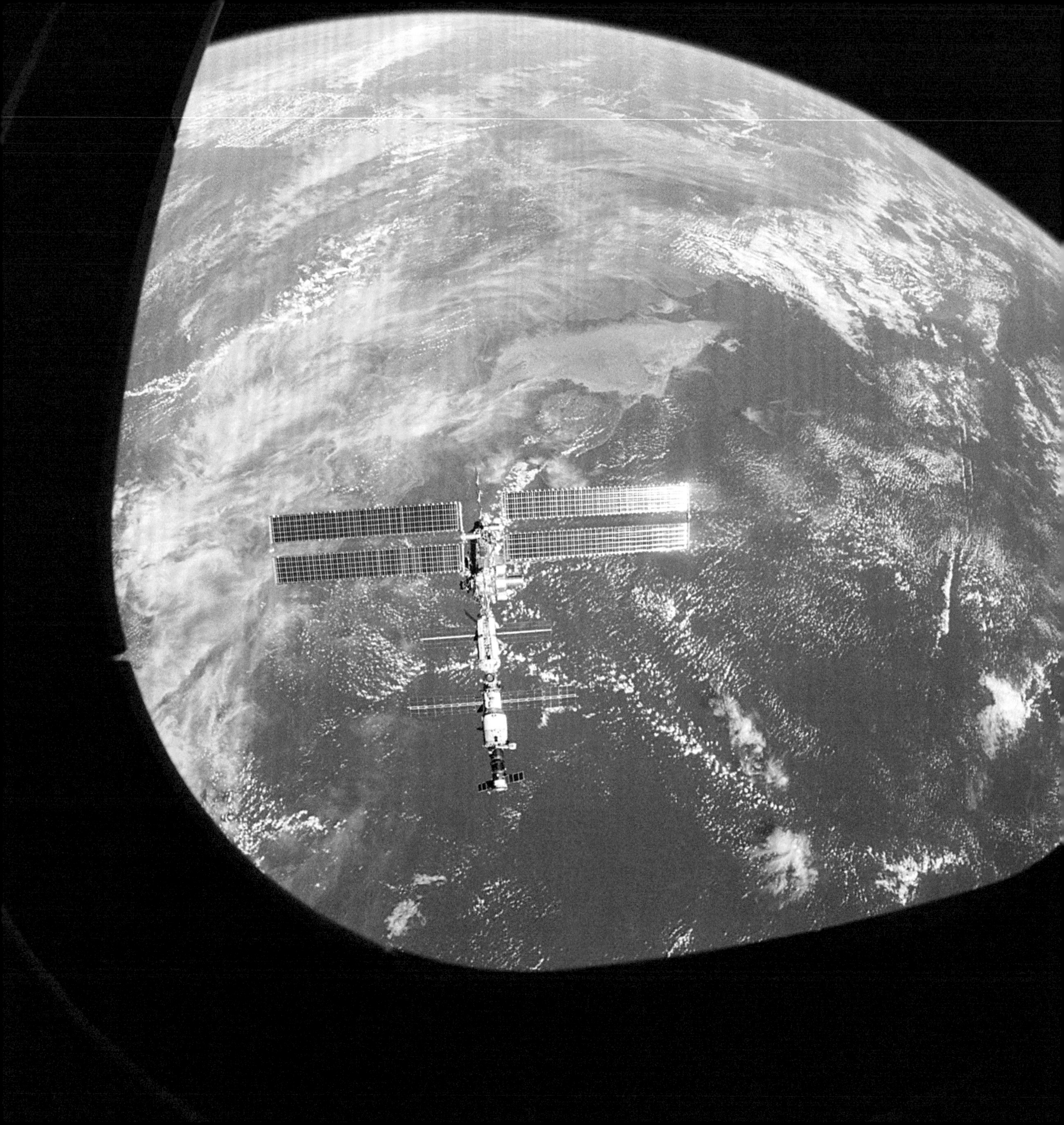

Crisis and resolution 1989–1993

International Space Station The Station with the Earth as a backdrop, photographed through an overhead window from the Orbiter Endeavour on Shuttle Mission STS-108 in December 2001.

Money troubles
Space Station Freedom, 1989-1991

In January 1989, George HW Bush succeeded Ronald Reagan as President. Reagan had given the Space Station the go-ahead in 1984 and now it was President Bush who would have to deal with Space Station Freedom's continuing cost challenges. The 1988 estimate for the Revised Baseline Station of $12.2 billion had now risen to $12.8 billion in 1984 dollars or $19 billion in current year dollars. As before, this was the Station's basic construction cost. It excluded launches, operations, personnel, ground support and international contributions. March 1995 would see the launch of the first Station building block, April 1996 would see its first permanent crew, and March 1998 the completion of its assembly after nineteen Space Shuttle flights. NASA now estimated the total Station programme costs for all phases up to the completion of its assembly at $30 billion in current year dollars.

NASA's fiscal year request for 1989 was $1.87 billion and the White House and Congress had together cut it down by a draconian $972 million or 52 per cent. The annual ritual pruning that Washington was performing on NASA's budget requests had become a handicap to the Station's progress. Its 1985 budget had been cut by 36.2 per cent, its 1986 budget by 28.5 per cent, its 1987 budget by 31.7 per cent and its 1988 budget by a brutal 62.8 per cent.[1] Every time this happened it risked a descoping of the Station's capabilities and a stretched-out timetable with higher costs to be picked up further on down the line. The 52 per cent cut in 1989 forced NASA to postpone the Phase 2 Dual Keel additions to the Revised Baseline design indefinitely. The polar orbiting platform that had been part of the Space Station programme since 1984 was transferred out of it and into NASA's Office of Space Science and Applications.[2] In 1984, NASA Administrator James Beggs had announced to Congress that the Space Station would comprise three major elements – a Station and two orbiting platforms – that together would perform eight key functions: a laboratory, an observatory, a transportation hub, a servicing facility, an assembly point, a manufacturing plant, a storage depot and a mission staging point. Now, the descoping of that plan over the five years since was shrinking the Space Station's role dramatically.

Space Station Freedom An artist's impression of the Space Station Revised Baseline design as it was from 1987 to early 1991 with the long truss structure built from individual struts. Its new name was Space Station 'Freedom.' President Reagan had chosen the name Freedom in 1987 from hundreds of suggestions.

NASA had issued the full development contracts for the Station to its four work package contractor teams in December 1987 and progress on those depended on stage payments that, in turn, relied on approval in Washington of NASA's annual budget requests. The 52 per cent cut to NASA's 1989 request caused NASA's Administrator James Fletcher to warn of the negative effects that the reductions were having on the nation's space programme. Congress increased the figure to $902 million, giving NASA almost what it had asked for. After further acrimonious exchanges with Congress, James Fletcher resigned in April to be succeeded in July by Admiral Richard Truly, a former astronaut.[3] To help ease the budget problem, NASA split the project into an initial phase with reduced capabilities and an 'assembly complete' phase when 'full capabilities' would be restored. In July 1989, NASA formed a review team to examine options for re-phasing the Station's development to respond to the new budgetary realities. To cut the short-term costs, some things in Space Station Freedom's Revised Baseline design would have to go. The team identified a basket of savings in the technical systems, including propulsion subsystem changes, the elimination of one of the two airlocks (giving the job to one of the node modules), scrapping a new spacesuit and eliminating a mechanical device called a flight telerobotic servicer intended for maintenance, servicing and inspection tasks.

There were some more serious reductions. The crew would shrink from eight to four and the solar power generation would be cut from 75kW to 37.5kW by omitting half the solar arrays. The life support system would change from a closed to an open loop, meaning that it would not recycle its waste water and the Space Shuttle would have to ferry up more clean water to resupply the Station.[4] The smaller crew of four would curtail the Station's ability to carry out scientific research as there were three laboratory modules full of experiments that needed attention – American, European and Japanese – and a large part of the crew's time would be devoted to housekeeping and maintenance duties. Equally serious was the halving of the power supply, which would impact the performance of the laboratory experiments. The 'one-two' punch of the reduced crew and the reduced power would weaken the productivity of all the Station's laboratories. An effective way to cut NASA's annual budgets would have been to reduce the Station assembly flights by the Shuttle to five a year, but this would have delayed the completion date and increased costs in the long term because of recurring operations overheads. NASA proposed to launch the first building block in March 1995 and then to match the subsequent assembly sequence to its annual budget, but the Station's completion would slip to 1999 – five years later than the original target.[5]

On 20 April 1989, President Bush established the National Space Council within the Executive Office of the President. Its role was to coordinate the development of a national space policy and a strategy for its implementation. Anything to do with space, whether civilian, military or commercial in nature, would now come under the Council's watchful eye, including Space Station Freedom.[6] Then on 20 July 1989, the President made a speech on space policy to mark the twentieth anniversary of the Apollo 11 Moon landing. After describing the Apollo success, he turned to the Space Station. Pointing out that the space race of 1961 had speeded up space exploration, he said, 'I'm not proposing a ten year plan like Apollo; I'm proposing a long-range, continuing commitment. First, for the coming decade, for the 1990s: Space Station Freedom, our next critical step in all our space endeavours ... It is to have Space Station Freedom up there, operational and underway, a new bridge between the worlds and an investment in the growth, prosperity, and technological superiority of our nation. And the Space Station will also serve as a stepping stone to the most important planet in the Solar System: planet Earth.'[7] The President went on to say that the Station was the next step in human exploration. He appointed Vice President Dan Quayle to lead the new National Space Council with the task of determining what the next round of space exploration should be. Following the President's speech, NASA commissioned a team at Johnson Space Center to carry out a study to provide input to the National Space Council while it deliberated on America's future in space.[8] The study was to last ninety days and it would cover both human and robotic missions under the umbrella title of the Space Exploration Initiative. It would examine five exploration targets of which Space Station Freedom, a return to the Moon and a journey to Mars were the first three, in that order.

Early in November 1989, and unconnected with NASA's '90-Day Study' as it came to be called, the President approved a national space policy that clarified American goals and activities in space resulting from the National Space Council's work.[9] It reaffirmed America's commitment to have a Space Station in place by the mid-1990s. The Station would contribute to America's leadership in aspects of spaceflight, support scientific research and technology development, generate materials and life sciences benefits, promote private research and commercial space, and enable evolution in keeping with long-term goals including human space exploration beyond

Earth. It would also encourage more international cooperation in civilian space programmes, including cooperation with the Soviet Union.

NASA published the results of the 90-Day Study in November 1989 to coincide with Admiral Truly's briefing of the National Space Council.[10] It made much of the role that the Station would play in supporting future exploration activities. It described the Station's capabilities at the completion of the Revised Baseline design. It was going to support research in many fields, including microgravity, materials processing, medicine, Earth observation, life sciences, astronomy and space physics. It would be a development testbed for automation and robotics, advanced structures and materials, power generation, space electronics and communications. With the addition of more building blocks, it would evolve into a support base for space exploration. 'First, it will serve as a transportation node for assembling, testing, processing, servicing, launching, and recovering lunar and Mars vehicles. It will also supply crew support, data management and communications systems, and logistics services to accomplish these activities,' said the report. To do all this the Station would evolve through four additional stages. The lower keel of the Dual Keel design would return. There would be a second habitation module and a lunar spacecraft hangar. There would be two more crew members and more solar arrays to bring the power up to 175kW. The upper keel of the Dual Keel would return with a Mars spacecraft assembly facility. Accompanying drawings showed all four phases. The Dual Keel design had returned, bigger and more powerful than before. Nowhere in the report was there a mention of what it was now going to cost. However, Admiral Truly gave the National Space Council a colossal figure of about $500 billion to cover everything over a period of twenty to thirty years. Vice President Quayle asked the National Academy of Sciences to assess the 90-Day Study. Though the Academy was supportive, reaction at the White House and in Congress to the budget was understandably negative and the plan failed to win support.[11]

By 1990, the Station's projected hardware cost stood at $13 billion in 1984 dollars or $19 billion in current year dollars up to its assembly completion point. This was the forward cost. NASA had already spent about $2.4 billion on the Station's design and development since 1985. NASA now estimated the total programme cost to America from beginning to end, including Shuttle launches, operations, personnel and ground support would add up to $38.3 billion. The dates for the first launch and a permanent crew were as they had been in 1989 and its completion now needed twenty Space Shuttle flights, one more than before.

NASA's ongoing cost reduction efforts resulted in a 1991 projection of a slight drop in the Station's cost to $12.3 billion in 1984 dollars or $18.5 billion in current year dollars, a permanent crew date slip to March 1997, assembly completion by August 1999 and the need for twenty-nine Shuttle flights, nine more than the previous year. NASA had achieved the cost reduction by simply moving the polar Earth observation satellite and its launch costs into the agency's science budget. By NASA's estimate, the programme costs up to 1999 – no longer the assembly completion year – stood at $38.3 billion. The Government Accountability Office, though, had by now estimated the Station's programme costs through thirty years of operation at $118 billion. NASA's budget request for 1990 had been $2.1 billion and Washington had again cut it, this time by 18 per cent.[12] Congress then refused NASA's request for $2.7 billion for 1991 and demanded the restructuring of the Station to shave off $6 billion between 1991 and 1996.[13] Congress gave NASA ninety days to present a new proposal.

The White House thought that it was time for experts outside NASA to look at the nation's space programme and the agency's modus operandi. In September 1990, it formed the Advisory Committee on the Future of the US Space Program led by Norman Augustine, the chairman and CEO of aerospace giant Martin Marietta Corporation, and told him to report back with recommendations in 120 days. In its report, published on 17 December, the committee was critical of NASA. It highlighted nine concerns. First was the lack of a national consensus about civilian space goals; second was NASA's inadequate resources to cover its programmes; and third was the problem of the agency's fluctuating budgets. Fourth, fifth and sixth dealt with NASA's institutional ageing, the civil service's fossilised personnel policies and the problem of 'mission creep' in project scope, complexity and cost. Seventh, eighth and ninth dealt with the starvation of NASA's technology base, the fault-intolerant nature of space technology and the dependency on the Space Shuttle for space access.[14]

The Augustine committee believed that the top priority was to fund the space science programme as so many benefits flowed from it. The specific recommendation on the Space Station was that 'NASA, in concert with its international partners, reconfigure and reschedule Space Station Freedom with only two missions in mind: first, life sciences experimentation (including the accrual of operational experience on very long duration human activities in space) and, second, microgravity research and applications. In so doing, steps should be taken to reduce the Station's size and complexity, permit greater end-to-end testing prior to launch, reduce transportation requirements, reduce extra-vehicular assembly and maintenance, and, where it can be done without affecting safety, reduce cost.' NASA should be given whatever time it needed to re-evaluate the Station with these objectives in mind and not be limited to Congress's ninety-day deadline. There should be a crew escape vehicle based at the Station before permanent crews arrived to enable crews to evacuate in an emergency. There should also be another means of crew travel to, and from the Station if the Space Shuttle system was unavailable. The committee reaffirmed that, in its judgement, the Station was the essential initial building block of the manned space exploration programme.[15]

Dan Quayle and the National Space Council By 1989, the Space Station was running into big trouble with a ballooning budget that was forcing major reductions in the Station's capabilities. President George HW Bush had succeeded President Reagan in January 1989 and in April he appointed Vice President Dan Quayle to chair a newly formed National Space Council to develop a national space policy that would include the Space Station and help to bring the project under control. This photograph shows Dan Quayle testing spacesuit gloves in a vacuum glovebox at Johnson Space Center in 1989.

1991 Space Station Restructured Design – Man Tended Capability Continued problems with Space Station Freedom's budget caused Congress in 1990 to instruct NASA to reduce the Station's costs. The result was the 1991 Space Station Freedom Restructured Design. It came in three distinct stages, each with its own price tag. NASA named the first stage 'Man Tended Capability' and it offered sufficient facilities for Shuttle Orbiter crews to make short visits to tend scientific experiments in the single laboratory module. They would not be able to stay on the Station but would have to sleep and eat on board the docked Orbiter. Image: author.

On 3 January 1991, following the publication of the committee's report, Chairman Norman Augustine and his Vice Chairman Laurel Wilkening appeared before the Space Science and Applications Subcommittee of the Committee on Science, Space and Technology in the House of Representatives. They summarised the findings and answered questions. When it came to the Space Shuttle they made a stark prediction: 'Although it is a subject that meets with reluctance to open discussion, and has therefore too often been relegated to silence, the statistical evidence indicates that another Space Shuttle is likely to be lost in the next several years ... probably before the planned Space Station is completely established on orbit. This would seem to be the weak link of the civil space program – unpleasant to recognise, involving all the uncertainties and statistics, and difficult to resolve.'[16] Events would prove that true twelve years later.

Slimming down

Restructured Space Station Freedom, 1991

While the Augustine committee had been pondering the future of the nation's space programme, NASA had been restructuring the Station design. It announced its revised programme in January 1991.[17] The new design reflected the realities of a reduced budget. Funding for 1991 was slashed to $1.9 billion from NASA's $2.7 billion budget request. This followed a White House cut of $240 million added to a Congressional cut of $550 million. Henceforth, this figure would not rise more than 10 per cent a year, reaching a peak annual funding commitment of $2.6 billion. The Station's total target expenditure from 1991 to 1996 – the year it would begin to host crew visits on orbit – was $14.2 billion: 29 per cent less than the previous $19.9 billion. However, this was just up to 1996 and there were a further four years of assembly missions for the Station to reach completion. Neither did the $14.2 billion figure include funds for the development of a crew escape vehicle, which the Augustine committee had asked for. The key to keeping the costs down was an incremental assembly approach labelled by NASA as 'go as you pay'. The nation would be able to buy the Station with the equivalent of instalment payments tied into a phased assembly timetable.

A NASA senior management review of this approach in January 1991 reads in places like a car-purchase deal that offers the prospect of stretched-out financing for a basic model at the outset with another round of financing if the customer wants to add various options to turn it into a luxury model. The customer, mostly the American taxpayer, would get the basic model at the first assembly milestone in June 1996, when crews would be able to make short visits. Sleeping in the Orbiter but working on the Station, they would be able to activate a few experiments. NASA called this the 'Man Tended Capability'. It would need six Shuttle assembly flights to achieve and would comprise an American laboratory module, a node, a docking

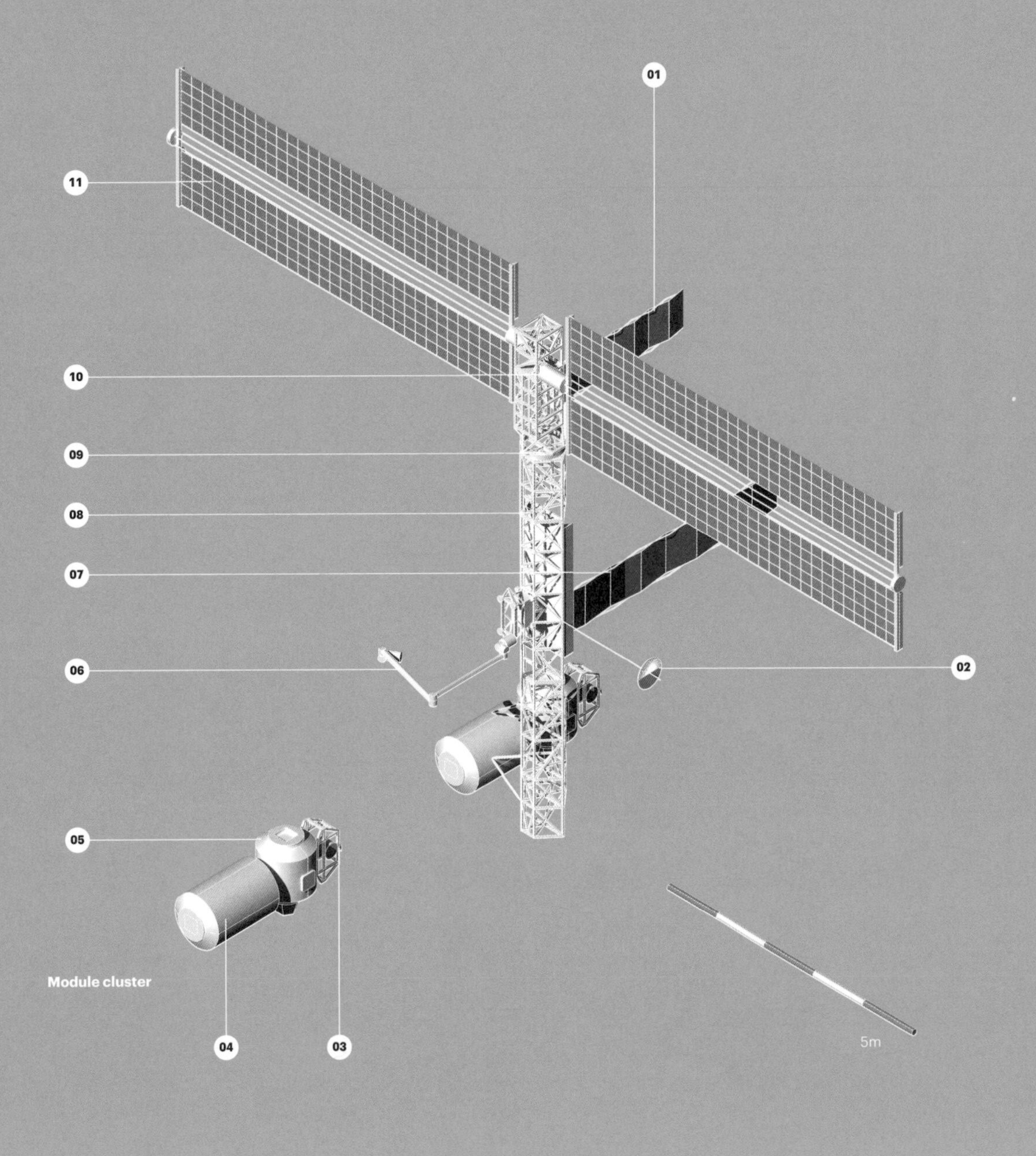

01 Power radiator
02 Tracking & data relay satellite antenna
03 Docking port
04 US short laboratory module
05 Node
06 Robotic arm on mobile platform
07 Thermal radiators
08 Prefabricated truss segments
09 Rotating joint (alpha joint)
10 Rotating joint (beta joint)
11 Solar array

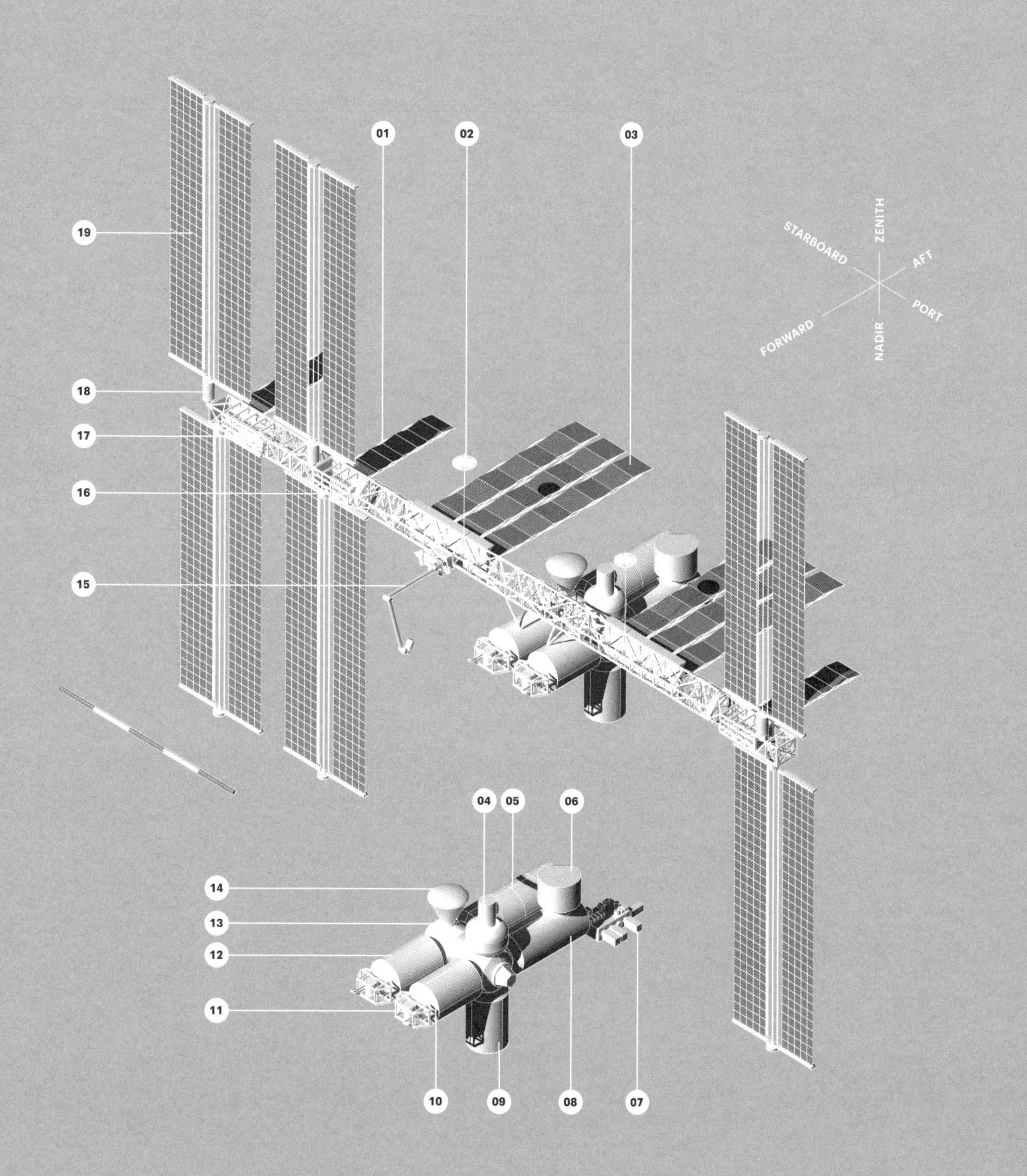

01 Power radiator
02 Tracking & data relay satellite antenna
03 Thermal radiators
04 Airlock module
05 European laboratory module
06 Japanese logistics module
07 Japanese experiment platform
08 Japanese laboratory module
09 US logistics module
10 US short habitation module
11 Docking ports (2)
12 U.S. short laboratory module
13 Nodes (2)
14 Emergency return capsule
15 Robotic arm on mobile platform
16 Rotating joint (alpha joint)
17 Prefabricated truss segments
18 Rotating joint (beta joint)
19 Solar array

port, one pair of solar arrays, the robotic arm and a portion of the truss structure. The next milestone in June 1999 was when the Station would be able to host permanent crews of four. NASA called this the 'Permanently Manned Capability' and it would need ten more assembly flights to achieve on top of the previous six. It added an American habitation module, the Japanese laboratory and logistics modules and experiment platform, the European laboratory module, an American logistics module, a crew emergency return capsule, an airlock, a second node, a cupola, a second docking port, more truss structure and two more pairs of solar arrays. Then there was a further evolution to a crew of eight with more modules and capabilities called 'Eight Manned Crew Capability' when the Station would reach an advanced state of completion. It added second American laboratory and habitation modules, a second emergency return capsule, additional truss structure and a fourth pair of solar arrays. This was the luxury model and it would need a further six assembly flights to achieve, on top of the previous sixteen. There would also be twelve more Shuttle flights during the assembly sequence for Station resupply using logistics modules as well as crew and equipment exchanges.[18]

So, there would be a total of thirty-four Space Shuttle flights between the end of 1995 and the end of 2000 at an average of seven flights a year. This was a highly optimistic flight schedule, despite NASA's 1987 go-ahead to Rockwell International for the new Orbiter Endeavour to bring the Station fleet back up to four (the first Orbiter Columbia was too old and heavy for Station deliveries). The most missions the Shuttle had flown in a single year so far were three in 1989.

NASA's senior management review did not include the Station's total price tag but NASA had separately estimated that total programme costs now stood at $30 billion in current year dollars through to 1999. Significant savings had been achieved over the previous figure of $38.3 billion.[19] The new Station phasing plan had a hidden benefit. At each milestone, assembly would be able to cease without operating penalties. So, if Congress killed off funding at $14.2 billion in 1996, for example, there would still be a usable American laboratory module for crew visits to tend experiments. However, there was a catch. Among the cost reduction measures, NASA had shortened the length of the American laboratory and habitation modules to just over half their former size. Now, the laboratory could house only twenty-four racks instead of forty-four. Though NASA's senior management review in 1991 referred to two short laboratory and two short habitation modules with twenty-four racks each, instead of the two longer modules of the Revised Baseline design, in the same review document there was no sign of the extra two modules in the detailed launch manifest up to the Station's completion in 1999. Nor did they appear in Station contractor drawings later in the year.[20] To all intents and purposes they did not exist, except perhaps in the minds of some of NASA's senior managers.

1991 Space Station Restructured Design – Permanently Manned Capability Congress's instruction to NASA in 1990 to lower the cost of the Station resulted in the Space Station Freedom Restructured Design with a proposal to build the Station in three distinct stages. 'Permanently Manned Capability' was the name of the second stage. It provided for permanent on-board crews with the addition of a habitation module. Also added at this stage would be the European and Japanese elements and an emergency return capsule. The aim of the plan to build the Station in three stages was to spread out the costs over a longer period. Image: author.

1991 Space Station Restructured Design - Eight Manned Crew Capability The third stage of the 1991 Space Station Freedom Restructured Design was called 'Eight Manned Crew Capability'. This added second American habitation and laboratory modules, a second emergency return capsule, more truss structure and another pair of solar arrays. The major change of the Restructured Design from its predecessor was the replacement of the strut-built truss structure by an entirely different design made of prefabricated truss pieces that would be joined together in space. This resulted in a heavier truss structure because the truss pieces would have to be sturdier to withstand the Space Shuttle's launch loads in prefabricated form. Image: author.

Though NASA was touting the restructured Station's ability to begin materials research just one year after the launch of the first building block, it had reduced American research facilities by 45 per cent. Not only that, but the shorter modules also carried a crew that was reduced from eight to four. As a large slice of their time would be spent on maintenance and housekeeping duties, they would have less time to spend tending experiments. NASA had cut the one thing that the Augustine committee had considered vitally important – scientific research. In contrast, the new assembly sequence included all the contributions of the international partners as NASA was committed to honouring its agreements. NASA, though, had reduced the four pairs of solar arrays from the Revised Baseline design that produced 75kW of power to three pairs, resulting in a power drop of 25 per cent to 56kW. American, European and Japanese laboratory modules, the American habitation module and various node and logistics modules would have to share this reduced power supply.

The restructured Station's overall design was much as it had been in the Revised Baseline version of 1987. As before, the Station's backbone was a long transverse truss aligned horizontally and perpendicular to the orbital plane with its ends pointing out to port and starboard. The truss's engineering design and construction method, however, had changed radically from the 5-metre strut-built truss with its provenance that reached all the way back to the Station concept studies of the early 1980s. For some time there had been growing concern about the heavy work burden the strut-built truss placed on astronauts and the number of spacewalks that would be needed to bolt or plug it together in space. Added to this were complications of preassembling it on the ground before launch to check it, then disassembling it into pieces for launch, then reassembling it again on orbit. After the Preliminary Design Review in June 1990 – that in NASA's definition was intended to show that the design met all requirements, that the correct design option has been selected, its interfaces identified and verification methods described[21] – the strut-built truss was scrapped and replaced by a completely different design devised by McDonnell Douglas. NASA approved the new design in January 1991.

The new truss structure was 96 metres long, made up of seven prefabricated pieces at the Permanently Manned Capability stage. They would plug together on orbit, end to end. Each truss piece was a three-dimensional open framework assembled on the ground from bulkhead frames and longitudinal and diagonal members. Inside was a mixture of equipment and subsystems such as utility raceways, electrical distribution gear and stationkeeping controls. Outside were rails for the Canadian mobile robotic arm, a mobile cart to assist spacewalks and attached equipment such as deployable solar arrays, thermal radiator panels and science experiments. Aluminium was the material of choice because it was relatively easy to machine into structural pieces of complex profile,

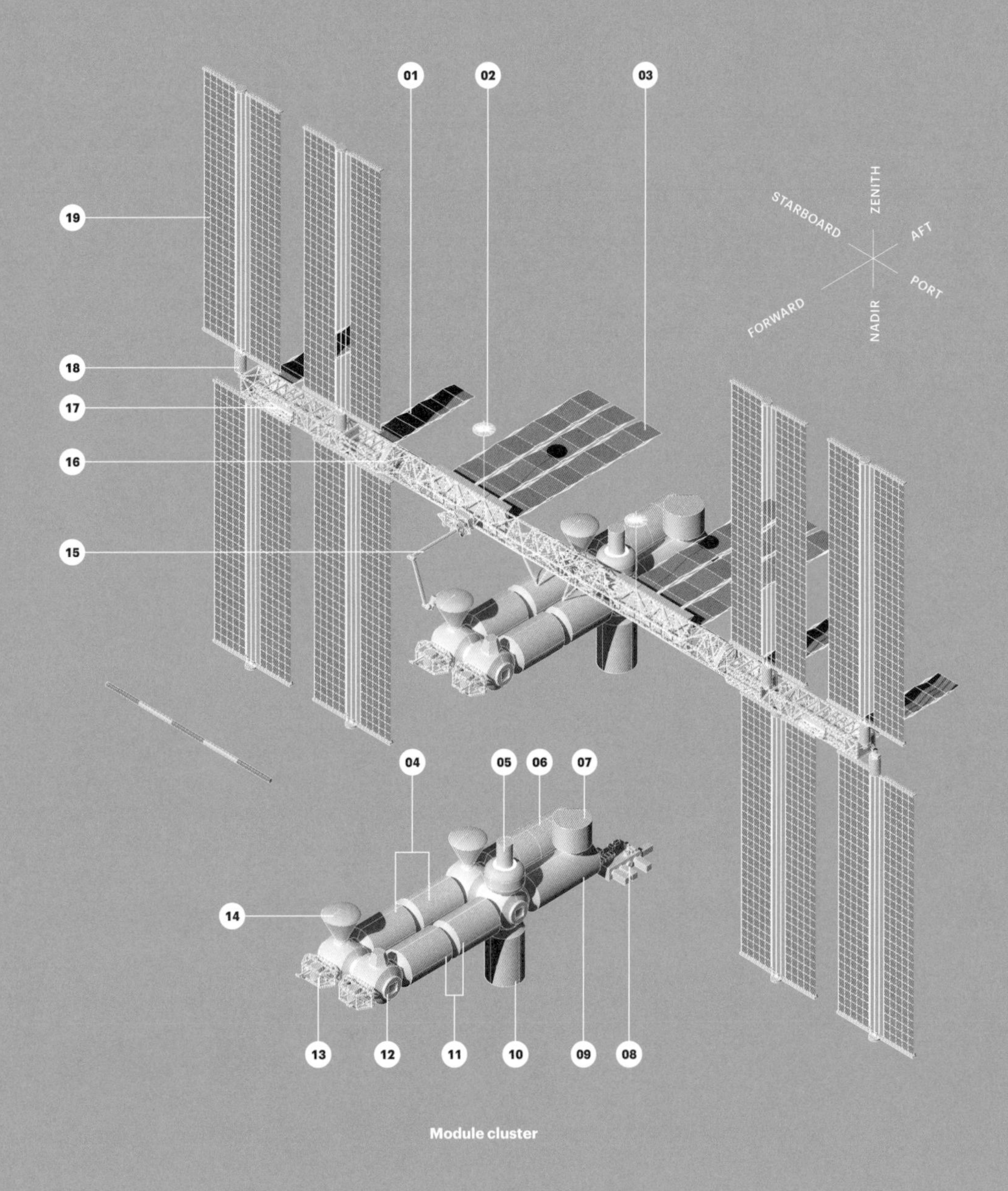

Module cluster

01 Power radiator
02 Tracking & data relay satellite antenna
03 Thermal radiators
04 US short laboratory module (2)
05 Airlock module
06 European laboratory module
07 Japanese logistics module
08 Japanese experiment platform
09 Japanese laboratory module
10 US logistics module
11 US short habitation module
12 Nodes (4) & Cupola
13 Docking ports (2)
14 Emergency return capsule
15 Robotic arm on mobile platform
16 Rotating joint (alpha joint)
17 Prefabricated truss segments
18 Rotating joint (beta joint)
19 Solar array

section or joint detail. The bulkhead frames would be machined from solid plate. Bolted to the bulkheads would be the longitudinal and diagonal struts with each framework assembled on a jig with great precision.[22] Fabrication, outfitting and checking on the ground before launch would greatly help to ease assembly on orbit, simplifying the installations using the docked Orbiter's robotic arm and eliminating the need for a lot of spacewalks.

Each framework's length of 13.7 metres was the maximum that the Orbiter's payload bay could carry. The payload bay's cross-section was basically circular and to take advantage of it, the framework's cross-section would follow the shape of a hexagon that would fit more effectively into a bounding circle than the square of the earlier truss. There was, however, a disadvantage. The struts of the earlier truss would have been ferried to orbit bundled together, with no structural loads to support during the Space Shuttle's launch and ascent other then their own weight. In contrast, each framework was a large skeletal structure filled with equipment and supported for its ride in the Orbiter by pins called trunnions that slotted into sockets along the Orbiter's payload bay door sills and the bottom of the hold. Though the trunnions secured each framework during transit, acceleration would weigh down on the framework's structure and it needed stiffening and strengthening to transfer the dynamic flight loads through the framework's load paths to the Orbiter. As a result, each framework would be four times heavier than its strut-built truss counterpart. All seven of them would add 3,200kg to the total truss structure's weight though there would be savings with the elimination of some support equipment.[23]

Each of the seven pieces of the new transverse truss had a specific job to do and had its own alphanumeric name. At the truss's centre was a piece with a hexagonal cross-section named M1. This functioned as the truss's physical connection to the module cluster positioned just below the truss's bottom edge and on its centreline. An arrangement of struts formed the connecting structure. The next two pieces at each end of the M1 were the P1 on the port side and S1 on the starboard side. These were almost identical, but mirrored on each side of the M1. Their common cross-section was cut back on its centreline from a hexagon to a trapezoid with symmetrically angled sides to accommodate three boxes containing stowed thermal radiators. The boxes had some thickness and that of the framework was halved to enable them to fit into the Orbiter's payload bay for launch. Once on orbit, the radiator panels would unfold like pantographs in the Station's aft direction. Also stowed on the S1 was part of the Station's robotic arm system. Next were the two P2 and S2 pieces attached respectively to the outer ends of the P1 and S1. The P2 and S2 were also almost identical and mirrored. Their inner ends matched the trapezoidal ends of their inboard neighbours but they reverted to hexagons to match the hexagonal cross-sections of their outboard neighbours next in the line. The conversion from one to the other occurred roughly halfway along each framework. The

P2 and S2 each contained four gyroscopes that would control the Space Station's attitude and station keeping. Beyond the P2 and S2 were two P3-P4 and S3-S4 pieces. These had twinned names because they comprised a section of framework and two pairs of solar arrays in one payload. The P3 and S3 frameworks were hexagonal to match their inboard neighbours. The P4 and S4 solar arrays stowed in long canisters set on a framework with a square cross-section. Once in space, the paired solar arrays would unfold vertically above and below the truss line.[24] The two sections of framework attached at a rotating joint called an alpha rotary joint that enabled the solar arrays' pitch angles to rotate around the port-starboard axis to track the Sun as the Station circled the Earth.

The assembly of the transverse truss on orbit would occur as a sequence working from starboard to port. The first piece to arrive in December 1995 would be the S3-S4, followed by the S2 in January 1996, the S1 in March and the M1 in April when the truss would reach its constructional halfway mark at the Man Tended Capability stage. After this, the truss's construction would continue with the P1 piece in December 1996, followed by the P2 in March 1997 and the P3-P4 in April, at which point the truss would reach the Permanently Manned Capability stage.[25] Beyond this, the Eight Manned Crew Capability stage showed a total of four pairs of solar arrays and two more pieces of truss structure, one at each end.

Though reductions to the lengths of the American laboratory and habitation modules enabled their full outfitting with racks before launch – not possible with the longer modules because of Shuttle payload launch load restrictions – they impacted the internal accommodation and life support systems. Habitability standards fell as there was now a lack of room for crew sleep compartments. The Station's crew size had been one of the casualties of cuts to NASA's 1989 budget. Now down to four persons from eight, the crew would have to camp in the module corridors until the Station reached its Eight Manned Crew Capability stage in 1999. However, there was no money to reach this eight-person crew stage.[26] Also delayed were a wardroom, a galley and personal hygiene amenities. Of the twenty-four racks in the habitation module, two were for the wardroom, two for a galley, two for life support system water, one for urine processing, one for a shower, one for communications, five for cabin air and thermal control, two for waste management and housekeeping, and eight for general or special storage. Four of these racks would be transferred to the laboratory module after the habitation module had arrived on orbit and their spaces filled with more stowage from a Shuttle resupply flight. In the laboratory module at the Man Tended Capability stage, only half of the twenty-four racks contained scientific experiments. Of the rest, five were for cabin air and thermal control, two for water, one for a workstation, one for communications, one for storage and two reserved for later use. Electrical power, thermal control, data management, vacuum, video and gaseous nitrogen would be available

1991 Space Station Restructured Design – Permanently Manned Capability An artist's impression of the Restructured Space Station as it would look after the 'Permanently Manned Capability' stage with an extra pair of solar arrays and truss piece added on the starboard side. The long transverse truss structure is made up of a set of prefabricated frameworks delivered by the Space Shuttle and ready to install on orbit with far fewer spacewalks and far less risk. This was the last major change in the design of the Space Station's truss structure. NASA incorporated the prefabricated framework approach in the International Space Station that began assembly on orbit later in the decade.

for the experiments. Rack frameworks changed from aluminium to graphite/epoxy to save weight.

Of the node modules, there would be one at the Man Tended Capability stage, another at the Permanently Manned Capability stage and two more at the Eight Manned Crew Capability stage. Their main functions were to connect the four laboratory and habitation modules into a unified pressurised volume, provide berthing ports for an airlock, crew emergency return capsules, a docking tunnel, a cupola and a logistics module, and provide distributed systems routing. Each node had six hatches – one in each endcap and four positioned at 90° intervals around the central cylinder. Each node would function as a six-way 'crossroads' for the crew and for the Station's distributed life support, electricity, communications and data systems. Each node had a small amount of space for rack-mounted communications, data management and crew healthcare equipment. Short Stand-Offs would support the racks off the pressurised hulls as they would do in the larger modules. There would be a 2.5m diameter centrifuge for life sciences research at the outer conical end of the last node.[27]

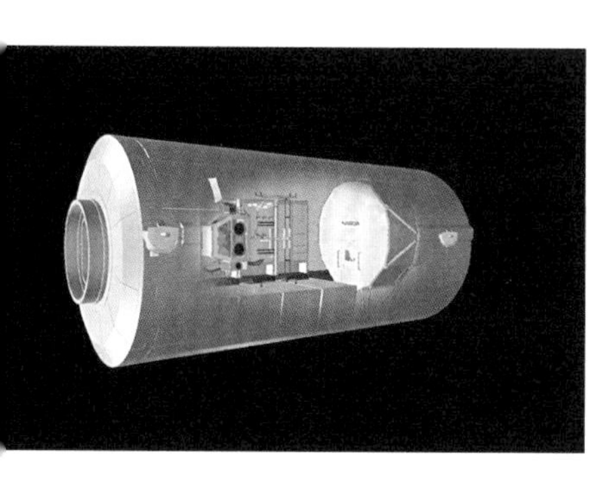

Centrifuge Module Despite the intractable problems with the Station's budget that continued to force reductions in its functions while stretching out its assembly, a large centrifuge remained a vital feature with its ability to generate a range of artificial gravity conditions for biological research. This artist's impression shows the 2.5m diameter centrifuge located at one end of a pressurised module called the Centrifuge Accommodations Module. It would be built by Japan but owned by NASA.

The restructured Station design showed one crew emergency return capsule at the Permanently Manned Capability stage and two at the Eight Manned Crew Capability stage. By mid-1991, the crew emergency return capsule's development had reached the end of Phase A. It was still only a concept and it still lacked funding for full development, but NASA now viewed it as essential. The Skylab laboratory outpost had always had a docked Apollo capsule and the Soviet Salyut outposts and Mir Space Station always had docked Soyuz vehicles. NASA called it the Space Station's 'lifeboat' and confirmed that it had three specific roles: return a disabled crew member in a medical emergency; return a Station crew (or part of one) from Station accidents or failures of on-board systems; and return a crew during interruption of Space Shuttle launches. NASA pointed out that each of these emergencies had occurred during the history of spaceflight. There had been system failures on the Gemini 8 and Apollo 13 missions while the Soviets had experienced two medical emergencies, a contaminated cabin atmosphere and a damaged window that had caused emergency crew returns. The Station's emergency return capsule would be able to seat four or six persons and would require no piloting. It would be able to leave the Station in a matter of minutes in an emergency and have enough food and water on board for the return journey to Earth. The prime contractors for the vehicle would be Lockheed and Rockwell. They would develop competing designs and Phase B would begin in 1992.[28] However, there was a misfit between intention and reality on what many viewed as a critical part of the Space Station from a crew safety standpoint.

In March 1991, NASA published the findings of a preliminary study on the environmental impact of the Space Station. Though the Station's site would be on orbit far above Earth, it was not exempt from a review of potential environmental hazards caused by its operation. There were some concerns about the venting of gases from the Station, the type of propellant it would use to maintain altitude and attitude, and the effect on the upper reaches of the atmosphere. There were questions about the preparation activities on the ground for launching the Station and their effects on the environment. The major concern, however, was the possibility of the Station's accidental descent and burn-up in Earth's atmosphere that would cause pollution of the air and a debris shower. Debris from Skylab's fiery destruction in 1979 had dropped over Australia when it should have sunk into the Indian Ocean. However it would be done, the eventual termination of the Space Station – a far bigger object than Skylab – would require careful planning.

The Station's operating timespan was to be at least thirty years. NASA's plan at that time was to disassemble it at the end of its lifetime and use the Space Shuttle system to ferry the pieces back to the ground. This assumed that the Shuttle system would still be operational. NASA was taking steps to ensure that the Station would not make an unplanned re-entry. If that happened, however, it would be either controlled or uncontrolled. If the former, it would keep to a defined entry and debris path. If the latter, a disaster was possible if it fell over a populated area. NASA was determined to ensure that this would not happen, though the agency had yet to do a detailed analysis of controlled and uncontrolled re-entry scenarios. Follow-on studies would examine these in detail.[29]

Gathering storm
Restructured Space Station Freedom, 1991-1992

On 15 May 1991, another Station crisis erupted in Washington when the House appropriations subcommittee turned down NASA's funding request for 1992. The subcommittee's chairman, Bob Traxler, wanted the $2 billion NASA had asked for to go to social and environmental programmes. The next day, NASA Administrator Richard Truly told the Committee on Science, Space and Technology that the cut would spell disaster for the agency's human spaceflight plans as well as its relations with its European, Japanese and Canadian partners. The House reversed the cut on 6 June and voted $1.9 billion for the Station. The Senate gave it another $100 million on 10 July, thus restoring NASA's full budget request.[30] It was more see-saw meddling – a symptom of the distrust that had grown between NASA and Washington and the political frustration that the Station's role was diminishing while its budget was increasing.

In the meantime, the National Space Council had endorsed NASA's restructured Station design on 21 March and in May it published the results of its study of American space exploration plans in a report entitled *America at the Threshold*.[31] The report was a distillation of numerous contributions from those who supported President Bush's space exploration quest and who had voiced their support through an outreach programme. The Synthesis Group, a spin-off of the National Space Council, had carried out the distillation under its chairman, Thomas Stafford, a former astronaut and retired Air Force general. Like its predecessor *Pioneering the Space Frontier* produced in 1986, it offered a choice of ambitious visions and measures aimed at boosting America's human space exploration programme.

Among its nine recommendations, the report mentioned the Space Station just once. Headed 'Conduct focused life sciences experiments' it advised NASA to 'implement a definitive life sciences program, along with the necessary experiments and equipment, on Space Station Freedom, consistent with the recommendation of the Advisory Committee on the Future of the US Space Program. These experiments are needed to reduce the uncertainties of long duration space missions.' On this one thing the report agreed with the findings of Norman Augustine's 1990 committee.

Sadly, there would be no part for the Station to play in assisting the operational preparations for human missions to the Moon and Mars. Its role would be to provide some research input but that was all. Unlike *Pioneering the Space Frontier*, which had made a brief reference to programme costs, there was no word on how much funding this new vision would require. As Congress bickered and NASA struggled to force down the Station's costs, it was perhaps not surprising that the report avoided the subject. President Bush had by now reviewed two studies by expert teams on America's long-range plans for its future in space, neither of which had price tags attached. Unless these grand visions were properly costed and funded, America's human space exploration plans would remain a fantasy.

Further concerns about the Space Station were voiced by the Space Studies Board of the National Academy of Sciences in its 1991 Annual Report.[32] Echoing the Augustine committee's recommendations that the Station's focus should be on life sciences and materials research, the Board judged that the Station's Restructured Design would not meet this objective. The quality and quantity of the anticipated research did not merit the projected investment. The aim of life sciences research on the Station would be to study the physiological consequences of long-term spaceflight, inevitably involving a long period of investigation on board. Life sciences research could not begin until permanent occupation occurred late in the decade and the facilities would be inadequate. Neither would the Station be much good for microgravity research which the Board felt could be carried out in the European Spacelab laboratory on the Space Shuttle or on free-flyer platforms. The Board considered that 'continued development of Space Station Freedom, as currently redesigned, cannot be supported on scientific grounds. If the present station design is implemented, this major national investment must be justified on the basis of considerations other than research in these two disciplines'. The report cited some specific shortcomings: there would be insufficient electrical power for many long-term biological experiments; planning of an essential centrifuge and animal-holding facility was insufficiently clear; the crew size was too small; there was no dedicated life sciences laboratory; crew movement, docking manoeuvres and other physical activities would disturb materials research; and upgrading computers, which became obsolete quickly, would be difficult. The Board suggested removing all materials microgravity research from the Station and devoting its research role to life sciences. Another long-term research role for the Station would fall by the wayside.

While NASA fended off criticism of the restructured Station, Canada was quietly making substantial progress with its contribution – the robotic arm and its mobile trolley system. By the end of 1991, Spar Aerospace Ltd., the Canadian fabricator, was well into prototype development and final design. The Mobile Servicing System was the umbrella title given to the combination of a teleoperated robotic arm, a supporting framework, a trolley car and a hand-like manipulator that would provide the Station with a tool to carry out exterior mechanical tasks as well as supporting astronauts on spacewalks.

The Space Station Remote Manipulator System was the name given to the robotic arm portion of the device.[33] Its genesis lay in the Shuttle's robotic arm, which Spar Aerospace had built as Canada's contribution to the Space Shuttle programme. The Shuttle's arm was fixed to a position on one side of the Orbiter's payload bay but the Station's robotic arm was going to be a much more sophisticated piece of engineering. Its role would be to carry out assembly, maintenance and repair duties and it would need to move about the Station's exterior. It was designed to travel the length of the transverse truss on a trolley car moving on rails, with the ability to detach itself if necessary and crawl over the Station's surface like a giant insect. It was 17.4 metres long and functionally similar to the human arm. It had the equivalent of a shoulder joint, an upper arm, an elbow joint, a lower arm and a wrist joint. Each end joint connected to a grappling device called an end effector. The arm's assembly was symmetrical about its elbow joint. It would grapple a support at one end, move its two long limbs in a somersaulting motion, grapple a support at the other end and then release the first grapple and perform the same somersault. By repeating this manoeuvre, it would be able to walk across the Station's surface, grappling on to predetermined anchor points as it went.

Towards the end of 1991 an event occurred that would change the Space Station's future course, though its significance may have escaped the notice of the NASA team, as it battled with the project's costs and critics. On 26 December, the Union of Soviet Socialist Republics was formally dissolved. Mikhail Gorbachev, the last Soviet President, declared his office extinct, and Russia, the largest nation in the new Commonwealth of Independent States, gained a new President, Boris Yeltsin. With the demise of the Soviet Union, NASA's traditional competitor and occasional colleague in the space race ceased to exist. In an extraordinary turn of events, within two years, this

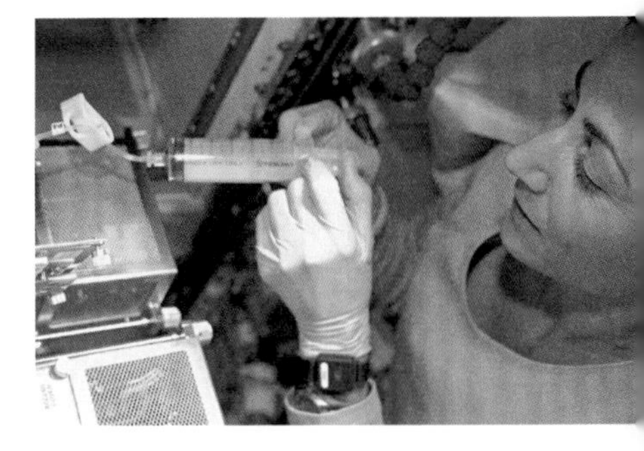

Life sciences research Life sciences research remained a top priority for the Space Station during its budget crisis years. The National Space Council published its report in May 1991 called *America at the Threshold*. A focus on life sciences research was one of its nine recommendations. In this later photograph taken on board the International Space Station during Expedition 21, astronaut Nicole Stott is at work on the Cell Biology Experiment Facility SPACE SEED experiment in the Japanese pressurised module. It is a typical example of the life sciences research carried out on the Station today.

quiet revolution would lead to the reincarnation of the Space Station, placing it on a firm financial and technical footing.

Dan Goldin succeeded Admiral Richard Truly as NASA Administrator in March 1992. Goldin had been General Manager at TRW Space and Technology Group. A passionate advocate of a streamlined approach to project management, he soon challenged his new agency as well as industry and academia to follow him. In a speech to NASA's staff, he urged them to: 'tell us how we can implement our missions in a more cost-effective manner. How can we do everything better, faster, cheaper, without compromising safety?' He was talking about forthcoming NASA missions to Mars and other destinations beyond Earth planned for later in the decade.[34] It was too late to apply this philosophy to the Space Station.

On 7 May 1992, NASA successfully launched the new Orbiter Endeavour on Shuttle Mission STS-49. Endeavour replaced the Orbiter Challenger lost in the 1986 disaster. The mission's main task was to capture and repair the crippled Intelsat VI communications satellite that a defective Titan rocket had launched two years earlier and left in a useless orbit. The crew retrieved the satellite, attached a new rocket motor and then released it to climb to its operating orbit.[35] The mission was a success and Endeavour and its crew returned to Earth without incident. NASA's fleet was now back to four Orbiters of which three – Endeavour, Atlantis and Discovery – would carry out the Space Station's assembly in orbit.

It was time to start selling the Station to the scientific community. In August 1992 NASA organised its first conference for prospective researchers. Called the Space Station Freedom Utilization Conference, it was to become an annual event. Over 700 people attended of whom about 200 were researchers and scientific experiment developers. The conference featured more than thirty exhibits sponsored by private companies, NASA offices and foreign space agencies.[36] In a keynote address, Dan Goldin explained that 'essentially, the Space Station should be thought of as an international research centre in orbit. Researchers from universities and the private sector, such as pharmaceutical companies, and our international partners will be able to share facilities on Freedom to facilitate basic research in materials processing, biotechnology and life sciences.' He defended NASA's budget. 'Many people don't realise that NASA only receives one per cent of the federal budget – literally two cents a day for every American citizen. When you consider the enormous return on investment a space station will yield, Americans will get far more than their two cents' worth. For that small amount, the dividends we pay are enormous.' He envisioned the future. 'Space Station Freedom isn't just a job, or a chance to make money. It's a mission to move the human species into breaking the chains of gravity and becoming a multi-planetary society. Pursuing a mission of this monumental importance will lift civilization on Earth to new heights of health, wealth and knowledge.'[37]

The Station's long-term objectives were now to spearhead cutting-edge research in three fields: microgravity, life sciences and technology. The importance of the Station in the life-sciences research field had been stressed by committees outside NASA and it was the first main item on the conference agenda. The life sciences goals were manifold: to ensure the health, safety and productivity of humans in space; to acquire fundamental scientific knowledge of space biological sciences; to expand mankind's understanding of life in the Universe; to develop an understanding of the role of gravity on living systems; to provide for the health and productivity of humans in space; and to promote the application of life-sciences research to improve the quality of life on Earth. The Station would enable long-term investigations into these areas under controlled laboratory conditions for the first time. American life-sciences facilities would comprise a biomedical monitoring and countermeasures facility of four racks, a gravitational biology facility of two racks, a 2.5-metre diameter centrifuge, a gas grain simulation facility of one rack and a controlled ecological life support system test facility of two racks. Specific areas of study would be the maintenance of long-duration crew health and performance, fundamental cell and plant biology, the influence of variable gravity on biological systems (the centrifuge again), submillimetre particle chemical and physical processes and crop growth and yield. The total inventory of life sciences equipment added up to eleven racks and, optimistically perhaps, the large centrifuge.[38]

The second major research area was microgravity. Here, American research would cover fluid dynamics and transport phenomena, combustion science, materials science, biotechnology and fundamental physics. Protein crystal growth was a particularly important subject and it would need a single rack. The aim would be to study the effects of gravity on crystals, the physics and dynamics of their growth and the role of macromolecular crystals in X-ray crystallography. Materials science would explore unique materials structures with specially crafted properties in a low gravity environment and measure their thermophysical properties. It would start out with two racks and evolve to three. A further two racks would serve both combustion and fluid experiments. Fluid science research would advance fluid physics theory, seek improvements in thermophysical property measurement and explore specific fluid phenomena. It needed a single extra rack as well as the two shared with combustion experiments. Containerless materials processing would take a rack. Biotechnology research would explore cell function and cultured tissue models for genetic data. It would take a rack. A group of small microgravity experiments called 'mid-deck class' would need two racks and a microgravity glovebox would need a rack. Altogether, American microgravity research on the Station would require twelve racks from about 1998 onwards.[39]

The third major area of research was simply called 'technology' and dealt with engineering experiments. NASA claimed that the Station itself

First launch of the Orbiter Endeavour
NASA launched Endeavour, the last of the Space Shuttle Orbiters, on 7 May 1992. Endeavour was built by Rockwell International as a replacement to the Orbiter Challenger lost in the 1986 tragedy. Endeavour's arrival increased the number of Shuttle Orbiters able to build the Space Station to four.

was going to be an engineering experiment: it would involve the first assembly of a seriously large structure in space and its dynamic behaviour on orbit would be of great interest. There would be research on the interaction of different materials with the space environment, the thermal control of a large structure, the performance of large solar arrays for electrical power, and propulsion for orbital reboost and station keeping. Then, at some point in the Station's lifetime, there might occur the assembly of spacecraft destined for the Moon or Mars, the repair and upgrading of satellites, and the in-space assembly, testing and approval of space parts and systems for commercial uses. None of these, however, was planned at that time.[40] What the Space Station could offer that no other space laboratory had before was a large structural truss for mounting experiments needing direct exposure to the space environment – a truss that could be visited by astronauts on spacewalks if need be. It would offer the science community a unique platform for research.

There were four possible sites for mounting experiments on the Station's truss, two to port and two to starboard. The experiment footprint area at each site – the area available for attaching experiments to the structure – measured 1.8 by 4 metres. 3kW of power and data connections would be available at each site, but there would be no centralised thermal control. Experiments could face zenith (-Z), nadir (+Z), forward (+X) or aft (-X) relative to the Station's flight path, depending on the site. The maximum experiment payload weight would be about 4,500kg at both port (-Y) sites and 4,100kg – 2,300kg at the two starboard (+Y) sites.[41] Experiments in the materials technology field would benefit most from the exterior sites. Long-duration exposure of materials and coatings to a mixture of vacuum, microgravity, day/night temperature swings, atomic oxygen, cosmic radiation, occasional solar particle radiation and occasional micrometeoroid impact – all of which are present in Earth orbit – would provide experimenters with a unique space environment that was impossible to replicate on Earth.

Towards the end of 1992, Buzz Aldrin, the former astronaut, launched himself into the Station's cost debate. He proposed alternatives to Space Station Freedom which he described as 'too little for too much, too late'.[42] He suggested relieving the Station of its dependency on the Space Shuttle to lift it into orbit. Coupling the Station to the Shuttle had been an expensive decision because the Shuttle was so costly to fly. Among alternatives, he mentioned using the Russian Mir Space Station for biomedical research, reallocating Space Station modules to expand Mir, borrowing the Shuttle's Spacelab modules, and salvaging an empty Shuttle External Tank and outfitting its cavernous interior on orbit to turn it into a habitat. He recalled the success of the Skylab missions of the 1970s. It had needed just one launch to loft the Skylab laboratory to orbit.

It was, of course, too late to address Aldrin's ideas seriously. America had spent billions of dollars getting thus far, its international partners were

Buzz Aldrin Former astronaut and second man to walk on the Moon, Aldrin entered the Space Station's cost debate in 1992 with some of his own ideas. He suggested scrapping the use of the Space Shuttle to build the Station, attaching American modules to the Russian Space Station Mir instead and outfitting a spent Shuttle External Tank on orbit as a habitat. This photograph taken by Neil Armstrong in July 1969 shows Buzz Aldrin inside the Apollo 11 Lunar Module before the lunar landing.

committed and contractors were poised to begin fabrication. Nevertheless, Aldrin had made a valuable observation. He pointed out that the Russian reformation offered an opportunity to join two great space powers in a noble common cause and that attaching an American-built habitation module to the Mir station could be a first step.

Survival test
Space Station Alpha, 1993

In January 1993, Bill Clinton was inaugurated as President of the United States. The outgoing Bush Administration had envisioned a robust human exploration programme to return America to the Moon and then continue to Mars, but it handed over a project in deep financial and political turmoil. Clinton took action. On 9 March, he directed NASA to explore lower cost alternatives to Space Station Freedom's design. It was time for NASA to retrench. Dan Goldin set up a Station Redesign Team based at Crystal City in Virginia headed briefly by Joseph Shea and then by Bryan O'Connor.[43] Then on 25 March, the President formed the Advisory Committee for the Redesign of the Space Station headed by Charles Vest, president of Massachusetts Institute of Technology. It would be the job of the Advisory Committee to review the Station Redesign Team's proposals. Clinton also directed NASA to consider inviting Russia into the existing international partnership, and to study the use of Russian space assets[44] – all with the objective of bringing the Station's costs under control. There was a political objective, too. The emergence of Russia presented the White House with an opportunity to propose a joint initiative on something the two countries were expert at – human spaceflight.

With NASA instructed to develop a new design, there was hope that from Freedom's ashes would arise, to Dan Goldin's mantra of 'better, faster, cheaper', a new space bird that would fly to orbit on time and on cost. Understandably, Clinton wanted to rename the project. Space Station Freedom had been conceived when the Cold War still existed. To some, the name evoked the spirit of a free West pitted against an oppressed East that President Reagan had captured when he spoke at the Brandenburg Gate in 1987 and exhorted Mikhail Gorbachev to tear down the Berlin Wall. The wall had come down, *glasnost* and *perestroika* had blossomed and 'freedom' was old hat. NASA chose 'Alpha' as the new name.

On 9 March 1993 – the day that the White House instructed NASA to redesign the Station – Dan Goldin wrote to NASA's field centre directors and senior staff. He set out concisely what NASA and the Station Redesign Team had to do to respond to the President's directive and report to the Advisory Committee. They had to: 'provide a cost effective solution to basic and applied research challenges whose merit is clearly indicated by scientific peer review, significant industrial cost sharing, or other widely accepted method; provide the capability for significant long-duration space research in materials and life sciences during this decade; bring both near-term and long-term annual funding requirements within the constraints of the budget; continue to accommodate and encourage international participation; and reduce technical and programmatic risks to acceptable levels.' The new design should meet the budget, absorb effort spent on the project so far, achieve a research start on orbit in 1997, be completed by 1998, focus on life science and materials research, include the international partners, involve Russia for the first time, be less costly to operate, require fewer Shuttle launches and fewer assembly spacewalks, use streamlined project management, maintain a budget reserve, have a ten-year orbital life extendable to fifteen years and stimulate spin-off technology. The Station was also to re-establish American leadership in space. The deadline for the final report was 1 June 1993.[45]

The Station Redesign Team had less than three months to come up with a new design for a project that would stand expert scrutiny. The team at Crystal City comprised forty-five NASA staff members and ten representatives from the Station's international partners. Assisting them were five consultants drawn from inside and outside government and three groups of NASA engineers and scientists from Johnson Space Center, Marshall Space Flight Center and Langley Research Center. The team received a precise set of design guidelines, which encompassed crew size, crew health, pressurised volume, microgravity environment, external experiments, life support, safety, electrical power, data, communications, propulsion, thermal control, spacewalks and robotics. NASA was mindful of the international partners' contributions and the potential impact of a new design on them and the team had more guidelines drawn from the international agreements. The new design would incorporate the Japanese modules, the European module and the Canadian mobile robotic arm system at the 'Permanent Human' (formerly called Permanently Manned) capability stage. The expectation was that all the partners would agree to this capability's timetable, which would not deviate much from its predecessor. There would be a crew of four astronauts at Permanent Human Capability stage with a potential for eight later on, as before.

The Station's solar arrays would deliver 75kW of electrical power. The Station's operational life would begin at its completion, not some earlier interim date. Systems and interface requirements for the international contributions would change as little as possible, A small Italian-built module would become the main pressurised resupply module in the early stages and equally important in the same role later on. NASA proposed new operating guidelines with the European Space Agency, the Canadian Space Agency, the Japanese Science and Technology Agency and the Italian Space Agency, under its own initiative, was providing small pressurised modules (later called multipurpose logistics modules). These were agreed quickly.

The team conducted a major outreach with briefings. It received comments on science, research and technology issues as well as numerous concepts from the NASA centres, the aerospace industry, the Station's programme office, the international partners and others. It briefed a delegation from the Russian space programme.[46]

The Station Redesign report proposed three basic design approaches: Option A – Modular Buildup; Option B – Space Station Freedom Derived; and Option C – Single Launch Core Station. The team based two of them on the earlier Restructured Space Station design and created one that was entirely new and surprising.

Option A – Modular Buildup, as its name suggested, followed a gradual and incremental approach to the Station's construction. It ensured that all the international building blocks remained in the assembly sequence. There were four phases in its assembly on orbit. Phase 1 was a simple platform named 'Power Station' to which Orbiters would dock and attach payloads. It needed three Shuttle flights to assemble and consisted of a piece of truss structure, one pair of solar arrays producing 20kW, thermal radiators, stationkeeping and communications systems and part of the mobile robotic arm system. Orbital stationkeeping demanded attitude control for correct flight orientation and a propulsive boost for maintaining altitude – both vital functions to ensure the platform remained in a safe and stable condition when it was left alone on orbit.

NASA Administrator Dan Goldin Bill Clinton was inaugurated as President in January 1993 and inherited the Space Station that was in financial and political turmoil. In March he directed NASA to come up with lower cost alternatives and explore the possibility of bringing Russia into the existing Station partnership. Dan Goldin, seen here in a later photograph, had become NASA Administrator in April 1992. He set up a redesign team and gave it three months to come up with a solution. Photo: NASA/Bill Ingalls.

Phase 2 offered 'Human Tended Capability'. It added a core laboratory module, docking ports and the rest of the mobile robotic arm after a fourth Shuttle flight. It would be able to support short crew visits. The new module design was a hybrid of an American-built common module and an Italian-built mini-module. Two cylindrical segments taken from the American module were mated to a multiple docking port segment taken from the small Italian logistics module. The result was an American laboratory capability even smaller than the previously descoped design, with just nine racks for experiments. The new module was called a Common Core/Laboratory Module. At its Italian end was a new tapered docking tunnel for visiting Orbiters, given the name of a Pressurized Mating Adapter.

A further eight Shuttle flights would achieve Phase 3 and what was now called 'International Human Tended Capability'. It would see the addition of more truss-structure building blocks, the European and Japanese laboratory modules, the Japanese laboratory logistics module, the Japanese external platform, another pair of solar arrays to double the power to 40kW, the multi-windowed Cupola and more thermal control equipment. Phase 3 would support expanded crew visits but still not permanent crews. Phase 4 eventually achieved 'Permanent Crew Capability' after another four Shuttle flights, making sixteen in all. It would see the addition of another solar array pair to increase the power to 60kW, a habitation module (another hybrid module), an airlock, a 'closet' module based on the small Italian logistics

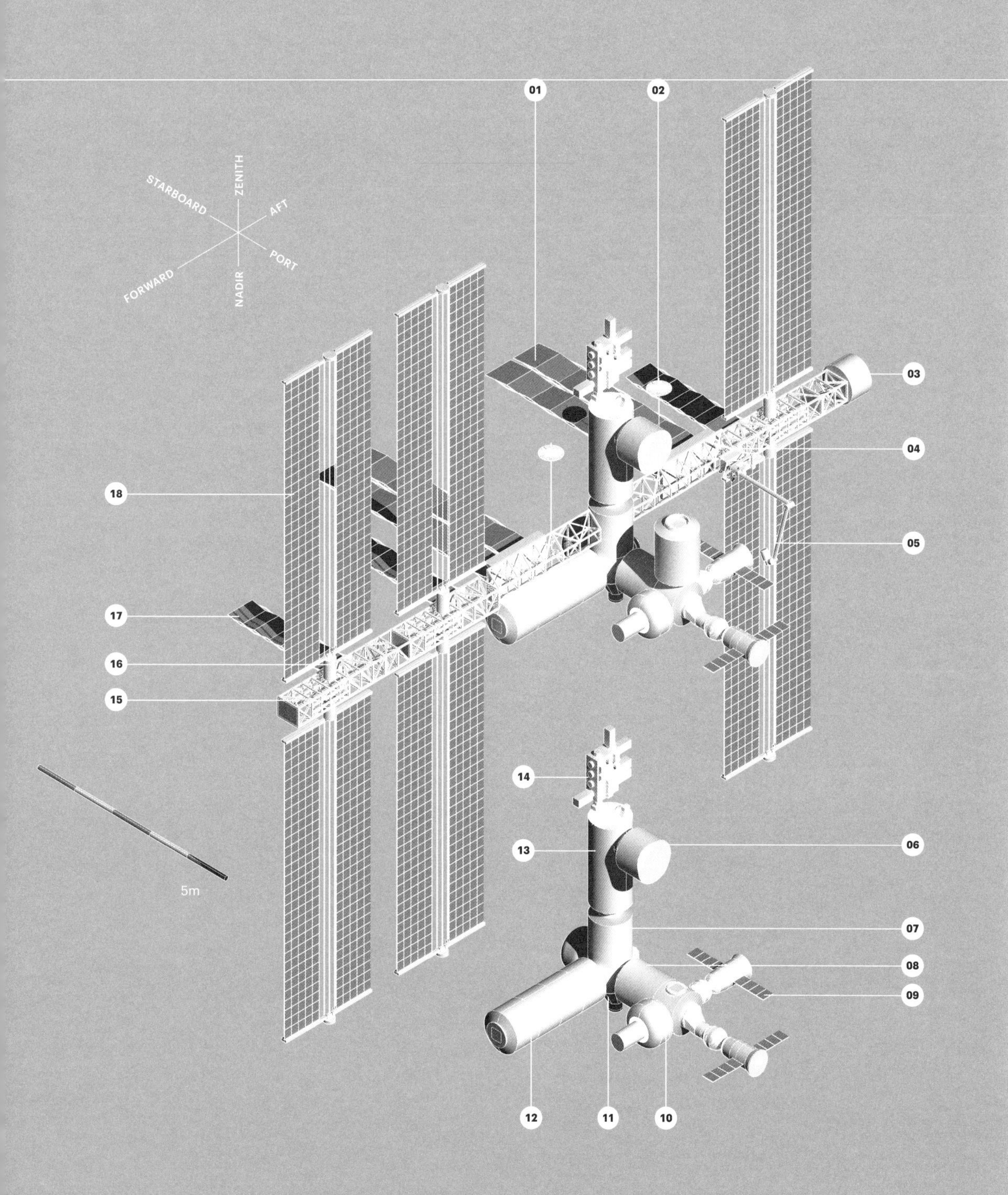

01 Thermal radiators
02 Tracking & data relay satellite antenna
03 Propulsion unit (DOD BUS-1)
04 Rotating joint (alpha joint)
05 Robotic arm on mobile platform
06 Japanese logistics module
07 US core laboratory module
08 US core habitation module
09 Soyuz return vehicle (2)
10 Airlock module
11 Docking port
12 European laboratory module
13 Japanese laboratory module
14 Japanese experiment platform
15 Prefabricated truss segments
16 Rotating joint (beta joint)
17 Power radiator
18 Solar array

module and, most significantly, two Russian Soyuz spacecraft docked at the Station as the crew return vehicles. The team had responded to Dan Goldin's instructions to include Russian elements. It was a breakthrough, for NASA could now abandon the American-built crew emergency return capsule, which had never proceeded beyond concept stage.

An advantage of Option A was its ability to function as a useful Station if assembly stopped at any phase, with some slight adjustments. This was an asset, given the unpredictable economic and volatile political environment that surrounded the project. The downside was that crews could not live on board permanently until the last phase of the assembly sequence. The quality of some of the scientific research would suffer, particularly in the field of human physiology that required long stays in weightlessness.

Option A came in two design versions. The first version designated Option A-1 eliminated the M1, P2, P3, S2 and S3 truss building blocks of the former Restructured Space Station and added a new short one called P5 at the forward end. Mounted at the aft end on a truss piece called S6 was a new propulsion element called a 'bus' – a word borrowed from the electrical word 'busbar'. Originating at the Department of Defense, the bus was a new propulsion unit that would have the job of orbital reboost and station-keeping. If necessary, the P5 at the forward end of the truss could take a replacement propulsion unit with the Station rotated through 180° to switch its bow to the stern. The solar arrays flew vertically with their tips pointing to nadir and zenith. The second version, designated Option A-2, eliminated the M1, P2 and S2. It avoided the use of the P5 and the new propulsion unit and retained the Station's existing propulsion hardware along the truss structure. Option A-2 also flew with the truss aligned on the flight path but with the solar arrays turned horizontally so that they pointed to port and starboard.[47] Also reoriented in Option A-2 was the module cluster.

Option B – Space Station Freedom Derived was similar to Option A in its overall design and assembly but in some respects was closer to the earlier Restructured Space Station design. It too ensured that all the international building blocks were taken into account. Phase 1 was the Power Station platform that would need two Shuttle flights to deliver it to orbit. As before, it had one truss building block supporting a pair of 20kW solar arrays, thermal radiators, attitude control, propulsion and communications systems to enable it to remain safely on orbit on its own. It had an Orbiter docking port and a power feed to the spacecraft to support experiments on the Orbiter missions that flew Europe's Spacelab laboratory. A further six Shuttle flights would deliver Human Tended Capability in Phase 2. This would include sufficient truss building blocks to complete the starboard side of the truss, the American-built laboratory module (in place of the hybrid American-Italian module of Options A-1 and A-2) and two experiment locations along the truss to give experimenters an early start. Also delivered and in Phase 2 were the complete mobile robotic arm, the airlock, the first

1993 Space Station Alpha Option A-1 - Modular Buildup Continued problems with the Space Station's costs in 1993 prompted the new Clinton Administration to instruct NASA to produce fresh proposals for an affordable design. NASA came up with four options for the Station, now named Alpha. The first of these named Option A-1 Modular Buildup was a gradual assembly in four phases beginning with a 'Power Station' and proceeding to 'Human Tended Capability', 'International Human Tended Capability' and finally 'Permanent Crew Capability', shown here. The Station's flight attitude was rotated through 90° to align the truss in the fore-aft direction so that a new propulsion module at its aft end could handle the orbital stationkeeping. The module cluster was now at the centre. Image: author.

1993 Space Station Alpha Option A-2 – Modular Buildup Option A-2 was similar to Option A-1 Modular Buildup except that it flew differently. It was the second of NASA's new proposals in response to the Clinton Administration's instructions to the agency in 1993 to produce an affordable Station design. In Option A-2, the Station was rotated 90° on its X axis so that the solar arrays pointed outwards to port and starboard.
Image: author.

node module, the Cupola and the tapered Orbiter docking tunnel. At Phase 2 of Option B it would be possible to dock two Orbiters simultaneously to provide a larger crew for experiment and assembly tasks. Phase 3 achieved International Human Tended Capability after another eight Shuttle flights. It included all the truss building blocks on the port side to finish off the truss structure (eliminating the P2 piece), two more solar array pairs to increase the power to 60kW, the European and Japanese laboratory modules, the Japanese laboratory logistics module and external platform, another node module, more thermal radiators and the port attitude control and propulsion system to match the one on the starboard side. There were two docking ports, one on the front of the American laboratory module and the other on one of the node modules. A further four Shuttle flights completed Phase 4 and Permanent Human Capability by delivering the habitation module and the two Soyuz spacecraft for crew emergency return. One of the docking ports would move from a node module to the front of the habitation module. The Option B team investigated three variations to the Option B design aimed at reducing costs but found either performance disadvantages or the need for redesign work in each case.[48]

Option C, Single Launch Core Station was utterly unlike Options A and B. It abandoned the 'kit of parts' approach of using the Shuttle to deliver a set of building blocks for orbital assembly and replaced it with the single launch of a new prefabricated habitat that was similar in concept and size to the 1970s Skylab. Not only did it reject much of fourteen years of work and billions of dollars spent on a Station tailored for construction over a series of Shuttle missions, it largely renounced the use of the Shuttle Orbiter fleet as the principal delivery vehicles and replaced it with an unwinged Orbiter mutation to lift the single Station to orbit in just one flight. Like Hughes Aircraft Company's rotating wheel design from the early 1980s, Option C was the joker in the pack. It is hard to imagine how NASA could have offered it as a serious proposal so late in the Station's development. Perhaps it was simply intended to prod the White House and Congress into viewing the other options in a more favourable light.

The core of Option C was a single pressurised module with conical endcaps. It had a cylinder length of 19.5 metres, a diameter of 6.7 metres and a habitable volume of 736m^3. Grafted on to one end of the module was the aft fuselage thrust structure and main engine cluster taken from the Space Shuttle system. Plugged on to the other end was a nose cone. These accessories transformed the pressurised module into a rocket. Mating the rocket to the standard Space Shuttle combination of the big External Tank and two Solid Rocket Boosters gave it the propulsion it needed to reach orbit. NASA called it the Space Shuttle Derived Vehicle. Both Johnson Space Center and Marshall Space Flight Center had been working on the idea, on and off, since 1975. Though its designers claimed it took advantage of much of Space Station Freedom's development work, it dispensed with a vast

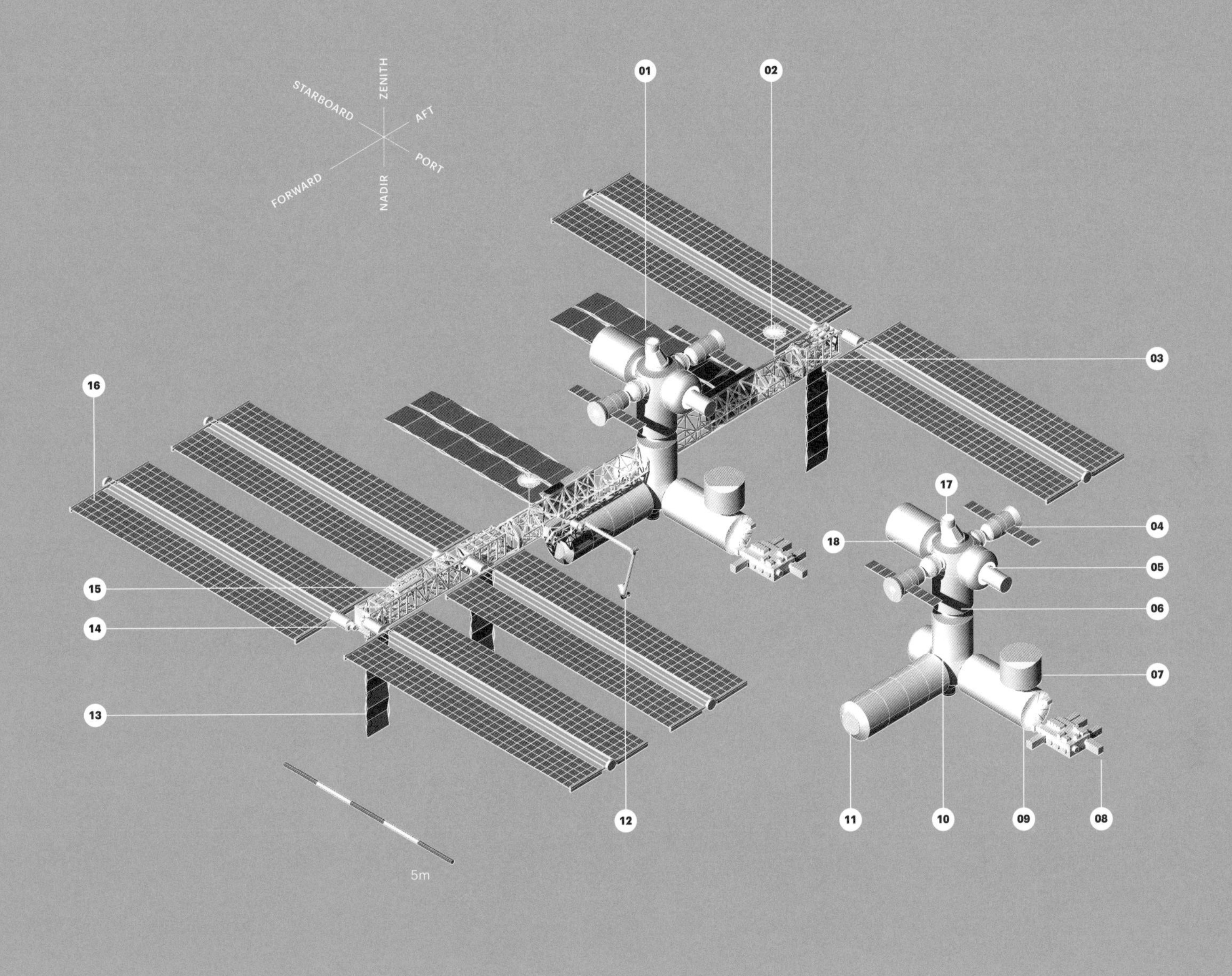

01 Thermal radiator
02 Tracking & data relay satellite antenna
03 Rotating joint (alpha joint)
04 Soyuz return vehicle (2)
05 Airlock module
06 US core habitation module
07 Japanese logistics module
08 Japanese experiment platform
09 Japanese laboratory module
10 US core laboratory module
11 European laboratory module
12 Robotic arm on mobile platform
13 Power radiator
14 Rotating joint (beta joint)
15 Prefabricated truss segments
16 Solar array
17 Docking port
18 Logistics module

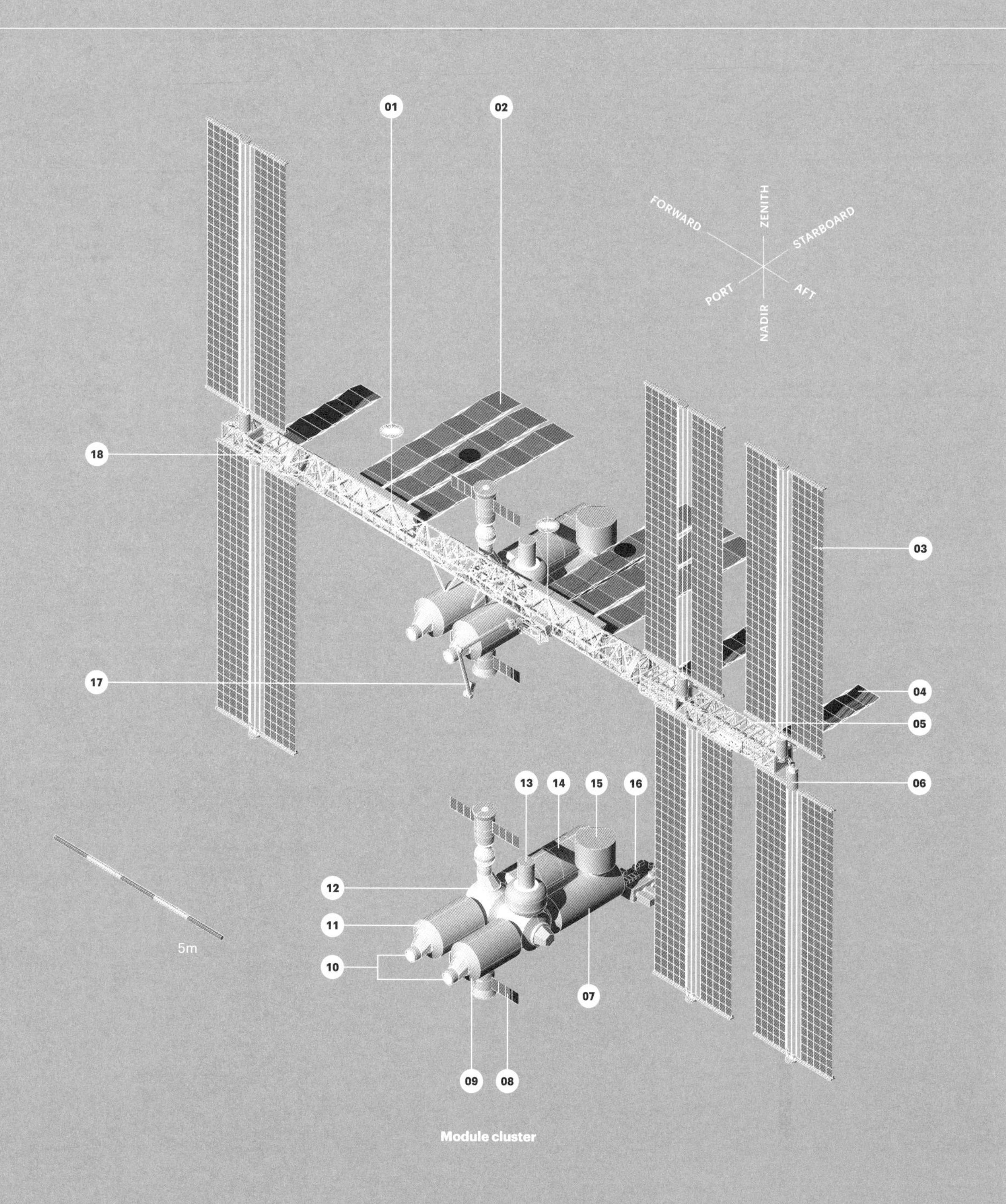

Module cluster

01 Thermal radiators
02 Tracking & data relay satellite antenna
03 Solar array
04 Power radiator
05 Prefabricated truss segments
06 Rotating joint (beta joint)
07 European laboratory module
08 Soyuz return vehicle (2)
09 US laboratory module
10 Docking ports (3)
11 US habitation module
12 Nodes (2)
13 Airlock module
14 Japanese laboratory module
15 Japanese logistics module
16 Japanese experiment platform
17 Robotic arm on mobile platform
18 Rotating joint (alpha joint)

quantity of hardware. Gone were the long truss structure, the American laboratory module, the American habitation module and the node modules. All their pressurised accommodation was now rolled together into the big cylindrical habitat. Retained were the European and Japanese modules, the Japanese experiment platform, the Canadian robotic arm and the two Russian Soyuz vehicles – all now mandatory international elements.

The first mission would deliver the core module, jettisoning the two boosters and the big tank on the way up in the normal Shuttle manner. Just before reaching orbit the core module would release the engines and the nose cone. Once it reached its orbit it would activate its thermal radiators and two pairs of solar arrays and become a fully operational Station. Then a Shuttle mission would deliver a Soyuz vehicle and a commissioning crew. Following missions would deliver another Soyuz, the Japanese laboratory module, the European laboratory module, the Canadian mobile robotic arm (less mobile than it was before because of the loss of the truss), the Japanese external platform and laboratory logistics module, and a third pair of solar arrays. Inside the cavernous module were seven decks of crew accommodation arranged vertically, like a lighthouse. Three decks were for laboratories, one was for habitation, another was for crew healthcare and module berthing and two more were for life support systems and supplies. A central crew circulation corridor ran through the decks from end to end.

Introducing the use of the Russian Soyuz vehicles to Space Station Alpha as the means of crew return raised a question mark over the Station's orbital inclination as it circled the Earth. The physics of launching to Earth orbit dictates that the most efficient launch trajectory is due east on the equator to take advantage of the Earth's natural rotation rate, giving an added boost to ascent velocity. This advantage reduces the more distant the launch site is from the equator because high and low latitudes produce more inclined orbits that turn trajectories away from due east. NASA had assumed that the Station would orbit the Earth at 28.8°, the lowest inclination orbit for launches from Cape Canaveral in Florida. The Soyuz vehicles meant that the Station's orbit would have to switch to 51.65° to enable Russia to launch them from its Baikonur Cosmodrome in Kazakhstan. Launching Station payloads from Florida to the 51.65° orbit would require refinements to the Space Shuttle system to reduce vehicle weight, increase propulsive energy or both. The first refinement was changing the metal used to manufacture the Shuttle's External Tank to an aluminium-lithium alloy, which is lighter than aluminium on its own. The original tank weighed 34,505kg. By the sixth Shuttle mission in 1983 the tank's weight had dropped by 4,540kg and it was called the Lightweight External Tank. With the new alloy it would become the Super Lightweight External Tank, with a further weight loss of 3,405kg.[49] The second refinement would have increased the propulsive energy of the Space Shuttle system by using a more powerful version of the strap-on booster called the Advanced Solid Rocket Booster. This, however,

1993 Space Station Alpha Option B – Space Station Freedom Derived Option B followed the gradual four-phase build-up of Options A-1 and A-2 but proved to be more costly. Carefully included in all these new designs were the module and equipment contributions of Japan, Europe and Canada. None of these changed despite NASA's continued tinkering with the Station's overall form. In Option B, the Space Station flew with its truss structure aligned along the Station's flight path like Options A-1 and A-2. NASA produced these new options in response to instructions from the White House to lower the project's costs. Image: author.

1993 Space Station Alpha Option C – Single Core Launch Station Utterly unlike any other Space Station design considered over the previous decade was Space Station Alpha Option C – Single Core Launch Station. It was more like the Skylab project of the early 1970s. It comprised a single large cylindrical habitat with the international elements, the two emergency return spacecraft and the solar arrays plugged on to its exterior surface. Adopting this new design for the Space Station would have wasted years of work on the Shuttle-delivered module, node and truss building blocks that the Station's contractors were about to begin fabricating. Image: author.

was cancelled in 1993 in favour of smaller improvements to the boosters. Without both refinements it was necessary to reduce Shuttle payloads. Options A, B and C all featured long pressurised modules. It meant shortening them or sending them up empty and installing equipment later, or some combination. The European and Japanese partners complained about this in their evaluations of the Space Station Alpha options.[50]

Included in the Station Redesign Team's report of June 1993 were new cost estimates. However, none of the new design options fell within the maximum budget expenditure of $9 billion between 1994 and 1998 that the Clinton Administration had instructed NASA to follow. Over this period, the two versions of Option A were costed at $13.3 billion and $12.8 billion, Option B at $13.3 billion and Option C at $11.9 billion. If there was any good news it was that the estimates for Options A and B – the serious options – compared favourably with the 1992 figures for Restructured Space Station Freedom over the same period of $15.8 billion and $17.5 billion produced respectively by NASA and an independent source. The report also gave life cycle costs from 1994 through Permanent Human Capability to the end of ten years of operation. For the two versions of Options A these amounted to $35.7 billion and $35.2 billion; Option B came to $38.3 billion; and Option C came to $29.4 billion. In closing, NASA and its redesign team admitted that the costs exceeded the target figure but observed that 'the three options do represent the most effective approaches that could be determined'. It was just not possible to build Space Station Alpha, deliver it to orbit, achieve its life sciences and materials research objectives, keep to the timetable and do it all within the given budget. The project had reached an impasse.

On 10 June 1993, the Advisory Committee, in turn, reported to the President on the Station's redesign work.[51] On general issues, it saw the Station as an evolving programme that was part of the nation's high-technology infrastructure. It also saw it as an international cooperative venture requiring long-term multilateral commitments. It recommended that research project selection should make the most of the Station's long-term presence on orbit and suggested shifting it to 51.6° inclination to make it more accessible for Russian launches. On the redesigned options, it recommended Options A and C for further consideration. It thought that the Power Station platform phase was not worth pursuing and that the Human Tended Capability phase was of marginal benefit. It also advised that firm funding and commitments from the Administration and Congress was the best way to control the Station's long-term costs. On the international partners, it pointed out that their reaction to Option C had been understandably negative because it was a new and untested design.

The Advisory Committee reiterated that emergency crew return vehicles were essential. It was concerned about development risk due to problems and delays and over-reliance on the Space Shuttle with its vehicle weight burden. It was also critical of NASA's management of the project. Improved

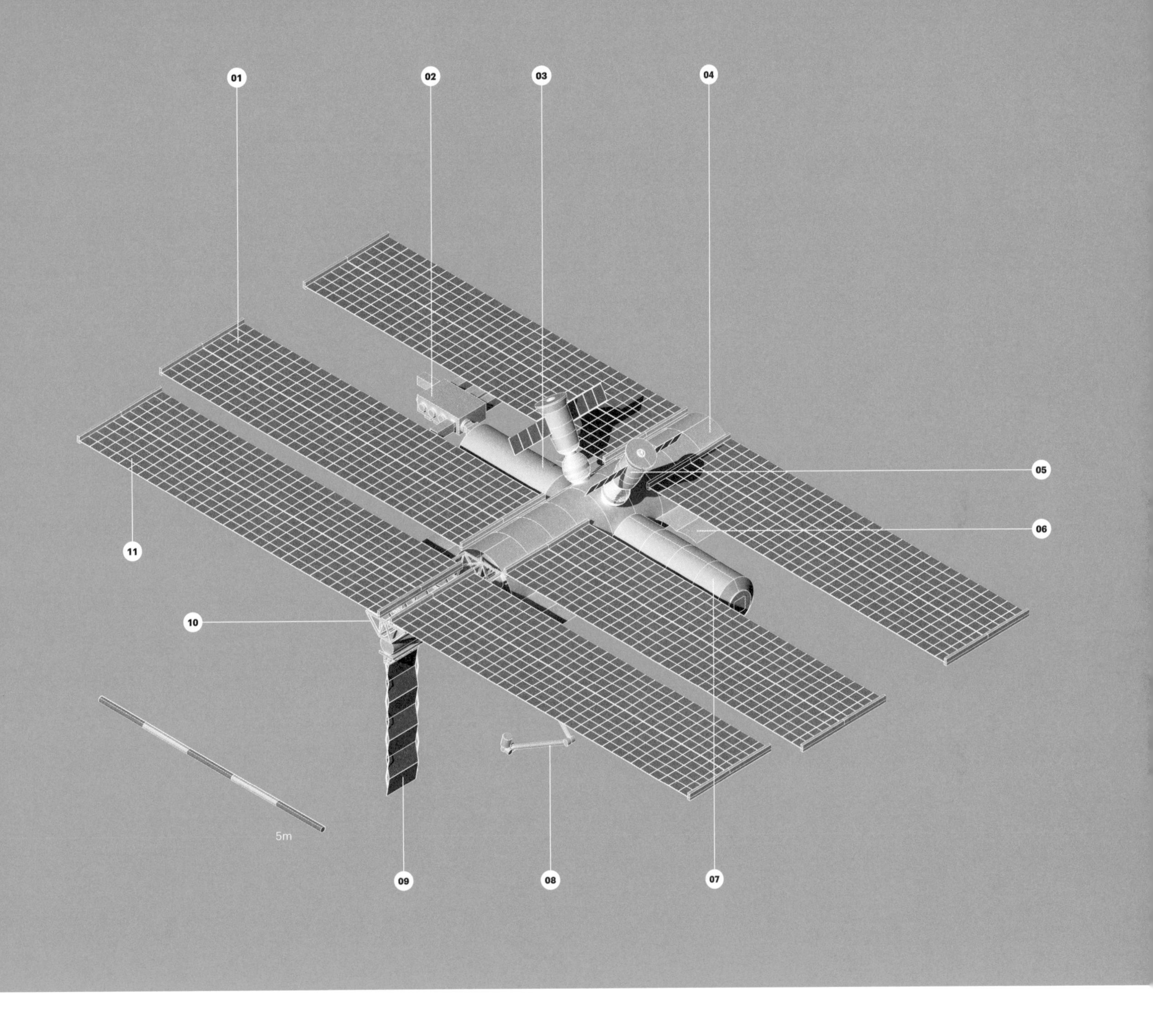

01 Body-mounted fixed solar array (4)
02 Japanese experiment platform
03 Japanese laboratory module (logistics module hidden from view)
04 US Shuttle-derived large cylindrical laboratory and habitat
05 Soyuz return vehicles (2)
06 Thermal radiator
07 European laboratory module
08 Robotic arm
09 Power radiator
10 External truss structure
11 Truss-mounted fixed solar array (2)

project management between NASA and its contractor teams was the key to cost reductions. The NASA Administrator must apply lean management techniques, overlapping and duplicated responsibilities must be eliminated, management layers reduced, programme responsibility centralised, staffing levels cut by 30 per cent and a single prime contractor should run the project. The report noted that: 'this will require great resolve on the part of NASA and the Administration.' Based on these recommendations, on 17 June 1993 President Clinton in his first major speech about space chose a version of Option A with the Station's Permanent Human Capability milestone delayed by two years in an effort to shave off more costs.[52]

However, the seemingly endless bad news about the Station's costs had given the project's Congressional opponents the ammunition to mount another attack. On 23 June 1993, the House of Representatives debated an amendment to an appropriations bill for NASA that would have eliminated all funding from the Station programme at a stroke. The amendment was sponsored by Congressman Tim Roemer from Indiana. Opening the debate, he spoke about the success of the Apollo missions, that 'in those days the nation was proud of NASA'. He said it was time to 'restore NASA's credibility' and that it was a mistake to cut off small science programmes for 'callous and cavalier' cost overruns. He wondered why NASA had to commit to the international partners and, in particular, Japan which had been unhelpful to America in other fields.

First to speak against the amendment was Congressman George Brown from California. He pointed out that the Station was the heart of the manned space programme and that to cancel it would amount to 'walking away from the legacy of Mercury, Gemini and Apollo'. Cancellation would result in a 'lost opportunity to establish a permanent presence in space'.' Congressman Tom Lewis from Florida accused the Station's opponents of a 'divide and conquer' campaign. He said that cutting the Station was equivalent to taking money from the education of future children. For just one-thousandth of the annual budget, the nation would have a space research capability that none could imagine. The debate wore on for three and a half hours before the House took a vote. There were 215 votes for the amendment and 216 against.[53] If the vote had gone the other way the consequences for America's space programme would have been catastrophic.

Rescue party
Russia Joins as Space Station partner, 1993

America and Russia had been cooperating on new space mission plans during 1992. In October that year, Dan Goldin and Yuri Koptev, his opposite number at the new Federal Russian Space Agency, completed negotiations for the flight of a Russian cosmonaut on the Space Shuttle, for the tour of American astronauts on the Russian Mir Space Station and for an Orbiter rendezvous and docking with Mir. A joint programme office was set up at Johnson Space Center to oversee the missions. In March 1993, Koptev wrote to Goldin with a proposal to incorporate the core module from the Mir 2 Station into Space Station Alpha's design. Russia had planned to replace Mir with a new Mir 2 Space Station after the end of Mir's life.[54] Mir 2 had been through several incarnations since first studied in 1989. Originally quite an ambitious design, it had been scaled back by 1992 after the demise of the Russian Buran space shuttle programme and the future of the replacement station was in doubt. It comprised a pressurised core module, a transverse beam (similar to NASA's Space Station truss), a docking module, a resource module and two laboratory modules – one for life sciences and another for materials processing. Progress vehicles would deliver the unpowered modules to orbit and Soyuz spacecraft would ferry crews up and down.[55]

The aim of Yuri Koptev's proposal was to help NASA with its Station budget headaches by suggesting that Russia should partner America in a new joint Space Station leadership initiative. Russia would contribute its Mir 2 modules as well as resupply flights using its Progress cargo vehicles. The net result would be major cost savings for NASA and the rebirth of a project that would benefit both nations.

In early April, President Clinton met Russian President Boris Yeltsin at a summit in Vancouver. Their discussions covered a range of issues including space cooperation. Then at a post-summit press conference Yeltsin declared that the two nations would join forces on a new Space Station initiative. The announcement was a complete surprise. The White House assured Congress and NASA that there was no firm decision. For one thing, there were political questions over Russian sales of missile technology and the White House wanted Russia to abide by an arms non-proliferation agreement in exchange for an invitation to join the Station project. As an incentive, America had offered Russia a $400 million contract for work on the Orbiter-Mir mission and Space Station Alpha. On 16 June, Russia signed two agreements in Washington covering non-proliferation. The following day, in his speech about space, President Clinton began to lay the groundwork for Russia to join the Space Station project. In July, a Russian delegation came to Washington to begin negotiations on how Russian hardware could be incorporated into the Station's design. The Station rose to the top of the agenda for a planned meeting between Vice President Al Gore and Russian Prime Minister Viktor Chernomyrdin in September and the American-Russian Station negotiations group came under pressure to achieve positive results for the meeting. On 2 September, Gore and Chernomyrdin signed an accord merging Space Station Alpha and Mir 2 to create the new Space Station.

To underscore the pivotal role that Russia would now play, the Space Station acquired a grander title. It briefly became International Space Station Alpha, then Alpha disappeared and it became the International Space Station – the name by which the world has known it ever since.

Dan Goldin and Yuri Koptev This photograph shows Dan Goldin (right) shaking hands with Yuri Koptev (left) at a signing ceremony in 1993. Yuri Koptev was Dan Goldin's opposite number at the Russian Federal Space Agency at the time that Russia joined the Space Station project. Koptev proposed to NASA that Russia would contribute two of its modules from its Mir 2 Space Station, no longer going ahead, as well as supply flights using Progress cargo vehicles. By this time, NASA had already chosen Russian Soyuz spacecraft for emergency escape vehicles for the Station. After Russia joined the project it was reincarnated and rebranded as the International Space Station.

In July, the Senate showed support for the new American-Russian partnership by voting forty-nine to forty in favour of continuing the project. Then in late November, Clinton and Gore met with key House and Senate members and obtained backing for the partnership. Support, however, was not universal and a clique in Washington continued to battle against the Station. Between 1991 and 2003, it survived twenty-two votes in Congress aimed at cutting off its funding. It was a measure of the strength of the new International Space Station with its obvious foreign policy benefits that motions to stop it were overcome. On 16 December 1993, NASA and the Federal Russian Space Agency signed a second accord with a detailed statement of work for the International Space Station. The next day Al Gore and Viktor Chernomyrdin issued a joint statement confirming Russia's acceptance of an invitation to join the project.[56]

It had been a tumultuous year for the International Space Station. When President Clinton entered the White House he inherited a project whose design was too ambitious, its budget too high, and its management disorganised. Its rescue came with Russia's involvement. Underlying this supreme irony was the message that, henceforth, teamwork by the leading space nations might be the only way to tackle the grandest space challenges as their complexity and cost would prohibit any nation from carrying them out alone. True, the Space Station had begun life as an international project, but with the exception of Canada whose robotic arm was vital, the partners could have dropped out and the Station would have carried on. The new American-Russian relationship marked a major policy shift. NASA had become an instrument of American foreign policy. It made possible an extraordinary example of détente when Sergei Krikalev became the first Russian cosmonaut to fly on the Space Shuttle in February 1994,[57] something that would probably have been unthinkable at the beginning of the previous year.

Unlike the single nations of Canada and Japan, Europe's participation in the Space Station involved ten nations led and orchestrated by the European Space Agency. Each of these nations to a greater or lesser extent had its own space agenda and some of them, such as France and Germany, were leading space nations in their own right. NASA kept the Europeans informed of what was going on with the Russians but that was about all. NASA told the European Space Agency that Russia's entry would not affect the intergovernmental agreement with the existing partners, or the bilateral memorandum of understanding between them. The Japanese and Canadians were given the same story. The existing partners saw the benefits of Russia's participation. They had witnessed the upheavals at NASA and there had been growing concern about the agency's ability to maintain control over the project. The Russians would bring stability.

Rising concerns about NASA and the Station had provided the impetus for European studies for an independent orbital infrastructure. The

The Russian module, Zvezda One of NASA's main arguments for bringing Russia on board the Station was that the lower fabrication costs of space hardware in Russia would help to bring down the Station's costs. NASA claimed that Russia's involvement would shave $2 billion off the Station's budget. This photograph taken in 1997 shows the fabrication of the Russian Zvezda module at the Khrunichev State Research and Production facility in Moscow. Zvezda was a critically important early building block of the International Space Station. In this photograph, Zvezda had just undergone a pressurisation test. Image: PD-USGov-NASA.

Apollo-Soyuz Test Project An artist's impression of the American Apollo spacecraft to the left about to dock with the Soviet Soyuz spacecraft to the right on the Apollo-Soyuz Test Project mission in July 1975. This highly successful mission paved the way for future cooperation between America and Russia on the Russian Mir Space Station and later the International Space Station. Image: R Bruneau and NASA.

European Space Agency had envisioned this comprising the Columbus Module berthed at the Station, the free-flying laboratory called the Man-Tended Free-Flyer that would periodically dock with the Station for servicing, and an independently orbiting Polar Platform. This triad would give Europe a considerable foothold on orbit. France had been developing and promoting a small spaceplane called Hermes that would reach orbit on the Ariane 5 launch vehicle and then fly back to Earth on its own like the Space Shuttle. As worries about the Station mounted and the idea of European independence in orbit gathered momentum, the European Space Agency decided that Hermes would service its Man-Tended Free-Flyer platform rather than the Station. Hermes, though, had disadvantages. Because it was so small, its payloads ferried to and from the European platform would be much smaller than the Space Station's racks and this would seriously restrict their value. The Ariane 5 launcher would need to be rated for human spaceflight to launch Hermes and this would seriously compromise its commercial capability. Then there was the Hermes spaceplane itself, which was proving to be technically marginal and expensive to develop. Both Hermes and the Man-Tended Free-Flyer were later phased out.[58] There was also concern about the Columbus Module's rising cost that later led to the shortening of its length, a fate that also befell NASA's Destiny Module. On the positive side, the Columbus Module's final cost of about €700 million came in not far over its original €658 million non-negotiable fixed price. Despite a delay of five years in the delivery of Columbus to the Station caused, in turn, by the delayed launch of the Russian Zvezda module and the Shuttle Columbia Orbiter disaster, the Columbus Module came in close to its cost target.[59]

A subsequent assessment of America's cooperation with Russia on the Space Station listed five benefits from the American viewpoint. First, the Station would provide employment on a civilian space project for engineers and scientists in Russia as an alternative to projects that may be military in nature. It would also supply incentives for the Russian government and industries to comply with non-proliferation measures such as missile technology control. It would help to build bridges between American and Russian institutions and industries in the aerospace sector. It would allow the American government to inject hard cash into the Russian economy to bolster stability and growth and support the new Yeltsin administration. Finally, it would help Russia to maintain its highly respected human spaceflight programme, a symbol of its superpower status.[60]

One of NASA's strongest arguments for bringing Russia on to the project revolved around the budgetary benefits that would result. Russia would provide Soyuz capsules that would release America from the need to develop its emergency crew return capsule and Progress vehicles that would periodically resupply the Station and relieve the Space Shuttle of some of that duty. Russia would provide a range of habitation, laboratory

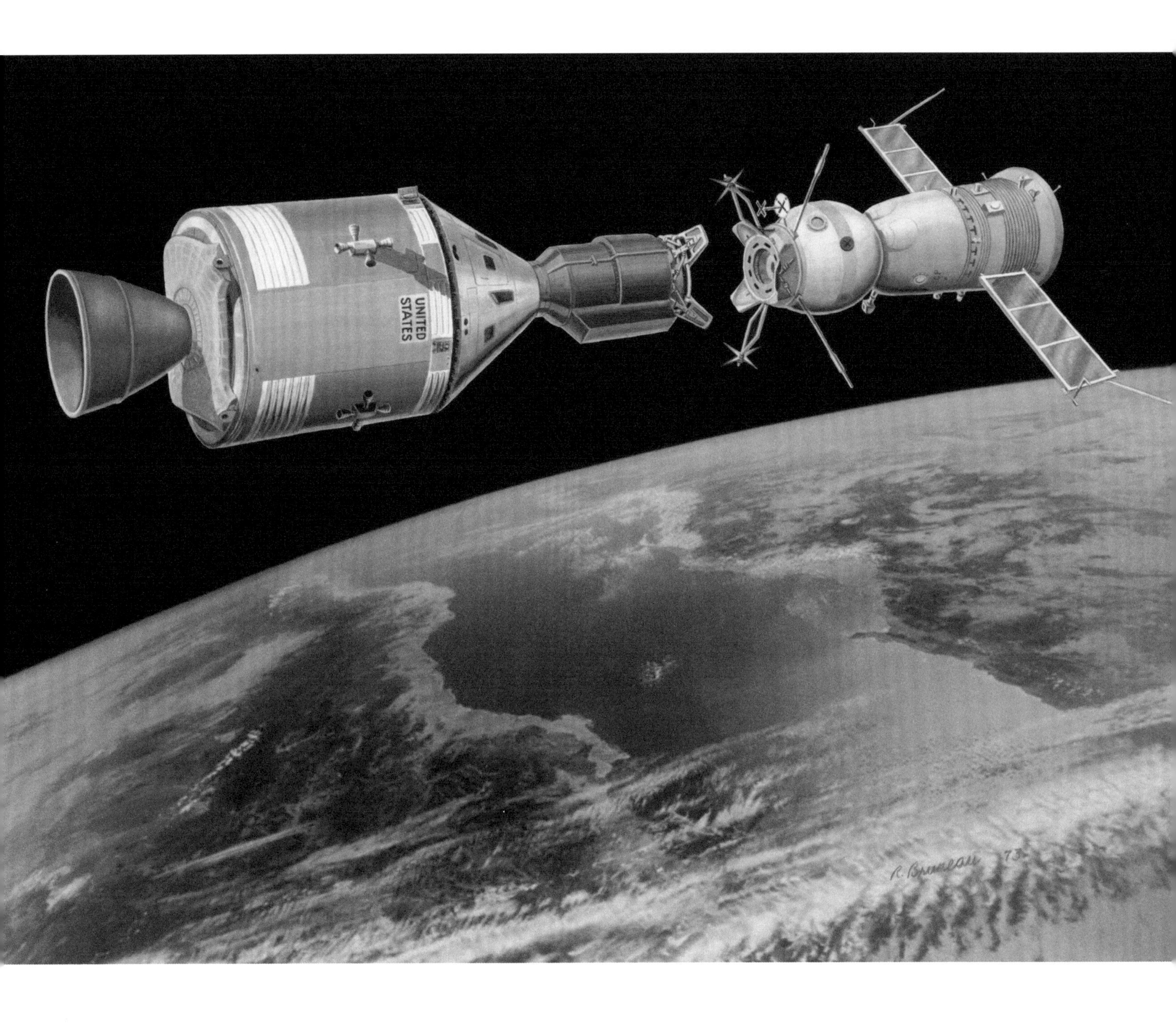
UNITED
STATES

and docking modules and solar arrays, all of which would help to take pressure off NASA's shopping list of Station building blocks. Russian space hardware was cheaper to fabricate than its American counterparts and NASA could pay Russia to supply some of it. NASA claimed that Russian participation would reduce the Station's funding needs by $2 billion. In September 1993, NASA estimated that under an annual funding cap of $2.1 billion imposed by the Clinton Administration, the chosen Space Station Alpha design would have cost $19.4 billion between 1994 and 2003. Russian participation would reduce this figure to $17.4 billion in 1993 dollars.

Appendix 4 shows the two figures of $17.4 billion and $19.4 billion in the context of the Station's changing development and fabrication costs from its go-ahead in 1984 up to the start of its assembly on orbit in 1998.

NASA listed other valuable benefits. Russian involvement would boost the crew size from four (reduced from the earlier eight) back up to six at completion of the Station's assembly, bring forward the launch of the first American building block to December 1997, advance Station assembly completion to June 1992, enable dual access to the Station by American and Russian spacecraft, improve the Station's research capabilities and increase on-board resources such as crew time, electrical power and pressurised volume. However, NASA's claimed savings were disputed by the Government Accountability Office, which countered that the $2 billion of savings would be mostly offset by $1.4 billion of increased funding needed to pay Russia for some Station hardware, pay for missions by the Space Shuttle to Russia's Space Station Mir, the impact of the Station's new orbit on the Shuttle and more Shuttle flights to support Station assembly. Further, $0.6 billion of the claimed savings was not necessarily due to Russian participation as it concerned reductions that, as yet, were unresolved. The Government Accountability Office concluded that 'when all Space Station elements are considered, current estimates would indicate that much of the savings NASA attributes to expanded Russian participation will not be achieved'.

Naturally NASA disagreed.[61] Set against the backdrop of past, present and projected costs for the Space Station, however, the contested $2 billion paled into insignificance. NASA's new $17.4 billion Station development estimate for 1994 to 2003 included neither Space Shuttle launch costs nor associated civil service salaries over the same period. Adding these raised the figure to $47.9 billion. Then there was the money spent by NASA on Station design and development work over the decade to 1993, which came to $11.4 billion. There were also the estimated operating costs for the completed Station over a ten-year period which came to $13 billion.[62] Summing all these figures up produced a new life-cycle bottom line cost for the International Space Station of $72.3 billion of American money in 1993 dollars. This did not include the financial outlays of all the other partners.

Appendix 3 shows the International Space Station's initial and final crew sizes in the context of the crew size fluctuations during the Station's design and development years. Though NASA's claim that Russia's involvement would increase the Station's crew size to six, it would not be until 2009 – nine years after the first three-person crew visit – that the Station would reach the six-person crew milestone.

The 1988 Intergovernmental Agreement among the Station's partners had provided for any of them to use the lack of availability of funds as a reason for reconfiguring the project despite each signatory's commitment 'to make its best efforts to obtain approval for funds to meet its international obligations'.[63] Though the unstated target of this provision might have been Europe and some of its participating nations, it was NASA that invoked the provision with the shambles of its Station budget and its invitation to Russia to save the project. If that provision had not been in the Agreement, the consequences for the International Space Station might have been dire indeed. In the end it was the project's salvation.

Détente and dominance

American and Soviet-Russian collaboration on Apollo-Soyuz and Shuttle-Mir, 1969-1997

One cannot overstate the importance to the International Space Station's fortunes of early American-Soviet collaboration on the Apollo-Soyuz programme in the 1970s and later American participation in the Russian Mir Space Station in the 1990s, up to the start of Space Station assembly missions. Without their successes and the experience, trust and goodwill that resulted, the Clinton Administration might not have had the confidence to invite Russia on to the project as a senior partner. An enduring miracle in the history of spaceflight is that America and the Soviet Union were able to brush aside their global differences and collaborate so successfully on the Apollo-Soyuz mission in the midst of the Cold War. Overcoming the political and cultural challenges of collaboration on both projects was an achievement equal to mastering the complexities of spaceflight that both demanded. The détente strengthened the dominance of both nations over the human spaceflight field up to the International Space Station and the present time, a shared supremacy that is now being challenged only by China. The significance of the Apollo-Soyuz and Shuttle-Mir programmes for the International Space Station was so great that it is instructive to turn from the Station's story to consider the achievements and the lessons learned.

The Apollo-Soyuz Test Project was the name given to the first jointly crewed space mission of the United States and the Soviet Union in 1975. Its genesis lay in contact that opened up in 1969 between the NASA Administrator and the President of the Soviet Academy of Sciences at that time. Exploratory talks in Moscow in 1970 led to the suggestion that the two agencies might develop a single docking mechanism for use on space stations planned by both nations. To get technical discussions going,

International Space Station human spaceflight heritage The human spaceflight heritage of the International Space Station began in 1961 with the first cosmonaut and astronaut flights to orbit by the Soviet Union and America respectively. The two nations then embarked on different space paths. The Soviet Union pursued a series of crewed orbital outposts beginning with Salyut 1 and ending with Salyut 7, launched respectively in 1971 and 1982. Meanwhile, America carried out a series of crewed exploration missions to the Moon beginning with the Apollo 11 landing in 1969 and ending with the Apollo 17 landing in 1972. America then followed these up with its first crewed orbital outpost Skylab in 1973. America and the Soviet Union collaborated on the Apollo-Soyuz mission in 1975. America then developed the Space Shuttle system that first flew in 1981 while the Soviet Union – later Russia – built the world's first proper Space Station, Mir, that first launched in 1986. Russia became a partner in the International Space Station in the early 1990s and launched the first piece of it to orbit in 1998.

Vostok I
1st human spaceflight
Yuri Gagarin
12 April 1961

Salyut I
1st Soviet orbiting laboratory
19 April 1971

Salyuts 3-7
3rd-7th Soviet orbiting laboratories
1974–1982

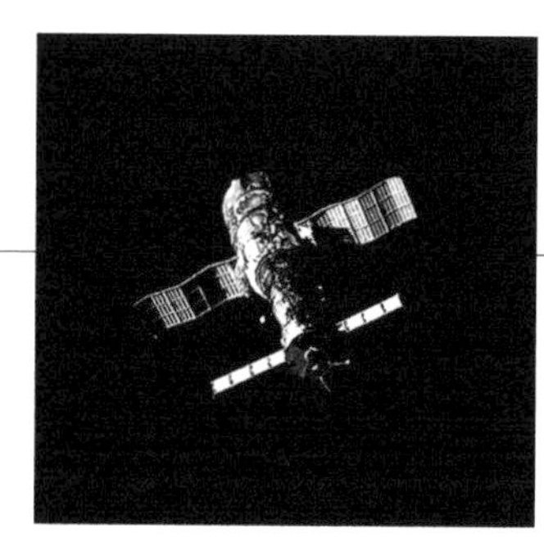

Freedom 7
2nd human spaceflight
Alan Shepard
5 May 1961

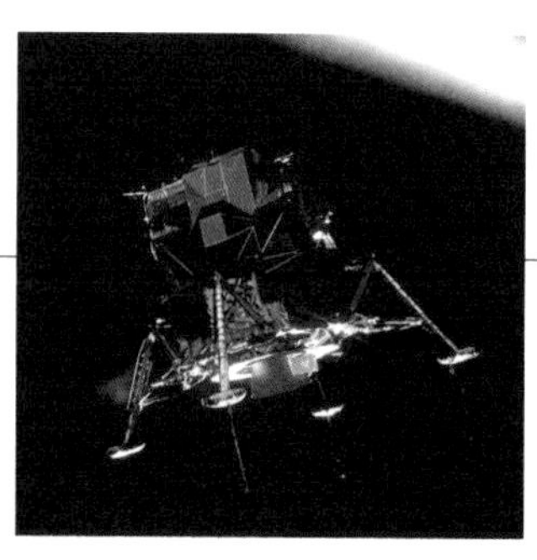

Apollo 11
1st lunar landing
20 July 1969

Skylab
1st American orbiting laboratory
25 May 1973

Mir
1st space station
1st launch 10 February 1986

Apollo-Soyuz
Joint USA-USSR mission
17 July 1975

Space Shuttle
1st mission with Columbia
12 April 1981

International Space Station
1st element launch
20 November 1998

NASA floated the idea that a Soviet spacecraft might rendezvous and dock with Skylab, planned to launch in 1973. NASA would install Soviet docking equipment on Skylab for this purpose. Then in a speech in July 1970, President Nixon voiced his interest in pursuing discussions on space cooperation with the Soviet Union. NASA began outline studies of a joint mission and a design team came up with two scenarios using a common docking mechanism. The first was a demonstration of a rescue mission using either an Apollo or a Soyuz spacecraft to assist a disabled spacecraft of the opposite type and the second was a mission to test rendezvous and docking procedures. Soyuz proved too small to accommodate a rescued American crew so the first scenario revolved around Apollo as a vehicle to rescue a Soviet crew, which halved the intended dual rescue capability. On the other hand, testing a universal system for rendezvous and docking on a joint mission could benefit Skylab or a future Space Station.[64]

USA-USSR space cooperation President Richard Nixon of the United States on the left and Premier Alexei Kosygin of the Soviet Union on the right sign an agreement on science and technology cooperation during a bilateral summit in Moscow in May 1972. This agreement paved the way for the successful Apollo-Soyuz Test Project of the mid-1970s and was the genesis of decades of space cooperation between the rival superpowers that eventually led to the International Space Station. It is a relationship that has endured and prospered despite the ebb and flow of political differences and diplomatic sparring between the two nations over the years.

Whatever the mission scenario, one of the first tasks was to compare the Apollo and Soyuz spacecraft. Soyuz was capable of either automatic or manual rendezvous and docking using radar and attitude control system participation by the target vehicle. It could fly unpiloted or with one, two or three cosmonauts. Crew transfer from one Soyuz to another was by spacewalk, a procedure that the Soviets had already tested. By contrast, Apollo rendezvous and docking was a manual operation without any target vehicle participation. The Apollo command module normally seated three astronauts but this number could be reduced or its cabin could be modified to accommodate a mixed crew of five astronauts and cosmonauts, though this had never been done. Transfer between Apollo command and lunar modules had been by an internal passageway after the removal of docking gear. There was also a difference in cabin pressures. Soyuz operated at one Earth atmosphere while Apollo operated at a lower ambient pressure. A conference in Moscow in October 1970 saw the exchange of technical information and the formation of working groups to study the compatibility of rendezvous and docking procedures, guidance and navigation systems, communications, docking mechanisms and docking tunnels. In January 1971, NASA representatives went to Moscow to move the project forward, having been given a free hand by the White House.

Proposed was the joint development of a rendezvous and docking system for use with Apollo and Soyuz spacecraft. NASA had produced two basic concepts for the docked vehicles. The first of these was minimal with the Apollo command module docking directly with the Soyuz orbital module. Soyuz would then be depressurised for atmospheric equalization before hatch opening and the entry of two astronauts into Soyuz. After connecting umbilicals, the Soyuz orbital module would repressurise to full atmosphere for the duration of the astronaut visit and then depressurise for their return to Apollo. Pre-breathing would have to occur before both depressurizations. The second concept called for the addition of a 'go-between' module to act as a variable pressure airlock. After rendezvous and docking, astronauts would enter the new module at the lower pressure. The pressure would then rise to a full atmosphere before hatch opening with Soyuz and the crew visit. The astronauts would pre-breathe before returning to the new module that would then depressurise to enable them to re-enter Apollo.

By June 1971, NASA had a clear idea of the hardware required to support an Apollo-Salyut link-up – Salyut having replaced Soyuz as the preferred Soviet vehicle. The agency began work on the cost and schedule for a single mission and gave a contract to North American Rockwell for a detailed study of all the elements needed. NASA specified five functions for the new airlock module. It would serve as a structural connection between the two docking mechanisms on each side of it; it would provide atmospheric adjustment between the Apollo and Salyut vehicles; it would offer a habitable environment to the crew for the time that they occupied it; it would house communications equipment operating at the Soviet frequency; and it would provide added volume for an Earth resources survey. Its overall length would be about 2.8 metres and internal diameter about 1.4 metres with a proposed hatch diameter of 0.9 metres.

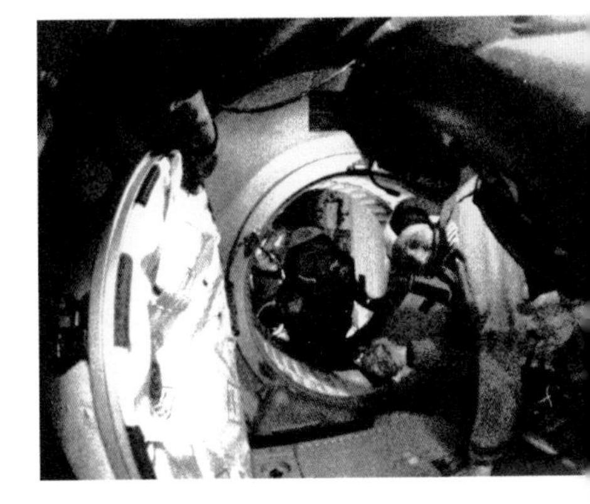

Apollo-Soyuz Test Project Apollo commander and astronaut Thomas Stafford on the right shakes hands with Soyuz commander and cosmonaut Alexei Leonov after the historic docking of their two spacecraft on the Apollo-Soyuz Test Project mission on 17 July 1975. This mission was the first time that America and the Soviet Union – two political adversaries on the world stage – had teamed up on a joint space mission. This photograph was taken from a 16mm film.

By November 1971, North American Rockwell had produced a detailed design. That same month a NASA team went to Moscow to confer with their Soviet colleagues. An important outcome was agreement in principle on a two-ring docking mechanism with a set of guides that intermeshed when the rings came together. Shock absorbers would dampen the impact and structural latches would lock the vehicles together. Each side had slightly different ideas about the detailed design but agreement was reached.

In February 1972, NASA imposed a budget ceiling of $250 million on the project. The two sides met again in Moscow and produced four sheets of American-Soviet engineering drawings of the docking mechanisms, which needed just minor fine-tuning before their handover to the contractor. NASA meanwhile began to focus on the necessary political approval. First, there was a need for Congressional authorisation and funding appropriation before fabrication could begin. A bilateral agreement between the United States and the Soviet Union would have to precede the Congressional process. By April, NASA's senior management was again in Moscow, poised to reach an agreement with their Soviet counterparts on a technical proposal, organisational plan and project schedule, items that were considered essential to support NASA's recommendation to the White House on the mission's feasibility. The Soviet team also reverted to a Soyuz rather than Salyut vehicle for the mission to which the American team had no objection. NASA proposed a document called *Summary of Results*, which contained seventeen points. Negotiations over these proved difficult at first but both sides again reached an understanding. All was now ready. At their Summit in May, President Nixon and Soviet Premier Kosygin signed a space agreement with the joint Apollo-Soyuz mission as its centrepiece.[65]

Apollo-Soyuz Test Project The American Apollo spacecraft as seen from the Soviet Soyuz spacecraft during the joint Apollo-Soyuz Test Project mission of 1973. The docking module at the front of the Apollo spacecraft in this photograph was specially developed by NASA to enable the two different spacecraft to dock together safely in space.

NASA formed a management team to run the project and sent a statement of work to North American Rockwell with authorisation to proceed under a sole source contract. The work covered modifications to an Apollo command module that was already built and the engineering, fabrication and assembly of a docking module, a docking system and a support structure, together with ground testing of the various elements. American and Soviet teams agreed that a joint training programme for the crews and flight controllers would be best so that both crews would be familiarised with each spacecraft. By the end of 1972, engineers in both countries were moving the project forward with an intensive schedule of reviews and meetings taking place in Houston and Moscow. By October 1973, all the shared design work on the new docking module and system was completed. North American Rockwell had begun fabrication of the docking module, completed fabrication of the first of five docking system sets for tests and carried out modifications to the Apollo command module. The first round of docking seal tests began the following month and went well.

NASA's public relations office in Washington now began to take an active interest in the project. NASA had a policy of full disclosure of all its space activities and with the Apollo-Soyuz mission there was a tremendous opportunity to show détente at work to maximum advantage. Live television coverage of the mission would be essential. The Soviets, however, were more secretive, normally only issuing press releases after the completion of a mission. They did not indulge in extensive television coverage. The public disclosure issue was added to the agenda for the mid-term review meeting. The two sides reached agreement just before the mission's launch.

With the mission launch date just two months away, a flight readiness review took place in Moscow in May 1975. It was vitally important that all went well. To begin with, the mission operations team had completed the principal flight plan for a 15 July launch as well as a later back-up plan in case of a delay due to weather or equipment problems. There would be twin control centres – one in America and the other in the Soviet Union – with each responsible for its own spacecraft and crew. Interpreters would be on hand at each location. The guidance and control group had mounted the American-designed docking equipment on the outside of Soyuz to provide the incoming Apollo module commander with his sighting target. There were final adjustments needed to procedures for controlling Apollo's attitude during the final few seconds before docking. The docking hardware team reported that tests had been carried out successfully. There was, however, a problem with the design of American-made alignment pins and sockets, which could jam under some circumstances during docking. American and Soviet hardware changes were made to deal with this issue. The communications and tracking team had completed its work and reported no problems. The life support and crew transfer group had concerns over the fire safety of equipment or materials transferred

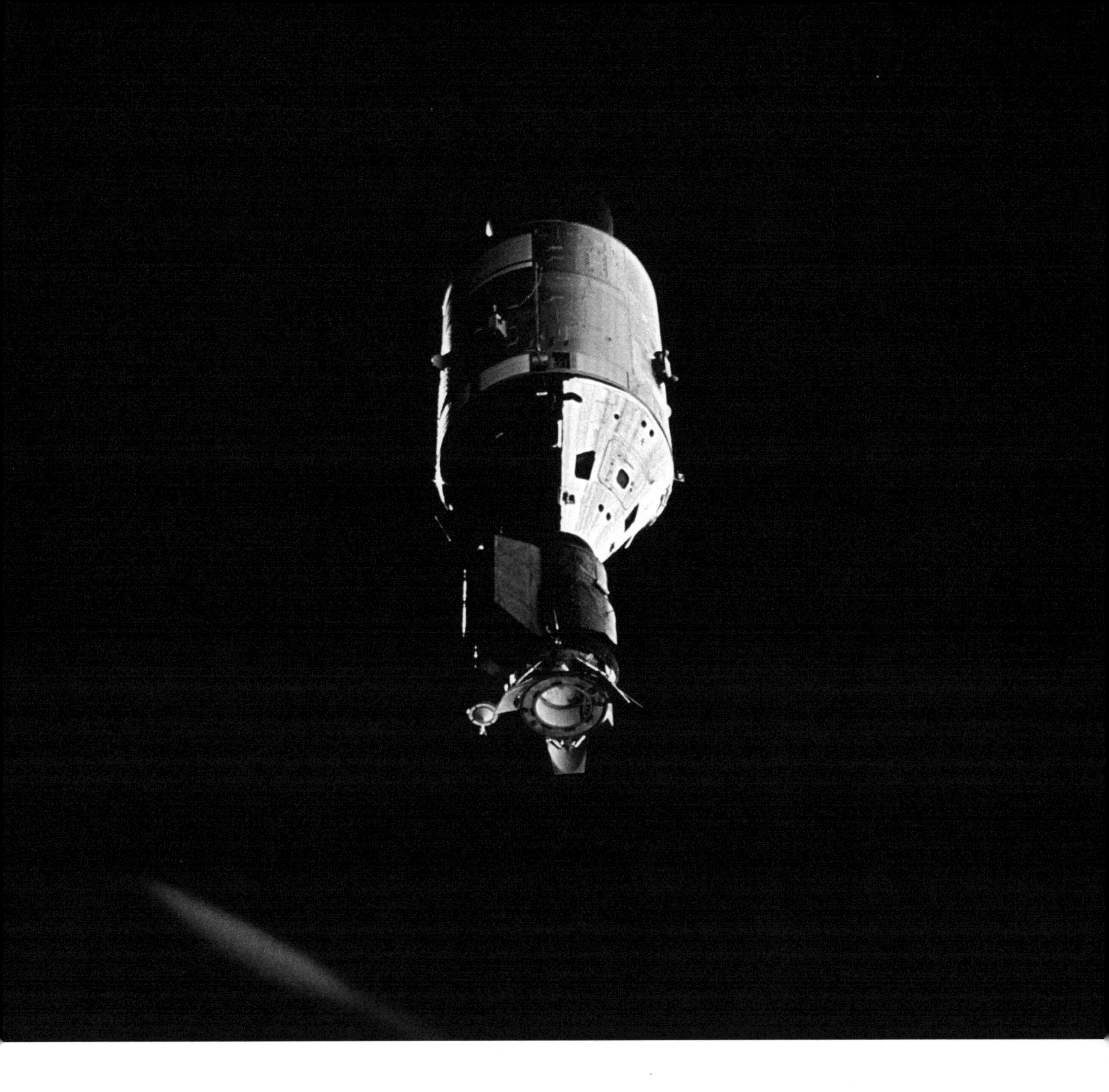

Atlantis

from one spacecraft to another because Soyuz had an 80 per cent nitrogen to 20 per cent oxygen atmosphere while Apollo had a 60 per cent oxygen atmosphere at a lower pressure. Soviet cosmonauts would have to wear clothing made from a flame-retardant material specially developed for the mission to eliminate the use of potentially risky wool and cotton garments in the high oxygen environment of Apollo's cabin.

Though the meeting had gone well, there was a residual concern about the lack of a single commander in space or a single flight controller on the ground with sole authority to act in an emergency. Each nation had its own commander and controller and there was no attempt by either side to exert its will over the other.

On 15 July 1975, shortly after 15.00 Moscow time, the Soyuz spacecraft with two cosmonauts on board lifted off from the Baikonur Cosmodrome. The Apollo spacecraft with three astronauts on board launched the same day at 14.50 East Coast time from Kennedy Space Center. Apollo now had to adjust its orbital flight path and velocity over several phases to place it on the right track to catch up for its rendezvous with Soyuz. The two spacecraft established radio contact early on 17 July when they were still over 200km apart. The gap reduced to 35km about one hour, forty minutes later and about one hour after that Apollo commander Thomas Stafford performed a perfect docking with Soyuz. Later, after passing through the docking module, Stafford opened the hatch into Soyuz and shook hands with its commander Alexei Leonov. It was to become one of the most famous handshakes in history. Shortly before, Soviet General Secretary Leonid Brezhnev had sent his congratulations to the cosmonauts and astronauts on the mission. He referred to the cooperative friendship between the Soviet and American designers, engineers, cosmonauts and astronauts who had participated in the project. He noted that 'one can say that Soyuz Apollo is a forerunner of future international orbital stations' and that 'the détente and positive changes in the Soviet-American relations have made possible the first international spaceflight'.[66]

American and Soviet crew visits to each other's spacecraft lasted a couple of days and undocking of Soyuz and Apollo took place on 19 July. Both crews returned to the ground safely. The mission had been a complete success. It had taken America and the Soviet Union just six years from the birth of an idea for a joint mission to the handshake of an astronaut and a cosmonaut in space for the world to see on television. It was an outstanding achievement and model of bilateral space cooperation that no other space programme has surpassed.

Leonid Brezhnev's remark that Apollo-Soyuz was a forerunner of future international orbital stations was prophetic. The Soviet Union went on to develop and launch its Space Station Mir beginning in 1986, a project that deserves the title of the world's first true Space Station (there is no easy translation for the Russian word 'mir' into English – it can mean world,

Mir Space Station On 29 June 1995, Atlantis became the first Shuttle Orbiter to dock with the Russian Mir Space Station on Space Shuttle Mission STS-71. The Mir-19 crew took the photograph of Atlantis docked to Mir on 4 July 1995. Mir-19 commander and cosmonaut Anatoliy Solovyev and flight engineer and cosmonaut Nikolai Budarin had temporarily undocked their Soyuz TM spacecraft from Mir to perform a brief fly-around. Photo: Anatoliy Solovyev and Nikolai Budarin.

01 Progress-M cargo vehicle
02 Mir core module
03 Priroda Earth sensing module
04 Spektr power module
05 Soyuz-TM crewed vehicle
06 Kristall technology module
07 Docking module
08 Kvant-2 augmentation module
09 Kvant-1 astrophysics module

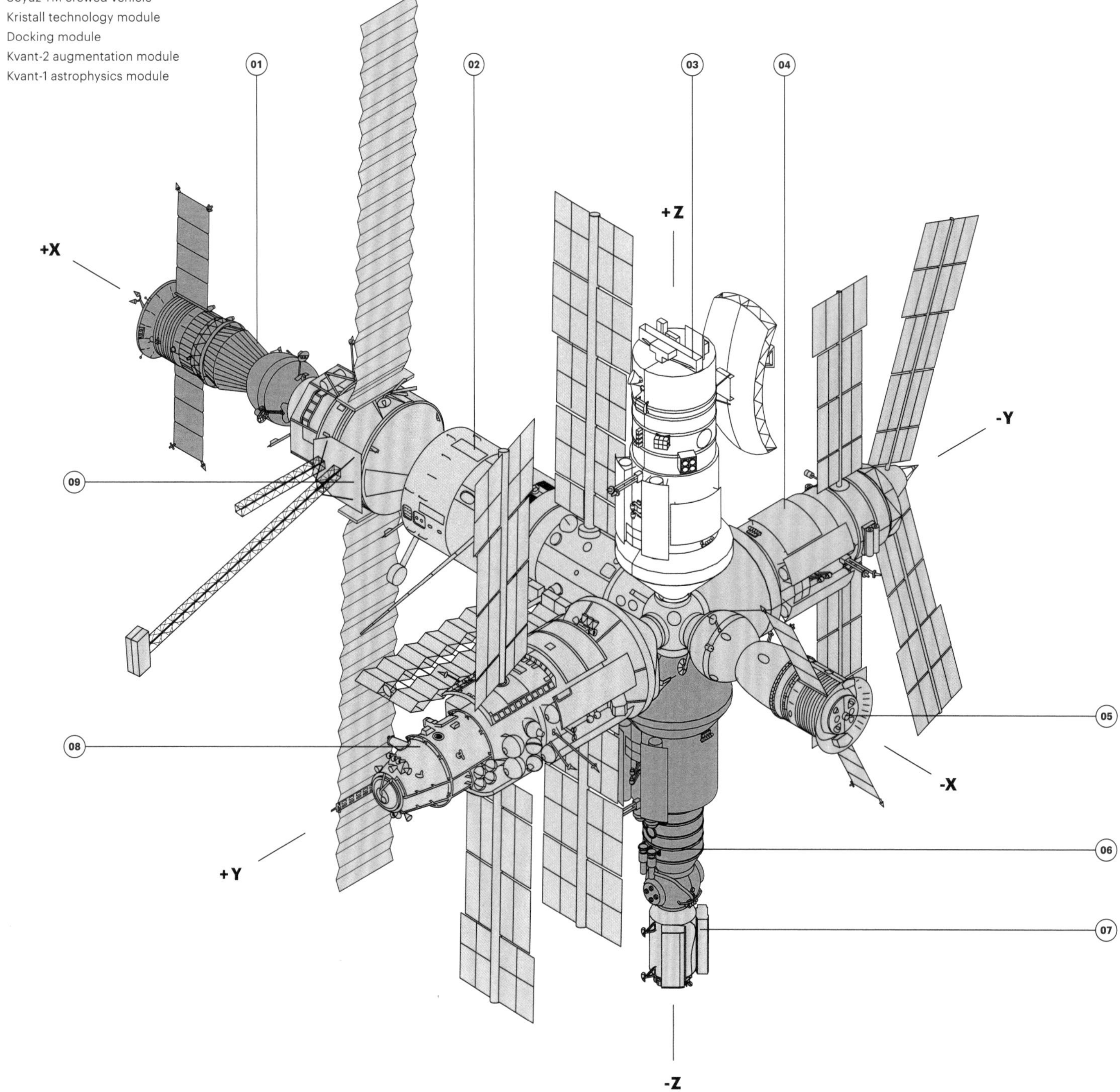

peace, community or gathering). The Soviet Union and America teamed up again on Mir, when America became substantially involved in flying missions and astronauts to it.

Like Apollo-Soyuz, Shuttle-Mir – as it was called in America – was a highly productive partnership. Following the success of Apollo-Soyuz, talks between NASA and the Soviet Academy of Sciences took place in October 1976 about follow-up projects. These led to another agreement between them in 1977 on more collaboration in the human spaceflight field with the main aim being joint Shuttle-Salyut flights. That same year, Secretary of State Cyrus Vance and Soviet Foreign Minister Andrei Gromyko renewed the 1972 agreement that had led to Apollo-Soyuz and endorsed the new agreement. Shuttle-Salyut talks continued and Johnson Space Center proposed a Shuttle-Salyut mission in 1978 as the Space Shuttle programme approached its first flight. However, American worries about technology transfer risks halted progress. NASA Administrator Robert Frosch wrote to President Alexandrov at the Soviet Academy of Sciences explaining that the complexity of the issues involved was the problem. Time passed and it was not until 1990 that contact opened up again when Vice President Dan Quayle discussed space cooperation with Soviet President Mikhail Gorbachev and representatives of the Soviet Academy of Sciences. In 1991, a leading Soviet industry minister named Oleg Shishkin met with Quayle to proposed an ambitious cooperative programme, including a Shuttle rendezvous with Mir, which was due to succeed the Salyut programme. Though this proposal was rejected, Presidents Bush and Gorbachev signed a space cooperation agreement in Moscow. It called for a flight by an American astronaut on a Soyuz spacecraft to the Mir Space Station for a stay of up to six months and, in exchange, a flight by a cosmonaut on a Shuttle-Spacelab mission. This led to the establishment of an American-Soviet joint working group on human spaceflight and opened up space cooperation contact at the highest government levels in both nations.

In 1991, the new Russian Federation emerged from the imploding Soviet Union. The following year America and Russia renewed a space cooperation agreement and issued a joint statement on space cooperation. It called for a Russian cosmonaut to fly on a Shuttle mission, American astronauts to undertake tours of duty on Mir and a Shuttle docking with Mir. The statement also paved the way for the American commercial purchase of Russian space services. In July 1992, America and Russia began discussions on the development of a Russian docking mechanism to enable Shuttle Orbiter visits to Mir. NASA Administrator Dan Goldin met with the Federal Russian Space Agency's General Director Yuri Koptev in July and October and they signed an implementing agreement on human spaceflight cooperation.[67] It was a simple document of just twelve pages that laid out concisely what both space agencies would do. First, a Russian cosmonaut would join an American crew to fly on Space Shuttle Mission STS-60 that

Mir Space Station The Mir Space Station comprised seven building blocks delivered to orbit and assembled in the following order: Mir core module, Kvant-1 module, Kvant-2 module, Kristall module, Spektr module, docking module and Priroda module. The Proton-K launch vehicle delivered them all except for the docking module which was ferried up in the Space Shuttle. Assembly of Mir began in February 1986 and ended in April 1996.

was scheduled for launch in November 1993. The selected cosmonaut would train at NASA as a mission specialist and a further cosmonaut would train as a back-up. Then an American astronaut would join a Russian crew on a Mir mission that would exceed ninety days to provide him or her with long-duration experience. There would also be a back-up astronaut. Next, the astronaut's tour on Mir would coincide with a Shuttle Orbiter rendezvous and docking with Mir in 1994 or 1995. The Orbiter would transport two cosmonauts to Mir and exchange them with two cosmonauts returning to the ground. Finally, American and Russian engineering teams would proceed with the development of an androgynous docking mechanism to provide the Orbiter with a new docking capability. There were also provisions for further crew exchanges, transportation of experiments and supplies and spacewalk activities. Each agency would fund its own participation.[68]

In April 1993, President Clinton met Russian President Boris Yeltsin at a summit in Vancouver, Canada, where both agreed to an expanded Shuttle-Mir programme. Further discussions between the two sides resulted in a plan for two three-month tours and four six-month tours of American astronauts on Mir up to 1997, with Soyuz spacecraft and the Space Shuttle delivering the astronauts to Mir as well as American experiments for installation in Mir's Spektr and Priroda modules. Also planned were joint spacewalks to help to extend Mir's lifespan to 1997. Then in September, Vice President Gore and Russian Prime Minister Victor Chernomyrdin chaired the first meeting of a new American-Russian commission on energy and space and agreed to begin a new era of Space Station cooperation. Phase 1 of this was the Shuttle-Mir agreement expanded to include up to two years of total astronaut time on Mir. Phase 2 was the International Space Station. America had opened the door of the International Space Station and invited Russia to enter.

Sergei Krikalev became the first Russian cosmonaut to fly on the Space Shuttle with the launch of the Orbiter Discovery on the STS-60 mission on 3 February 1994. The same day, NASA announced the names of the first two astronauts to go to Russia to train to join the Russian crew of Mir. Norman Thagard was the primary astronaut and Bonnie Dunbar was his back-up. American-Russian groups organised along the lines of those formed for Apollo-Soyuz and began work on joint mission management, cargo and scheduling, public affairs, safety, science, integration, spacewalks and medicine. NASA approached the scientific community for ideas for experiments to be carried out by its astronauts on Mir and appointed a director of operations to be based in Russia. The first meeting at Johnson Space Center of a task force on Shuttle-Mir rendezvous and docking took place in May. In June, Vice President Gore and Russian Foreign Minister Chernomyrdin met again in Washington and announced a new space contract between America and Russia that had just been signed by their agency heads Dan Goldin and Yuri Koptev. It provided for $400 million in

American expenditures, up to about two years of additional astronaut flight time on Mir and up to nine more Shuttle Orbiter dockings with Mir facilitated by a new docking module. It also provided for American use of Mir's Spektr and Priroda modules for experiments, a joint American-Russian research programme on Mir and the extension of Mir's lifetime to enable all these activities to take place. It was to be a highly ambitious undertaking.

Shuttle-Mir became a formal and independent space programme in October 1994, a testament to the attention that Shuttle-Mir had received at the highest political levels in both nations and the importance that each had attached to its success. Not all went smoothly, however. The collapse of the Russian rouble in October 1994 disrupted the financial health of the Russian space programme and the launches of the Spektr and Priroda modules were delayed until 1995 and 1996 respectively after the first two astronauts had become Mir residents. Meanwhile on 3 February 1995, the Orbiter Discovery lifted off on Shuttle Mission STS-63 and became the first American spacecraft to visit Mir. Its purpose was not to dock with the Russian Space Station but to carry out a rendezvous and perform a fly-by to test approach techniques and communications and navigation systems. Despite problems with a propellant leak, Discovery came to within 11 metres of the Russian Space Station. It was the Orbiter Atlantis that docked with Mir on STS-71. Launched on 27 June 1995 with six astronauts and four cosmonauts on board, Atlantis made a virtually perfect docking on 29 June.[69]

The Shuttle-Mir partnership went on to encompass a further eight Space Shuttle visits to Mir. A total of seven American astronauts carried out tours of duty on Mir and accumulated nearly 1,000 days on board until the programme ended in 1998, the year of the launch of the first International Space Station building block to orbit. As Phase 1 of America's long-term partnership with Russia ended, Phase 2 began and the International Space Station took over where Shuttle-Mir left off.

There were several incidents on Mir during its lifetime. American astronauts were on board when it experienced two such emergencies – a fire and a collision. In February 1997, an oxygen-generating canister that was part of the life support system caught fire in one of the modules, filling the interior with smoke. It was serious because the fire's location was on the route to one of the Soyuz modules that comprised one half of Mir's crew escape capability. The canister burned itself out and the fire was extinguished. Then in June, an undocked Progress supply spacecraft collided with Mir. An arriving Progress vehicle was about to replace its counterpart docked at the Station. Moscow ground controllers had instructed Mir's commander to use the old vehicle to test a new teleoperated remote docking system. It used a camera on the vehicle to project a changing image of Mir's docking port on to a video screen for the remote pilot as he guided the vehicle in. However, Mir's image

on the screen was poor and lagged behind the closing rate of the vehicle as it came in to dock. By the time that the commander realised that the speed of Progress was too great, it was too late to take corrective action. Progress collided with a solar array on Mir's Spektr module and then hit the module and punctured it. This was dangerous because Mir's internal atmospheric pressure began to drop as air escaped through the hole. The crew readied itself for another possible evacuation but it managed to close Spektr's hatch to seal it off. Mir was now out of power because the crew had disconnected electrical cables to close Spektr's hatch. The collision had also knocked Mir into a spin. The repairs needed to bring Mir back to normal after this accident were arduous and time-consuming.[70]

The programme manager for Apollo-Mir, Frank Culbertson, later summed up the main lessons learned. 'You're going to have things happen that are going to require problem-solving continuously. The best-laid plans, the best-designed systems – you're still going to have difficulties. Long-duration spaceflight is hard and the most valuable lesson we learned from that is to expect it to be difficult. Plan for that. Train for that. And then, be prepared to handle the unexpected.'

From an American viewpoint, the Shuttle-Mir partnership achieved several goals. It gave NASA experience of working with an international partner on Space Station operations on Earth orbit. This involved American astronauts and Russian cosmonauts training in each other's languages and taking part in each other's operations, much as they had done on Apollo-Soyuz in the 1970s. Lessons learned during the Shuttle-Mir cooperation helped to reduce risks. The Russian-built and American-integrated docking system made dockings easier and safer and NASA gained its first experience of docking its Shuttle Orbiters with a Space Station. American crew member experiences on Mir also led to the introduction of emergency controls, better fire precautions, stronger lighting and other improvements on the International Space Station. Shuttle-Mir also gave American astronauts experience of long-duration space missions for the first time since Skylab in the 1970s. They accumulated more time in space than all the Shuttle missions added together up to that time. Lastly, astronauts were involved in scientific research on Mir that covered over 100 experiments in eight scientific fields, another valuable prelude to the work they would be doing on the International Space Station.

Mir Space Station The Russian Space Station Mir with Earth as a backdrop taken from the Shuttle Orbiter Atlantis after undocking from Mir at the end of the STS-71 Shuttle mission on 4 July 1995.

International Space Station on its Y axis
This side-on, Y axis view of the Station shows a docked Shuttle Orbiter on the right and the European Automated Transfer Vehicle at the far end on the left. Each solar array pair has two rotating joints (alpha and beta joints) that attach it to the Station and enable it to track the Sun as the Station revolves around the Earth on orbit.

Close-up of the Russian Zvezda and Zarya modules This photograph of the Station from an aft-above position was taken from the Orbiter Discovery during its fly-around of the Station after undocking on Shuttle Mission STS-114 in August 2005. The view shows the two large Russian modules on the Station in full sunlight. At the top is the Zvezda Service Module. At the bottom is the Zarya Control Module. Zarya was the first Station building block launched to orbit in November 1998.

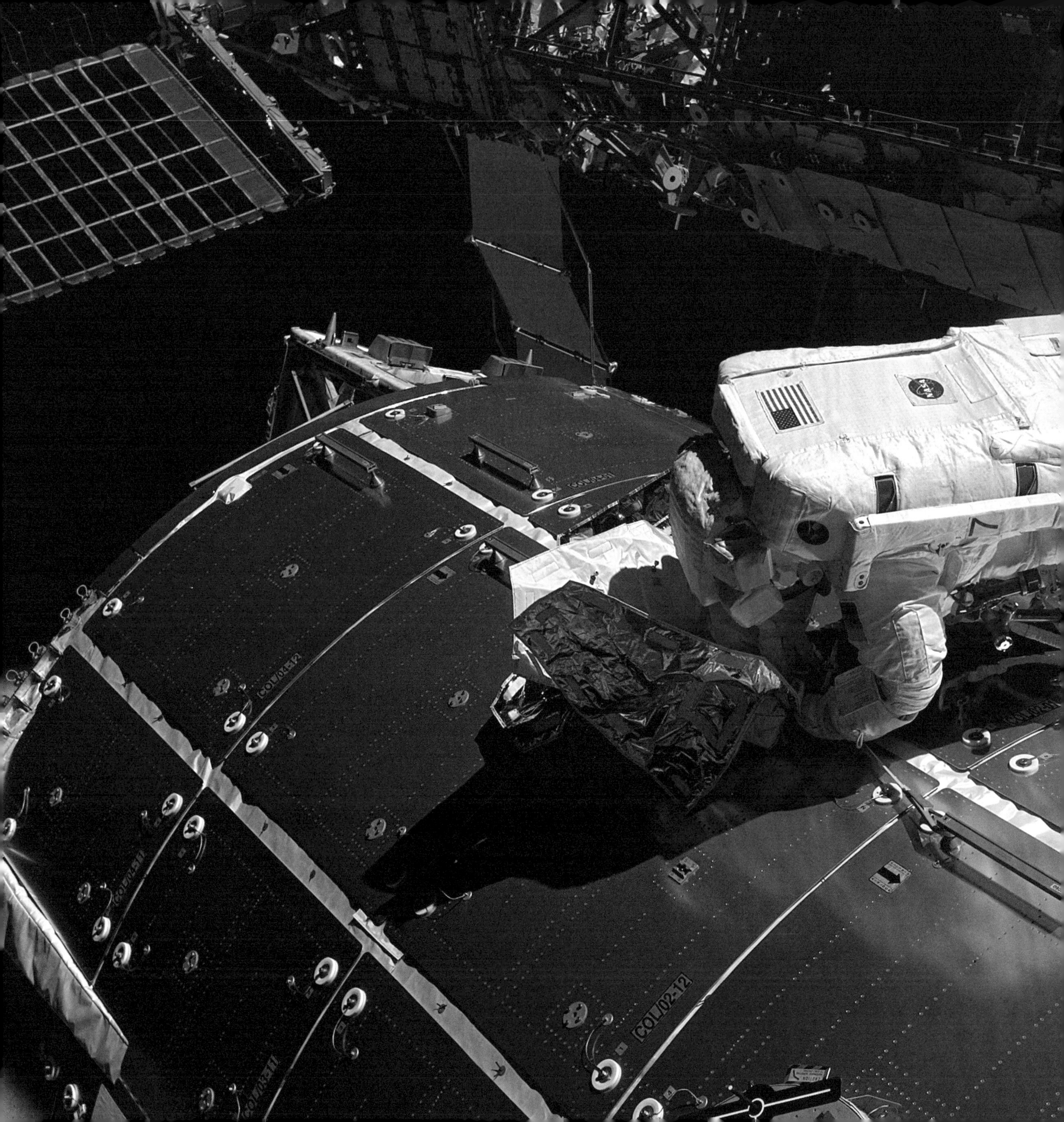
COL/03-12
COL/02-12

The European Columbus Module Astronaut Hans Schlegel works on the exterior of the European Columbus Laboratory Module after its delivery and attachment to the Station on Shuttle Mission STS-122 in February 2008.

Anticipation and preparation 1994–1998

Project renaissance
International Space Station, 1994

The Space Station that President Reagan had given the go-ahead to a decade before had survived. It had been an eventful ten years during which NASA had kept the project on track through three White House Administrations and eight of its own Administrators. From a design viewpoint, the Station's evolution had been a roller-coaster ride. After a cautious start in 1979, it had picked up speed and climbed to precarious heights with the 1984 Power Tower and 1985 Dual Keel designs. From there the tracks led downward. The 1987 Revised Baseline design gave way to the 1988 Space Station Freedom design and then to the 1991 Restructured Design as, pushed and pulled out of control by its heavy cost burden, the project's descent into a budgetary abyss had accelerated. Hard braking with the Space Station Alpha designs had not worked and the Station had almost crashed in the Senate in 1993. Russia had sent out a rescue party and had helped to save the project.

Work on the new International Space Station began almost immediately after Al Gore and Viktor Chernomyrdin signed their accord in September 1993. Boeing started metal forming and welding work on the first node module just days after the announcement that Russia was joining the project.[1] By early 1994 a new Station design had emerged. Bigger and

The International Space Station with the Shuttle Orbiter Endeavour and the Automated Transfer Vehicle Johannes Kepler This photograph of the Station was taken in May 2011 by astronaut Paolo Nespoli from a departing Soyuz spacecraft that was returning to Earth. On the left at the forward end of the Station is the docked Shuttle Orbiter Endeavour. Endeavour's robotic arm extends over the vehicle's underside for an inspection procedure. Projecting out at the bottom right and appearing bright white with an 'X' pattern of solar arrays is the docked European Automated Transfer Vehicle named Johannes Kepler.

1979–83
Space Station concept studies begun by NASA and industry

1981
First flight of the Space Shuttle

1984
NASA Space Station grand concept studies

1984
President Reagan gives project go-ahead and invites international participation

1984
'Power Tower' reference Space Station design

1985
'Dual Keel' Space Station design succeeds the 'Power Tower' design

1984
Scientists criticise the 'Power Tower' design for poor research conditions

1986
Loss of Space Shuttle Challenger and its crew

Design Evolution of the International Space Station The Station's design took fifteen years to evolve from various initial concepts to the International Space Station. NASA began to explore designs in 1979 just before the start of the Space Shuttle era and then widened these to include aerospace industry studies up to 1984 when President Ronald Reagan announced the project. Two highly ambitious designs – the Power Tower and the Dual Keel – emerged between 1984 and 1985. From then on, gradual descoping of the project caused by constant budget overruns resulted in the Revised Baseline design in 1987 that became Space Station Freedom, the Restructured Design in 1991 and the Space Station Alpha designs in 1993. Despite all these changes, the Station retained its cluster of international modules and long transverse truss that supported the solar arrays and other equipment. Its range of capabilities, though, diminished over the fifteen years to the single role of a research laboratory. This role expanded after Russia joined the project in 1993.

more impressive than before, it was now full of Russian building blocks. NASA began contracting out the fabrication work and provided Boeing with a temporary funding stream until the company was appointed as prime contractor for the American portion of the Station.[2] In January 1995, NASA signed a $5.65 billion contract with Boeing. This enabled the Station to pass the Phase C milestone with its design complete and ready to fabricate, integrate, verify and proceed to full fabrication in Phase D.

One of the first issues that NASA and the Russian Federal Space Agency had to agree on was the Station's orbital inclination relative to the equator. It was a concern that had been under discussion between the two agencies since the Soyuz spacecraft had been chosen as Station crew escape vehicles. Russia had originally planned to operate Mir 2 in a 65° inclination orbit. This high orbit would have carried Mir 2 over much of Russia's land area and avoided overflight of most of the newly independent former Soviet republics to the south. NASA, on the other hand, wanted a 28.5° inclination orbit. This low orbit was ideally suited to launches from Cape Canaveral in Florida and made the most of the Space Shuttle's lift capabilities, being closer to the equator. It was virtually impossible for Russian launches to reach the 28.5° orbit because changing inclination would require a lot of propellant to push their spacecraft out of one orbital path into another. The 28.5° orbit would also cause Russia problems with tracking its spacecraft as

1987
'Revised Baseline' descoped Space Station design with keels omitted to reduce assembly effort and costs

1988
Space Station 'Freedom' design development contracts signed with Industry

1991
Space Station 'Restructured' design to reduce costs

1993
White House instructs NASA to perform 90-day cost-cutting redesign resulting in Space Station Alpha alternatives

1993
Russia joins the project at President Clinton's invitation and the project is reborn as the International Space Station

1990
Project is well over budget

1992
Project is still well over budget

1993
Bill in Congress to kill the Space Station defeated by one vote

their tracking stations were so much further north. NASA and the Russian Federal Space Agency settled on 51.6° for the Station's orbital inclination as it could be reached by launches from both countries. Nevertheless, it was a disadvantage for Russia as the Station would overfly only 7 per cent of Russia's land area and it would miss Moscow. The Space Shuttle too would be at a disadvantage because its maximum payload weight would be reduced.[3] However, to mitigate the negative impact on the Shuttle's performance, NASA had reduced the weight of the External Tank and this would help to maintain the Shuttle's payload lift capability to the 51.6° orbit.[4]

The new International Space Station design born in 1994 was somewhat similar to Option B of the Space Station Alpha studies of early 1993, but bigger. The truss was horizontal and perpendicular again to the flight path with its ends pointing out to port and starboard. The modules were generally horizontal, at right angles to the truss and in line with the flight path. The first major change was the increased length of the truss with a fourth solar array framework added at the starboard end. The truss and solar array frameworks were now symmetrical in position on each side of the truss's centreline. The second major change was the number of modules and their arrangement. The Station now had seven primary modules and eleven secondary modules; five of these connected end to end to form a horizontal spine in line with the flight path. The spine of modules intersected with the

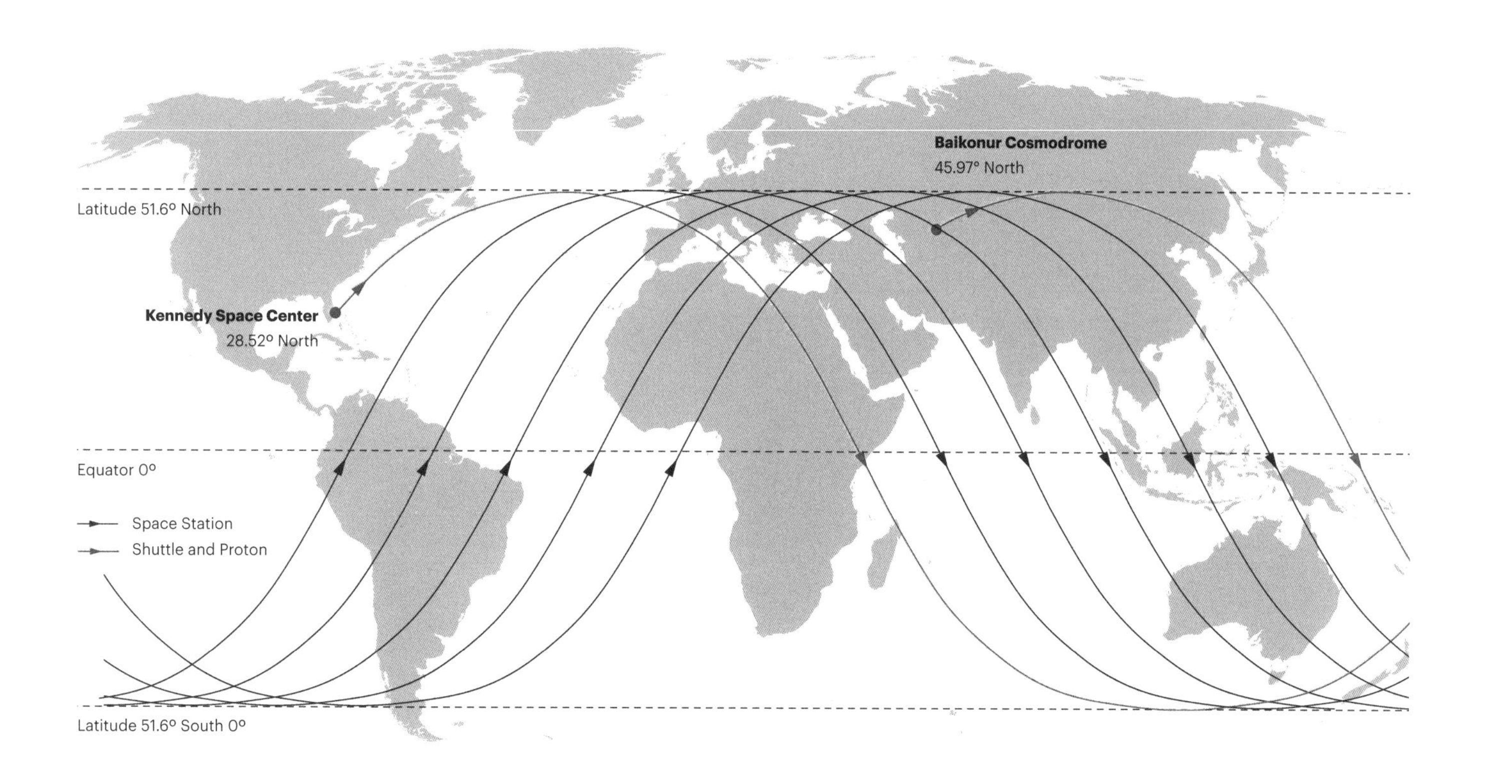

International Space Station's orbit America would launch its Station building blocks from Kennedy Space Center in Florida. Kennedy could launch vehicles into a 28.5° inclination orbit that was reasonably close to the equator. This low inclination orbit was advantageous as easterly launches benefited from the boost to vehicle ascent performance caused by the Earth's eastward rotation rate that peaked at the equator. Russia, on the other hand, would launch from Baikonur Cosmodrome that was much further north than Kennedy. Its vehicles could not reach low orbital inclinations without using up considerable quantities of propellant to shift their orbital paths sideways. This would be impractical and wasteful. NASA and the Russian Federal Space Agency agreed on a 51.6° orbital inclination for the International Space Station. It was accessible to launches from both sites and the Station would overfly large swathes of the Earth's surface. The Station's 51.6° orbit would prove to be very useful for Earth observation. Image: author.

truss's centreline to resemble the letter T if viewed from above with the T's post extending forward a little beyond its crossbar. Connected to the spine modules and pointing out in different directions were the other primary and secondary modules, hardware elements and the crew return vehicles. At its Assembly Completion stage the Station would have a pressurised volume of 1,200m³, a mass of 419 tonnes, maximum electric power of 100kW and overall dimensions of 108.4 metres wide by 74 metres long, looking down on it from above.

Most of the new modules were Russian and two of them together formed the core of the new design. These were the Zarya (Russian for sunrise) Control Module and the Zvezda (Russian for star) Service Module. It was vital that the core modules could, first, place themselves on the correct orbit and be able to maintain position and manoeuvre there and, second, provide basic habitation and life support systems to enable early crew visits. The Zarya Control Module would be the first Station building block to arrive on orbit in 1998. Its final location at the Station would be towards the aft end of the module spine.

Zarya's heritage dated back through the Mir Space Station to the beginning of the Soviet Space Station programme in the early 1960s and the development of a common module called the Universal Block Module (referred to in Russian as the FGB). The block module had a standardised design with a module structure, a pressurised chamber with a breathable atmosphere, a rendezvous and docking system, a main engine, thrusters and an attitude control system, propellant tanks, a power system with solar arrays, a guidance, navigation and control system, and a thermal control system with radiator panels. A cylindrical structure encircled the pressurised chamber with the deployable solar arrays, thermal radiators and propellant tanks arranged around the outside of the cylinder and a docking mechanism at one end. It was capable of adaptation into three categories of vehicle. These block modules were used to build transport logistics spacecraft (referred to in Russian as TKS), space station modules and space tugs. Modifications included engine location, size and number of propellant tanks, power system capacity, solar array design, internal layout, pressurised volume and the type of modules attached at the aft end. The hardware was capable of mating on the ground before launch with other module types and it also provided a rendezvous and docking capability to arriving spacecraft once on orbit. The Universal Block Module became the foundation of a generation of Soviet spacecraft, beginning with the development of the Almaz military space outpost in 1964 through the Salyut space outpost series and ending thirty years later with the Zarya module on the International Space Station. [5] The Zarya Control Module was highly capable and versatile, and the ideal building block with which to begin the Station's construction.

After the Soviet Almaz outpost came the transport logistics spacecraft based on a Universal Block Module twinned with a conical capsule for three cosmonauts, called Merkur (Mercury in Russian). Merkur was attached to the block module's aft end through the capsule's heat shield with an access hatch at its centre. Launched together on a Proton rocket, the block module and capsule could be controlled automatically and independently from the ground. The capsule could separate from the block module and return to the ground while the block module could remain on orbit or dock with a space station. This arrangement led to a series of spacecraft missions named Cosmos that continued until the mid-1980s. Cosmos spacecraft began to dock with the Soviet Salyut outposts in the mid-1970s and continued up to Salyut 7 in 1991 after which the Salyut programme ended, to be succeeded by the much larger Mir Space Station.[6] The Universal Block Module formed the basis of the designs of the Kvant-2 (quantum in Russian) and Kristall (crystal) modules on Mir. Kvant-2 was a scientific and airlock module that supported research in fields of biology and Earth observation as well as spacewalks via the airlock. It supplied Mir with drinking water and oxygen and performed attitude control and power distribution functions. It had a crew shower and washing facilities. Kristall's focus was on astrophysical and biological research as well as materials production technology. It had a radial docking port and was used to dock the first visiting Shuttle Orbiter on STS-71 in 1995.[7]

NASA funded Zarya's development through Boeing as prime contractor under Boeing's $5.65 billion contract. Boeing subcontracted its construction

1994 International Space Station design
This artist's impression of the 1994 International Space Station shows the influence on the Station's design of the Russian additions. In this view, the new Russian sector lies just behind the transverse truss and between the port and starboard solar arrays. The dominant Russian element is a zenith-pointing tubular power module with fanned solar arrays at its upper end.

to Khrunichev space centre in Russia. Zarya was a fully self-contained spacecraft equipped for independent orbital operations in accordance with its Universal Block Module origins. Controlled from the ground, it would have its own propulsion and orbital stationkeeping capabilities enabling it to place itself precisely on to the Station's 51.6° orbit. Berthed to Zarya's nadir port was a Docking and Stowage Module, also Russian, with internal stowage volume, a docking port and solar arrays.

Permanently docked at Zarya's aft port was the Zvezda Service Module at its docking node forward port. Zvezda was a major module that had its origins in the base block module on the Mir Space Station (known as DOS-7 in Russian) that, in turn, was descended from the Salyut space outposts. The Salyut missions began in 1971 with Salyut 1, the first crewed outpost on orbit around the Earth. Following Salyut 1 was Salyut 2 in 1973 that failed, then Salyut 3 and Salyut 4 in 1974, Salyut 5 in 1976, Salyut 6 in 1997, and Salyut 7, which was operational from 1982 to 1991.

The Salyut module had a conical transfer compartment at its forward end that was replaced in the Mir base block module by a spherical docking and berthing node with five ports to be used for the docking or berthing of other Mir modules. The forward longitudinal port of the spherical node was equipped with utility plug-ins for transferring propellant and water from Progress supply spacecraft. The longitudinal port at the aft end of the module provided the docking point for the shorter Kvant module. Twin solar arrays projected on each side of the module and were much bigger than those on the earlier Salyuts. The base block module's pressure hull was made up of two cylindrical chambers of different diameters attached to each other end to end. The hull was made of chemically milled aluminium sheet, about 2mm thick, welded to webs that were 4mm thick. The hull thickness increased to 5mm in the spherical node. Attached to the curved outer surface of the large chamber was a thermal radiator. The module was clothed in a thermal blanket of about twenty-five layers of aluminised Mylar protected on the outside by a material similar to Kevlar. Inside the module was a pressurised living and working volume in two sections that corresponded to the large and small cylindrical chambers. The cosmonauts had two private sleeping compartments that were formed as pockets in the curved hull of the larger chamber. Next to these was a personal hygiene compartment with a lavatory. Storage compartments and drawers occupied most of the residual inner curved hull surfaces of the chambers. There were several portholes including one in each of the crew cabins and one in the floor for Earth observation. There was also a physical exercise area with a velo-ergometer (exercise bicycle), a treadmill and a medical cabinet.[8]

Zvezda was known as the Service Module. Like Zarya, it was a highly capable module, but one that would provide the Station's early living quarters and its initial life support, communications and data processing, electrical power distribution, flight control and propulsion systems.

Like its predecessor on Mir, its accommodation included sleeping quarters, a personal hygiene compartment, a galley with a refrigerator-freezer and a table for meals. Spacewalks using Russian spacesuits would be possible using Zvezda's transfer compartment as an airlock. Zvezda had an aft port for visiting spacecraft that would approach from the rear of the Station.

Berthed to two of Zvezda's other ports in its docking node was the Science Power Platform, pointing to zenith, and the Universal Docking Module pointing to nadir, both also Russian. The Science Power Platform comprised a pair of fan-shaped solar arrays mounted on top of a thick mast with a thermal radiator mounted on the side. The Space Shuttle would deliver it to orbit where it would provide additional electrical power and attitude control on the Station's X-axis. The Universal Docking Module was a node module with five ports for the berthing of other modules or the docking of visiting spacecraft approaching from beneath the Station. Berthed to it were two Research Modules pointing out to port and starboard and a Docking Compartment facing aft, also Russian. The twin Research Modules had externally mounted payload points and would provide Russia with internal and external experiment facilities. The Docking Compartment would enable spacewalks by Russian crew members. Zarya, Zvezda, the Science Power Platform, the Universal Docking Module, the twin Research Modules, the Docking Compartment and the Docking and Stowage Module together comprised the new Russian sector of the Station.[9] It was a very impressive contribution. The arrangement of the Russian sector would provide considerable flexibility for docking visiting Soyuz and Progress spacecraft, enabling several of them to visit the Station at the same time. This flexibility was essential as two three-seat Soyuz spacecraft would always be docked at the Station to be ready for the evacuation of its six-person crew in an emergency.

Berthed to Zarya's forward port on its docking node was the Pressurized Mating Adapter-1. This was the long name given to a short American-built tunnel piece shaped like a truncated cone that acted as the link between the Russian sector and the American, European and Japanese sectors of the Station. Moving along the spine, the forward end of Pressurized Mating Adapter-1 connected to the aft port of Node-1, the first node module to arrive on orbit. Node-1 (named as Unity) was built by Boeing and was one of three node modules on the American side of the Station. It had six ports – one at each conical end and four located at 90° radial intervals around its cylindrical shell. Berthed to Node-1's port on the port side was the Cupola, whose faceted windows would offer direct observation of the Earth and the Station's robotic operations on the port side. Berthed to Node-1's starboard port was the Joint Airlock (named as Quest). The Joint Airlock was a Boeing-built element that would enable spacewalks by the Station's crew members wearing either American or Russian spacesuits. Berthed to Node-1's nadir port was Node-3. Like Node-1, the Node-3 module (named as Tranquility)

Pressurized Mating Adapter The American and Russian sectors of the International Space Station were like separate wings of a mansion built in different styles: there was a need for a passage between them that would be architecturally compatible with the unique characteristics of each. The problem was solved on the Station with a small building block called a Pressurized Mating Adapter that mated with different hatches on American and Russian modules, providing an access passage for the crew between the two sides. There are three of these building blocks on the International Space Station. This photograph shows one of them undergoing pre-launch processing at Kennedy Space Center.

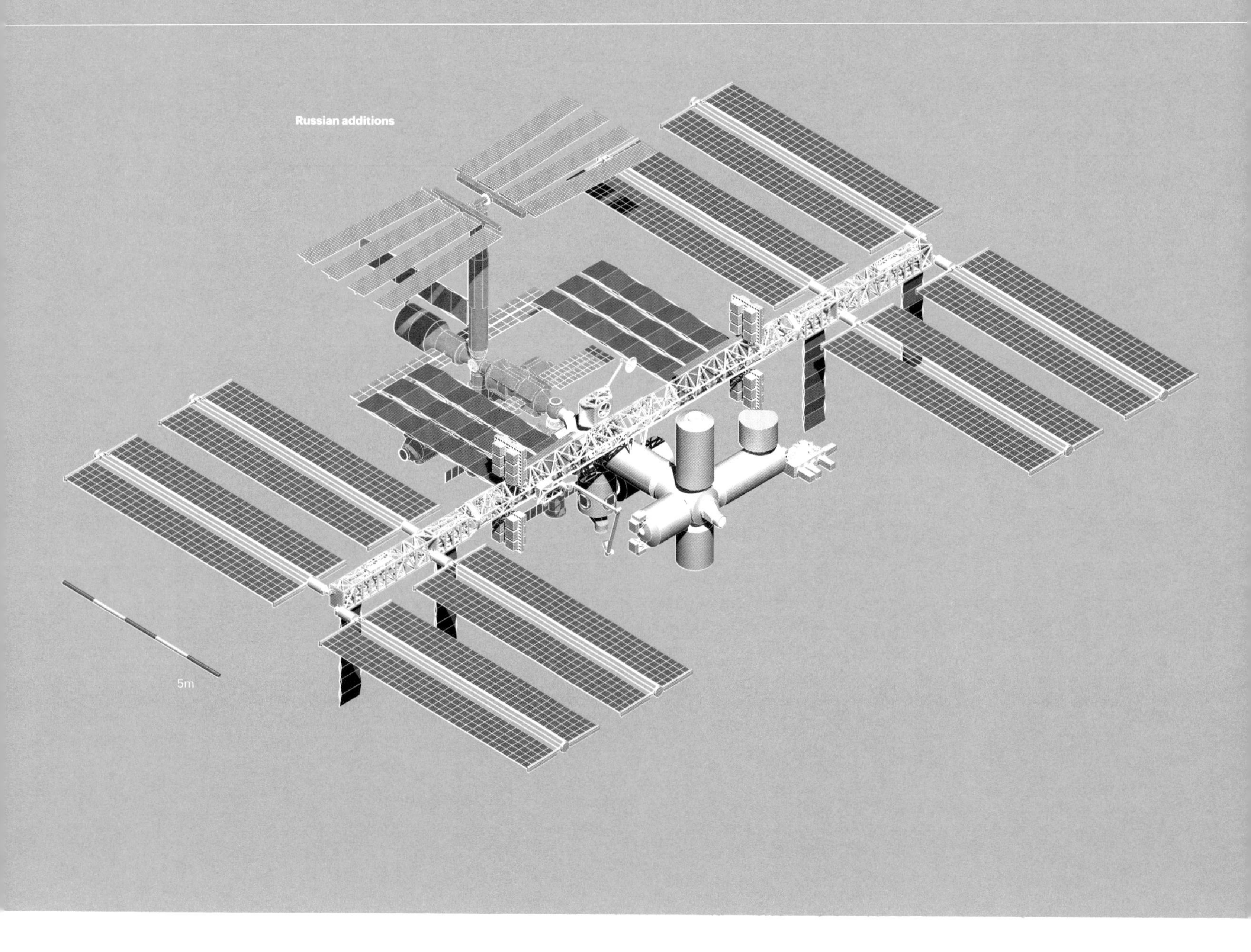

1994 International Space Station design – assembled and disassembled The first major change with the new International Space Station's design concerned the truss and the solar arrays. The new design returned the long transverse truss to its earlier alignment on the port-starboard axis. In the now-discarded Space Station Alpha Options A and B, NASA had aligned the truss on the flight path. NASA also added a fourth pair of solar arrays so that these were now symmetrical either side of the Station's fore-aft centreline. The Station now resembled the Eight Man Crew Capability configuration of the 1991 Restructured Design. The second major change concerned the addition of the new Russian sector at the aft end of the Station's module cluster. Russia's planned additions would comprise eight new building blocks including one American-funded pressurised module. There was now a single fore-aft spine of modules with the American and Russian sectors joined together at about the spine's midway point. In the illustration showing the assembled Station, the Russian additions are coloured red. The impact of these on the Station's overall design was considerable. Images: author.

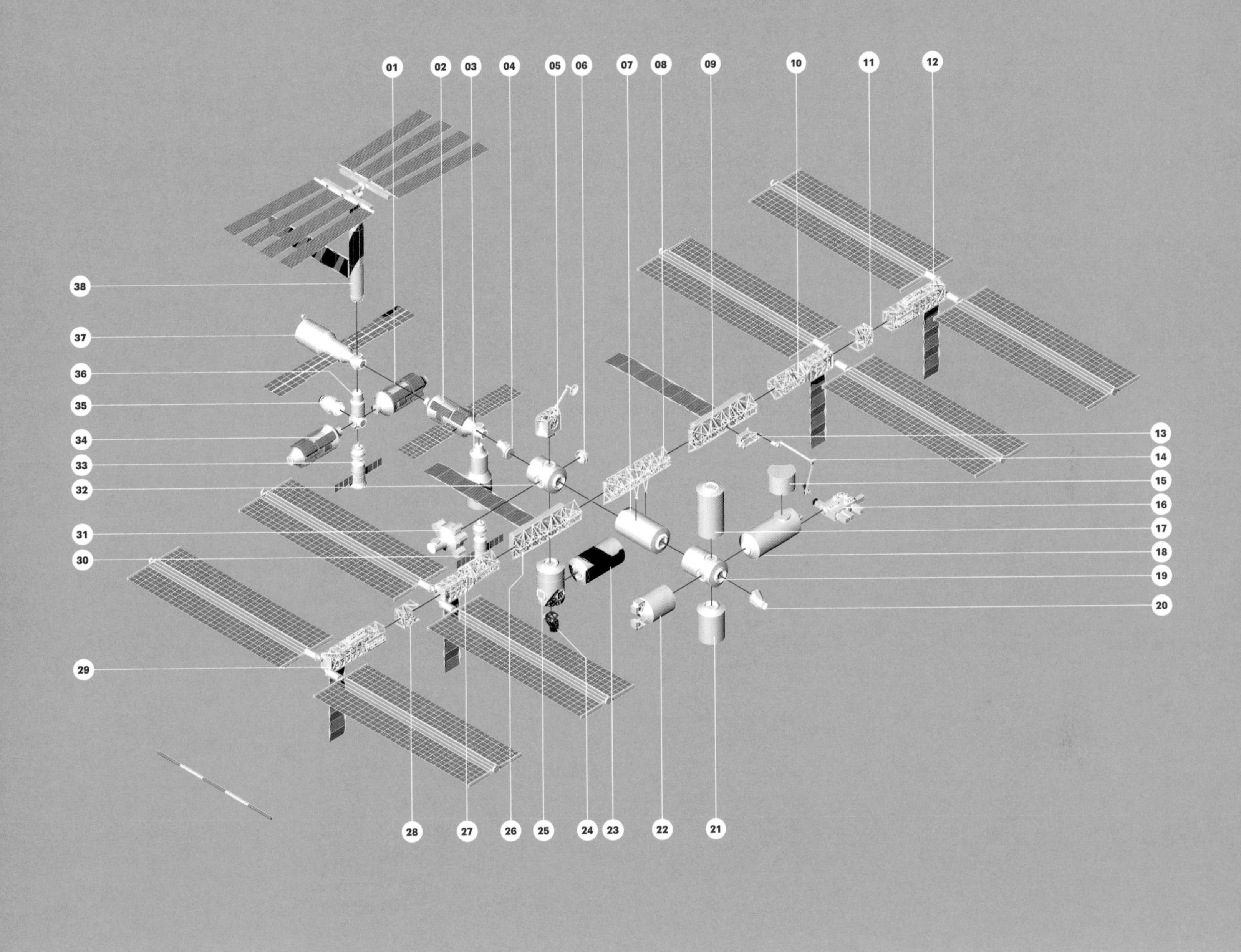

- **01** Research Module
- **02** Zarya Control Module
- **03** Docking & Stowage Module
- **04** Pressurized Mating Adapter-1
- **05** ZI Truss
- **06** Cupola
- **07** Destiny Laboratory Module
- **08** SO Truss
- **09** P1 Truss
- **00** P3/P4 Truss
- **11** P5 Truss
- **12** P6 Truss
- **13** Mobile Transporter
- **14** Canadarm2 robotic arm
- **15** Kibo Experiment Logistics Module
- **16** Kibo Exposed Facility
- **17** Centrifuge Module
- **18** Kibo Pressurized Module
- **19** Node-2
- **20** Pressurized Mating Adapter-2
- **21** Multipurpose Logistics Module
- **22** Columbus Laboratory Module
- **23** Habitation Module
- **24** Pressurized Mating Adapter-3
- **25** Node-3
- **26** S1 Truss
- **27** S3/S4 Truss
- **28** S5 Truss
- **29** S6 Truss
- **30** Soyuz TMA Spacecraft
- **31** Joint Airlock
- **32** Node-1
- **33** Soyuz TMA spacecraft
- **34** Research Module
- **35** Docking Compartment
- **36** Universal Docking Module
- **37** Zvezda Service Module
- **38** Science Power Module

had a port at each end and four ports placed at 90° intervals around its circumference. Node-2 would be provided to NASA by the European and Italian Space Agencies and built by Alenia Spazio.

Berthed to Node-3's port on the port side was the Boeing-built habitation module, now with its length reduced. On Node-3's starboard port was a place for an American-built crew return vehicle.[10] Initially designated as a capsule, a NASA illustration of the Station configuration in 1995 showed it as a small vehicle with a lifting-body form and winglets. NASA had assumed it would be similar to the X-38 vehicle. It would have seven seats to permit the Station's crew size to increase by one and a fully automated, deorbit-landing mode. This vehicle had been under consideration by NASA for several years but had never been funded beyond concept stage. Its job of returning the Station's crew to the ground in an emergency had now been taken over by the docked Soyuz spacecraft. It soon disappeared from the Station's new configuration. Berthed to the nadir port of Node-3 was Pressurized Mating Adapter-3.

Crew delivery and escape vehicle concept
By the late 1980s, the need for an escape vehicle of some kind to assure the safe return of the Station's crew to the ground in an emergency had emerged as an essential requirement. This design called the HL-20 Personnel Launch System dating from around 1990 was a concept developed by NASA Langley Research Center for a lifting-body vehicle that would deliver a crew to the Station, remain docked there during the crew's tour and return it to Earth, while providing a means of evacuation in an emergency. It was one of a number of similar studies that responded to the need for a Station 'lifeboat' but none of them ever proceeded beyond concept stage due to lack of funding. NASA's final choice to do the job was two docked Russian Soyuz spacecraft.

Moving forward along the spine, berthed to the forward port of Node-1 was the Boeing-built Laboratory Module (named Destiny), also shortened. It was positioned at the intersection of the Station's T shape and functioned as the structural link between the spine of pressurised modules and the truss structure. It was, as before, the twin of the Habitation Module but filled with experiment racks and workstations. Berthed to the forward port of the Laboratory Module was Node-2 (named as Harmony), to be provided to NASA by the European and Italian Space Agencies. Berthed to its upper port and pointing to zenith was the Centrifuge Module. This was the same length as the Habitation and Laboratory Modules and another American element. It would contain the centrifuge for variable-gravity life sciences research and payload racks and would be built by the National Space Development Agency of Japan. Mounted to the forward port of Node-2 at the front of the spine was Pressurized Mating Adapter-2. This completed the collection of American-built or American-funded modules.

The European and Japanese contributions were much the same as before. Berthed on Node-2's port side were the Japanese Experiment Module pointing out to port, the Japanese Experiment Logistics Module that sat on top of it and the Japanese Experiment Exposed Facility – Japan's external experiment platform – that perched on the end. Berthed to Node-2's starboard port was the shortened European Columbus Laboratory Module, now with external experiment mounting frames attached to its endcap. A recent addition to Europe's contribution was the Multipurpose Logistics Module berthed to Node-2's nadir port. The Italian Space Agency would provide this module system to NASA under a bilateral agreement. It was not part of the European Space Agency's contribution. Italy planned to build three of these modules to deliver supplies to the Station and return experiments to the ground on future Shuttle missions.[11]

The transverse truss structure made up from a set of prefabricated building blocks had replaced the stick-built truss system in the Restructured Space Station Freedom design of 1991. Now it had reached a state of consolidation in the International Space Station's design, having also been featured in Space Station Alpha's Options A and B. After beginning life at 95 metres long in the 1991 Restructured Freedom design, NASA trimmed it to 74 metres, 71 metres and 83 metres in the 1993 Space Station Alpha Options A1, A2 and B respectively as part of the budget-cutting campaign. Now it had grown again to 95 metres. At its centre was the newly named S0 Truss framework that provided the connection between the truss and the spine at the intersection of the Station's T post and crossbar.

Positioned just above the Destiny Laboratory Module, the S0 was connected to it by a cluster of struts. Beyond the P1 and S1 were the P3-P4 and S3-S4 frameworks (NASA had cut out the P2 and S2 of the Restructured Freedom design but maintained the truss numbering system). Each of these carried two pairs of solar arrays. Then there were two short truss pieces called the P5 and S5 mounted on the ends of the P3-P4 and S3-S4 respectively. These enabled the inboard ends of the last truss frameworks with their twin solar arrays – the P6 and recently added S6 – to clear the solar alpha joint rotation mechanisms of the P4 and S4. This completed the transverse truss, now comprising nine structural building blocks.

There was now, however, a tenth piece of frame called the Z1 that was physically separate from the transverse truss. In NASA's illustrations at the time, it appeared to exist by itself, attached to the zenith port of Node-1. The purpose of the new Z1 framework would become apparent later on during the Station's assembly on orbit. Also new to the Station's team was Brazil which was joining the project as a participant. Brazil was going to provide several small elements including a Shuttle-delivered external experiment pallet mounted on the truss structure called the Express Pallet.

This was the state of affairs with the Station's design in the years immediately following Russia's entry to the project. It did not mean that the Station would not experience further changes up to the launch of its first building block later in the decade and, indeed, up to the Station's assembly completion on orbit whenever that would occur. Ten years after the project's launch, the Space Station had matured into a physically ambitious but technically confident piece of space architecture, although it was now a very different design.

After all the rescoping and restructuring that had happened over the years with one design succeeding another, how did NASA now explain the project's mission? As before it was manifold. First, in NASA's words it would 'create a permanent orbiting science institute in space capable of performing long-duration research in the materials and life sciences areas in a nearly gravity-free environment'. Second, it would 'conduct medical research in space'. Third, it would 'develop new materials and processes in collaboration with industry.' Fourth, it would 'accelerate breakthroughs in technology and engineering that would have immediate, practical applications for life on Earth – and create jobs and economic opportunities today and in the decades to come'. Fifth, it would 'maintain US leadership in space and in global competitiveness and serve as a driving force for emerging technologies'. Sixth, it would 'forge new partnerships with nations of the world.' Seventh, it would 'inspire our children, foster the next generation of scientists, engineers and entrepreneurs, and satisfy humanity's ancient need to explore and achieve'. Eighth, it would 'invest for today and tomorrow. Every dollar spent on space programmes returns at least $2 in direct and indirect benefits.' Ninth, it would 'sustain and strengthen the United States' strongest export sector – aerospace technology – which in 1995 exceeded $33 billion.'[12]

These were bold words but only time measured in decades would show if NASA's vision would become a reality. The Station was going to take years to assemble on orbit and only when a full crew of six astronauts and cosmonauts had a full set of experiment facilities to work with would it begin to reach its real research potential.

Testing the Automated Transfer Vehicle – Europe's resupply spacecraft Europe's Automated Transfer Vehicle was a complex and sophisticated resupply spacecraft that would rendezvous and dock automatically with the International Space Station. Simulating this manoeuvre on the ground for testing purposes under full gravity was always going to be difficult. These two photographs taken in 2006 in a high-bay building at Astrium at Val-de-Reuil in France show how it was done. The circular docking interface, shown in the photos in white, simulated the aft end of the Station's Zvezda module with which the vehicle would dock. The Automated Transfer Vehicle's guidance, navigation and control avionics are simulated in equipment mounted on the robotic arm inside the miniature hangar opposite. Slowly moving the circular piece on a mobile platform towards the equipment on the arm across the floor simulated the final closure of the vehicle with the Station on orbit. Photos: Astrium and ESA.

To enable crews, experiments and equipment to perform efficiently on orbit and fulfil the project's mission, the Station was going to depend on regular visits by spacecraft. They would deliver new crews, supplies, consumables, spare parts and new experiments to the Station and return crews, completed experiments and waste material to the ground as well as evacuate crews in an emergency. Regular resupply of pressurised cargo was vital, for it included the water, food, oxygen and nitrogen essential for crew life support, as well as clothing and personal articles. A major benefit of the broad internationalisation of the Station was that five types of spacecraft – one American, two Russian, one European and one Japanese – would share these duties and work together as a fleet.

First and foremost was the Space Shuttle system with its three Orbiters – Atlantis, Endeavour and Discovery. Their main task would be to ferry up most of the Station's building blocks, including all the American modules and truss pieces, all the Japanese and European modules and at least one of the Russian elements. The Orbiters would also exchange crews, and deliver water, pressurised and unpressurised cargo and gases. Using the Multipurpose Logistics Module as a shipping container, the Orbiters would bring up nearly 9 tonnes of pressurised cargo on a supply mission. The Orbiters would also offer the only means of returning intact items to the ground such as completed experiments. No other spacecraft was able to do this.

Russia's three-seat Soyuz spacecraft would deliver and return crews and provide the means of emergency escape and evacuation for six-person Station crews. Russian Progress spacecraft would supply pressurised cargo and attitude control propellant and, docked to the Station, provide occasional reboost to keep it at the correct altitude on orbit. Progress spacecraft would transfer unused reboost propellant to storage tanks on Zarya or Zvezda for use in the Station's attitude control. After transferring supplies from Progress, crews would fill the empty compartments with trash after which Progress would undock, deorbit and burn up in the Earth's atmosphere.

Europe's Automated Transfer Vehicle would be about three times as large as the Progress spacecraft. Launched on the Ariane 5 rocket, it would deliver pressurised cargo and propellant to the Station. Controlled from the ground, the Automated Transfer Vehicle would approach and dock with the aft port of Zvezda, right at the rear of the Station. There, it would be able to boost the Station's orbital altitude, just like the Progress spacecraft.

Completing the fleet was Japan's Transfer Vehicle called the HTV. It was the only spacecraft of the fleet unable to dock with the Station under remote control. Launched on Japan's H-2A rocket it would rendezvous and approach the forward end of the Station but stop short to be grappled by the Canadarm robotic arm and manoeuvred in for berthing at Node-2's nadir port. The Japanese vehicle would carry a mixture of pressurised and unpressurised cargo.[13]

The fleet of vehicles would be busy. Building and servicing the Station was projected to involve more than eighty flights by the fleet over a four to five year period. The Space Shuttle system would fly thirty-one of these of which twenty-four would be for the delivery of major building blocks and seven for supplies, consumables and experiments. On the Russian side, ten unmanned Russian launches would deliver modules and other elements. About eleven Soyuz flights were needed for crew rotation and escape capability. About thirty Russian Progress flights were needed for supplies, consumables and experiments.[14] Not included in the eighty-flight figure were five flights planned for the European vehicle and up to eight flights for the Japanese vehicle.

In July 1995, the Committee on Science of the House of Representatives approved the International Space Station Authorization Act, which was intended to give NASA sufficient funding to complete the Station's development. Representative Robert S Walker had introduced the bill in May. The Committee claimed that it had 'conducted rigorous oversight of the space station program since first proposed by the President in 1984'. In its view, funding stability and predictability were now essential to complete the Station, a project that was in the national interest. Congress authorised the total amount at a little over $13.1 billion 'to remain available until expended, for complete development and assembly of, and to provide for initial operations, through fiscal year 2002 of, the International Space Station'' NASA would be held to the Clinton Administration's spending limit of $2.1 billion in any single year, beginning in 1996.[15] The Washington antipathy to the Station and its price tag that had forced NASA to shoulder a constant burden of budget reductions, stretched timetables and hardware cancellations, would become a thing of the past. Or would it? In September 1993, NASA had estimated that the International Space Station as then defined with Russia as a new partner needed $17.4 billion between 1994 and 2003. The General Accounting Office thought this sum was too low and had disputed the way that NASA had done its estimating. The figure took into account savings resulting from Russia's involvement but overlooked payments to Russia for associated space activities, the cost of upgrades needed by the Space Shuttle system and the cost of additional Shuttle flights and Shuttle launch window constraints.[16] The budget figure was also tied to an optimistic assembly timetable. The Authorization Act contained International Space Station Assembly Sequence Revision A, which showed twenty-eight Shuttle flights for assembly, utilisation and logistics over a fifty-six month period from November 1997 to June 2002. This would average six flights a year – well beyond the Shuttle's record at the time. It was not the first time that wishful thinking had driven the Shuttle's Station assembly timetable. Any Station assembly delay caused, for example, by hardware fabrication, payload or Shuttle problems would push the Station's completion beyond 2002 and probably break the budget.

A fabrication problem had already occurred. The first high-precision weld on the Node-1 module had failed. The Committee noted that though the problem had been solved, the lack of contingency funds to deal with such problems was an ongoing concern. Then there were still some dissenting voices. Led by Representative Tim Roemer who had nearly succeeded in killing off the project before, a subcommittee proposed an amendment to terminate the Station. This time, however, it was convincingly defeated. Congress had moved on and its mood was upbeat. Fabrication of the Station's parts was ahead of schedule and Congress noted with satisfaction that '48,200 pounds of hardware had been produced (of which) 19,900 pounds was scheduled to be completed by this time'.[17]

Bending metal
Manufacture of modules and trusses, 1994 onwards

The task of producing much of the hardware was Boeing's as prime contractor for the American portion of the Station. Boeing would produce or oversee the production of major Station building blocks including several pressurised modules and the transverse truss structure that it had inherited from McDonnell Douglas and Rocketdyne when it took over those two companies in the 1990s. Contracts for the full development and fabrication of the truss frameworks were in progress at those companies before NASA appointed Boeing as prime contractor in 1995. McDonnell Douglas had the S0, S1 and P1 frameworks and Rocketdyne had the P3-P4, S3-S4, P5, S5, P6 and S6 frameworks. An exception was the small Z1 truss piece that NASA added late and Boeing developed quickly. The Z1 was originally going to support a potential Russian propulsion unit to help boost the Station's orbit from time to time and contain four control moment gyroscopes. It ended up as a temporary site for the P6 solar array truss piece to provide more early electrical power before assembly of the main truss structure began. When Boeing took over McDonnell Douglas and Rocketdyne, it assumed responsibility for the Station's entire truss structure. Rocketdyne's experiences in designing their truss pieces had been a major factor in NASA's decision to scrap the strut-built truss and replace it with a prefabricated version. The P3-P4, S3-S4, P6 and S6 pieces incorporated solar arrays and a lot of other equipment that needed installation before delivery to orbit and assembly there. Rocketdyne realised that the task of attaching all this extra equipment to a lightweight strut-built truss on orbit and then running the electrical cabling down the truss to the modules would be difficult and involve lengthy amounts of astronaut spacewalk time. It was just not feasible.

Changing the truss design to a series of prefabricated building blocks with the stowed solar array assemblies already in position would greatly ease the astronaut workload. The P6 and S6 building blocks, for example, each had just twelve utility routes that would need connecting together in-line by astronauts on spacewalks after their structural attachment by the Station's mobile robotic arm.[18] However, the solar arrays, though folded up and stowed very efficiently in canisters and boxes for their ride up in the Shuttle Orbiters, still took up volume that impinged on the shape and size of the truss pieces that supported them. Everything had to fit within the Orbiter payload bay's overall dimensions. There were two solar array assemblies on each of the outboard portions of the P3-P4 and S3-S4 building blocks. Their presence forced a change in the cross-section of the pieces from a hexagon to a rectangle. The smaller rectangle's corners lay within the hexagon's boundary with the stowed arrays occupying the leftover volume between the two shapes. This change is clearly visible in technical drawings of the P3-P4 and S3-S4 building blocks.[19] It was a similar story with the fold-out triple thermal radiators on the S1 and P1 pieces that resulted in a reduction of the hexagon to a trapezoid half its width. There were also projecting rails for the robotic arm's mobile trolley that would run along the edges of the transverse truss and projecting trunnion pins that kept the truss pieces in place in the Orbiter payload bays for their ride up to orbit. Then there were the large diameter rotating joints on the P3-P4 and S3-S4 pieces to enable the solar arrays to track the Sun on orbit. The impact of all the equipment on the truss structure is today most evident in photographs of the Station from above. There are few continuously smooth lines. The truss edges move in and out with transverse cuts like a roughly whittled stick. Despite the truss's curious overall appearance, experience has shown that it does its job very well.

Each truss piece comprised a series of bulkhead frames connected at their vertices by longerons with diagonal struts across bulkheads and outer faces to add stiffness. The bulkheads, longerons and struts were individual

Transverse Truss fabrication A close-up view of an aluminium bulkhead that formed part of the S0 Truss building block just after its machining out of thick plate at Boeing facilities in Huntington Beach, California, in October 1997. The bulkheads were coupled with longerons and diagonals to make up the 13.1m long truss piece.

components machined from aluminium alloy plate and then anodised. Boeing placed all the components on a giant assembly jig, then positioned and aligned them and bolted them all together. Over 99 per cent of the bolt holes were drilled manually using templates and the match-drilling process in which the drill penetrates multiple parts to ensure the holes align properly. In the aerospace field it is common to fabricate an initial full-size version of the final product and put it through its paces by testing it under a range of simulated performance conditions. It is called qualification hardware and if it passes the tests the fabrication of the final flight hardware can proceed. Some qualification versions of the truss pieces were built, but not all. Originally there were four flight units and one qualification unit for the outer truss pieces but Boeing scrapped the qualification unit to save money. In another case, a prototype went through qualification and ended up as the S6 building block. Testing covered thermal vacuum, acoustic, modal and static structure parameters. The thermal vacuum test included functioning utility systems and live wiring; the acoustic test examined the launch sound pressure on installed equipment; the modal test monitored structural movement from an impacting object using accelerometers; the static structural test used hydraulic actuators to apply push and pull loads while strain gauges measured the response. Verification of the deployment of the large solar arrays employed a system of balanced weights to suspend the stowed array faces as they unfolded. Much of the testing was done at Kennedy Space Center after delivery of the truss pieces and before launch processing. Final touch-up of scratches on the anodised surfaces was done by hand using Alodine, a chromate-based chemical coating that resists corrosion. Touch-up work continued on the truss frames until they were finally dressed up in white thermal insulation blankets shortly before launch.[20]

Transverse Truss fabrication The two photographs show the S0 Truss piece – fabricated, assembled and finished – after its removal from a tooling fixture in Boeing's clean room at Huntington Beach, California. The S0 subsequently became the first piece of the Station's truss structure that the Space Shuttle launched to the Station.

The flagship of the American-built modules was the Destiny Laboratory Module. Despite the fact that Boeing had used a product called CATIA exclusively as the computer-aided design and engineering tool in the development of the 777 aircraft, it used another program called Intergraph for the International Space Station as the company's space division was not quite ready to convert to CATIA. Destiny's pressurised hull comprised a structural aluminium alloy skeleton of ring and bulkhead frames, longerons, conical endcaps, end hatches, berthing mechanisms and curved, pressure-resisting panels that formed the modules inner skin. Outside this were thermal insulation blankets made of Kevlar and Nextel and debris shield panels called bumpers made of aluminium alloy.[21] The three-layer hull construction was known as a stuffed Whipple shield[22] invented by American astronomer Fred Whipple. There were gaps between the shield panels, the blankets and the pressurised hull.

The fabrication of Destiny and the other American modules took place at Marshall Space Flight Center in Alabama in buildings that dated from the Saturn V rocket era. Assembly and welding of the module hulls occurred in one area while final assembly and integration of insulation, shielding, outfitting, utilities and harnessing occurred in another area upgraded to clean room standards. Construction began with the fabrication of the family of components that made up a module. The circular frames of the structural skeleton had sections shaped like the letter H and were machined from solid aluminium alloy plate.

A vertical milling machine produced complex components such as the end hatches. Each hull panel was machined from flat alloy plate with a grid of waffle-shaped stiffeners and then put into an autoclave to form its curve and anneal it. Module assembly took place on a rolling cradle that held the module structure securely and rotated it to simplify welding operations. Welding and outfitting took place inside the hull with access through the two end hatches. This made fabrication and installation procedures difficult. The welding method used was the electron beam type, carried out robotically by a mandrel inserted through one of the hatches. The mandrel's shape contracted during insertion through the end hatch and then opened out inside like an umbrella. It welded together the curved pressurised hull panels and conical endcap panels to form the pressure-resistant hull. It produced continuous 3mm full-depth welds between adjacent curved panels to an accuracy of 1/800 of a millimetre. X-ray scans were used to check the welds.[23]

Boeing produced no full-size qualification versions of the American-built modules apart from one earlier node prototype. For one thing, the major structural challenge for the module design was the loads imposed by the launch and ascent on the Space Shuttle and accurately simulating these loads on a full-size module under test conditions was not possible. Thorough structural analysis qualified the modules for loads such as these. After arrival on orbit, the modules would experience constant pressure, dynamic loads and thermal cycling throughout their lifespans, set at thirty years for qualification purposes. Boeing subjected the modules to structural, pressure and thermal cycling tests to simulate these conditions. There were problems during pressure testing when pressure hot spots developed at some points. With the first node module it became necessary to refine the NASTRAN-PATRAN computer program that performed the structural analyses. Some gussets failed and had to be cut out by hand. Boeing tested the module primary structures to one-and-a-half times their maximum design pressure – 157kPa – in ambient temperatures but not under thermal cycling conditions, and apertures and openings underwent leak tests. Also used were static flight loads tests to simulate Shuttle launch and ascent effects. Some ionising radiation testing was done on selected components and debris impact tests were performed on the hull elements.

Boeing considered that its hull design offered better protection than the Russian modules because the Nextel and Kevlar thermal insulation blankets behind the shield panels helped to resist penetration. Weighing of the modules took place frequently during their fabrication and assembly to ensure they remained within Shuttle payload weight limits. Some fabricated components that were too heavy were scrapped. After Boeing had installed and integrated electrical, communications and data cabling and gas and fluid lines within the hull, it carried out live testing on them.[24] The aluminium alloy used was 2219, a grade ideal for space applications. It had a high copper content, was capable of welding and had high stress corrosion resistance after age hardening.

Perhaps the most complicated component in the construction of the Destiny Laboratory Module was the window. The first time that NASA had installed a window in an American-built module was on Skylab in the early 1970s and there is a story connected to it. The window had led to a major disagreement between Leland Belew, manager of the Skylab office, and his team at Marshall Space Flight Center on the one hand and NASA's Office of Manned Space Flight director George Mueller and his industrial design consultants Loewy/Snaith on the other. The Marshall team was responsible for designing Skylab's pressurised habitat while Loewy/Snaith were tasked with recommending habitability improvements. The Marshall team opposed a window in Skylab, arguing that it would weaken the hull structure, was too costly, would take too long to develop and test and was not essential to mission success. Raymond Loewy, the leading industrial designer, countered that it was unthinkable not to include a window as its recreational value alone on a long mission would justify its cost. Loewy's argument won the day. The window proved popular with the Skylab crews and valuable for Earth observation.[25] NASA had evidently learned a lesson, as windows became a constant feature of the Station's pressurised modules from the earliest design work in 1979. Destiny would get a single window and the Station as a whole would get a multi-paned Cupola to provide astronauts with panoramic views of Earth and the Station's exterior.

Destiny Laboratory Module – hull fabrication Boeing fabricated the American laboratory module named Destiny in the Space Station manufacturing facility at NASA Marshall Space Flight Center in Alabama. Destiny was made of aluminium alloy. It was 8.5m long and 4.3m in diameter. It comprised three cylindrical sections and two endcaps with hatches. Inserted on one side of the central cylindrical segment was a 509mm diameter window. Destiny was able to accommodate twenty-four racks. The photograph shows Destiny's pressurised hull after its fabrication. A rolling cradle securely held the module and rotated it during welding work.

Destiny's window was circular and 509mm in diameter. Boeing designed it to have very high light transmittance over a wide spectral range to make it suitable for scientific Earth observation. It contained four panes of glass, each with an anti-reflective coating. The outer pane was 9mm thick and provided micrometeoroid debris protection. Inside this were two pressure-resisting panes – one active and one redundant – each 32mm thick. These three panes were made of high strength, fused silica glass. The fourth innermost pane was a scratch pane to protect the two pressure panes. It was made of 11mm thick borosilicate glass. Holding the panes in place with gaps between them were two thick and two thin structural alloy hoops with interlocking Z-shaped sections. The window's overall thickness from outer surface of outer pane to inner surface of inner pane was 143mm. A ring of thirty captive bolts fixed the window against the hull like a ship's porthole. A set of seals maintained the pressurisation integrity of the hull interior and across some of the glass panes while a system of valve ports maintained pressure levels in the gaps. The window had an external shutter controlled internally by a hand wheel, and came with inside and outside covers to maintain pressurisation in an emergency if a pane cracked or a seal leaked. It was put though a variety of pressure, leak, temperature, vibration and optical tests for its qualification.[26]

As Boeing pressed ahead with the fabrication of the truss structure and the modules, the other Station partners were doing the same, albeit with some belt-tightening. In October 1995, Europe's space ministers agreed to fund a smaller version of the Columbus Module. Columbus had been the name adopted in 1985 for a triad of European space elements – an Attached Pressurised Module, a Man-Tended Free-Flyer and a Polar Platform. The pressurised module was the lone survivor after the other elements had been cancelled and it had inherited the title of the Columbus Module. Columbus, however, had shrunk to a third of its original size and Germany was bearing 41 per cent of its costs. France was committed to paying 27 per cent of the costs of the Automated Transfer Vehicle that would resupply the Station with propellant, water, air, and experiments every eighteen months or so.[27]

Though all appeared to be going well with development progress on the American, European and Japanese building blocks, a problem was emerging with the Russian Zvezda Service Module. The Russian aerospace company Energia was responsible for building Zvezda, funded by the Russian government, and it was due for launch in April 1998. Zvezda was a critically important module, as it would provide the basic accommodation necessary for early Station crews to take up residence. In 1995 though, funding problems emerged. In December, the Russian Federal Space Agency announced that Zvezda could not proceed unless the Russian government released the equivalent of $43 million then overdue. Possible solutions included using the core module from the Mir Space Station instead or altering the International Space Station's assembly sequence. The White House began to take notice and Vice President Gore pursued the matter with Russian Prime Minister Chernomyrdin.[28]

In June 1996, NASA and Roscosmos (the new name for the Russian Federal Space Agency) signed a protocol that summarised the contributions and obligations of each nation to the Station, which went far beyond earlier commitments to build, launch and assemble just the hardware. The protocol's focus was on issues in which one party was dependent on the other or they were equally involved in achieving an objective. Beginning with the obvious – each party would be responsible for the launch of its own elements, spares and logistics unless there were specific agreements to the contrary – the agreement ranged far and wide. The parties would 'keep what they bring' and have preferential rights to their own facilities. NASA would be in charge of Station systems engineering and integration and operational and utilisation integration with the support of data and personnel supplied

Destiny Laboratory Module – hull fabrication The photograph shows the module with thermal blankets covering its hull after its shipment to Kennedy Space Center in Florida. Protecting the thermal blankets and the pressurised hull from micrometeoroid penetration were exterior bumper shields (not yet added in this photograph).

by Roscosmos. There would be an exchange and sharing of electrical power generated on the Station and a sharing of propellant deliveries for maintaining the Station on its correct orbit. During the assembly phase, six Shuttle flights and eleven Soyuz flights would rotate fifty-one crew members with numbers fairly balanced between each nation. Each crew would include at least one representative of each nation and both nations would be equally responsible for delivering crews to orbit. Crews would train and work together as a single team on the ground and on orbit but the main operations language would be English. Russia would be responsible for emergency return of three-person crews to Earth during the assembly phase with a Soyuz spacecraft permanently docked at the Station. Both parties would host the training of joint crews free of charge and each would provide the other with office space. Each party would provide food, supplies and personal items for its own crew members. There were two exceptions to each nation's responsibility for launching its own hardware: the Russian Science Power Platform, which an Orbiter would deliver to the Station, and the Zarya module, which was Russian-built and launched but paid for by NASA. There was no mention in the protocol of Zvezda's delay, and its launch date still showed as April 1998.[29] What was evident in this protocol was the camaraderie that had developed between America and Russia over the Station.

The delay with the Zvezda module continued. In August, Energia warned NASA that it could not launch Zvezda on time in 1998. Forced by the White House to take action, NASA responded by beginning the development of an interim control module that would be able to take over some of Zvezda's duties if necessary. In February 1997, Prime Minister Chernomyrdin went to Washington and signed a document promising the equivalent of $100 million by the end of the month and a further $250 million by the end of the year, but the money continued to trickle in. There was a renewed economic crisis in Russia the following year and the rouble collapsed, creating more financial headaches for the Russians.

To help Russia out and speed up Zvezda's completion, NASA negotiated a deal to pay the Russians $60 million for thousands of hours of cosmonaut time to help run American experiments on the Station and to store American experiments in Russian modules.[30] NASA, however, was having problems of its own as some of the American contractors were running behind schedule. Then in 1999, against a backdrop of political disagreements between America and Russia over war in the Balkans that caused Russia to think again about its Station involvement, two Proton rockets failed shortly after launch prompting Roscosmos to announce an indefinite postponement of Zvezda's launch and leading NASA, in turn, to think again about Russia's involvement. It would not be until February 2000 that Russia would return Proton to flight status, eventually launching Zvezda successfully in July of that year.[31]

Destiny Laboratory Module – window Boeing installed a window in the curved hull of the Destiny Laboratory Module. It was 509mm in diameter and made of four panes of glass. From outside to inside, a 9mm thick outer pane provided micrometeoroid protection, two 32mm thick panes (one active, one back-up) retained the module's internal air pressure and one 11mm thick pane provided scratch-resistant protection to the pressure panes. The window had an overall thickness of 143mm.

Columbus Laboratory Module – test prototype Seen here at the European Space Research and Technology Centre (ESTEC) at Noordwijk in the Netherlands is a test prototype of the European Space Agency's Columbus Module. By 1995, its length had shrunk to about one third of its original size as Europe's space ministers sought savings. The test prototype was used to evaluate the installation and fit of racks and experiments. It was also publicly accessible at certain times through guided tours as part of ESTEC's policy of encouraging visitors. Photo: ESA.

esa

Vital functions
Environmental control and life support, power and thermal systems development, 1994 onwards

Like the modules and truss structure, the International Space Station's life support system had been under development well before the Russians joined the project as a senior partner. The task of keeping six crew members – more during crew exchanges – alive, healthy and comfortable over three or more decades of continuous operation would be a huge challenge. Never before had such a system had to sustain so many humans in space for such a long time. Life support was such an essential feature of the Station that it is important to review the basic functions it was designed to perform. It would provide oxygen and nitrogen for crew respiration and water for crew drinking, food preparation, crew hygiene and oxygen generation; it would recycle waste water into clean water; it would remove carbon dioxide, toxic gases, particles and micro-organisms from the air; it would monitor and control the module air gas mixture; and it would maintain internal air pressure, temperature, humidity at comfortable levels and circulate the air throughout the Station. Two important objectives on this list for the Station's life support system would be to improve its technology to supply oxygen by the electrolysis of water, and clean water by the recycling of waste water. This would help to reduce a reliance on resupply deliveries by the Space Shuttle, Progress vehicles and the European and Japanese spacecraft.[32]

Thorough testing of all the life support systems on the ground and in some instances in space on Shuttle missions would be a vital part of the development process to ensure their safety, reliability and durability. A major change to the Station's life support system occurred with Russia's participation. Russia had accumulated a great deal of experience in long-duration spaceflight and life support systems with its series of seven Salyut orbital outposts in the 1970s and 1980s and then with the Mir Space Station from 1986 up to 2001. American experience was limited to the 1973-74 Skylab missions and the Space Shuttle. From the outset, the division of the Station's pressurised accommodation into two sectors – American and Russian – drove the need for separate environmental control and life support systems that would function independently. The two nations had slightly different definitions of what a life support system was. To the Americans it comprised air control and supply, air regeneration, water recovery and management, crew body waste management, temperature and humidity control, fire detection and suppression, and experiment air control. From this list the Russians excluded thermal control but included food storage and preparation, food refrigeration and freezing, spacewalk support, crew personal hygiene and housekeeping. Fire detection by smoke detectors and suppression by portable extinguishers was a feature common to all pressurised modules.[33]

Columbus Laboratory Module – Stand-Off and utilities installations The shortened version of the Columbus Module was able to accommodate sixteen racks for experiments and equipment. The interior outfitting followed that of the American modules with four Stand-Offs used to support and restrain the racks and ensure that they were interchangeable with other racks on the Station. This photograph shows the interior of the Columbus Module during utilities installations. Clearly visible are the four Stand-Off structures at the 45°, 135°, 225° and 315° radials around the hull's circumference. Initial laying and harnessing of partly wrapped fluid and gas lines and electrical and communications cables through the Stand-Offs and on the conical endcap inner surfaces has occurred. Some of these have been stubbed off, ready to receive the racks with which they will connect. Image: ESA.

The Russian sector went through many changes from the time that Russia joined the project up to the Station's assembly in space, but two Russian building blocks – the Zarya Control Module and the Zvezda Service Module – stand out and remained vital parts of the Station from beginning to end. The Zarya Control Module's life support system included air control and supply based on gas and pressure sensors, temperature control based on fans and heat exchangers and trace contaminant removal by air filters. Zarya had no air regeneration and operated in two modes: first, when unoccupied with pressure and ventilation control; second, docked to Zvezda when, occupied, it would rely on Zvezda for air regeneration. Zarya also had blowers to help to circulate air between the Russian and American sectors. The Zvezda Service Module offered a self-contained life support system for up to six crew members during the Station's early assembly phases. It incorporated all Zarya's features and added air regeneration with carbon dioxide removal by lithium hydroxide or regenerable sorbents, air filtration to remove airborne particles, oxygen air resupply by electrolysis of recovered humidity and waste water, water supply for potable or crew hygiene use, crew body waste management and waste water disposal, and spacewalk life support consumables.

Environmental control and life support test facility A view of the facility at NASA Marshall Space Flight Center for testing the environmental control and life support system and the internal thermal control system in the Station's American sector (Russian modules had their own life support and thermal systems). In the foreground is the life support simulator for testing the performance of three Station modules: the Destiny Laboratory Module, Node-1 and Node-3. At the centre left is the internal thermal control system simulator for testing the temperature of equipment in the Station's racks. Thorough testing of the Station's life support and thermal systems was essential to confirm their durability and reliability for operation on the International Space Station before launch.

The ability of the Russians to fit so much equipment into Zvezda was testament to their experience in designing, fabricating and operating their earlier fleet of Salyut and Cosmos spacecraft. Up to about 1997, the Russian sector also included a separate Life Support Module – later eliminated – that replicated most of Zvezda's functions for a six-person crew and added a personal hygiene facility. Other Russian building blocks such as the Research, Universal Docking, Docking and Stowage Modules had various basic life support functions such air pressure sensing, contaminant removal, temperature control and air circulation.[34]

The building blocks that comprised the American sector on the Station up to the end of the 1990s included the Destiny Laboratory Module, the Habitation Module, the three Node Modules, the Joint Airlock, the three Pressurized Mating Adapters, the Centrifuge Module and the Cupola. Life support functions in the Laboratory and Habitation Modules would comprise air supply and return, air mixture regeneration, and temperature and humidity control. Destiny also had a water supply, a gas supply and vacuum for experiments while the Habitation Module had waste management and water recovery and management capabilities. Life support functions in the Nodes comprised air circulation and ventilation, air filtration and pressure monitoring. Racks would house most of the life support equipment in the Laboratory, Habitation and Node Modules. The Joint Airlock was designed for depressurisation and repressurisation to enable crew members to exit to conduct spacewalks and then re-enter. Its life support capabilities included air regeneration, air control and supply, temperature and humidity control, potable water storage and spacesuit consumables.

The main design objective for the life support system in the American sector was to minimise the use of consumables and expendables by using regenerable technology wherever appropriate and cost effective. There were two systems developed for this – water recovery and oxygen generation. The water recovery system would provide clean water by recycling waste water from a body shower (if provided), handwash and oral hygiene uses as well as crew urine. It would also recover condensate water from humid cabin air and spacesuit system water. The water recycling hardware comprised a urine treatment unit and a waste water treatment unit. Urine treatment involved low-pressure vacuum distillation in a rotating chamber in which the distilled liquid would migrate to the perimeter for collection. It would then go to the waste water treatment unit, where along with other waste waters it would pass through preliminary filters to remove solids such as hair and lint and then through a series of multifiltration beds for purification. A high-temperature catalytic reactor would then remove any remaining contaminants or micro-organisms. Finally, electrical sensors would check the purity of the water. Acceptable water would go to a storage tank for crew use and unacceptable water would be reprocessed.

Recycling would reduce water resupply needs from the ground by 2,760kg per year. The oxygen generation system would produce oxygen for the breathable air for the crew and any laboratory animals and to replace oxygen lost to module leakage, airlock depressurisation, carbon dioxide venting and experiment use. The heart of the unit was a cell stack to electrolyse water into oxygen and hydrogen, the latter being vented overboard. A power supply unit would provide the electrical power needed for electrolysis. The oxygen generation system would be able to supply 5.4kg per day during normal use cycles or up to 9kg if operated continuously.[35] A further refinement of the system envisioned for later addition to the Station was a carbon dioxide reduction unit called a Sabatier reactor. The process involves reacting hydrogen with carbon dioxide at high temperature using nickel as a catalyst to produce methane and water. On the Station the water would be added to the water supply and the methane vented overboard.[36]

As well as life support, several other finely engineered mechanical and electrical systems were essential to maintain the Station as a safe living and working environment. As the project's design had changed after 1993 with the addition of the Russian sector, it is worth reviewing what these had become as NASA later defined them after the Station's completion.[37]

First, there was the electrical power system to provide the Station and its systems, occupants and experiments with an adequate and reliable power supply. Zarya and Zvezda each had a pair of solar arrays and a distribution system to provide power during the early assembly phases before the arrival of the large American-built arrays that formed part of the truss structure. Zarya's and Zvezda's arrays produced up to 3kW and 13.8kW respectively.

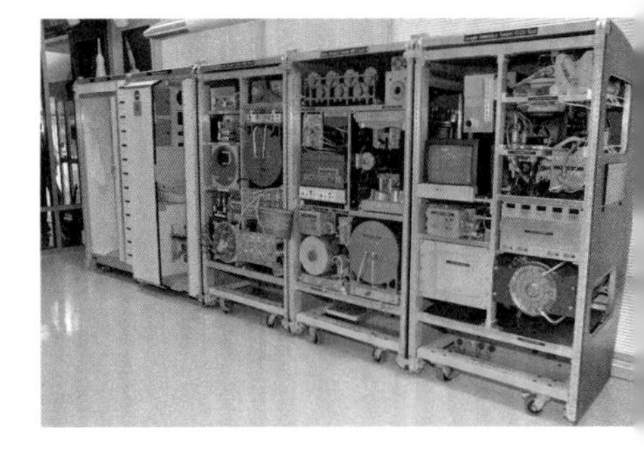

Life support system racks and compartments This photograph from NASA Marshall Space Flight Center shows the test set-up of five life support racks intended for Node-3 on the Station. From left to right are the shower compartment, the waste management compartment, two water recovery system racks and the oxygen generation rack. The water recovery racks provide clean water through the recycling of waste waters. Included is a urine processor and a waste water processor. The oxygen generation rack produces oxygen for crew and laboratory animal respiration and tops up oxygen loss in the system. It comprises a cell stack that electrolyses clean water into oxygen and hydrogen molecules and separators that harvest the gases from the water after electrolysis.

Environmental control and life support system diagram Oxygen for crew respiration is generated by the electrolysis of processed water drawn from crew personal hygiene use, urine and cabin air humidity condensate. Processed water is purified and returned for crew drinking and hygiene. The long-term objective for the life support system on the International Space Station is to increase its independence from Earth resupply and to recycle waste gases as far as possible rather than venting them overboard. This diagram shows the evolution of the Station's life support system in pursuit of this objective. It has a carbon dioxide reduction subsystem (named Sabatier after its inventor). This generates water from carbon dioxide removed from the cabin air by combining it with hydrogen taken from the oxygen generator. The methane (CH_4) produced is vented overboard. In this diagram, trace contaminant control and fire detection and suppression functions are omitted for clarity. Image: adapted by the author from a NASA diagram.

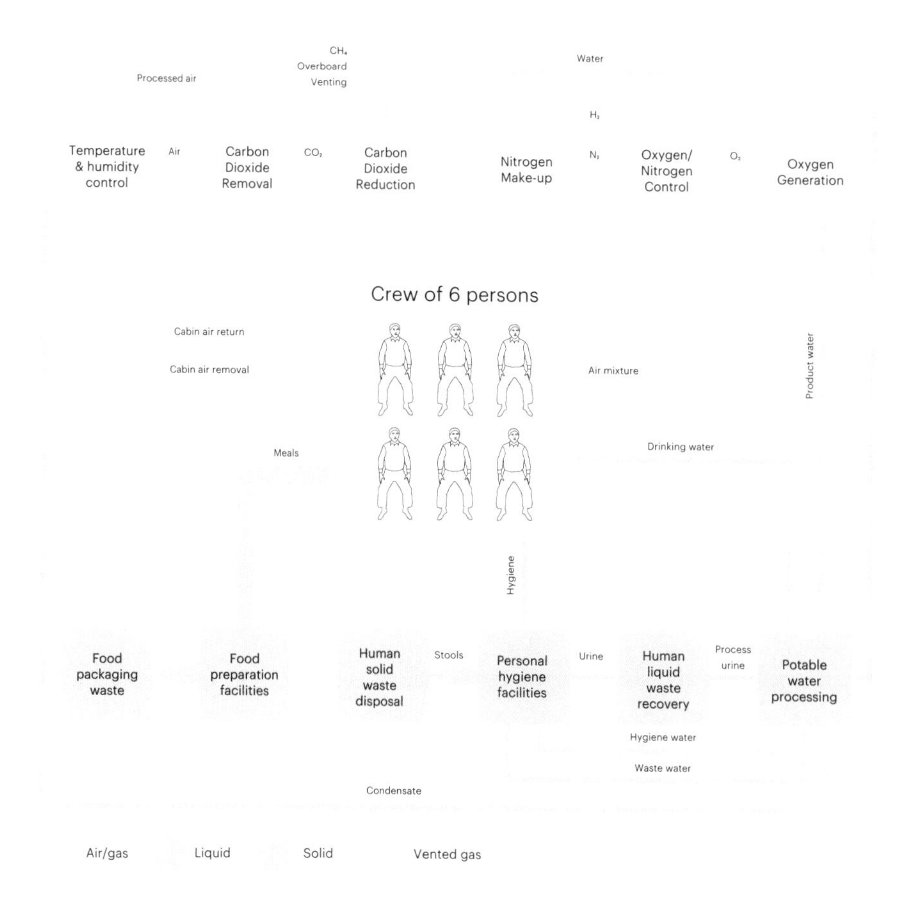

Later, and for the rest of the Station's life, the eight pairs of solar arrays along the truss would become the main power source, generating up to 84kW. The circular joints called Alpha joints on truss pieces P3-P4 and S3-S4 and the gimbal joints called Beta joints at the base of the eight masts of the solar arrays would enable the arrays to track the Sun effectively as the Station orbited the Earth. Nickel-hydrogen batteries on the P3-P4, P6, S3-S4 and S6 truss pieces would store the generated power for use during the night portion of each orbit. Radiators on the same truss pieces would provide cooling for the power system. Electronics control units would steer the solar arrays to face the Sun. Primary power from the port and starboard solar arrays would come in 160VDC through cables along the truss to converter units that reduced it to 124VDC. The power then went to main bus switching units and remote power controllers on the S-0 truss building block that distributed it throughout the Station. Once the large solar arrays were in action, Zarya's solar arrays would be folded up out of the way. Russia later planned to provide the Solar Power Platform to generate more power for the Russian sector.

Next, there was the thermal control system to deal with all the heat resulting from electrical power use inside and outside the Station, from the crew's metabolism and other sources. In space, heat can only be disposed of by radiation, as convection is impossible in a vacuum. There were two means of thermal control – passive and active. Passive thermal control included insulation, surface coatings, heaters and heat pipes. Active thermal control dealt with heat that exceeded passive thermal control capabilities. In the American sector, the thermal control system used mechanically pumped water and ammonia in closed-loop circuits to collect the heat from inside the modules, convey it to the exterior and radiate it out into space. Pumped water coolant circuits operated inside the modules controlled from the Destiny Laboratory Module. A low temperature circuit running at 4°C-10°C collected heat from the life support system and some experiments. A medium temperature circuit running at 12°C-17°C collected heat from avionics and other experiments. These circuits fed the heat-to-heat exchangers on Destiny that connected with external ammonia coolant circuits. The ammonia circuits conveyed the heat along the truss structure and delivered it to large radiators pointing aft on truss building blocks P1 and S1. The external coolant circuits also collected heat from coldplates on experiments and payloads mounted along the truss and disposed of it in the same way. The cooling system on the Russian modules utilised triol and polymethyl siloxane fluids.

Next, there was the propulsion system to maintain the Station at a safe orbital altitude and enable it to manoeuvre out of the way of orbiting debris. The Station would orbit at an altitude of 370-460km with 400km as its normal planned altitude. The higher the orbit, the less atmospheric drag there would be to slow the Station down and the less propellant needed to keep it in position. The slowing of the Station caused by drag was a particular concern at times of high solar activity when the Earth's atmosphere expanded outwards and the braking effect increased on orbiting spacecraft. To combat the drag, the Station had propulsion capabilities to reboost itself from time to time and raise its altitude. Both Zarya and Zvezda had propulsion systems using nitrogen tetroxide and hydrazine as propellants. As the first building block to arrive on orbit, Zarya had three types of engine to help it to position itself precisely on the correct orbital path. Zvezda had main engines and attitude control engines to provide roll, pitch and yaw control and would take over orbital station-keeping duties from Zarya after docking. Both the Russian Progress spacecraft and the European Automated Transfer Vehicle would be able to reboost the Station when docked at Zvezda's aft port, as would the Shuttle Orbiters from a different location. The Station would also have its own thrusters that used its own propellant supply, occasionally topped up by the visiting vehicles. The Station, therefore, would be able to rely on several sources for periodically raising its orbit.

Russian oxygen generator units Russian cosmonauts Andrey Borisenko, Alexander Samokutyaev and Sergei Volkov are pictured with three Russian Elektron oxygen generator canisters in the Zvezda module on the International Space Station. Similar in size to industrial vacuum cleaners, these highly efficient machines were first used on the Mir Space Station. They electrolyse water into oxygen and hydrogen. The oxygen is vented directly into the cabin atmosphere and the hydrogen is vented overboard. Each canister can process 1kg of water to generate 25 litres of oxygen per hour. This is enough to support the respiration of one crew member for one day. The Elektron canisters were not without operating problems on the Station.

Then, there was the guidance, navigation and control system that managed the Station's attitude in flight and orientation to the Sun and to the Earth. This system tracked the Sun, American and Russian communications and navigation satellites, and ground stations around the world. The Station used the tracking information to point its solar arrays, thermal radiators and communications antennae in the right directions. Attitude control relied on control moment gyroscopes powered by electricity from the Station's solar arrays. These were steel wheels weighing 98kg each and spinning at 6,600 revolutions per minute. They stored considerable energy through angular momentum. Gimballing each wheel enabled it to change its attitude and the resulting change of force direction caused the Station to move. Multiple wheels mounted on the Z1 truss piece gave the Station the ability and flexibility to change its orientation or maintain a constant attitude. Though the control moment gyroscopes offered smooth, continuously adjustable attitude control, they were limited in the degree they could move the Station. Large movements required the use of the propulsion system.

Lastly, there was the Station's communications system that kept the astronauts and cosmonauts in constant contact with control centres on the ground and with visiting spacecraft. It also enabled ground control to monitor and maintain the Station's systems, operate payloads and enable flight controllers to send system commands. The system provided two-way audio and video communications between crew members inside or outside the Station on VHF or UHF bands as well as two-way audio, video and file transfer between the Station and its ground control centres and payload scientists. It also handled transmission of system and payload telemetry between the Station and the ground, circulated experiment data to scientists and ensured overall flight control of the Station from the ground. Because the Station was continually orbiting the Earth, constant line-of-sight communications to the control centres was not possible and tracking and data relay satellites in geosynchronous were needed to redirect transmissions on a continuous basis on S, Ku or Ka bands.[38]

Control moment gyroscopes Technicians working on a control moment gyroscope in the Space Station Processing Facility at Kennedy Space Center in July 1998. The International Space Station has four of these gyroscopes. Their job is to control and maintain the Station's attitude with regard to Earth. Each gyroscope has an internal rotor spinning at high speed that exchanges momentum with the Station, enabling slight adjustments in the Station's attitude in the X, Y and Z axes.

Unfinished business
ISS cost, schedule, retrofit and life-cycle issues, mid-1990s

While NASA and Roscosmos were working to move the International Space Station forward, others at NASA were pondering the day that the Station's life would come to an end and it would come tumbling back to Earth. NASA had produced the first environmental impact statement for the Station in 1991[39] and it was time to update it. Under normal conditions, the greatest impact would occur at the beginning and end of the Station's lifespan. At the beginning, there would be the effects of the launches from Kennedy Space Center. Shuttle launches resulted in predicable, short-term air, water and biological effects in the area and NASA had studied and recorded them over the years. The main impact was a decrease in the diversity of vegetation close to the launch site caused by the deposition of products from solid rocket exhaust. There were some upper atmospheric effects that included a localised decrease in stratospheric ozone density and possible decreases in the ion and electron concentration in the space environment surrounding the Station. Venting of non-hazardous liquids and gases would be permitted and cover helium, argon, neon, carbon dioxide and oxygen. Some outgassing and leakage of the Station's internal atmosphere through seals and joints would be unavoidable. All solid waste products and hazardous substances would be returned to Earth in sealed containers and disposed of in accordance with established procedures for such materials.[40]

At the end of its life, the planned method of dispatching the Station would be by controlled and targeted re-entry and burn-up in the Earth's atmosphere. From 6 per cent to 19 per cent of the Station's hardware would survive re-entry as fragments falling into a remote ocean area. The expected footprint of the surviving fragments was 43,000km² in area but could extend to about 960,000km² due to variables such as fragment aerodynamic form, wind speed and wind patterns. The number of fragments reaching the Earth's surface could vary from about 600 to over 2,000 with a total weight of 24.3 to 78.6 tonnes. Some pieces might have enough energy to cause damage to people, structures and ships. Once the fragments splashed down they would sink to the ocean floor. Some leakage of sealed containers that survived re-entry was possible, though unlikely. There was also a remote possibility that some Station parts could re-enter the atmosphere accidentally during assembly, operation or disposal with some fragments falling on land. Causes of accidental or uncontrolled re-entry included lack of propellant to keep the Station on orbit, multiple and major on-board systems failures or a disabling collision with orbital debris, meteoroids or other spacecraft. Failures such as these could render the Station's attitude and reboost functions inoperative or could disable the ability of propulsive vehicles to dock. Though the Station would break up, burn and produce fragments in a similar manner to a controlled and targeted re-entry, the fragment footprint area was unpredictable. NASA's analysis showed that the risks to life beneath the Station's overflight area were small – just one in 78 billion for any given individual.[41]

In 1998, with the Station on course for the beginning of its assembly phase, NASA turned its attention to human space exploration missions beyond Earth. Mars was a leading destination candidate. From the earliest Station studies, aluminium or an alloy of it containing lithium had been the preferred material for the pressurised modules, sized to fly in the Space Shuttle cargo bay. The aerospace industry understood the material thoroughly, its fabrication techniques were well known and its performance was predicable. For space exploration missions beyond Earth, however, it had disadvantages. It was heavier than other materials that might be able

Space Station Processing Facility NASA built the huge Space Station Processing Facility at Kennedy Space Center in Florida to carry out the pre-launch processing and preparation of all the Station's building blocks launched from the United States mainland. The cavernous interior had a floor area of 42,455m² with high and low bays designed to 100,000-Class Clean Room standards. In the foreground, this photograph shows a Station robotic refuelling experiment being prepared for launch in the Space Shuttle. Carrying the experiment to orbit in the Space Shuttle's payload bay is a white structural multipurpose carrier framework. At the far end of the high bay are two pressurised modules also being processed for launch. As well as the high and low bays, the building contains laboratories, control rooms, offices and logistics areas. While undergoing processing, Station crews were able to use the modules and other equipment for familiarisation and training.

to do the same job of containing pressure, such as high-performance membranes. It also required fabrication into complete modules before launch and was impossible to collapse into compact form for efficient stowage in the payload compartments of rockets or spacecraft. While this was tolerable for International Space Station modules, it was a disadvantage for habitats intended for the surface of Mars where payloads had to be as small and lightweight as possible to reduce their burden on launch and interplanetary vehicles. Flexible and lightweight habitats that could fold up into compact form for launch and then expand and open up into travelling accommodation for outbound and return voyages would be a great asset. In 1997, a team at NASA Johnson Space Center began work on an interplanetary habitat for a crew of six on a Mars mission. They called it the 'TransHab' project and began to develop it as a future addition to the Space Station as a possible back-up to the Station's Habitation Module which had run into budget trouble.

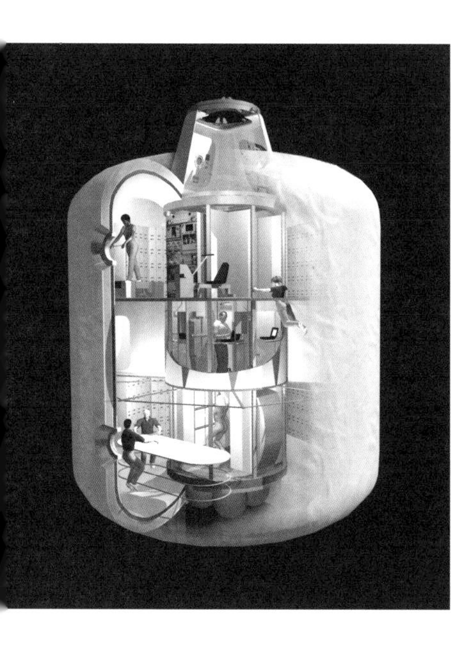

TransHab inflatable module design An artist's cutaway view of the inflatable TransHab module that NASA began to design and develop in the mid-1990s as a potential habitat addition to the International Space Station. The design had three accommodation decks. In this image orientation, the top deck was for crew healthcare and hygiene with exercise equipment, a medical examination station and bathroom facilities. The middle deck contained six private sleep compartments located in the core and surrounded by a water jacket for solar radiation shielding. The bottom deck was for crew communal activities such as meals, meetings, leisure and conferences with ground control.

The word TransHab was a contraction of transit habitat. It was a hybrid structure with an inflated habitable volume wrapped around a rigid central core in much the same way that a car tyre encircles a metal wheel. 10.5 metres long and 7.3 metres in diameter, it contained 342m^3 of pressurised volume. There were three levels of accommodation. Level 1 was the main communal activity level with a galley, wardroom, stowage, a window for Earth viewing and enough room for twelve people during a crew exchange. Meals, meetings, conferences with ground control, planning sessions, socialising and public relations video events would all take place on Level 1. Level 2 was the central level and housed six crew compartments arranged radially around a core that contained mechanical equipment. The crew compartments were surrounded by a water jacket to provide radiation shielding during solar flare events. Each crew compartment had a sleep restraint, a workstation, personal effects stowage and environmental control. Level 3 was for crew healthcare and more stowage. This level contained medical equipment, a medical examination area, treadmill and bicycle exercise machines, personal hygiene facilities and another Earth viewing window. Connecting the three levels was a corridor along the central core. At one end would be a pressurised vestibule that connected TransHab to its berthing point on the side of Node-3 on the Station.[42]

TransHab deployed the technology of inflatable structures, which for years had found applications in architectural enclosures on Earth. It advanced the state of the art of highly loaded composite structures, inflatable pressurised skins and the integration of both into a space habitat. Wrapping the flexible skin around the rigid tubular core to form a compact payload ensured that TransHab made the most of the Orbiter's payload bay volume but had enough structural strength to survive launch loads. Once on orbit, partly dismantling and reassembling the core structure would convert it into partitions to help to form the accommodation. TransHab's weight was about equally divided between the core and the skin. The flexible skin had three functions: to contain pressure, to act as a thermal insulator and to resist the penetration of debris. The skin comprised four functional layers: an abrasion-resistant and pressure containment membrane on the inside, a structural restraint layer, a micrometeoroid and debris shield, and a thermal insulation blanket on the outside.

TransHab was based on unproven technology and NASA carried out a test programme at Johnson Space Center in 1998 with three goals to validate the habitat's performance and safety. The first goal was proof that TransHab could survive the impact of a high-velocity projectile. For this, NASA assembled a multilayer sample of TransHab's skin and fired small metal balls at it at 7km per second. The sample comprised a triple-layer sandwich of open cell foam and Nextel sheets with a Kevlar membrane on the inside with an overall thickness of 300mm. The balls used were 1.7mm in diameter and made of aluminium. The test sample survived shot after shot. The second goal was proof that TransHab's pressurised membrane could contain atmospheric pressure safely. For this, NASA carried out a pressurisation test on a 7-metre diameter, 3-metre tall, doughnut-shaped prototype using a safety factor of four, which meant subjecting it to four times the normal atmospheric pressure of 101.3kPa. The prototype survived this test without leaks or failures. The third goal was proof of TransHab's ability to compress and fold up into a small volume for Shuttle launch. For this, NASA used the pressurisation test prototype, added a thick debris shield to build it up to its 300mm thickness, folded it up and then inflated it in a vacuum chamber to show how it would deploy in space. This test was also successful.[43]

However, the speed and success of TransHab's development ultimately led to its demise. It became a potential rival to the Station's cylindrical aluminium Habitation Module, which was at risk of being cut out of the Station due to its high cost. Congress had approved the Habitation Module and Boeing was due to fabricate it. Replacing this module with TransHab would lead to all sorts of contractual and political difficulties. Congressional language in NASA's appropriations in the year 2000 directed NASA to stop work on the project and NASA complied. From the beginning, one of the roles planned for the Station had been to act as a proving ground for the technologies needed to enable humans to carry out deep space exploration missions. The International Space Station offered an ideal site for testing technologies like those embodied in TransHab. Having invested much time and effort in developing and testing the habitat on the ground, NASA was unable to move its development forward. An opportunity to evolve the technology of space habitats was lost. In view of the later cancellation of the Station's Habitation Module, TransHab would have been an immensely valuable addition, providing successive crews with the comprehensive living accommodation that they deserved.

Concern was again mounting over the Station's costs and its assembly timetable. In 1993, NASA had committed to an annual funding limit of $2.1 billion and a total cost to completion of $17.4 billion. By 1998, NASA and its contractors had fabricated most of the American building blocks and the project was well into its test and integration phase. Cost growth and delivery delays, however, had been making news. In May 1997, the launch of the first building block had to be delayed from November that year to June 1998 while the completion date slipped by a full year. In September 1997, NASA Administrator Dan Goldin formed the Cost Assessment and Validation Task Force to carry out a review of the cost, schedule and partnership performance on the Station. The objective was to make recommendations on improving the project's business structure and cost management and to determine the life-cycle cost. Congress was asking for detailed information on Boeing's monthly staffing levels, hardware delivery dates, hardware launch plans, and post-assembly contract negotiations and it wanted an independent cost and schedule analysis. The task force would provide this.

TransHab inflatable module design This artist's view shows the TransHab inflatable module as it would have been attached to the Station. TransHab would have been berthed to one of the side ports of Node-3 of the Station's 1993-94 design configuration. Two circular windows on TransHab's underside look down towards Earth.

With assembly on orbit about to begin, the International Space Station's partners took an important step on 29 January 1998 and signed new agreements. On behalf of the United States, Acting Secretary of State, Strobe Talbott, signed the 1998 Intergovernmental Agreement on Space Station Cooperation with representatives of Russia, Japan, Canada and participating European Space Agency nations of Belgium, Denmark, France, Germany, Italy, the Netherlands, Norway, Spain, Sweden, Switzerland and the United Kingdom. NASA Administrator Dan Goldin also signed three bilateral Memoranda of Understanding with his overseas space agency counterparts Yuri Koptev of Roscosmos, Antonio Rodotà of the European Space Agency and William Evans of the Canadian Space Agency (signing with Japan took place later in February 1998). The new agreements superseded those from 1988 between the United States, Europe, Japan and Canada. They took into account the changes to the Station's programme resulting from significant Russian involvement and design changes made in 1993.

Article One of the 1998 Intergovernmental Agreement confirmed that the object 'is to establish a long-standing international cooperative framework among its partners, on the basis of genuine partnership, for detailed design, development, operation, and utilization of a permanently inhabited Civil International Space Station for peaceful purposes, in accordance with international law'. The vaguely defined 'genuine partnership' was open to interpretation. There was a vital partnership between the Station's leaders, America and Russia, and their responsibility to deliver the building blocks that formed the Station's 'foundations'. There was another partnership with Canada, whose robotic arm was an 'essential part'. There was a third partnership with Europe and Japan who would contribute building blocks that would significantly enhance the Station's capabilities. Presiding over all the partners' contributions was NASA, which had the lead role for overall programme management and coordination.[44] The Agreement provided a general description of the Station and its constituent parts and specified the roles of NASA and Roscosmos in particular. It established a management structure, clarified operations costs sharing, specified various crew issues and emphasised the use of all the Station's resources in a safe and efficient manner.

Though the document dealt with safety aspects, there were no descriptions of roles and responsibilities in emergencies. 'The Space Station will enable its users to take advantage of human ingenuity in connection with its low-gravity environment, the near-perfect vacuum of space and the vantage point for observing the Earth and the rest of the Universe.' Echoing the words of NASA Administrator James Beggs at the time that President Reagan gave the project the go-ahead in 1984, the Agreement envisioned that the Station's capabilities could grow to enable it to become a laboratory, an observatory, a transportation node, a servicing base, an assembly base, a research and technology centre, a storage depot and a staging base. Major milestones were the first building block launch in June 1998, permanent crews on board from January 1999 and, optimistically, assembly completion in December 2003 and full utilisation from December 2004 onwards.[45]

Dan Goldin's task force published its findings in April 1998. It was a bleak assessment that listed a range of problem areas. NASA's $2.1 billion annual spending limit had resulted in stretched-out procurements, deferred and untimely work and inadequate contingency planning, all of which had contributed to cost increases and assembly delays. Cost and assembly plans had been optimistic from the outset and continued to be so, while reserve budget levels were inadequate. NASA's attempt to reduce project staffing was unrealistic and did not reflect the need to maintain the appropriate mix of skills necessary to see the project through. Management challenges would continue among the Station's participating space agencies and contractors. The task force highlighted many areas of uncertainty and risk with the Station's development. The testing and approval of hardware was running late; the scale and complexity of the assembly phase on orbit was beyond current experience; a planned American-built crew return vehicle programme was far behind schedule and inadequately funded (it had already been scrapped); pre-launch integrated testing of multiple hardware elements had minimal reserve time to tackle unexpected problems, especially those concerning software; the Destiny Laboratory Module was running several months late; oral and written language complexities and international cultural differences were affecting training readiness; and spare parts inventories were being reduced to save money.

There was another set of risks that concerned Russia. The task force was critical of NASA's original claim that inviting Russia on to the project as partner would result in significant savings because Russia would build the Zarya Control Module and provide a crew return vehicle capability at its own

TransHab inflatable module prototype undergoing tests A TransHab inflatable prototype during tests in a large vacuum chamber at NASA Johnson Space Center in the late-1990s. After a highly successful prototype development and testing programme, Congress instructed NASA to drop TransHab as it was becoming a rival to the pressurised aluminium Habitation Module that already formed part of the International Space Station's design.

1998 Intergovernmental Agreement between the Station partners On 29 January 1998, representatives of fifteen International Space Station partner nations signed a new Intergovernmental Agreement on Space Station Cooperation in Washington DC. The new Agreement paved the way for the Station's assembly – then shortly to begin on orbit – and its subsequent operations. After signing, the international representatives went to Kennedy Space Center to view preparations for launching the Station's building blocks. This photograph shows them in the Space Station Processing Facility with the Unity Node-1 module in the background. Mounted in Unity's endcap is a Pressurized Mating Adapter (conical black object covered in cables). Dan Goldin, NASA Administrator, who forged the new International Space Station alliance with Yuri Koptev of Roscosmos, is standing at the front, sixth from the left.

expense. In fact, schedule savings from Russia's involvement were becoming a schedule slippage because of withheld Russian funding that was needed to complete its Station elements. 'Proceeding forward with full knowledge of the past' without contingencies, as the task force put it, was 'tantamount to accepting a level of risk that could drive US costs significantly higher' and NASA's contingency planning was deficient. There was the continuing funding shortfall in Russia over the Zvezda Service Module that was already running late and Russia had ongoing obligations to support its Mir station, not yet de-orbited. The task force also questioned NASA's claim that the Station's annual operations expenditure could be held to $1.3 billion. There was also equipment obsolescence and deterioration and NASA had not quantified replacement costs.[46]

The task force concluded that the Station had insufficient funding to cover its normal development and programme growth, its cost and scheduling risks, and measures needed for risk reduction. It forecast that the Station would be one to three years late in completion on orbit. There were nine recommendations of which the first – the most important – recommended that: 'the present program plan should be revised so that it is achievable within the financial resources available. Realistic milestone dates should be established as the basis for development of the program plan and internally defined target dates should be used for execution. If necessary, program content should be eliminated or deferred to fit within funding constraints'.[47] The message was clear. NASA had to bring the Station's costs and timetable under control. If something had to give way to make this happen, then so be it.

Then, in June 1998, the General Accounting Office jumped again into the Station's cost debate. It turned to the theme of the Station's final price tag in testimony given before the Committee on Science of the House of Representatives. The General Accounting Office had commented negatively in previous years on the Station's assembly budget. It was now concerned with life-cycle funding requirements. In April it had estimated that the Station would cost nearly $95.6 billion to develop, assemble and operate. This, however, was a base price that could increase with launch delays and specifically with the yet-to-be-developed (and never-to-be-developed) American crew return vehicle. The new stratospheric figure of $95.6 billion had increased from the office's June 1995 estimate of $93.9 billion. Development costs had grown by 20 per cent, while in-house staff costs had more than doubled. Funding reserves for the project had dropped from $3 billion in 1995 to $1.1 billion. Prime contractor Boeing was headed for a $817 million cost overrun. There was no funding allocated to cover the costs of tracking orbital debris that could pose a hazard to the Station during its lifetime. More Shuttle flights might be needed to support orbital assembly. The General Accounting Office's testimony was short and to the point. It offered no conclusions or recommendations.[48]

In 1998, Congress passed legislation aimed at stimulating the commercial use of space. The 1998 Commercial Space Act was intended 'to define policies and processes encouraging industrial investment in the development of space, and to remove regulatory barriers that might hinder commercial ventures'.[49] Mindful of the new Act, in November that year NASA published a commercial development plan for the Station. It had two objectives. In the short term, it would 'begin the transition to private investment and offset a share of the public cost for operating the space shuttle fleet and space station through commercial enterprises in open markets'. In the long term, it would 'establish the foundation for a marketplace and stimulate a national economy for space products and services in low-Earth orbit, where both demand and supply are dominated by the private sector'.

NASA's approach was to pursue a set of 'pathfinder' business opportunities that aimed at becoming profitable in the long run without public subsidies. The agency had chosen nine business areas and had begun business development activities with industry. The plan proposed to involve a prominent business school, private space investment consultants, two major aerospace companies, a major accounting firm and the agency's own commercial space centres. It raised several new ideas: dedicate a single Shuttle mission a year called 'Commerce Lab' to commercial research until the Station was complete; explore commercial flight access on expendable launch vehicles; transition from a cost-based to a value-based Shuttle flight pricing policy; apply brand management to advertising and customer relations; frame the intellectual property issue in a way that would not discourage commercial research. All this would be orchestrated by a new non-governmental organisation.[50]

A big question, though, was what, if anything, NASA could do to reduce Shuttle flight costs to and from the Station to stimulate commercial interest. These costs typically ran in the range of $25,000 per kilogramme for commercial payloads. Added to that base figure would be pre-launch costs for reviewing and approving the payloads and experiments for safety and reliability, and operating costs for managing and running the experiments at the Station. While large corporations might be able to cover these costs, they could well be an insurmountable barrier to start-up and small businesses that needed affordable space access for a new technology venture. There were also concerns about the quality of the scientific research that the Station would generate. The international science journal *Nature* – an opponent of human spaceflight – declared in a hostile article in its February 1998 issue that: 'there is potential harm to be done using questionable science to sell a project that might otherwise not pass muster. It confuses a public already struggling to distinguish between scientific consensus and junk claims. It makes it more difficult for space-based experimenters ... to be taken seriously by their colleagues'.[51] The journal did not believe that the Station was a science project. It was 'being undertaken for many reasons, including national prestige, employment for aerospace workers in many industrialised nations, global technopolitics – particularly a desire to help prop up the Russian aerospace industry – the unwillingness of Americans to abandon an astronaut programme to which they have a deep emotional attachment and, perhaps most important, the project's own political momentum'. There, the journal was right. From the Station's beginning in 1979, it had been promoted as a project in the national interest that would bring multiple benefits to the United States and its partners – technological, diplomatic, economic, educational and cultural as well as scientific. There was much to commend that line of thought.

Baikonur Cosmodrome, Kazakhstan
Night view of Soyuz launch complex No 6 at Baikonur Cosmodrome, Kazakhstan. The structural framework arms that support the rocket on the launchpad have been swung out of the way of the rocket's launch position at the centre of the view. These structures are essential to support Soyuz until just before launch because of the high winds that blow across Baikonur from the Steppe. Swung to horizontal positions to the far left and right are the access platforms. Photo: ESA.

Departure points
International Space Station assembly mission launch sites

By now, the focus of activity was shifting to the launch sites around the world where International Space Station payloads were arriving by land, sea and air and undergoing processing to prepare them for launch. There were four launch sites that would serve the Station. Baikonur Cosmodrome in Kazakhstan and Kennedy Space Center in Florida would launch the Station's building blocks as well as crews, equipment, supplies and consumables. The Guiana Space Centre in French Guiana and the Tanegashima Space Center in Japan would launch additional equipment, supplies and consumables but no building blocks.

Baikonur Cosmodrome would launch the Station's first building block to orbit. Baikonur was in the midst of the Eurasian Steppe to the east of the Aral Sea about 2,100km south-east of Moscow. The closer a launch site is to the equator, the greater the range of orbital inclinations it can serve. Baikonur was chosen as it is as far south as it was possible to get inside Soviet territory at the time. Baikonur could not launch to inclinations less than 45.6°. It was the only one of the Station's four launch sites located inland and concerns about overflight of China normally raised the minimum inclination to 51°. Well away from any populated areas were impact drop zones set aside for spent rocket stages and launch azimuths were limited to ensure that they did not fall outside these zones. Baikonur was the second busiest launch complex worldwide, just behind the combined launch complex of Kennedy Space Center and Cape Canaveral. The local climate has seasonal extremes. Temperatures can rise as high as 50°C in summer and fall to -40°C in winter with blizzards.[52]

The Cosmodrome is a place of considerable significance in the history of spaceflight. It dates from 1955 and is the oldest launch complex in the world. It was the site of the launch of the first satellite into space, Sputnik, on 4 October 1957 and Yuri Gagarin, the first human to travel to space, was launched from there on Vostok 1 in 1961. Since then, it has been the

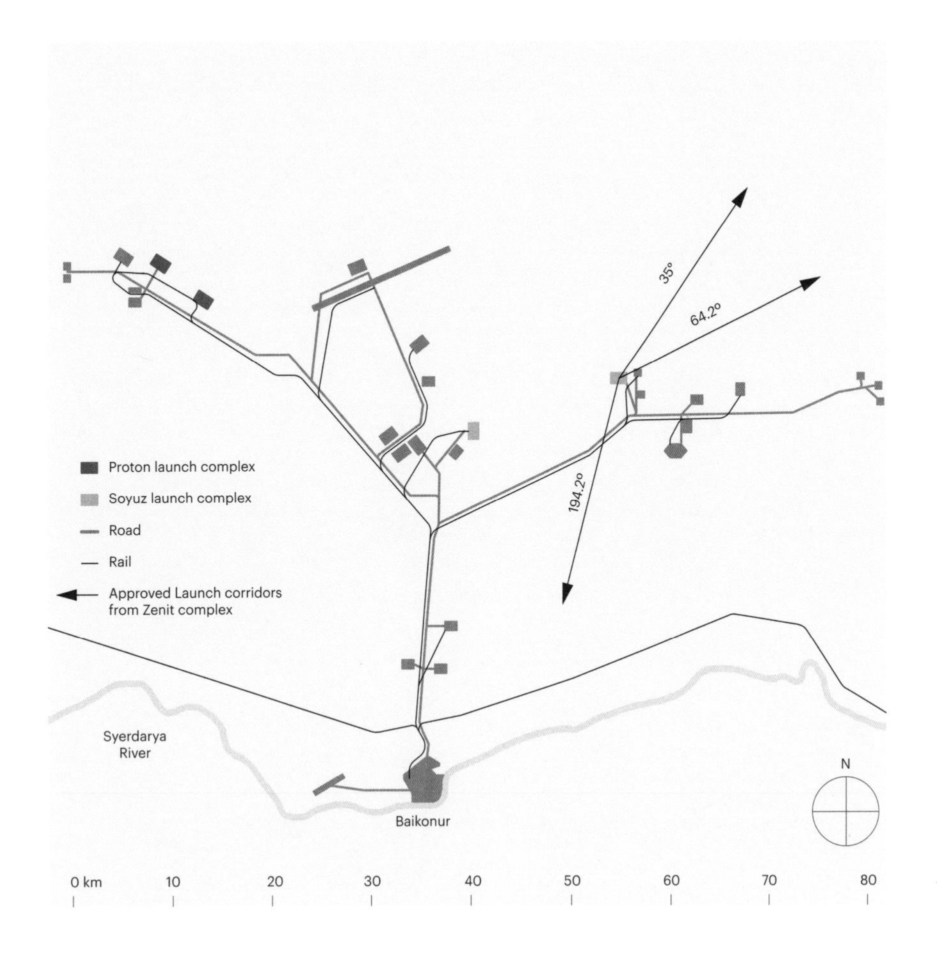

Baikonur Cosmodrome, Kazakhstan
This map shows the main features of Baikonur Cosmodrome in Kazakhstan. Baikonur was the only one of the International Space Station's launch sites that was inland. Land to the east of Baikonur on the Steppe is reserved for drop zones for spent rocket stages. Dating from 1955, the launch complex is the oldest in the world and was the site of the launch of the world's first satellite Sputnik in 1957. Image: author.

launch site for all Soviet and Russian human missions, lunar and planetary missions and geostationary satellite missions. Many Russian rockets began their flights at Baikonur over the years, including Proton, Rokot, Soyuz, Molniya, Tsyklon and Zenit.[53] In the late 1950s and 1960s, the Soviet Union built a 'science city' named Leninsk to the south of the launch complex to accommodate the thousands of workers based at the Cosmodrome. In 1995, its name was changed to Baikonur. The Soviets took the name from a small mining town about 322km to the north-east in an attempt to mislead the West as to its precise geographical location.[54] After the demise of the Soviet Union, the Cosmodrome came under the control of the Republic of Kazakhstan. Today, it is operated by Russia under multi-year leasing arrangements. There are ten operational launchpads and all major rocket parts and hardware arrive by rail on three tracks – one for rockets and spacecraft and two for rocket fuel.[55] The tracks fan out into a shape like the letter Y on plan leading to the numerous launch complexes. The Cosmodrome has a total surface land area of 7,360km^2.

Kennedy Space Center, Florida, USA This map shows the main features of Kennedy Space Center in Florida. All launches from the Center are to the east, out over the Atlantic, between azimuths of 35° and 120°. There were two Space Shuttle launchpads at the Center designated 39A and 39B. Image: author.

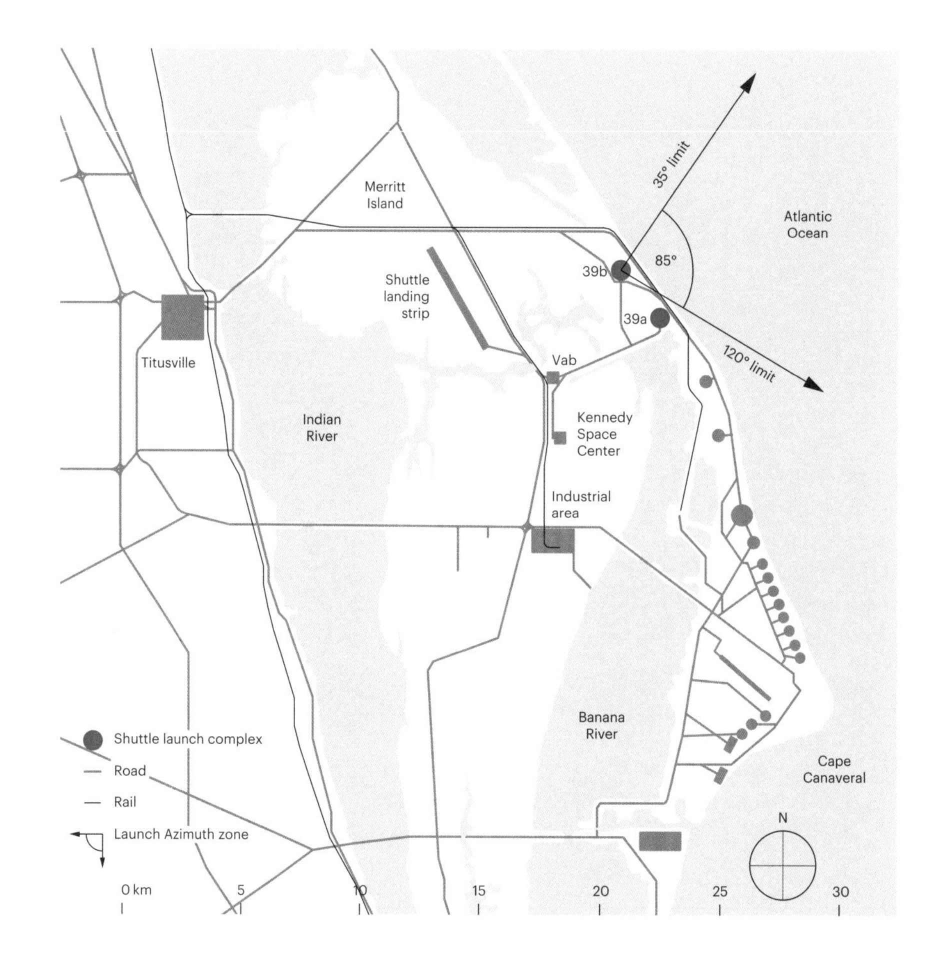

Kennedy Space Center is on Florida's central Atlantic coast. It is based at Merritt Island, a flat area of land separated from the mainland by the Indian River. This strip is studded with launchpads operated by Kennedy Space Center and Cape Canaveral Air Force Station. Together they form the principal American spaceport for all human spaceflight launches, all launches to geostationary orbit and most planetary and lunar missions.[56] Like Baikonur, the site was originally chosen to be as far south as possible on national territory and, in Kennedy's case, consistent with safe launches without any overflight of land. All launches are to the east, with launch azimuths within an arc that sweeps from 35° to 120°. Spent rocket stages and recoverable boosters fall into the Atlantic. The combined operations of the NASA Center and the Air Force Station make it the busiest launch facility in the world. The climate is warm in winter and hot in summer with frequent thunderstorms and occasional hurricanes between June and November.[57]

Like Baikonur Cosmodrome, Kennedy Space Center is a place of considerable importance in the history of spaceflight. Unlike Baikonur, it has also become a major visitor attraction. The nearest major city is Orlando and there are numerous resort towns nearby. Launches from Cape Canaveral began in 1947 when the US Air Force established the Eastern Test Range there for testing ballistic missiles. It is now known as the Cape Canaveral Air Force Station. Up to the Apollo missions, all NASA launches were carried out from US Air Force launchpads. In 1963, NASA established Kennedy Space Center on Merritt Island next to the Air Force Station and built its own launchpads for the Apollo and later the Space Shuttle programmes.[58] The first Apollo mission, Apollo 7, was launched to Earth orbit from Kennedy in 1968 and the first Shuttle Mission, STS-1, in 1981. The two Shuttle launchpads 39A and 39B were in regular use up to the end of the Space Shuttle era, except in the aftermath of the Challenger and Columbia Orbiter disasters.

Kennedy occupies a land area approximately 16km from east to west and 55km from north to south. Many of the buildings constructed for the Apollo programme were refitted for the Shuttle. Notable among these is the tall Vehicle Assembly Building, which is visible for miles around in the flat Florida landscape. Major facilities added for the Shuttle included an Orbiter landing runway 4.8km long, the Orbiter Processing Facility and the Solid Rocket Booster Assembly and Refurbishment Facility. In the early 1990s, NASA added the Space Station Processing Facility – a vast new building for processing all Space Station hardware elements launched on the Shuttle to the Station, as well as all payloads returned to the ground.

The Guiana Space Centre known in French as *Centre Spatial Guyanais* is Europe's main spaceport in French Guiana on the north-east coast of South America. The spaceport extends along 18km of the Atlantic coast from Kourou in the south-east to Sinnamary in the north-west. French Guiana was settled by France in the seventeenth century and is an overseas department or *département d'outre-mer*. As part of France, it is also part of the European Union and its currency is the euro. The Guiana spaceport was the best placed of all the Station's launch sites due to its proximity to the equator just 540km to the south. It can launch rockets directly out over the Atlantic in an azimuth arc from 349.5° to 91.5°, thus serving both equatorial and polar orbits. Its near-equatorial location gives its easterly launches a 17 per cent payload mass advantage over Kennedy.[59] France began rocket firings there in 1968 and satellite launches in 1970. In 1966, the European Launcher Development Organization moved its launch operations there from Woomera in Australia.[60]

The European Space Agency replaced the earlier launcher organisation in 1975 and the French Government offered to share the launch site with the new agency. In return, the European Space Agency funded the upgrade of the launch facilities for the Ariane rockets, then under development. Launch vehicle elements, payloads and propellants are shipped from France to a seaport at Kourou and then transferred to the spaceport. Ariane 5 was and

Kennedy Space Center, Florida, USA
Kennedy was the launch site for all Space Shuttle missions to the International Space Station. This photograph shows the Shuttle on its way to the launchpad complex on its Crawler-Transporter. Seen here, the Shuttle comprises the large brown external propellant tank, a white Solid Rocket Booster on each side of it and the Shuttle Orbiter behind with its wingtips just visible. Just to the right of centre in the middle distance is the huge Vehicle Assembly Building, originally built for the Apollo programme.

still is the largest rocket launched from the Guiana Space Centre. It was used to launch the Automated Transfer Vehicle, Europe's delivery spacecraft, to the International Space Station. Ariane 5 was joined in 2008 by the Russian Soyuz rocket with its new launch complex built as an alternative to Baikonur, and by Europe's new Vega rocket intended for the small satellite market. Ariane 5 launch operations are managed by Arianespace, a company largely controlled by the *Centre National d'Etudes Spatiales*, the French space agency. The operation of the spaceport as a whole is shared between the French and European space agencies.[61]

The fourth launch complex to launch International Space Station missions was Tanegashima Space Center in Japan. The launch complex lies on the south-eastern coastal tip of Tanegashima island and is the largest Japanese launch complex. Its launch capabilities comprise the Yoshinobu complex with two launchpads for the H-IIA and H-IIB rockets and the Osaki launch complex and the Takesaki range for smaller rockets. Its facilities include vehicle assembly, spacecraft test and assembly buildings, a spacecraft and fairing assembly building, firing test facilities, a launch control centre and tracking stations.[62]

Guiana Space Centre, French Guiana
The Guiana Space Centre or *Centre Spatial Guyanais* is on the north-east coast of South America and is the site of Europe's principal launch complex. This photograph shows Europe's Ariane 5 rocket on its way to the launchpad. On the left is the Ariane 5 Final Assembly Building. The Ariane 5 moves to its launchpad on a mobile platform that travels on two sets of rails. The launchpad is in the distance at the top right of the photograph. Image: ESA.

Fleet review
International Space Station assembly mission launch vehicles

Russia's main rocket for launching its International Space Station building blocks was the Proton-K. Produced by the Khrunichev State Research and Production Space Center, it was the largest rocket in Russia's fleet[63] and would deliver the first building block – the Zarya Control Module – to orbit. Development of the Proton began in the early 1960s, directed by Soviet academician, VN Chelomei. Conceived to fulfil as a dual-role as a ballistic missile and a satellite launcher, it evolved to become a space launch vehicle. The rocket first flew in a two-stage configuration in July 1965 and delivered a scientific satellite called Proton to orbit, from which it took its name. Exploration missions launched by Proton included the Zond, Luna, Venera, Mars, Vega and Phobos missions.[64] Proton is launched from Baikonur where it has achieved a long-term average of about twelve launches per year.

The Proton-K that launched the Russian building blocks – the Zarya and Zvezda modules – was a three-stage vehicle with an overall height of 57.2 metres and a gross lift-off mass of 691.5 tonnes. Each stage was fuelled by a mixture of nitrogen tetroxide oxidiser and dimethylhydrazine – storable chemicals that require no cryogenic conditioning of the type needed for liquid oxygen or hydrogen. The Proton-K's first stage comprises a central oxidiser tank surrounded by six smaller fuel tanks, each with an RD-253 engine at its base. The second and third stages combine fuel and oxidiser tanks within their cylindrical volumes. The second stage is powered by four RD-0210 engines and the third by a single RD-0210 engine. On top of the

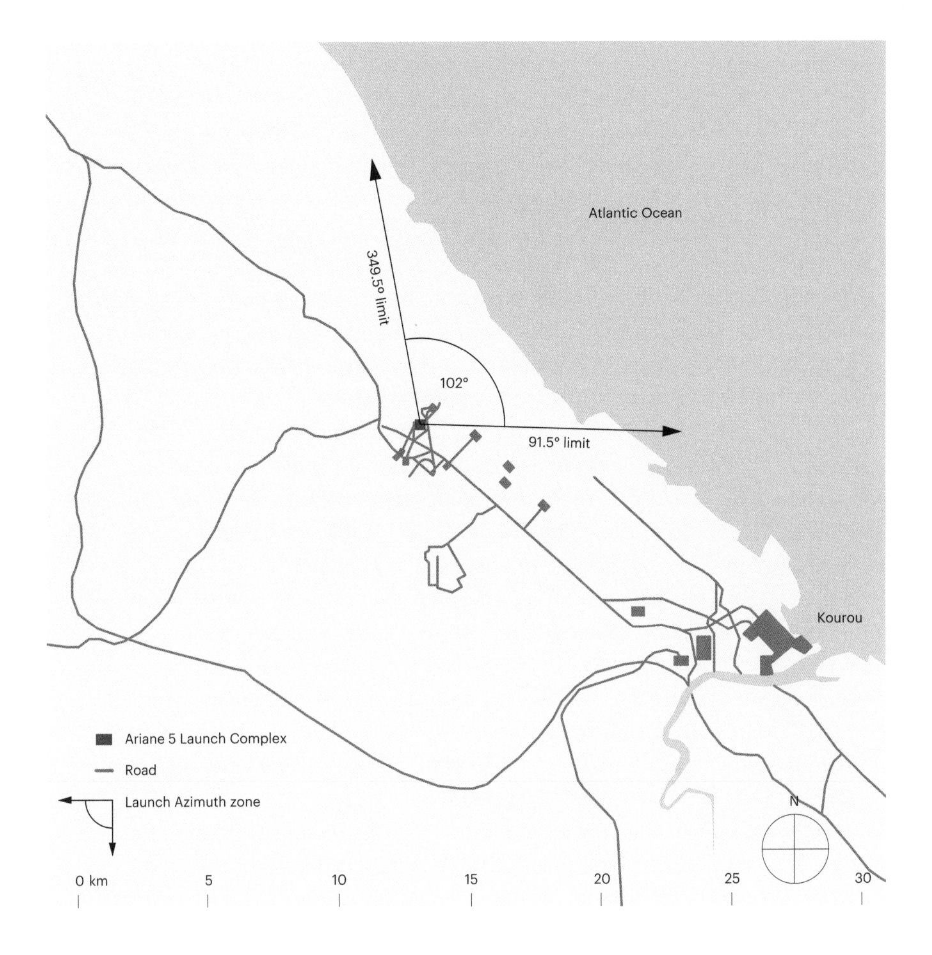

Guiana Space Centre, French Guiana
Launches from the Guiana Space Centre occur out over the Atlantic in an azimuth arc from 349.5° to 91.5°. Because of its proximity to the equator, the Centre is an excellent place to carry out launches to the International Space Station. This is due to the beneficial effect of the Earth's rotation rate (highest at the equator) on easterly launches.
Image: author.

third stage is the two-part payload fairing that housed the Russian-built Station modules. Proton vehicles are fabricated from aluminium alloy in sections and then shipped by rail to Baikonur for assembly before launch.[65]

Russia's Soyuz rocket launched the Soyuz and Progress spacecraft to the Station as well as two small Russian building blocks. Soyuz evolved from the Soviet R-7A intercontinental ballistic missile, first flown in May 1957. A variation of the R-7 missile delivered the first space satellite to orbit, Sputnik, in October 1957 and in April 1961, Yuri Gagarin flew on another R-7 variant to become the first human in space.[66] Soyuz launches take place from two locations in Russia – the Baikonur and Plesetsk Cosmodromes. Soyuz rockets are built by the TsSKB-Progress State Research and Production Space Rocket Center in Samara, Russia.[67]

A modern Soyuz rocket has four stages. The first stage comprises four strap-on boosters of tapered cylindrical shape assembled around a central core (the second-stage) that supports them. Each booster contains a kerosene fuel tank, a liquid oxygen oxidiser tank, a liquid nitrogen pressurisation tank, a hydrogen peroxide tank and four RD-107A engines with three-axis flight control provided by aerofins and movable vernier thrusters. The second stage is of similar construction to the boosters with similar tanks – kerosene, liquid oxygen, liquid nitrogen and hydrogen peroxide – and a single RD-108 engine with four combustion chambers and four movable vernier thrusters that provide three-axis flight control after booster separation. The third stage is mounted above the second stage and attached to it by a cylindrical lattice frame. This stage comprises kerosene and liquid oxygen tanks either side of an avionics bay and has a single RD-0110 engine with four combustion chambers and four vernier movable thrusters. The fourth stage, named Fregat, functions as an independent orbital vehicle, enabling Soyuz to deliver payloads to different types of orbit. Fregat consists of six spherical tanks – four for propellants and two for avionics – and has its own guidance, navigation, control and communications capabilities. It is incorporated in the Soyuz payload fairing together with the payload and payload dispenser.[68]

Soyuz is the most successful multipurpose launch vehicle ever built and has been used for just about every type of job in space including delivering satellites and research spacecraft for telecommunications, Earth observation, weather monitoring and scientific missions and crewed and cargo spacecraft to both the Russian Mir station and the International Space Station. The basic Soyuz configuration in use today was introduced in 1966 and has flown over 800 missions.[69]

The Space Shuttle was the most famous launch vehicle system in the history of spaceflight and the only reusable spacecraft system designed for both crew and cargo transportation. Though it needs little introduction, it is useful to review its chief features here, as it was by far the most important space vehicle during the Station's assembly years on orbit. It launched most of the major building blocks and many smaller payloads besides. The Space Shuttle was often described as a launch system as it comprised four distinct elements brought together at launch: the Orbiter spacecraft, the External Tank containing liquid oxygen and liquid hydrogen propellants and the pair of Solid Rocket Boosters containing a solid propellant mixture of aluminium powder, iron powder, ammonium perchlorate and polybutadiene. The Orbiter was a fully reusable fly-back spacecraft comparable in size and weight to a commercial passenger aircraft. The twin Solid Rocket Boosters were recovered after each launch and reconditioned for reflight. Jettisoned on orbit, the External Tank burnt up during re-entry. Development of the Space Shuttle began in the early 1970s at a time when the Apollo lunar missions were well under way and America was turning to the next major space goal – the design of a vehicle that would provide regular access for crews and cargo to Earth orbit. As President Nixon put it at the time, the Shuttle would 'transform the space frontier of the 1970s into familiar territory, easily accessible for human

Proton launch vehicle A Proton launch vehicle on the launchpad at Baikonur Cosmodrome. The Proton was a critically important launch vehicle for the International Space Station. It delivered the first and fifth Station building blocks – the Zarya and Zvezda modules. These two building blocks provided the physical foundations and life support functions to support the assembly of the rest of the Station. The Proton has flown over 300 missions since its introduction in the mid-1960s. It can fly in four launch configurations: with three stages, four stages and two more four stages with different upper stages. Image: RSC Energia.

endeavour in the 1980s and 1990s'.[70] The Shuttle was tasked to perform multiple roles covering civilian, military, research and commercial missions. There was a need for a payload bay size and mass lift capability that would permit the delivery of large satellites to orbit; and there was a requirement for a reusable vehicle design to reduce the high cost of missions using expendable launchers, which were in common use at the time. NASA's original plan was to develop a fully reusable spacecraft but this gave way to a design from the company North American Rockwell for a partly reusable vehicle system to reduce development costs. The project was given the go-ahead in 1972 and NASA chose Rockwell's Space Transportation Division as main contractor for the development of the fleet of Orbiters.

Contracts for the development of the External Tanks and the Solid Rocket Boosters went respectively to Martin Marietta in 1973 and Morton Thiokol in 1974. Rockwell also had the contract for overall system integration.[71] Development and flight testing followed a cautious sequence. The first Orbiter Enterprise was rolled out in 1976 at Palmdale in California and moved about 50km north to Edwards Air Force Base in the Mojave Desert for initial flight tests to demonstrate its unpowered glide approach and landing abilities. The Orbiter Enterprise was essentially an unpowered prototype intended for launch processing, approach, landing and ferrying tests. The first flight tests were carried out with Enterprise mounted on top of an adapted Boeing 747 aircraft. These captive tests led to free flight tests in which astronaut crews flew Enterprise back to the ground and were followed, in turn, by ferry tests with the Orbiter mounted on the Boeing 747 for long distance point-to-point delivery. In 1978, the Boeing 747 flew Enterprise to Marshall Spaceflight Center in Alabama for mating with the External Tank for vibration tests. 1979 was a milestone year in which Enterprise was ferried to Kennedy Space Center where it was mated with the External Tank and Solid Rocket Boosters and moved to the launchpad for pre-launch tests.

Meanwhile, the first Orbiter to fly to space, Columbia, was rolled out and ferried to Kennedy to begin preparations for the first spaceflight. NASA also gave Rockwell the go-ahead to manufacture two additional Orbiters, Discovery and Atlantis, and convert a prototype used for testing the Orbiter's structural design to another Orbiter, Challenger. A fleet of four Orbiters was in the making. The first launch of Columbia took place in April 1981, nine years after the project's go-ahead. Columbia spent two days on orbit and returned to Edwards Air Force Base. Columbia was followed by the Challenger, which first flew to space in April 1983. Discovery made its maiden flight in August 1984 and Atlantis in October 1985. Following the loss of the Challenger in January 1986, NASA ordered the replacement Orbiter Endeavour, which first flew in May 1992.[72]

The Space Shuttle became the backbone of the America human spaceflight programme and spanned three decades from its first flight

Soyuz launch vehicle A Soyuz TMA-9 launch vehicle is prepared for launch at Baikonur Cosmodrome. With the retirement of the Space Shuttle fleet in 2011, Soyuz became the only launch vehicle able to ferry astronauts and cosmonauts between Earth and the Space Station. Without this vehicle, the Station would be unable to continue operations. The Soyuz rocket and spacecraft series is regarded as the most successful in the history of spaceflight. Over 1,700 launches of it in various forms have taken place since its ancestor, the Soviet R-7A intercontinental ballistic missile, first flew in 1957. Soyuz is launched from Baikonur, Plesetsk Cosmodrome in Russia and now Guiana Space Centre in French Guiana. Image: RSC Energia.

Ariane 5 The Ariane 5 is Europe's principal launch vehicle and the latest in a family of European Ariane rockets dating back to the early 1980s. Built by Arianespace in France, the Ariane 5 can accommodate single or double payloads. Launched from the Guiana Space Centre, it delivered the Automated Transfer Vehicle – Europe's resupply spacecraft – to the International Space Station. Image: ESA.

in 1981 to its last in 2011. Hundreds of astronauts and cosmonauts flew on Shuttle missions. Without it, the Hubble Space Telescope, the Magellan and Galileo probes, the Chandra X-ray Observatory, the European Spacelab laboratory, the International Space Station and a host of smaller research and commercial missions would not have been possible. By the millennium in 2000, the Shuttle had carried out ninety-five successful missions. However, the Shuttle's record was marred by two tragedies – the losses of the Orbiters Challenger in January 1986 and Columbia in January 2003. Investigations into the causes of those losses found that human error, technical defects, and timetable pressures converged to trigger hardware failures. In the case of Challenger, it was the rupture of an assembly joint in the Solid Rocket Booster casing.[73] In Columbia's case it was the shedding of foam insulation around the External Tank's exterior.[74] Neither loss was the fault of the Orbiters – the most complex and risk-prone elements of the Space Shuttle system. They all performed superbly during their lifetimes.

The Space Shuttle system could transport cargo with a mass of nearly 25 tonnes to low Earth orbit and back in a payload bay measuring 18.3 by 4.6 metres; it could carry multiple cargo payloads of widely different shape and size on the same mission; it could ferry a crew of up to eight to orbit and back in a shirtsleeve environment and ten in an emergency; its acceleration loads during ascent or descent never exceeded three times the force of gravity; it could remain in space for up to sixteen days on its own or attached to the Station; it had a cross-range manoeuvring capability during return, enabling it to land on either east or west coasts of the American mainland; and it could land on a long runway of the type found at major airports.[75] At the time of writing, no other spacefaring nation has developed a launch vehicle system that begins to approach the multiple capabilities of the Space Shuttle. Neither has NASA replaced it with anything as capable. It will remain in a class of its own for a long time to come.

Ariane 5 would be Europe's rocket for transportation to the International Space Station. It was the latest in a line of increasingly powerful and capable rockets that began development in Europe in the early 1980s. Europe's rocket programme had its roots in the European Launcher Development Organization established in 1964, using the British Blue Streak intercontinental ballistic missile as the first stage of a new rocket named Europa. Europa was not successful and was abandoned in 1973.[76] The European Space Agency – an amalgamation of the European Launcher Development Organisation and the European Space Research Organisation – was approved by convention in May 1975, at which point the new fifteen-nation space agency effectively came into being.[77] France spearheaded the Ariane programme in 1972 to provide Europe with an independent satellite launch capability to reduce its reliance on America and the Soviet Union. The inaugural flight of Ariane 1 took place in December 1979. It was replaced by progressively more powerful versions – Ariane 2, Ariane 3 and Ariane 4.

The Ariane 4 came in six versions, giving it the potential to launch different types of payload. Between 1995 and 2003, there were seventy-four successful Ariane 4 launches.[78]

Ariane 5, the rocket that would deliver Europe's Automated Transfer Vehicles to the Station, is produced by Arianespace of Evry-Courcouronnes, France. It began operations in 1999. The first stage consists of two Solid Rocket Boosters and engines, powered by a mixture of ammonium perchlorate, powdered aluminium and polybutadiene. These are attached to opposite sides of a central core stage consisting of in-line liquid oxygen and liquid hydrogen tanks powering a single Vulcain engine. Above the core stage is the third stage comprising storable propellants, a third stage engine and an equipment bay for avionics and other subsystems. Ariane 5 can deliver single or double payloads to orbit and the payload fairing is produced in different lengths with compartments named Speltra and Sylda that can cater for a single payload or two stacked payloads of different height. Some fifty companies across Europe were involved in the vehicle's development with the largest participants being France, Germany and Italy.[79]

Japan's means of transportation to the Station was the H-IIB rocket. The H-IIB was an upgraded and more powerful version of its H-IIA rocket and it had two main purposes. It would deliver Japan's H-II Transfer Vehicle to the Station with its cargo of astronaut supplies as well as consumables, experiments, spare parts and other equipment. Then it would launch large satellites. The H-IIB was a two-stage rocket using liquid hydrogen and liquid oxygen as propellants in its first and second stages. Four strap-on solid propellant boosters were attached to its first stage.[80]

Payload loading at the Shuttle launchpad
Loading of Station payloads at the Space Shuttle launchpads used a structure called the Rotating Service Structure. This giant steel framework provided protected and shielded access directly to Orbiter payload bays. The main structural element was a vertically pivoted bridge that contained an environmentally controlled chamber called the Payload Changeout Room. A Station building block, such as a pressurised module, arrived at the launchpad in the vertical canister and was transferred to the Payload Changeout Room. The Rotating Service Structure then swung it around to close and mate with the Orbiter's fuselage. Opening the Orbiter payload bay doors enabled the payload to move the short distance into the spacecraft's payload bay.

Cargo terminals
Launch facilities and payload processing, mid-1990s onwards

The fabrication of payloads destined for launch to the International Space Station generally occurred far from the launch centres and the first step in their journey to orbit was to transport them from their places of origin to their places of launch. The Khrunichev State Research and Production Space Center built the Russian Zarya and Zvezda modules in Moscow. There, they underwent pre-launch ground testing before being shipped to Baikonur by rail – a distance of about 2,500km.

Of the Space Shuttle system, the bulky External Tanks were the only launch elements that needed regular transport to Kennedy Space Center. The Orbiters and Solid Rocket Boosters were reusable elements, which were serviced and maintained at Kennedy. The External Tanks were manufactured by Lockheed Martin at the Michoud Assembly Facility, east of New Orleans. From there, they were loaded on to barges and towed to Kennedy Space Center along the US Intracoastal Waterway, which runs along the Gulf of Mexico and Atlantic coasts. Each tank travelled about 2,000km.

Two NASA-owned vessels carried out the towing – the M/V Freedom Star and M/V Liberty Star. They were originally built in the late-1970s to recover the Solid Rocket Boosters from the Atlantic after separation from the Space Shuttle during ascent and continued in that role. Long-distance shipping was, and is, also involved in delivering the Ariane 5 to French Guiana. The rocket's parts are shipped from France to the port at Kourou, the transatlantic distance being about 6,700km.

In America, the Station building blocks and outfitting hardware were ferried from the prime contractor, Boeing, to Kennedy Space Center by a specially built aircraft. Boeing's Station fabrication facilities were based at Huntington Beach and Canoga Park near Los Angeles and at Huntsville in Alabama. Los Angeles and Huntsville were respectively about 3,500km and 1,000km from the launch site. The aircraft used was the NASA Super Guppy, originally built in America and later bought by Airbus Industrie in Europe. The European Space Agency supplied NASA with the aircraft as an offset to the cost of NASA's launch of the European Columbus Module. The Super Guppy had an exceptionally wide fuselage that could accommodate the Station's modules and truss pieces and it could carry payloads of up to 24 tonnes. The aircraft's nose was hinged to reveal the complete fuselage cross-section for loading, and the cargo hold had a system of rails with the loading process assisted by electric winches.[81] Europe has its own airlifter for large payloads called the Beluga, or A300-600ST Super Transporter, originally built to ferry Airbus aircraft parts around Europe. The Beluga delivered the European Columbus Module to Kennedy in 2006. The aircraft design is similar to the Super Guppy with front-loading via a hinged nose.[82]

Of the four sites that would send rockets and spacecraft on their way to the International Space Station, Kennedy Space Center was by far the most important. It had the task of processing and launching most of the Station's building blocks – modules, nodes and truss pieces – as well as internal racks and experiments, life science specimens, supplies and consumables, and a host of other items. The challenge was to carry out the processing and launching of all this material as efficiently, economically and safely as possible over a decade or more of Station assembly missions.

By the time of the first Station mission in 1998, Space Shuttle launches from Kennedy had been in progress for nearly two decades. International Space Station missions would make full use of this infrastructure and its iconic landmarks: the huge Vehicle Assembly Building for the integration of the Space Shuttles, the tank-like Crawler-Transporter for their transfer to the launchpad and the intricate Rotating Service Structure for the installation of payloads in the Orbiter bays before launch. The Station, however, needed its own processing capabilities for the building blocks that would pass through Kennedy. Chief among the newly constructed buildings was the Space Station Processing Facility, a vast industrial structure with a floor area of 42,455m^2 built between 1991 and 1994. Its task was to prepare the majority

Space Shuttle The Space Shuttle Orbiter Discovery undergoes maintenance in the Orbiter Processing Facility at Kennedy Space Center in August 2005 after its return from Mission STS-114. This mission marked the return to flight of the Space Shuttle fleet after the Shuttle Columbia disaster in 2003. More insulation material fell off the Shuttle's big propellant tank shortly after this mission's launch. This had been the cause of Columbia's loss. NASA again put the Shuttle fleet on hold while it continued to work on the tank's insulation problem and it was not until July 2006 that missions resumed.

Comparison of International Space Station launch vehicles These diagrams compare the launch and ascent characteristics and trajectories of four launch vehicles used to deliver building blocks, crews and supplies to the International Space Station. The Space Shuttle was American, the Ariane 5 is European and the Proton and Soyuz are Russian. The Ariane 5, Proton and Soyuz are conventional expendable launch vehicles that discard all their stages during their ascent to orbit. The Space Shuttle discarded only the External Tank. The twin Solid Rocket Boosters were recovered after splashdown for reconditioning and the Orbiter spaceplanes flew back for runway landings. Images: author.

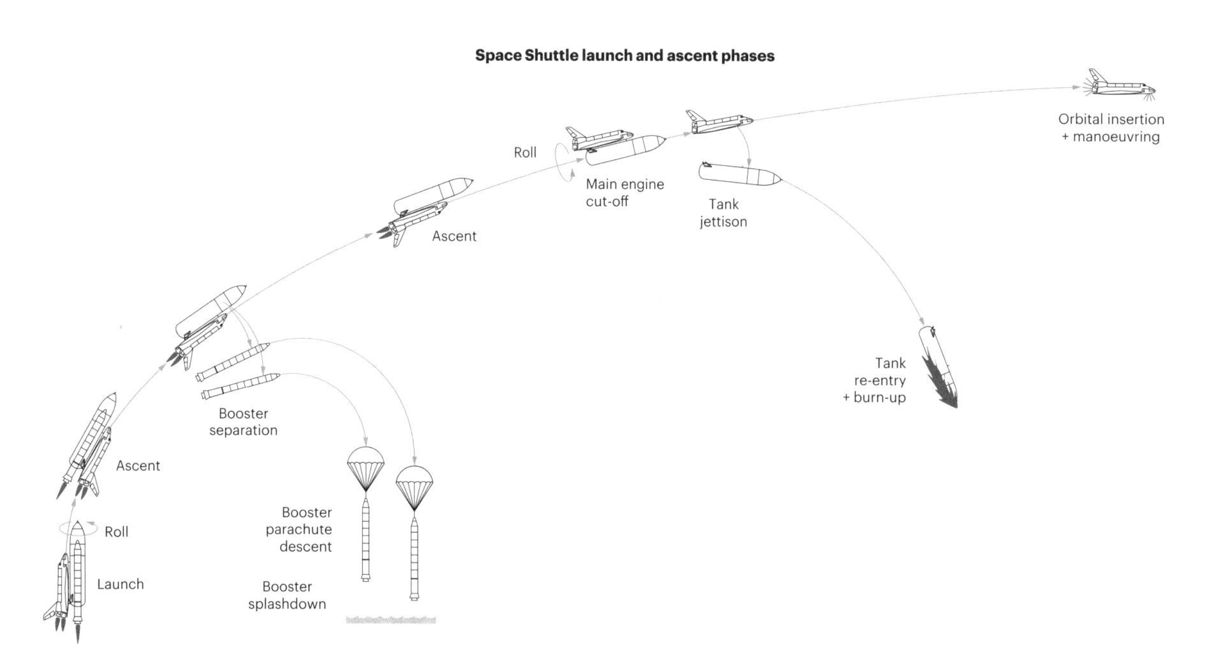

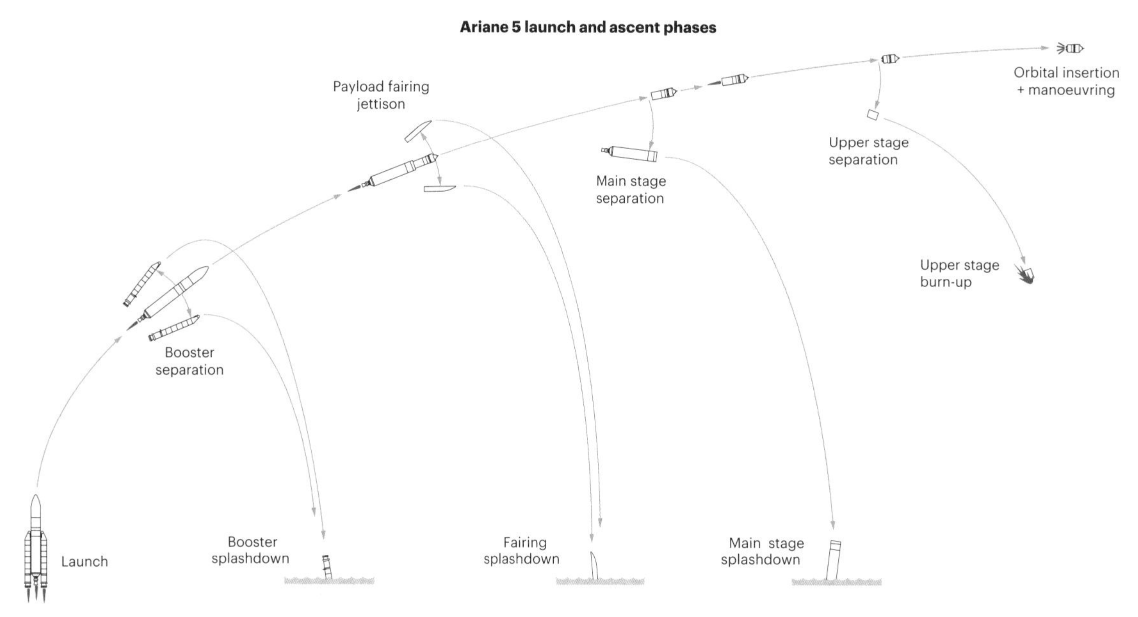

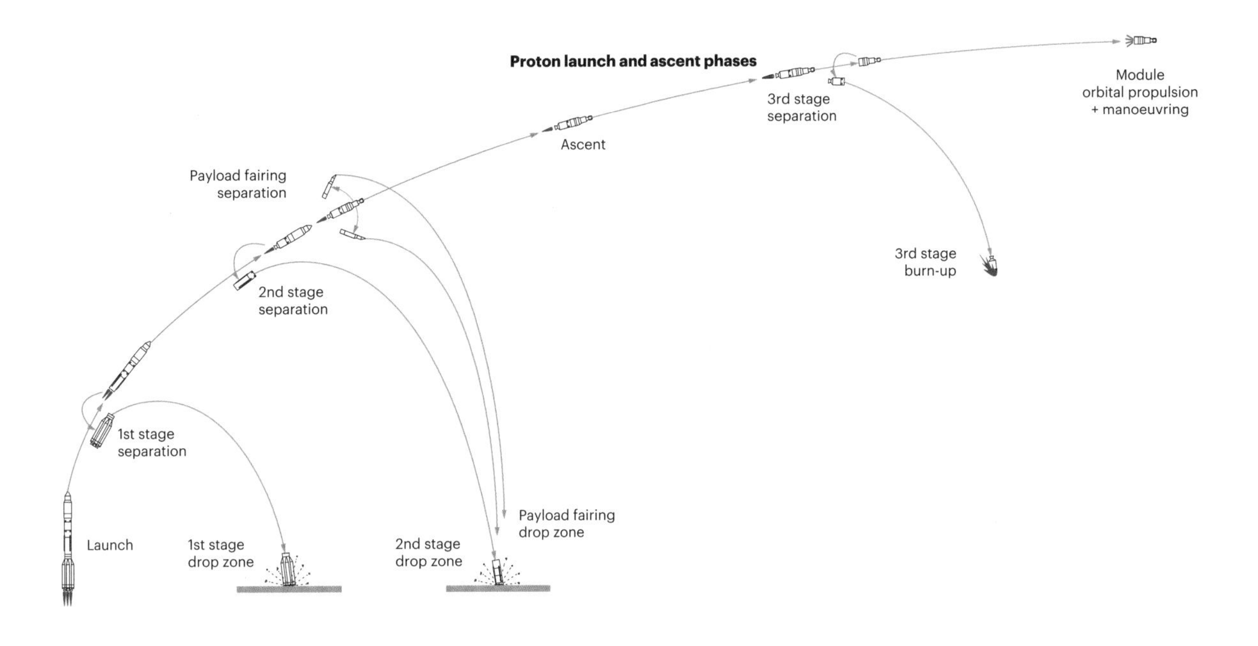
Proton launch and ascent phases
Module
orbital propulsion
+ manoeuvring
3rd stage
separation
Ascent
Payload fairing
separation
3rd stage
burn-up
2nd stage
separation
1st stage
separation
Payload fairing
drop zone
Launch
1st stage
drop zone
2nd stage
drop zone

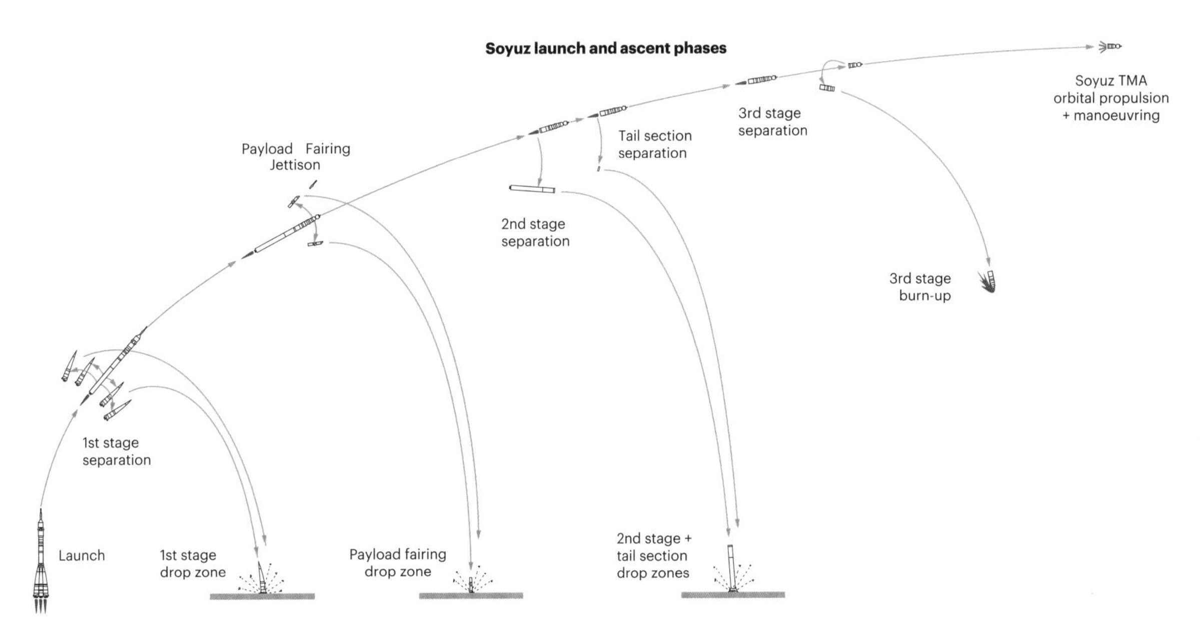
Soyuz launch and ascent phases
Soyuz TMA
orbital propulsion
+ manoeuvring
3rd stage
separation
Tail section
separation
Payload Fairing
Jettison
2nd stage
separation
3rd stage
burn-up
1st stage
separation
Launch
1st stage
drop zone
Payload fairing
drop zone
2nd stage +
tail section
drop zones

Columbus Module aircraft transportation
The European Columbus Module arriving sealed, sheathed and carefully encased at Kennedy Space Center in Florida, in 2006, after its flight from Europe in the A300-600ST Super Transporter aircraft named Beluga. The aircraft was originally built to ferry Airbus aircraft parts around Europe.

of the Station's building blocks for launch. The first of these to arrive was Node-1, named Unity, in June 1997. The Space Shuttle was due to deliver Unity to orbit on the first American assembly mission. The Space Station Processing Facility contained a cavernous high bay and a smaller intermediate bay for handling all the hardware, as well as a large airlock and multiple laboratories, all designed to 100,000-Class Clean Room standards. Each incoming building block entered the facility by means of an airlock. It would spend several months inside while being outfitted with equipment. As it was not possible to test the physical assembly and interconnection of these large payloads before launch, simulated assembly with a computer model was used to discover any misalignment problems or interface incompatibilities, giving engineers an opportunity to address any problems on the ground. During testing and processing, the large payloads were accessible to Station crews for training purposes.[83]

The high bay was 32 metres wide by 110 metres long by 19 metres high. It contained eight processing sites arranged in two parallel lines served by overhead travelling bridge cranes. The high bay's main purpose was to process the Station's modules and truss pieces. The intermediate bay was 15 metres wide by 103 metres long by 9 metres high. Placed along one side of the high bay, its purpose was to process racks and experiments and then transfer them to the high bay for installation in the modules and on the truss pieces.[84]

Once completed, each building block payload continued towards launch on the Space Shuttle. Loaded into an air-conditioned payload canister the size of the Orbiter payload bay, it would leave the facility from the same airlock at which it had arrived. Moved on a wheeled transporter, the canister would take one of two paths: either it would travel horizontally to the Orbiter Processing Facility for loading into the Orbiter's payload bay before the Orbiter's vertical rotation and integration with the Space Shuttle stack; or it would travel horizontally to the Canister Rotation Facility where it would be swung upright for moving directly to the launchpad on its transporter for vertical loading into the waiting Space Shuttle. The Canister Rotation Facility was a tower-like building also on a site at Kennedy's industrial area. Built in 1993, it had a high bay, 43 metres tall, and its own bridge crane.[85]

The facilities and equipment at Kennedy were brought to a high state of readiness well before the launch of the first Station payload. Everything was thoroughly tested, verified and approved before being put into use. The world would be watching as Kennedy and the other launch sites would swing into action as the daunting task of assembling the Station in space began.

BELUGA
Link-Belt
690

D
E
F
CETA
C
G
B
H

P1 Truss Astronauts Michael Lopez-Alegria and John Herrington work on the P1 Truss piece just after its installation on the Station in November 2002. They appear on either side of one of the wishbone-shaped keel pins that secured the P1 in Atlantis's payload bay during flight to the Station. They are in the process of removing the keel pin from the truss.

P3/P4 Truss Space Shuttle Mission STS-115 brought up the P3/P4 Truss piece, seen in this photograph, in September 2006. It was the first mission after the Columbia disaster to resume Station assembly operations. Clearly visible in this view is the Canadarm2 robotic arm mounted on its railcar. Just below it towards the end of the truss is astronaut Daniel Burbank with his legs pointing downwards. Difficult to spot in this photograph is a second astronaut, Steven MacLean, who is just above Burbank's head. The T-shaped object projecting out to the right at the end of the truss is a folded solar array stowed in its blanket box.

Canada

Canada
CETA 1
CETA 2

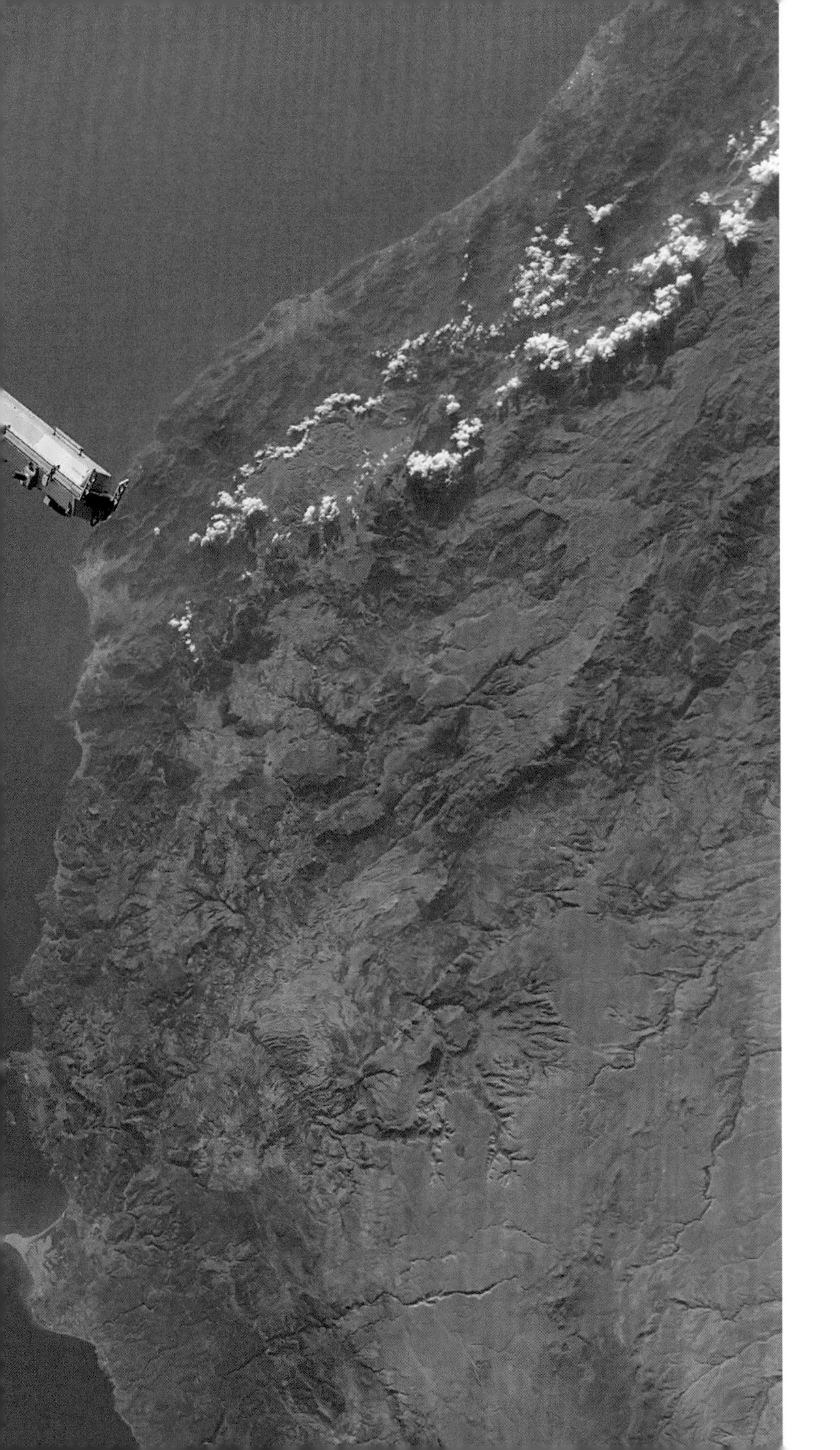

P3/P4 Truss Another photograph of the P3/ P4 Truss piece during its installation on the Station in September 2006.

Endurance and achievement 1999–2011

Firm foothold
First cluster of modules, 1998-1999

'It's like building a ship in the middle of the ocean from the keel up. You've got to float and you've got to sail. All this has to occur while you're actually building the ship, and that's what [the] Station is like.' Those words from an International Space Station programme manager seemed to sum up the huge challenge of building the most complex engineering project to be carried out beyond the Earth's surface.

On 20 November 1998, the assembly of the International Space Station began on orbit. The first building block on the launchpad on Mission 1 A/R was the Russian Zarya Control Module. Russia had chosen the name to signal a new era of cooperation between the Station's participating nations but especially between America and itself.[1] Zarya lifted off from the Baikonur Cosmodrome on the Proton rocket. Launching the Proton and putting Zarya on to a precise orbit was complex and involved many carefully planned steps over several days. After launch, ascent and ejection by the Proton, Russian ground controllers manoeuvred Zarya by remote control to the correct orbit using the spacecraft's built-in propulsion capabilities – not something needed on the American, European or Japanese modules as the Space Shuttle would deliver these as passive payloads. On the second day after launch, Zarya's engine was tested with a ten second burn, followed

Zvezda Service Module The International Space Station photographed after the addition of the Zvezda Service Module. On the far right is Unity Node-1 with a Pressurized Mating Adapter attached to each end of it. One adapter connects Unity to Zarya in the middle. The Zvezda Service Module is attached to the left end of Zarya. On the far left is a docked Soyuz spacecraft.

The first assembly flight – Zarya Control Module Photograph taken from the Shuttle Orbiter Endeavour of the Zarya Control Module on orbit as it approached the Shuttle for docking. At its left end is a port for berthing Zarya to the next Station building block. At its right end is a cluster of docking ports on a spherical node. These would enable Russian Soyuz and Progress spacecraft to dock with Zarya.

by another burn that put it onto a 346 × 246km orbit. On the fourth day, two further burns raised its orbit to 383 × 310km, followed on the fifth day by final burns that put it onto a circular 390km orbit – the orbit needed for rendezvous with the Space Shuttle coming up on the next mission. On the sixth day, ground controllers checked out Zarya to make sure it was in full working order and ready for the Orbiter's rendezvous, with tests performed on its on-board computer, docking control, solar arrays and batteries.[2]

Following closely Zarya's arrival on orbit was the launch of the first American building blocks on the Shuttle on 4 December 1998. Mission 2A with the Orbiter Endeavour (STS-88) delivered three separate elements – Unity Node-1 and two Pressurized Mating Adapters.[3] It was vital that NASA followed up Zarya quickly to establish an American presence and demonstrate that the assembly timetable had got off to a healthy start.

As its name implied, Unity with its six berthing ports would provide the link between the Russian and American operational sectors of the Station. Johnson Space Center described it as the 'cornerstone for a home in orbit'.[4] Unity arrived on orbit with the two Pressurized Mating Adapters already attached to it at each end. One would supply the physical attachment of Unity to Zarya and the other would provide a docking port for Orbiters arriving on subsequent missions. Endeavour did not dock physically with Zarya after rendezvous. Standing off at a short distance and flying in a close formation with Zarya, Endeavour's robotic arm first lifted Unity out of the payload bay and rotated it to a perpendicular position, pointing outwards to align with Zarya above. Then the robotic arm reached out to grapple Zarya to pull it in slowly to dock with the outward pointing Pressurized Mating Adapter. There was a misalignment problem at the first attempt and the arm had to ungrapple Zarya, swing it out and bring it back in again for a better berthing fit. This mission saw the first spacewalks on the Station. After the secure berthing of Zarya to the Pressurized Mating Adapter in the payload bay, two astronauts carried out three separate spacewalks to attach cables, connectors and handrails to exterior surfaces.[5]

In February 1999, sparring over the Station's cost between NASA and Congress continued when Congress refused the $600 million budget request to pay for Russian Soyuz and Progress spacecraft missions to the Station. Congress ignored an analysis by its own consultant who advised that, though there were cost and schedule problems, the first pieces of the Station were already overhead and the time had come to get on with it.[6] Reacting to Congress's refusal, NASA now wanted to add a new propulsion module to the assembly sequence to provide essential orbital reboost of the Station in place of the unfunded Progress spacecraft that were to perform that task. The next Station revision, Revision E, was issued in June 1999, seven months after the first assembly mission.[7] There were significant changes from Roscosmos. It gave a new launch date for the delayed Zvezda Control Module and the deferral of most of the other Russian building blocks until late in the assembly sequence. NASA, in contradiction of the recommendations of the April 1998 independent task force report to reduce the number of Shuttle missions to bring the Station's budget under control, added three more – two for logistics and equipment and one for the newly planned propulsion module.[8] The completion date slipped again to November 2004 and the number of assembly missions rose to thirty-seven. On 8 September 1999, the Station's opponents in Congress again mounted an attack on the project. They proposed through Bill Number HR 2684 to move $2.08 billion of the Station's funding over to other NASA and non-NASA programmes and to debt reduction.[9] It was soundly defeated by vote.

In July and October 1999, two Proton rockets failed after take-off. Russia's much-delayed Zvezda Service Module was the next block in the assembly sequence and it was due for launch on a Proton. Russia announced a postponement of Zvezda's launch after the failures. The resulting investigation saw a great deal of cooperation between NASA and Roscosmos[10] but the Station's assembly came to a halt for several months until Zvezda's eventual launch in July the following year.

On 14 March 2000, President Clinton signed the Iran Nonproliferation Act.[11] This Act was 'to provide for the application of measures to foreign persons who transfer to Iran certain goods, services, or technology ...' and it cited the International Space Station. In effect, it forbade NASA from paying for more Progress or Soyuz spacecraft missions to the Station until the President had determined that Russia had not exported weapons technology to Iran over a twelve-month period. It was a piece of legislation that would have serious consequences for the Station.

A few days later, on 16 March, the General Accounting Office in a letter to the Committee on Science of the House of Representatives, announced that the Russian Zarya Control Module and Zvezda Service Module did not meet certain American health and safety standards. Lack of debris shielding, sound insulation and effects of depressurisation were mentioned. The letter stated that 'Zarya and the Service Module are too noisy'.[12] NASA and Roscosmos responded with proposed remedies. However, the idea that terrestrial health and safety regulations could be applied to living and working facilities in space was naïve. High noise levels, in particular, present a challenge in space modules as noise cannot dissipate in the vacuum of space beyond the hull and tends to reverberate around the interior.

The hiatus caused by the delay in the completion of Russia's Zvezda Service Module came to an end on 12 July 2000 with its launch by a Proton rocket on Mission 1R. Zvezda was critical to the start of crewed occupation of the Station. It would be the focal point of the Russian operational sector and it would supply the first liveable quarters for a two-person crew in a compact, self-contained habitable environment.[13] Johnson Space Center, fond of the word 'cornerstone', used it to describe Zvezda's importance to

Zarya Control Module Astronaut Thomas Jones floats through the Zarya Control Module carrying IMAX camera equipment during Shuttle Mission STS-98 to the Station in February 2001.

Unity Node-1 and two Pressurized Mating Adapters The first Space Shuttle mission to the International Space Station brought up three building blocks: Unity Node-1 and the first two of the conical Pressurized Mating Adapters. This drawing shows the assembly sequence. In **I**, Zarya **01** approached the Orbiter from above **A**, perpendicular to the cargo bay, with its solar arrays at 90° to the Orbiter centreline to keep them clear of assembly operations. Unity **02** with its attached adapters **03**, **04** was rotated vertically out of the cargo bay **B** by the Orbiter's robotic arm **05** to await Zarya. In **II**, Zarya closed in on the Orbiter **C** and was grappled by the Orbiter's robotic arm **05** that moved it forward and attached it to the outer adapter **D**. Two astronauts then carried out three separate spacewalks to attach cables, fluid lines and other hardware to the exterior surfaces. After assembly work was complete, the Orbiter released the Station **E**. Image: author.

the Russian sector, but it was a vital part of the Station as a whole.[14] Like Zarya, Zvezda was put through its paces before adding it to the Station as a new building block. It was launched with many of its systems dormant. Once safely on orbit, a series of preprogrammed commands awakened the spacecraft and deployed a communications antenna and solar arrays. Russian ground controllers then manoeuvred Zvezda on to the same orbit as the Station and fully checked it out to confirm its readiness for rendezvous and docking. The docking of Zvezda with Zarya's aft port was entirely automated. Zvezda was the passive element and the ground controllers guided the embryonic Station comprising Unity and Zarya towards it and then brought the pieces together using Russian automated systems.

The next assembly sequence update, Revision F, was released in August.[15] It showed the addition of three Shuttle missions for delivery of logistics, equipment and experiments and two Russian missions moved forward in the sequence. The Station's completion date, however, had moved from November 2004 to June 2006, resulting in projected total assembly duration of seven years and seven months. The number of assembly missions now stood at thirty-eight.

The Space Shuttle launched the next Station building blocks on Mission 3A (STS-92) on 11 October 2000. This marked the hundredth Space Shuttle mission. The Orbiter's payload bay contained the Z1 Truss piece at the forward end and the third Pressurized Mating Adapter at the aft end. The third adapter would provide a docking point for future Orbiters at the forward end of the Station's American sector. The Z1 Truss was a transitional building block that NASA had added late in the Station's development. It supported an S-Band communications antenna system to enable the Station to communicate with the ground via the Tracking and Data Relay Satellites and it contained power distribution equipment to feed electrical power from the solar arrays. Its open structural framework housed four control moment gyroscopes that would be vital for orbital stationkeeping. After rendezvous, the Orbiter Discovery docked at the Pressurized Mating Adapter at the forward end of Unity Node-1. In this position, the Station was perpendicular to the Orbiter's fore/aft axis. Discovery's robotic arm lifted the Z1 piece out of the payload bay and mounted it on a berthing port on Unity facing towards the Orbiter's tail. The next arm action lifted the arriving Pressurized Mating Adapter out of the rear of the payload bay, moved it around Unity and mounted it on Unity's port facing the Orbiter's nose. Eight spacewalks were needed to complete work on this mission, divided between four astronauts making two spacewalks each. Their tasks included connecting electrical cables, deploying antennae, installing toolboxes and direct current converters, testing a berthing mechanism, removing a grapple, deploying a tray, attaching the third adapter and testing spacewalk rescue backpacks.[16]

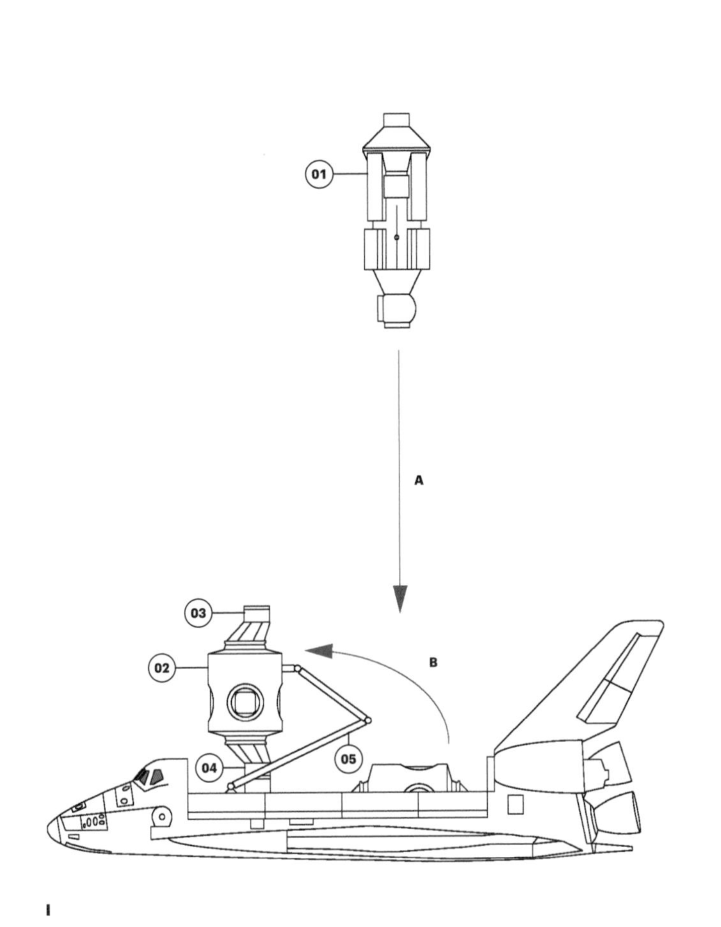
01
A
03
02
B
05
04
I

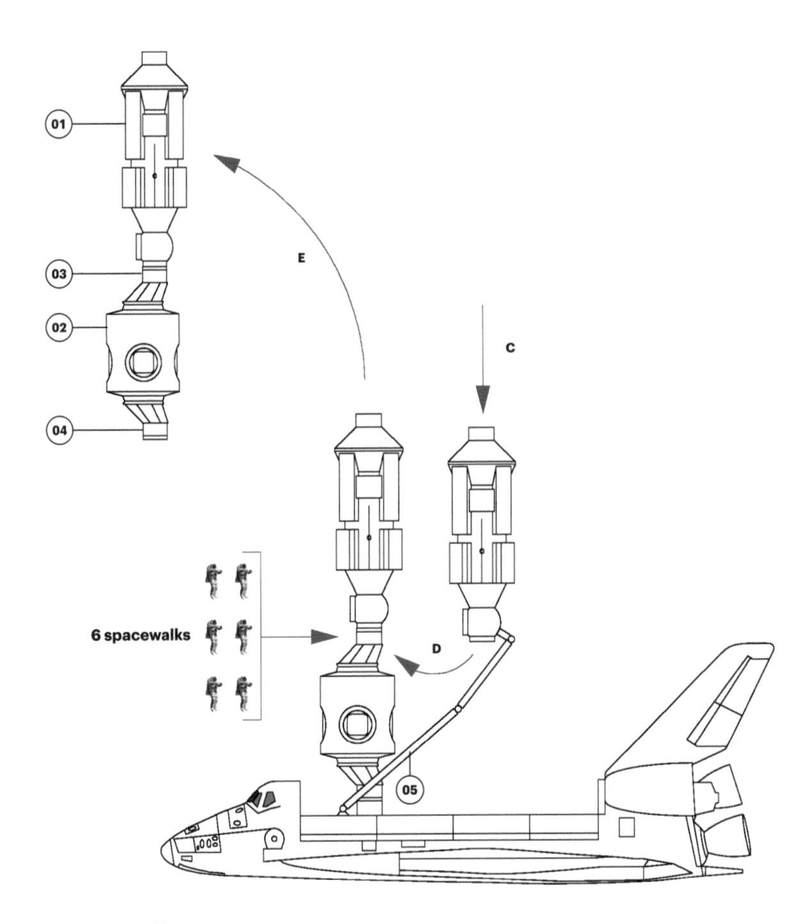
01
E
03
02
C
04
6 spacewalks
D
05
II

Permanent residency
First Crew Expeditions, 1999-2002

On 30 October 2000, the first International Space Station crew was launched on a Soyuz TM spacecraft from Baikonur, docking with the Station on 2 November. It was a major milestone that was formally labelled '3-Person Permanent Habitation'. The first crew named Expedition 1 comprised its commander, astronaut Bill Shepherd, a former US Navy SEAL who had flown on three Shuttle missions, flight engineer and veteran cosmonaut Sergei Krikalev and cosmonaut Yuri Gidzenko, the Soyuz pilot. Both cosmonauts had spent time on Mir, over one year in Krikalev's case. NASA proclaimed that 'a new chapter in human spaceflight history will unfold with the launch of the first permanent occupants of the new International Space Station' and described the crew's job as a 'shakedown mission'. In their first few days on board, the crew turned on Zvezda's computers to control the Station's systems and activated the air purification, oxygen generation and water supply. They set up exercise equipment and a medical centre and unpacked components, clothing, laptop computers, office equipment, cables and electrical gear.[17]

The *New York Times* was less sanguine than NASA about the arrival of the first crew. The newspaper had been 'lukewarm about the station from the beginning, feeling that it would be enormously costly and yield only limited scientific returns, most of which could be gained by satellites and unmanned probes to the Solar System. Many of the claims advanced for the station – such as the opportunity it provides to grow perfect crystals in the absence of gravity and to perform other exquisite manufacturing tasks – have been hyped beyond reason and debunked by expert committees. But that does not mean there is no value at all in the enterprise. The fact that sixteen nations are collaborating on the American-led project is a diplomatic achievement, as is the participation of Russian space and rocket scientists who might otherwise be tempted to put their skills to malevolent use.' The newspaper cautioned that a 'pressing task will be to devise a high-quality research agenda for the station lest it remain a structure in search of a mission'.[18]

The Expedition 1 crew spent 136 days at the Station. With the first crew up, the Station could begin the process of evolving into a scientific research facility. Soyuz, though just a crewed spacecraft, was a vital part of the Station. Soyuz spacecraft would ferry up new crews to exchange with the old, as well as some crews making short visits. A three-seater Soyuz would always be docked at the Station to provide a means of escape in an emergency. When the crew size increased to six, there would be two Soyuz spacecraft docked at the Station; when the crew size increased to nine during crew exchanges, there would have to be three. Soyuz vehicles at the Station had to be replaced with a fresh spacecraft every six months. Crew exchanges would be timed to coincide with these replacements.

The Soyuz vehicle is the longest serving human spacecraft in the world with its origins in Vostok and a remarkable history of missions in the Russian and former Soviet space programmes. Work began on a replacement for Vostok, the Soviet Union's first manned spacecraft, in 1958. Launched on the Soyuz rocket, the new spacecraft was to have a multipurpose role, including missions to the Moon and to space stations on Earth orbit. Over time, Soyuz has achieved many spaceflight objectives: sustained flight in manned or unmanned modes, orbital manoeuvring, docking, scientific research and assembly of crewed stations. It served the Soviet/Russian Salyut and Mir Space Station programmes and, after the loss of the Orbiter Columbia in February 2003, it provided the only means of Station access for crews up to the resumption of Space Shuttle flights in July 2005. Soyuz has undergone many modifications and upgrades since its introduction. The latest version, known as Soyuz TMA, incorporates cabin volume and seat improvements to ensure that it can accommodate tall crew members comfortably.[19] The first Soyuz TMA flight visited the Station in October 2002 with its three-person crew. During the Station's early assembly stages, Soyuz approached it from behind and docked with the aft docking port on the Zvezda Service Module.

Mission 4A (STS-97) with the Orbiter Endeavour was the next to launch on 30 November 2000. It carried the first building block of the long transverse truss – the P6 Truss piece with the first set of solar arrays. The arrival of this building block set the scene for a series of missions that would bring up the entire truss structure, piece by piece. Until now the Station had been running on fairly limited electrical power, generated by solar arrays on Zarya and Zvezda. The P6 piece would deliver more power for the growing number of module building blocks arriving on orbit and the life support equipment and scientific experiments incorporated in the racks inside. The electricity generation capability of the P6's giant solar arrays was sufficient to meet the needs of thirty homes of average size.[20] The truss piece occupied Endeavour's entire payload bay. The solar arrays, technically known as the Photovoltaic Array Assemblies, were folded up like accordions and stowed for spaceflight against the side of the P6's structural framework, as were three smaller thermal radiators. The positioning of the P6 on the Station during this mission was temporary. Placed on top of the Z1 Truss on the Station's centreline, it would be moved to its final location at the port end of the truss on a future mission.[21] On this mission, Endeavour docked with Station at the Pressurized Mating Adapter installed at the underside port of Unity on the previous mission. In this position, the Station's fore-aft axis was parallel to that of Endeavour. The Orbiter's robotic arm lifted the P6 piece out of the payload bay and parked it close by the Station overnight. The next day, the arm attached it to the top of the Z1 on Unity's upper port. Astronauts on spacewalks then released the stowed solar arrays and thermal radiators. The arrays rotated to horizontal and then unfurled themselves full length in port and starboard directions. The three thermal radiators did the same.

The International Space Station's first crew The Expedition 1 crew – the International Space Station's first permanent crew – pictured playing with their food inside the Zvezda Service Module in December 2000. From the left they are: cosmonaut Yuri P Gidzenko, Soyuz commander; astronaut William M Shepherd, mission commander; and cosmonaut Sergei K Krikalev, flight engineer.

With its completion, over seventy-seven hours of spacewalks had been spent working on the Station. Unfolding the solar arrays transformed the Station into the largest structure yet built beyond Earth. It was another marker as the Station continued to grow, mission by mission.

Back on the ground, the Station's budget problems continued. NASA had ignored previous task force recommendations to eliminate some Station elements and assembly missions. However, in January 2001, NASA announced a $4 billion budget deficit and President George W Bush, recently elected, instructed the agency to cut the Station back before its 2002 budget request. There would be long-term consequences for the Station's operations. First to be cut was the Habitation Module. This was serious as it would deprive Station crews of proper living accommodation. Second to go was the planned American-built crew return vehicle. This vehicle had been a chimera for years and responsibility for providing crew emergency escape had already been taken over by Soyuz spacecraft. Third to go was the propulsion module, designed to help keep the Station at the correct orbital altitude. As the Space Shuttle, Automated Transfer Vehicle, Soyuz and Progress spacecraft all had the capability to reboost the Station's orbit, this was no great loss either.

The diminished Station became known as the Core Complete design. It defined the financial end point of NASA's assembly missions as the delivery of the second node module, Harmony Node-2, beyond which further Shuttle-delivered hardware would be guillotined. However, the assembly schedule showed that the European and Japanese modules came after Harmony – a potential diplomatic time bomb as NASA had not consulted the Europeans or Japanese about its Core Complete decisions. The implication seemed to be that NASA's responsibility for the Station's assembly might stop at Core Complete, potentially leaving its partners without the means to deliver their building blocks to the Station. Roscosmos was due to deliver ten more Soyuz vehicles under its agreement with NASA without payment and NASA was prevented from ordering more under the Iran Nonproliferation Act. This implied that the Station's crew escape capability might cease once the ten Soyuz vehicles had been used up, effectively preventing the Station from being permanently occupied. NASA, evidently, had not thought through the Core Complete version properly.

Of particular concern was the loss of the Habitation Module. The notion of operating a circa $25 billion (as the price tag then was) international research facility without adequate living and sleeping accommodation for the crew made no sense. Crews were using the Zvezda Service Module for sleeping and it had only cramped compartments for two crew members; the third was camping out elsewhere on the Station. There would be no proper facilities for a six-person crew and one of the most fundamental habitability needs of the Station would be shelved. To soften the blow, NASA proposed to fit out a future Node for habitation purposes.

Soyuz TMA Spacecraft A Soyuz TMA spacecraft departs from the International Space Station in September 2010. On board were the three crew members of Expedition 24. A Soyuz spacecraft consists of three main elements. At the bottom with the solar arrays extended on each side is the Instrumentation and Propulsion Module. At the top is the crew's Orbital Module with its nearly spherical shape. Connecting them in the centre is the crew's Descent Module.

Mission 5A (STS-98) launched on 7 February 2001, with the Orbiter Atlantis's payload bay occupied by the Destiny Laboratory Module. Dedicated to scientific research, command and control, and operation of the Station's robotic arm, Destiny was by far the most important American pressurised module on the Station. Its experiments would cover human life science, materials technology, fundamental biology, ecology, space science, Earth observation, Earth photography and commercial applications.[22] It was vital for the Station to begin to show off its research capabilities and research work inside Destiny could not start soon enough. Its delivery to the Station in February 2001 expanded the Station's internal habitable volume and gave the crews more freedom to move around.

Due to launch weight restrictions, Atlantis delivered Destiny to the Station with a limited number of racks installed. The remainder would follow on subsequent missions. The Station flew on orbit with its central axis through the Zarya, Zvezda and Unity modules orientated along the flight-path line with Unity ahead of the others. Atlantis approached for rendezvous from beneath the flight path with its nose pointing aft and its payload bay open to dock with the Pressurized Mating Adapter on Unity's underside. Attaching Destiny to the Station required Atlantis's robotic arm to perform three major tasks. First, it detached the adapter on Unity's forward port and temporarily mounted it on the Z1 Truss just above. This freed up Unity's forward port. Next, the arm lifted Destiny a short distance out of Atlantis's payload bay and berthed it to Unity's exposed forward port. Then, the arm retrieved the adapter from its temporary position on the Z1 and reberthed it on Destiny's forward port. This adapter would provide the docking location for future Orbiter assembly missions. A total of six spacewalks supported this mission's objective of attaching Destiny to the Station.

Modules and spacecraft are plugged together in space in two main ways – by docking and by berthing. Docking involves correct alignment of both elements and then control of the closing velocity of the moving element for a smooth mating, usually automatic but sometimes piloted. Berthing involves a robotic arm or other mechanical device that grapples an arriving element that is standing apart, motionless, and brings it in to mate with the stationary element, a process that may involve manipulations in three axes. Sometimes there is more than one mechanical device and the arriving element is passed between them until it reaches its destination, as would be the case during the Station's assembly.

By now, Station assembly had involved several berthing and unberthing operations using Orbiter robotic arms. Unity and its two Pressurized Mating Adapters had been berthed to Zarya; Destiny had been berthed to Unity; an adapter had been unberthed and then reberthed. The key to the success of these operations was the means by which the pressurised modules came together at their end ports through a universal mechanical connecting device. It was rather like pushing a ski boot into a binding where the binding

Assembly P6 Truss The P6 Truss was the first piece of the transverse truss structure delivered to the Station. In **I**, the Shuttle Orbiter docked with the Station at the Pressurized Mating Adapter **01** on the underside of Unity Node-1 **02**. In this position, the Station was on the same line as the Orbiter's fore/aft axis. An astronaut in the Orbiter operated the robotic arm **03** to lift the P6 Truss **04** out of the payload bay, parking it on the arm's end overnight. The following day, it was mounted on top of the Z1 Truss **05**. In **II**, astronauts released the stowed solar arrays and thermal radiator panels after the P6 Truss was in place, enabling them to deploy to their final positions. The solar array wings rotated to the horizontal **A** and then unfurled to their full length in opposite directions **B**. The three thermal radiators extended to their full length in different directions **C**. There were six spacewalks on this assembly mission. Image: author.

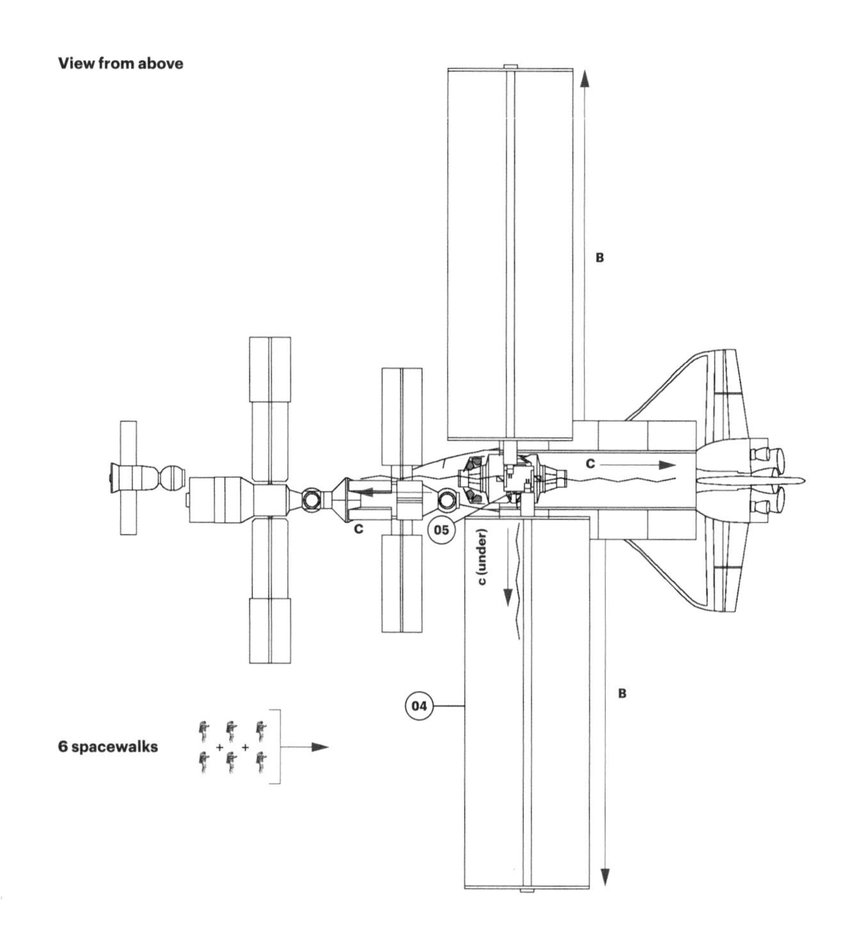

Views from side

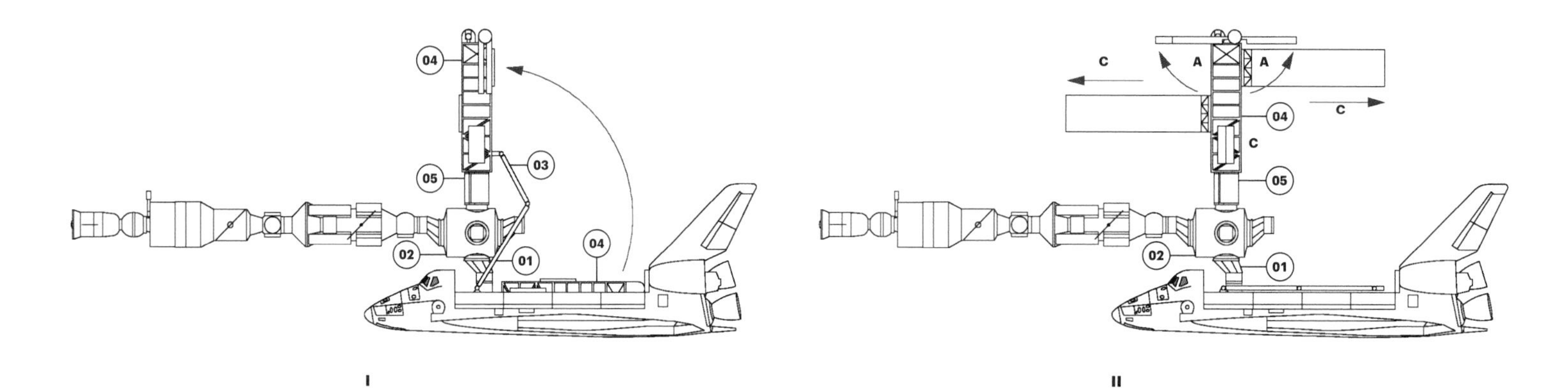

captures the toe and heel and clicks closed. The device used was called the Common Berthing Mechanism and it was installed on all Station modules – whether American, Russian, European or Japanese in origin – that would berth to each other, temporarily or permanently (Russian-to-Russian modules and spacecraft used a different automatic docking system).

The Common Berthing Mechanism came in two pieces – one active and the other passive – that locked together mechanically when they met.[23] The active half comprised a structural ring, forked alignment guides, capture latches, control panels and powered bolts. The passive half comprised a matching structural ring, capture latch fittings, forked alignment guides and nuts.[24] The procedure followed a step-by-step sequence. First, a robotic arm brought the two rings together until they were within a few centimetres of each other, with their alignment guides interlocked. A sensor on the active ring detected that the passive ring was close and that they were ready for mating. The active ring then extended four capture latches outwards. These latches were positioned between each of the four pairs of alignment guides around the ring's perimeter. Each latch was of an articulated design and operated by an electrically driven actuator that imparted rotational motion. Operating the actuators were four controller panel boxes arranged round the inside of the active ring. Next, lips on the end of the latches engaged with matching lips on the passive ring. The latches then retracted to draw the passive ring into the active ring in a coordinated movement. Finally, a set of sixteen powered bolts spaced around the active ring's perimeter screwed automatically into matching sockets in the passive ring to secure the structural connection. The bolts also used electrically driven actuators. The bolts initially screwed themselves partially in and then waited for half a day to enable the temperatures of the two rings to equalise. They finally screwed in tight in a prescribed sequence to their final load levels.[25] The Common Berthing Mechanism fulfilled stringent performance requirements. It ensured airtight seals between the modules; it provided reliable structural connections along the Station's length while absorbing dynamic loads caused by Station movements on orbit; it permitted crew circulation through module interiors; it acted as a bridge for utilities connections between modules; and it did all these things while enabling relatively effortless berthing or unberthing of its two parts.

Mission 6A (STS-100) launched on 19 April 2001. The Orbiter Endeavour brought up a vitally important Canadian building block, the Canadarm2 robotic arm. It was an evolutionary advancement of the Canadarm that had been used successfully on the Space Shuttle Orbiters.[26] With its ability to manipulate and move large items of hardware around the Station's exterior, Canadarm2 was indispensable. Its seven motorised joints gave it a wider range of motions than a human arm, which it resembled.

Canadarm2 has two long limbs and seven rotating joints that rotate up to 540° each, giving the arm seven degrees of freedom. The arm is

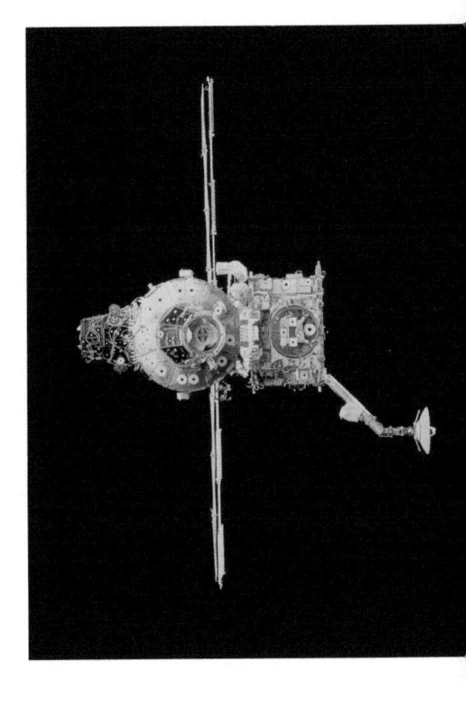

Z1 Truss and one Pressurized Mating Adapter
This is an end-on view of the Station after the addition of the Z1 Truss and a third Pressurized Mating Adapter. Slightly left of centre is Unity Node-1 with a foreshortened adapter facing towards the photograph viewpoint. The third adapter is attached to a port on Unity's left side. The box-like Z1 Truss with deployed antenna is on the right, attached to a port on Unity's right side as seen in this view. The white-and-black circular and square targets that cover the exterior of Unity and the Z1 are situational awareness and navigational aids for the cameras on the Canadarm2 robotic arm that would arrive on a future mission. They were used to calculate angles and assist the building block assembly process.

symmetrical about its mid-span joint, which is equivalent to the human elbow. Because it is free at both ends, the arm can somersault from point to point across the Station using its mechanical grappling devices called Latching End Effectors at each extremity to plug into sockets on the Station's exterior surface. Called Power Data Grapple Fixtures, these sockets provide the arm with power, data and video services wherever it plugs itself in. Force sensors provided a sense of touch and the arm has four colour cameras and an automatic collision avoidance capability. It is a highly sophisticated and remarkably effective tool.

Canadarm2 was capable of moving and manipulating a maximum mass of 116 tonnes in weightlessness. It could move at up to 370mm per second unloaded and 20mm per second when performing Station assembly duties with large payloads.[27] Built by MacDonald Dettwiler Space and Advanced Robotics Ltd of Brampton, Ontario, Canadarm2 was 17.6 metres long overall and weighed 1.8 tonnes at launch. Its tubular limbs were 350mm in diameter and made of a carbon fibre/thermoplastic material. Canadarm2 was part of the Mobile Servicing System for the Station that also included a Canadian dexterous manipulator named Dextre and a Mobile Base Platform. These arrived on later missions.[28]

The Space Station's computers failed during Shuttle Mission 6A. Canadian astronaut Chris Hadfield, who had arrived on Endeavour with the Canadarm2 to help to install it, later explained what happened: 'All [the computers] had an inherent flaw and started overwriting their own hard drives. It meant that the Station was effectively dead; it couldn't control its attitude, point its antennae, run its own diagnostics – all kinds of capabilities were gone, and the ground could barely communicate with us. If we hadn't had the Shuttle docked and ready to control the entire combined structure, we would have been in serious trouble. Fortunately, we could use the Shuttle's communications and thruster systems, and we still had oxygen, food and water, so the crew's attitude was just to keep working the problem.'[29] The problem was later resolved and the computers returned to normal.

Mission 7A (STS-104) launched on 12 July 2001, with the Orbiter Atlantis bringing up an airlock. It was another assembly milestone known as Phase 11 Complete. The Joint Airlock, built by Boeing and named Quest, would provide the Station with the means to carry out spacewalks with crews wearing either American or Russian spacesuits.[30] Up to the arrival of Quest, cosmonauts had to use Zvezda for spacewalks while astronauts had to carry them out from a docked Orbiter. Quest had two compartments with a total volume of 34m^3. One compartment was for storing American and Russian spacesuits and equipment and the other led to an exit hatch. Astronauts and cosmonauts scheduled to carry out spacewalks would be able to camp overnight in the large compartment and pre-breathe a higher oxygen air mixture to avoid decompression sickness in the low-pressure oxygen

P6 Truss The International Space Station seen from the Shuttle Orbiter Endeavour in December 2000 after the Station had received the P6 Truss building block with its solar arrays and thermal radiators. To get a series of photographs including this one, Endeavour moved down towards nadir after undocking and then began a complete tail-first circle of the Station at a distance of about 150m, taking about an hour for the manoeuvre.

atmosphere of a spacesuit. Quest's exterior surface had mountings for high-pressure oxygen and nitrogen gas tanks used to replenish the Station's internal atmosphere.

Atlantis's rendezvous with the Station was from forward of the Station's flight path, nose up, tail down with its open payload bay facing aft and perpendicular to the Station's fore/aft axis. After docking with the forward Pressurized Mating Adapter on Destiny, Canadarm2 – perched on Destiny's underside – reached down into the payload bay, grappled Quest and swung it up and around to berth it on to Unity's starboard port. Six spacewalks were needed on this mission. It was the first time that the Canadian robotic arm had performed a major assembly task on the Station and it worked well.

Cancelled Habitation Module The American Habitation Module being lifted by a crane during fabrication at Marshall Space Flight Center in Huntsville, Alabama, in 1997. The Habitation Module was to have provided living quarters for the Station's crew. NASA scrapped it in 2001 amid further budget reductions forced on the project by the incoming Bush Administration.

Back at NASA, the measures the agency had taken in February to solve its latest budget headache were proving inadequate. To help tackle the problem, in July 2001 NASA appointed the 'ISS Management and Cost Evaluation Task Force' to review the Station's management. Roscosmos's financial difficulties were also ongoing, leading to more changes to Russian elements. The Russian Universal Docking Module was replaced by another back-up module; a Docking and Stowage Module was superseded by a proposed module named Enterprise; a second Docking Compartment was cancelled; and a solar array platform was reduced in performance. Most significant of all, a Life Support Module and two Research Modules were dropped, substantially reducing the research role of the Station's Russian sector. Revision G had been issued in May 2001, showing the elimination of the propulsion module, a completion date in 2006 and more assembly sequence reshuffling.[31] Revision G was soon overtaken by these new changes, most of which found their way into a Revision G Update. Revision G had shown thirty-two assembly missions and the Revision G Update would show thirty-one.

The next mission was Russian. On 15 September 2001, Mission 4R launched from Baikonur with a Progress spacecraft ferrying up a docking module for the Station. The Pirs Docking Compartment (Pirs means 'pier' in Russian) was a small building block, built by RSC Energia in Korolev, Russia.[32] Its role was to provide an airlock for spacewalks from the Russian sector through a l-metre diameter hatch. Pirs also had another docking port for visiting Russian spacecraft, increasing the number on the Russian sector to three. Pirs would provide the Station with improved planning and operational flexibility. It would allow an arriving Soyuz to dock with the Station before a departing Soyuz left it, at the same time that a Progress resupply vehicle was docked, much like arriving, waiting and departing trains on adjacent platforms at a railway terminus.

In November 2001, the Task Force issued its report on the Station's management and costs.[33] It found that 'existing deficiencies in management structure, institutional culture, cost estimating, and program control must be acknowledged and corrected for the program to move forward in a

Impact of cost reductions on the American-built modules The International Space Station's American-built Habitation and Laboratory Modules were among the building blocks that bore the brunt of cost savings forced on NASA during the Station's development years as the agency grappled with constant Station budget overruns. When President Reagan announced the project in 1984, the Station was due to have two long Habitation Modules and two long Laboratory Modules. By 1993 when Russia joined the project and it became the International Space Station, the four American-built modules had been cut to two short modules. In 2001 in the face of more cost problems, NASA had to scrap the Habitation Module under instructions from President George W Bush. The result was that until the arrival of a later node with American-built sleep compartments, the crew had to manage as best they could. Zvezda had two cramped private compartments. The third crew member had to camp out elsewhere on the Station. Image: author.

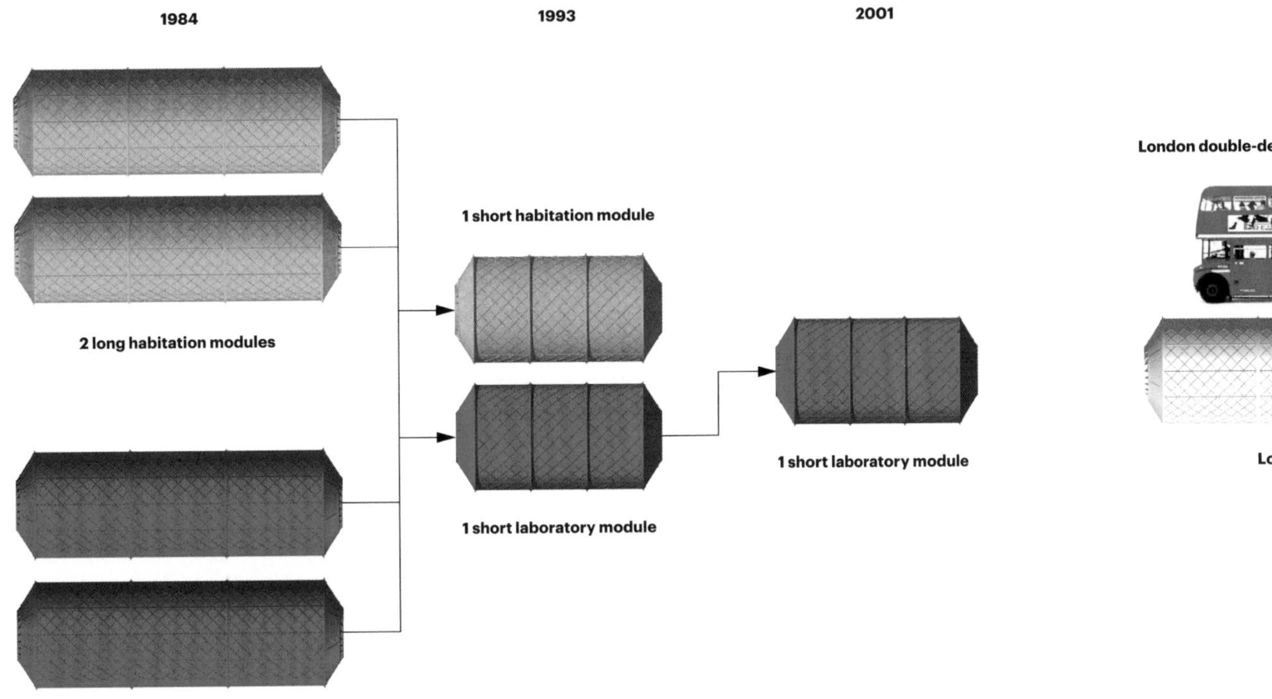

London double-decker bus at the same scale

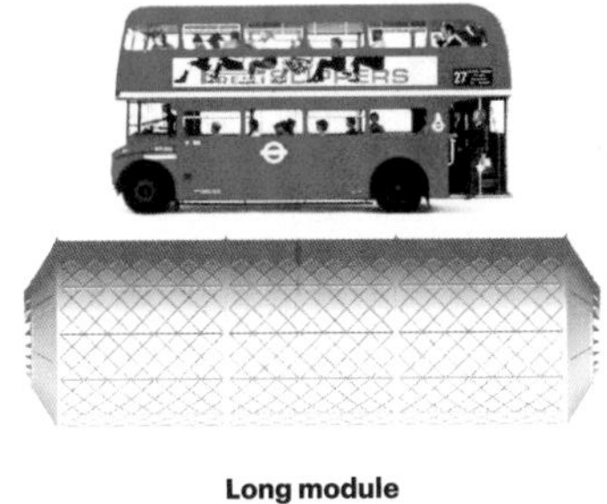

Long module

credible fashion'. It intimated that NASA's lack of financial management skills and cost underestimating were responsible for the budget overruns, that the forward budget from 2002 to 2006 was 'not credible' and that the project was being managed by NASA as an 'institution' with the emphasis on 'budget management rather than total program cost management'. It concluded with a stern requirement for NASA to 'demonstrate credibility ... to proceed beyond the Core Complete program'. It was NASA's job to find a solution to the Station's chronic cost problems while striving to ensure the launch of all the European and Japanese building blocks on the Space Shuttle. These launches would be exposed to some risk if NASA did not act swiftly. The effect was to disrupt NASA's relationships, which had been generally good so far, with its Station partners in Russia, Europe, Japan and Canada. NASA also upset the scientific community, which complained about the loss of the Station's research capabilities.

Destiny Laboratory Module The robotic arm on the Orbiter Atlantis has grappled the Destiny Laboratory Module and moved it out of the Orbiter's payload bay during operations to attach it to the Station. Taking this photograph was Thomas Jones, one of the astronauts on a spacewalk at the time to support Destiny's berthing.

Building site
Assembly of the Transverse Truss, 2002–2003

Mission 8A with the Orbiter Atlantis (STS-110) launched the first piece of the Station's transverse truss structure on 8 April 2002. From now on, the Station would begin to resemble a terrestrial building site where cranes lift pieces of steelwork into place and workers bolt them together into a frame. The big difference was that two robotic arms – the Canadarm on the Orbiters and the Canadarm2 on the Station – would carry out the lifting and bolting operations. The S0 Truss was the first of nine truss building blocks that the Space Shuttle would bring up over the next few years, all built by Boeing.[34] The S0 piece was the central element that would hold the whole structure together and transfer the various static and dynamic structural loads back and forth between the long truss and the line of pressurised modules. The S0 also provided routing of distributed utility systems between the modules and the transverse truss, including electrical power, thermal control, communications, navigational and data systems. The S0 had a three-dimensional rigid framework made from aluminium with five structural bays and a hexagonal cross-section that fitted snugly into Atlantis's payload bay for launch. Its length was 13.4 metres and its maximum girth was 4.6 metres. Astronauts on spacewalks would be able to enter the framework's interior to reach internal worksites. The S0 was launched already outfitted with equipment, including the Mobile Transporter for the Canadarm2 robotic arm, a trailing umbilical system, a portable work platform, four global positioning system antennae, two attitude control gyros and a charged particle detection system. Also delivered and installed on this mission were electrical bus switching units, circuit interrupters, crew and equipment movement aids and airlock equipment.[35]

Views from starboard side

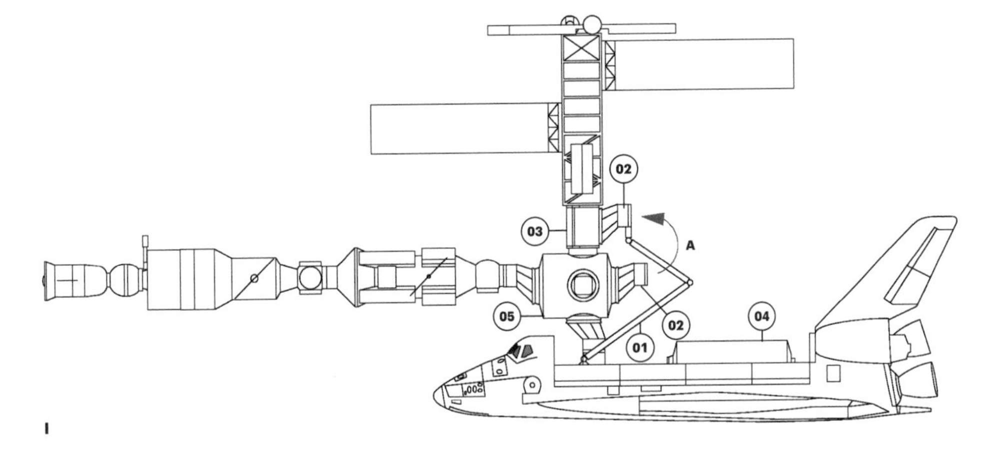

I

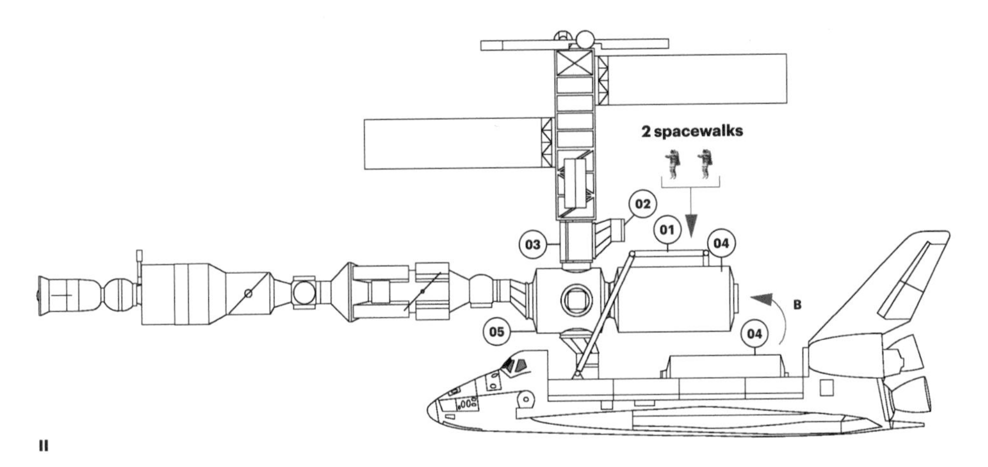

II

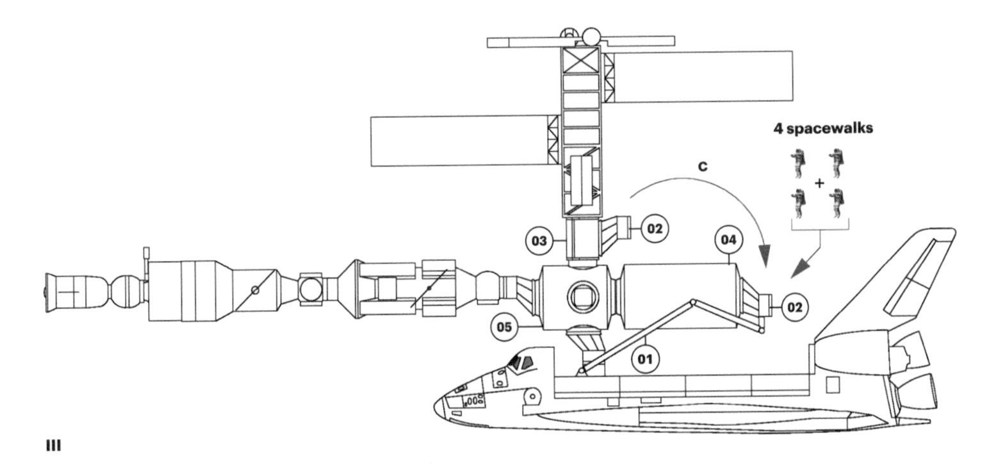

III

Destiny Laboratory Module The Shuttle Orbiter's robotic arm was used to great effect during the attachment of the Destiny Laboratory Module to the Station. In **I**, the robotic arm **01** detached and moved **A** the Pressurized Mating Adapter **02** to a temporary position on the Z1 Truss above **03**. In **II**, the arm lifted the Destiny Module **04** out of the Orbiter payload bay **B** and mounted it on the end of Unity **05**. Following up were two spacewalks. In **III**, the arm detached the adapter **02** from its temporary position on the Z1 Truss **03** and berthed it on the end port of Destiny **04**. Four more spacewalks completed the assembly operations. Image: author.

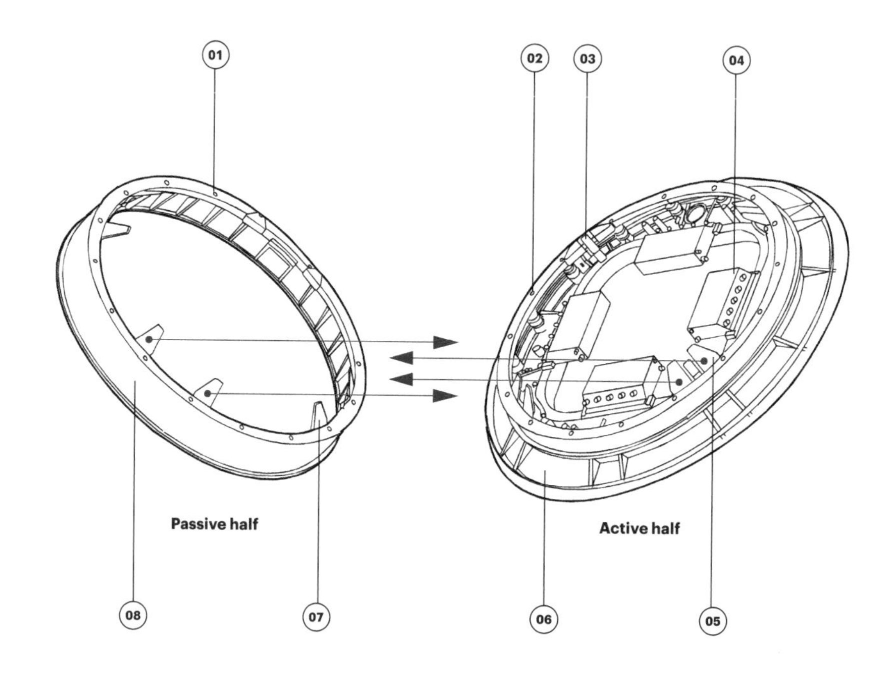

Common Berthing Mechanism The Common Berthing Mechanism comprised the vital structural and mechanical rings between most of the Station's modules, enabling them to plug together firmly and securely on orbit. Ports in the endcaps of modules due for mating were fitted with matched active and passive ring halves. These came together in a sequence using alignment guides, capture latches and powered bolts respectively. The red arrows show how a typical pair of fork-like alignment guides on one ring interlocked with a pair on the other ring when they came together. The photograph shows the passive ring half of the Common Berthing Mechanism. Image: NASA, with author additions.

01 Nuts (16 places)
02 Powered bolts (16 places)
03 Capture latches (4 places)
04 Controller panel assembly
05 Alignment guide (4 places)
06 Active structural ring
07 Alignment guide (4 places)
08 Passive structural ring

The Mobile Transporter, also built by Boeing, was the railcar that ran along the transverse truss. It was one element of the multi-element Mobile Servicing System that also comprised the Canadarm2, its work platform called the Mobile Base, a highly capable end manipulator on the arm called Dextre, and the robotic workstations inside the Station for astronauts to control and operate the system.[36] The Mobile Transporter railcar travelled on a roller suspension system along rails mounted on the leading edges of the truss frame pieces. It could lock down on to the rails automatically at any point to enable the Canadarm2 arm to handle large payloads. The railcar plugged into its own data, power and video systems using a trailing umbilical cord. Put simply, the Mobile Servicing System was the first mobile crane in space. Atlantis arrived at the Station in the same way it had done on Mission 7A in July the previous year – forward of the Station's flight path, nose up, tail down with its open payload bay facing aft and perpendicular to the Station's fore/aft axis. After Atlantis docked with Destiny's forward Pressurized Mating Adapter, Canadarm2 on Destiny grappled the S0 piece in the Orbiter's payload bay, swung it out and manipulated it around to mount it on top of a cradle-like group of struts pointing upwards from the aft portion of Destiny's cylindrical surface. Eight spacewalks were needed on this mission.

Mission UF-2 (STS-111) with Endeavour, launched on 5 June 2002 and brought up the Mobile Base element of the Mobile Servicing System as well as supplies for the crew. Once this part was in place, Canadarm2 and its railcar system were ready to build the rest of the truss structure that would arrive on later missions.

01

02

03

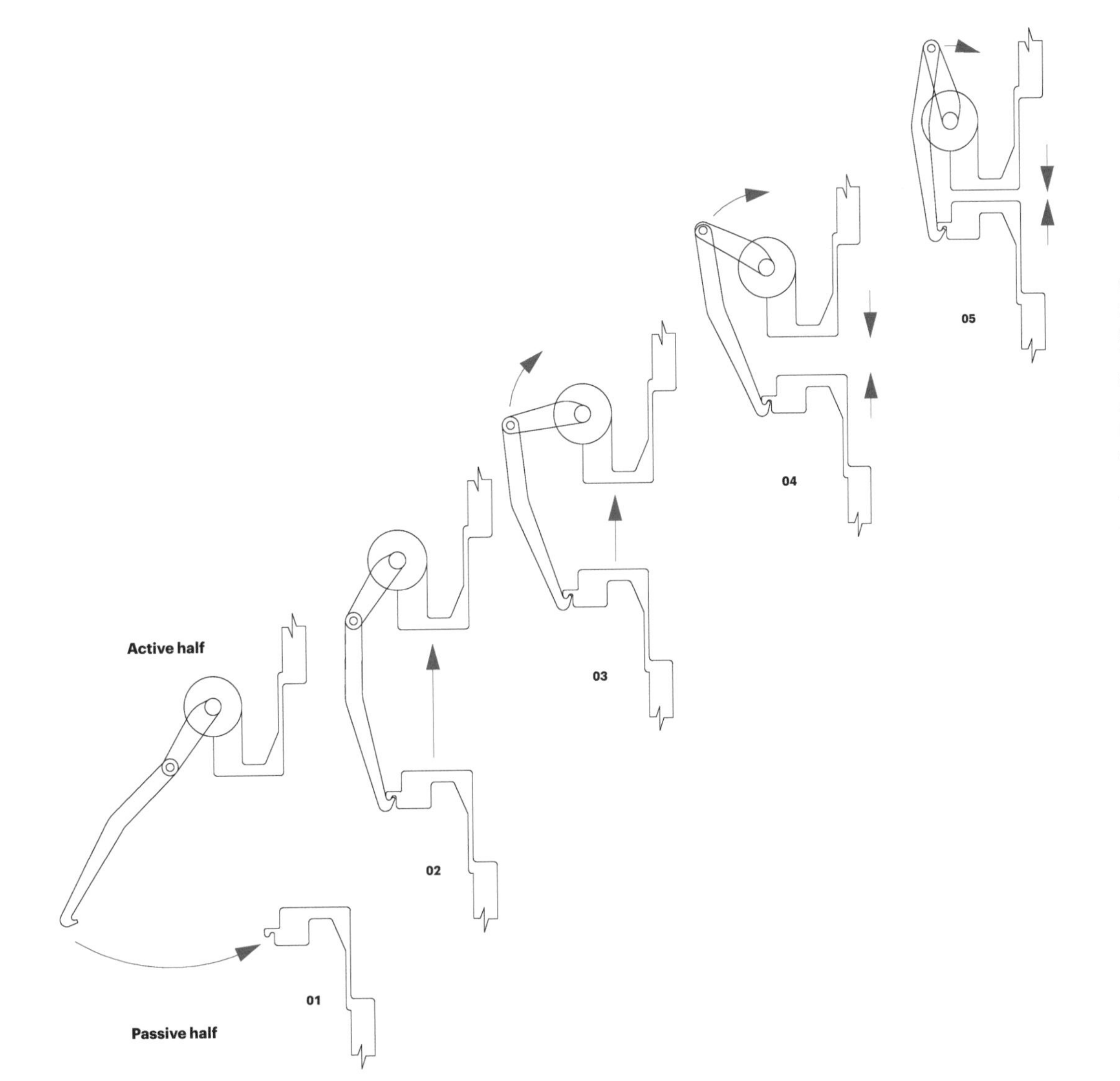

Common Berthing Mechanism Close-up photographs, above left, of a pair of alignment guides and a latch mounted on the active ring. The latch has an articulated design and works in a similar manner to latches found on some metal briefcases. In **01** the extended latch points inwards. In **02** it points downwards as though in the act of grappling. In **03** it retracts as though pulling the grappled piece towards it. Above, an electrically-driven latch actuator mechanism.

The diagrams, left, show the latch operation. In **01** and **02** the latch outer arm on the active half swings round and engages in a projecting lip on the inner edge of the passive half. In **03** and **04** the latch retracts towards closure. In **05** the two halves, correctly aligned, complete closure. Image: author.

In July 2002, the General Accounting Office reported to Congress on progress made by NASA in controlling the Station's costs. The Office observed that the cost cutbacks would limit research on the Station.[37] NASA had been aware of this for years. That same month, a NASA task force named 'Research Maximization and Prioritization' concluded that the Station should cease to be characterised as science-driven unless it proceeded beyond the Core Complete stage.[38] It was a warning about a project that now had space-based scientific research as its sole *raison d'être*.

Negotiations followed in 2002 between NASA and its partners, resulting in agreement over a revised 'International Core Complete' configuration that included the partners' modules but confirmed cancellation of the Habitation Module, which meant holding Station crew sizes down to three until other habitability measures were put in place. NASA issued the Revision G Update embodying these changes in October 2002.[39] Additional changes involved a delay to the delivery of the Japanese Experiment Module by a year and the moving forward in the assembly sequence of the Centrifuge Module, which was now considered a vital research tool. Two new logistics and equipment missions for the Space Shuttle were added and Europe's and Japan's resupply vehicles were featured in the sequence for the first time. More mission reshuffling occurred and the completion date slipped again to early 2008. By December, the Station's assembly planning was back on track.

Canadarm2 Robotic Arm The Canadian-built robotic arm Canadarm2 arrived at the Station in April 2001 folded up on a U-shaped Spacelab pallet in the Orbiter Endeavour's payload bay. After the Orbiter's robotic arm deployed the payload, Canadarm2 unfolded from its stowed position on the pallet and attached itself to the Station. Once securely in place, the arm grappled the empty pallet it had arrived on and handed it back to Endeavour's arm for re-stowage in the payload bay and return to the ground. This photograph shows the empty pallet held by both robotic arms – each operated by an astronaut – during the transfer procedure. Endeavour's robotic arm is on the left. Canadarm2 is on the right.

The next building block to arrive at the Station was the S1 Truss on Atlantis on Mission 9A (STS-112), launched on 7 October 2002. The S1 was the second piece of the truss and its first on the starboard side. The SI extended the truss's frame by 13.7 metres, providing similar functions to the centre S0 piece with railcar tracks for the Mobile Servicing System, as well as electrical power, thermal control, communications, navigational and data systems routes.[40] The S1 had a trapezoidal cross-section and six structural bays that were also accessible for spacewalks. It incorporated three folded thermal radiator panels stowed in boxes. The centre panel was deployed after the spacewalks, extending out like a pantograph in the aft direction. Atlantis's rendezvous with the Station was from forward of the flight path and it docked at the forward end of Destiny in what was now becoming a regular docking procedure. There were six spacewalks on this mission. After these spacewalks, the thermal radiator panel on the S1 piece also completed its deployment.

The truss building blocks were connected using an advanced version of what in the construction industry is known as a butt joint. The truss building blocks attached to each other end to end, by the Canadarm2 robotic arm. The procedure for attaching the S0, P1, S1, P3/P4 and S3/S4 Truss building blocks began with the robotic arm slowly bringing the active piece in close to the passive piece. The active piece performed the capture process. There were four tasks to be carried out sequentially. First, the two truss framework

Previous page: Quest airlock The airlock has two cylindrical compartments. The larger one contains spacesuits and associated equipment, with overnight accommodation for spacewalk pre-breathing for two crew members. The smaller one provides an exit tunnel and hatch. The photograph shows astronaut Garrett Reisman inside the larger compartment in May 2010.

faces were coarsely aligned using three sets of pins and sockets mounted on the framework's end faces. Second, four sets of fine alignment cones and cups – one at each vertex of the end face – engaged with each other. Third, with the faces correctly aligned, a centrally positioned capture latch shaped like a hook on the active face grappled a capture bar on the passive face, pulling the two truss pieces together. Fourth, the activation of motor-driven bolts at each vertex of the active face screwed them into sockets in the passive face and the two truss pieces were locked firmly to each other. The procedure for attaching the outer P5, S5, P6 and S6 truss building blocks to each other and to the rest of the truss was slightly different. The first two alignment tasks were repeated but the automatic capture and bolting tasks were carried out by spacewalking astronauts.[41, 42]

Pirs Docking Compartment View of the Pirs Docking Compartment after its berthing to the nadir port of the Zvezda module. Pirs contained an airlock to enable spacewalks from the Russian sector of the Station. It also provided another docking port for visiting Soyuz and Progress spacecraft. Visible on either side of Pirs in this photograph are two robotic cranes named Strela. The purpose of these telescopically deployable cranes is to move cosmonauts and equipment around the exterior of the Station's Russian sector.

Though the use of the Canadarm2 robotic arm to assemble the Station's truss helped to minimise the astronaut construction workload, there were some situations where astronauts had to carry out the truss bolting work themselves. To make this job easier and quicker for them in their bulky spacesuits, the bolting process used a special fast-action nut called a ZipNut® developed by a company called ZipNut Technology through NASA's spin-off technology programme. Unlike ordinary nuts that thread on to bolts by continuous turning, ZipNuts® are pushed on to bolts and then twisted to engage the bolt and nut threads. The operation is much faster than the normal threading procedure. The pushing action is made possible by a segmented nut inside an outer sleeve. The sleeve splits into separate segments that, first, splay outwards when the nut is pushed over the bolt head and, second, retract to engage when the nut is pulled slightly.[43] The technology was also used to repair the Hubble Space Telescope and has found many applications beyond space such as nuclear reactor repair, underwater pipe-laying, tyre manufacturing, medical equipment manufacturing and building construction.[44]

Mission 9A had begun to extend the Station's transverse truss on the starboard side and the next mission would do the same on the port side. Mission 11A (STS-113) that launched with Endeavour on 23 November 2002 brought up the P1 Truss. The P1 was the mirror image of the S1 and the third truss building block to be added to the Station. Its addition extended the truss's length by a further 13.7 metres to 40.8 metres. The P1 provided similar functions to the S0 in the middle and the S1 on the opposite side. These included more track for the railcar and robotic arm and extended power, thermal control, communications, navigational and data systems routes. Like the S1, the P1 had a trapezoidal cross-section and six accessible structural bays.[45] Endeavour followed the normal rendezvous and docking procedure at the forward end of the Station. Canadarm2 on top of Destiny grasped the P1 piece, moved it clear, rotated it and manipulated it into the correct alignment, swung it inboard and bolted it to the port end of the S0. A further six spacewalks were needed on this mission. After these, there

was another deployment aft of a thermal radiator panel. This time it was the centre panel of the three on the P1 piece that matched those on S1 the starboard side.

In 2002 a high-definition film of the International Space Station was screened in IMAX cinemas worldwide. Supported by NASA, Lockheed Martin and the Imax Corporation, it was simply entitled *SPACE STATION* and featured camera work by the Station's astronauts and an accompanying narrative by Tom Cruise.[46] Until now, the public had only seen photographs of the Station on orbit in newspapers, magazines and books or occasional movie clips on television news or science programmes. For the first time, audiences gained a sense of the complexity and scale of what already was by the far the biggest structure built beyond Earth. With its superb panoramas of the Station from the Orbiters, fascinating camera shots of crew members floating through the modules and Tom Cruise's narrative style, the IMAX production was a huge success, playing to packed houses.

Capped by the success of the IMAX movie, 2002 ended on a high note for the International Space Station. Supported by visiting Orbiters and Soyuz and Progress spacecraft, it was growing into a sophisticated scientific laboratory and habitat. The first five crews – Expeditions 1 to 5 with three astronauts or cosmonauts each – had come and gone and between them had already accumulated a total of 792 days of occupation. The Station's assembly was proceeding smoothly with arriving Orbiters delivering their building block payloads for plugging on to the Station, piece by piece, in a constructional procedure that was on the verge of becoming routine. It was all going very well. There was no hint of another catastrophe about to befall the Space Shuttle fleet and the impact it would have on the Station's prospects.

Construction work in space Astronaut Lee Morin holds a V-shaped keel pin that he has just removed from the S0 Truss piece after its delivery to the Station on STS-110 in April 2002. The keel pins were temporary restraints to hold the large truss pieces securely in the Orbiter payload bay during launch and ascent.

A second disaster
Loss of the Orbiter Columbia, February 2003

On 1 February 2003, on its way back home from orbit, the Shuttle Orbiter Columbia disintegrated during re-entry and its crew of seven perished. The Space Shuttle fleet was immediately grounded and assembly of the International Space Station halted. Columbia had not visited the Station, nor even come close to it. Of the four Orbiters in service – Atlantis, Columbia, Discovery and Endeavour – it was the oldest and too heavy to reach the Station's orbit with a big payload. Columbia was on a completely separate research mission.

As with the Challenger disaster of 1986, an investigation team was rapidly formed and given the title of the Columbia Accident Investigation Board. NASA quickly met with Roscosmos to plan how the Station could be kept resupplied by Russia's Progress spacecraft and crewed by Soyuz spacecraft for the foreseeable future. Abandoning the Station was unthinkable. A few

SO Truss In the diagrams to the right and below, the Shuttle Orbiter approaches the Station and closes in **A** with its payload bay doors open to dock with the forward end of the Destiny Module **01**. The Canadarm2 robotic arm **02** mounted on Destiny's exterior grasps the SO Truss **03** in the Orbiter payload bay and moves it away **B** from the Shuttle and the Station to rotate it to the desired orientation **C**. The Canadarm2 then brings the SO Truss inboard **D** and mounts it on a cradle-like group of struts **E** on top of the Destiny Module. Eight spacewalks were needed on this mission. Image: author.

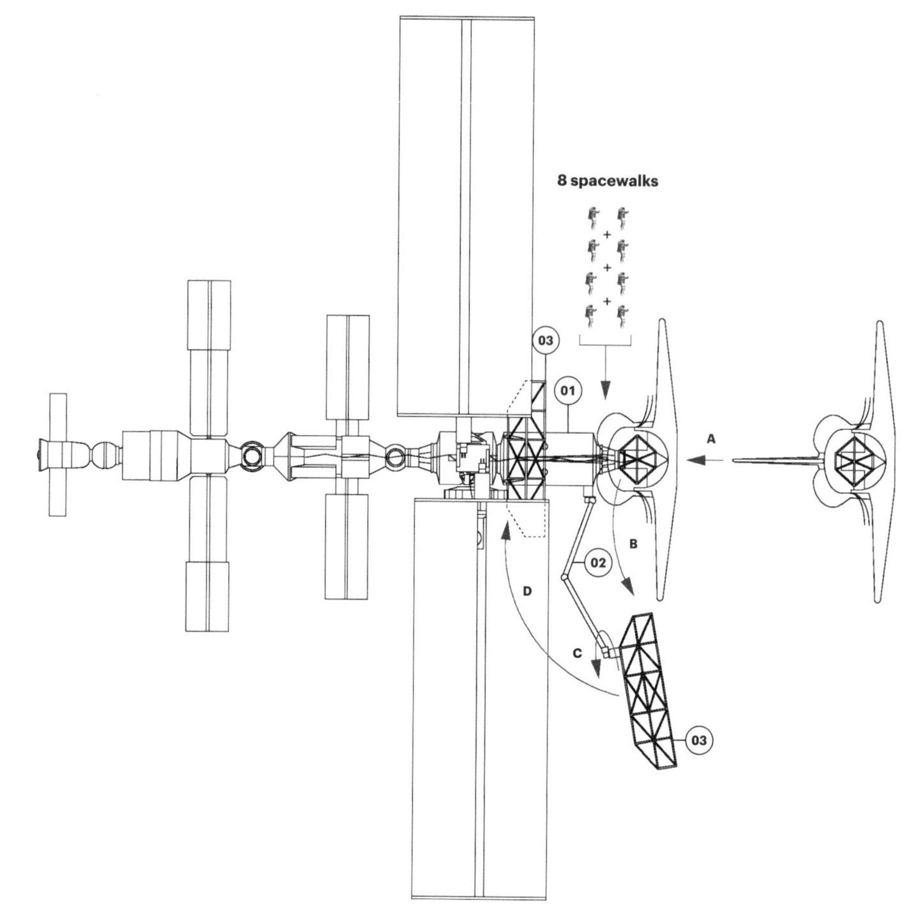

SO Truss The Canadarm2 robotic arm has the SO Truss building block in its grasp as it manipulates and moves it to its mounting place on top of the Destiny Laboratory Module. The SO Truss has a trapezoidal cross-section. The wide-angle photograph shows the two rails along the truss edges. Mounted on the rails towards the far end is the box-shaped railcar called the Mobile Base. Projecting outwards between the rails are two wishbone-shaped keel pins that secured the SO in Atlantis's payload bay during launch and ascent.

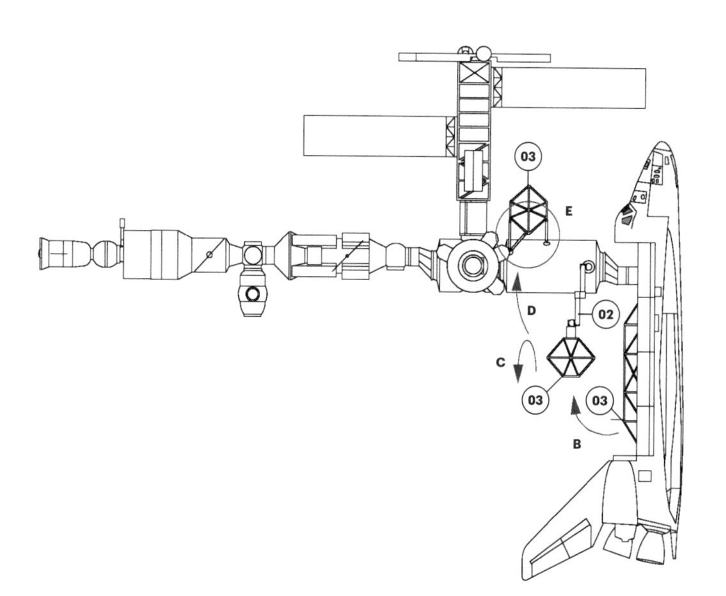

Mobile Servicing System The Mobile Servicing System is the name given to the International Space Station's travelling crane system. It comprises five main elements: the Canadarm2 robotic arm, a work platform called the Mobile Base that provides four grappling points for the arm and other equipment, the Mobile Supporter which is the travelling railcar that supports the arm on its work platform, a highly capable end manipulator for the arm called Dextre, and the astronaut workstations inside the Station for operating the whole system. This photograph shows the Mobile Base framework undergoing processing at Kennedy Space Center prior to launch.

days after the disaster, the weekly newspaper *Space News* wondered about the effect of the Iran Nonproliferation Act on Soyuz flights that were now so vital to keep the Station alive. It noted that 'the three-person, Russian-built Soyuz spacecraft currently is the only means of returning the Space Station's crew to Earth absent Shuttle flights and the only way to get new crews into orbit as long as the Shuttle fleet is grounded'.[47] The Station would become entirely dependent on Roscosmos and Russia for its survival until the Space Shuttle was operational again and no one had any idea when that would be.

The Columbia Accident Investigation Board released its findings in August 2003.[48] It had carried out a painstaking piece of detective work in a very short time and delivered a very thorough report. Over 100 staff had supported the Board's thirteen members, aided by 400 NASA engineers. They had examined 30,000 documents and conducted 200 formal interviews. According to NASA, more than 25,000 people from 270 organisations spent 1.5 million hours searching an area almost as big as the State of Connecticut looking for debris from the Orbiter's disintegration. Together with members of the public, they found 84,000 pieces.

In its Executive Summary, the report explained the cause of the accident in the following manner. 'The physical cause of the loss of Columbia and its crew was a breach in the Thermal Protection System on the leading edge of the left wing, caused by a piece of insulating foam which separated from the left bipod ramp section at 81.7 seconds after launch, and struck the wing in the vicinity of the lower half of Reinforced Carbon-Carbon panel number 8. During re-entry this breach in the Thermal Protection System allowed superheated air to penetrate through the leading edge insulation and progressively melt the aluminium structure of the left wing, resulting in a weakening of the structure until increasing aerodynamic forces caused loss of control, failure of the wing, and break-up of the Orbiter. This break-up occurred in a flight regime in which, given the current design of the Orbiter, there was no possibility for the crew to survive.'

The Executive Summary continued with comments on the Shuttle and NASA's management. 'The organisational causes of this accident are rooted in the Space Shuttle Program's history and culture, including the original compromises that were required to gain approval for the Shuttle, subsequent years of resource constraints, fluctuating priorities, schedule pressures, mischaracterization of the Shuttle as operational rather than developmental, and lack of an agreed national vision for human spaceflight. Cultural traits and organisational practices detrimental to safety were allowed to develop, including: reliance on past success as a substitute for sound engineering practices (such as testing to understand why systems were not performing in accordance with requirements); organisational barriers that prevented effective communication of critical safety information and stifled professional differences of opinion; lack of integrated management across program elements; and the evolution an informal change of command and decision-making processes that operated outside the organisation's rules.'[49] It was a damning critique.

At the end of the report were three pages of recommendations covering the External Tank's thermal protection system, launch imaging, Orbiter sensor data, wiring, bolts, close-outs, micrometeoroid and orbital debris, foreign object debris, launch scheduling, staff training, organisation, Shuttle recertification beyond 2010, and drawing and photograph records. In particular, NASA must 'initiate an aggressive program to eliminate all External Tank Thermal Protection System debris-shedding at the source with particular emphasis on the region where the bipod struts attach to the External Tank'. The Board considered that 'good leadership can direct a culture to adapt to new realities' and that 'NASA's culture must change'. The findings seemed to echo comments in the Young Task Force report of November 2001. NASA responded by appointing a Return-to-Flight Task Group to oversee the corrections and changes recommended in the Columbia report. The task group periodically reported its progress in 2004.[50] The result was that NASA agreed to carry out all the recommendations and make technical and management changes to the Shuttle programme with the focus on correcting the defect in the propellant tank's foam insulation. In parallel, NASA issued a series of implementation plans beginning in October 2003 describing its intentions for continuing the operation and occupation of the Space Station.[51]

As work continued to rectify the Space Shuttle's problems and return the Orbiter fleet to flight status, another space exploration plan was being hatched. In January 2004, President Bush announced a new space exploration vision and created a Commission chaired by Pete Aldridge to make recommendations on how it should be implemented. The Commission produced its report entitled *A Journey to Inspire, Innovate and Discover* in June 2004 in which it laid out a plan for future space exploration, recommending that, once it returned to flight, the Shuttle's focus should be to complete the International Space Station after which it should be retired.[52] The Aldridge Report recommendations found their way into a new national space transportation policy issued on 6 January 2005.[53] It sealed the Space Shuttle's fate.

NASA's focus was to shift away from the Space Station towards lunar and then Mars exploration with a new crew exploration vehicle. Meanwhile, the impact of grounding the Shuttle fleet worked its way throughout 2004 into the Station partners' future assembly planning and the partners issued an updated Station design configuration in July that reflected the reduced Russian hardware presence and various Russian and American assembly modifications.

On 26 July 2005, thirty months after the loss of Columbia, Space Shuttle flights to the International Space Station resumed with the launch of the

The Canadarm2 Robotic Arm, Mobile Base and Mobile Transporter The cutaway illustration shows the Canadarm2 arm, the Mobile Base work platform and the Mobile Transporter railcar. The Dextre manipulator does not appear in this view. Protective sheathing and insulation are omitted for clarity. The Mobile Base is depicted as an open framework with four grappling fixtures that look like small circular stools projecting out at angles. The robotic arm is attached to one of the grappling fixtures. Providing the foundation for the Mobile Base and the means of mobility along the rails is the box-like Mobile Transporter, coloured grey in this illustration. Image: CSA/MDA.

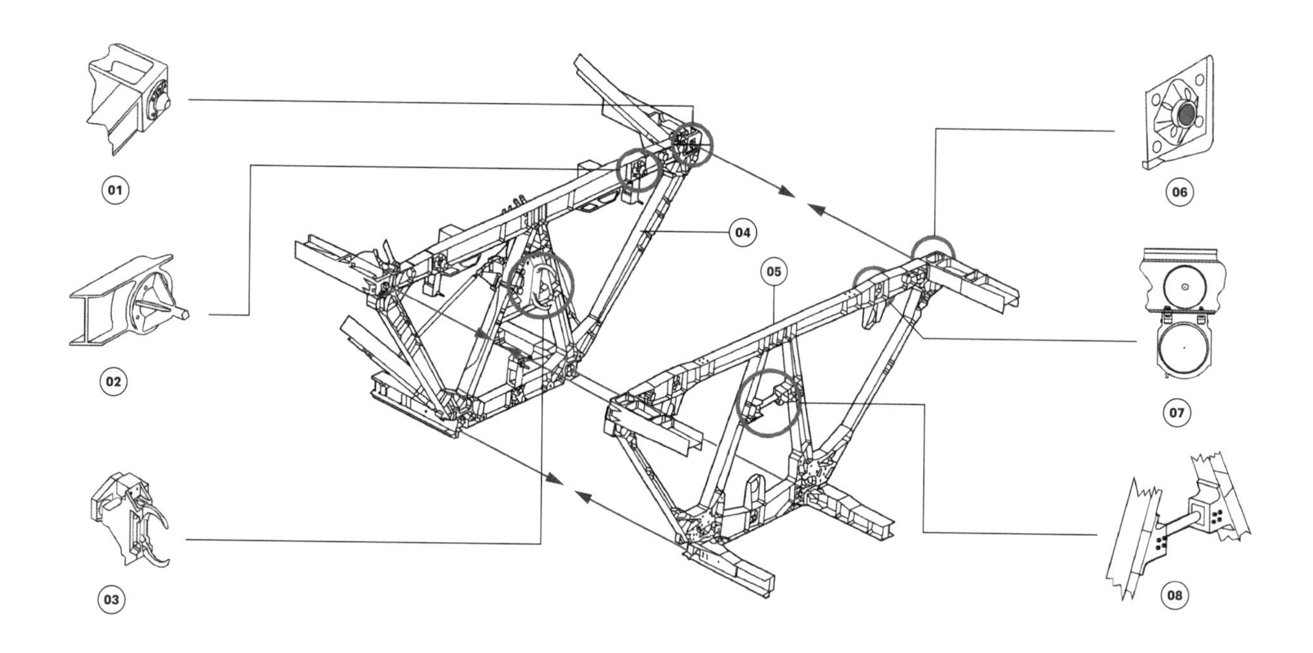

Truss interconnection technique The Canadarm2 robotic arm was used to connect the truss building blocks. This illustration shows the interconnection equipment. The trapezoidal truss end structure on the left was the active side and on the right it was the passive side. As the trusses were brought together, three coarse alignment pins on the active side engaged with matching sockets on the passive side. As closure continued, four fine alignment cones on the active side engaged with matching cups on the passive side. Then a hook-like capture latch on the active side grappled a matching bar on the passive side. Finally, motor-driven bolts secured the truss pieces. Image: NASA, with author additions.

01 Fine alignment cone (4 places)
02 Coarse alignment pin (3 places)
03 Capture latch (1 place)
04 S0 Truss
05 P1 or S1 Truss
06 Fine alignment cup (4 places)
07 Coarse alignment socket (3 places)
08 Capture bar (1 place)

Orbiter Discovery on Mission LF1 (STS-114).[54] NASA called the mission 'Return to Flight'. NASA had spent two and a half years implementing safety improvements for the Orbiters and their External Tanks. Discovery's launch and ascent was extensively monitored by cameras at the launch site, airborne cameras on aircraft and radar systems. Imagery of the Return to Flight launch showed a piece of foam falling off the tank and smaller tile and foam defects. There were two areas with gap fillers protruding. As Discovery neared the Station on 28 July, The Orbiter performed a Rendezvous Pitch Maneuver – a kind of 360° spacecraft somersault – to flip the Orbiter over and enable the Expedition 11 crew on the Station to photograph Discovery's underside. After looking at images of the damage, NASA's team on the ground decided that Discovery needed repairs before it came back. Astronaut Stephen Robinson went out on a spacewalk to deal with the protruding gap fillers, equipped with forceps and a hacksaw.

Another spacewalk involved testing repair techniques on intentionally damaged tile specimens to provide feedback for future missions. Discovery also delivered the Italian-built Multipurpose Logistics Module named Raffaello on this mission and attached it to Unity Node-1. Discovery returned to the ground safely but the Space Shuttle system was grounded again for further investigation and Space Station assembly missions went back on hold, for despite two and a half years of redesign and retesting, NASA and its contractors had still not solved the twin problems of the tank's foam insulation and the Shuttle Orbiter's sealed tiles. Spraying on foam fireproofing, bonding insulation panels and ceramic tiles to a substrate, grouting tiles to resist heat or water penetration are tasks well known to the building construction industry on Earth and usually carried out without defect. It seemed inconceivable that the Shuttle's problems so far down the space technology scale had still not been resolved by the scientists, engineers and materials specialists that NASA had at its disposal.

At about the same time, the issue of the impact of the Iran Nonproliferation Act on the Station was becoming critical. On 12 July, President Bush submitted to Congress an amendment to allow NASA to purchase goods and services from Russia to support the Station.[55] The need for more Soyuz and Progress spacecraft missions to keep the Station and its crews going was urgent. The House and Senate passed a modified version of Bush's amendment, allowing NASA to make purchases from Russia for Station purposes up to 1 January 2012. With Shuttle missions to the Station still on hold, this was welcome news to NASA.

In October 2005, NASA issued the results of its Shuttle/Station Configuration Options team study on the completion of the Station's assembly by 2010, in accordance with President Bush's directive.[56] The new assembly schedule showed nineteen more Shuttle missions of which eighteen would be for the Station's assembly and one for servicing the Hubble Telescope. NASA, in continued financial difficulties, announced it needed an extra $5.6 billion to carry out the Space Shuttle's remaining flights. The new reduced flight schedule meant that some hardware elements would be lost. The Japanese-built Centrifuge Module, previously considered to be of the utmost importance as a research tool on the Station, was scrapped. At a stroke, a whole branch of potentially valuable scientific research using the simulated partial gravity conditions of a centrifuge was wiped out. So too were the Russian Science Power Platform and the third American node module. The assembly sequence was reshuffled yet again to deliver infrastructure ahead of modules and, perhaps optimistically, NASA confirmed that the Station would be outfitted for a six-person crew. NASA's budget problems with the Station and Shuttle programmes persisted, forcing it to move funds over to them from its 2007 science and exploration budget to cover the shortfall, not a popular decision with the Station's scientists and experimenters.

In another cost-cutting measure, NASA announced that the Orbiter Atlantis would be retired early in 2008, allowing it to be scavenged for spare parts for the Orbiters Discovery and Endeavour. This would eliminate the cost of a major maintenance programme on Atlantis planned for that year. The end of the Space Shuttle programme had begun early and vultures were beginning to gather. In February 2006, the Space Station partners met and the following month they released a revised assembly sequence.

Truss interconnection technique This photograph shows the passive side of the truss end structure. Clearly visible are two coarse alignment sockets and strike plates along the top edge and one at the bottom midpoint. Positioned at each of the four vertices of the trapezoidal face is a fine alignment cup. At the centre of the face is a single capture bar.

Revisions identified by alphabet letter were no longer used – perhaps there had been too many – but are maintained in this account for continuity. This was the equivalent of Revision H.[57] The latest manifest showed that Station assembly flights had been reduced further – there were now only sixteen plus two additional contingency flights – but these would be able to deliver all remaining assembly hardware other than items deleted since the loss of Columbia. A third node module was back at the very end of the manifest in 2010: it was on the last mission. The final Space Shuttle missions were at risk of being cancelled if there were further delays. The third node module was the one NASA planned to outfit for improved crew habitation. Once again, crew living conditions on the Station seemed to be pushed to the bottom of the list of priorities. Also included in Revision H were the final Russian launches on the Proton rocket of a multipurpose laboratory module and a Research Module listed for delivery after the Shuttle's retirement date. Beyond these final Russian missions there was nothing. With Revision H, the total number of hardware missions required to assemble the International Space Station on orbit numbered twenty-eight from beginning to end, excluding American and Russian logistics and equipment missions and Soyuz replacement missions. The date of the very last Space Shuttle flight was set for July 2010.

Columbia disaster investigation A grid is laid out on the floor of a hangar building as members of the Columbia Reconstruction Project Team bring in and position pieces of Columbia's debris from recovery in the field. The team was reconstructing the bottom of the Orbiter as part of the investigation into the accident that led to the loss of Columbia and its crew as they returned to Earth on Mission STS-107.

Construction restart

Continued assembly of the Transverse Truss, 2006-2009

On 9 September 2006, Mission 12A (STS-115) with the Orbiter Atlantis launched to the Station carrying the next truss building block and the first since the Columbia disaster.[58] Forty-five months had passed since the last assembly mission in November 2002 and now the Station's assembly was up and running again. Atlantis brought up the P3/P4 Truss. Its installation would allow the complete Mobile Servicing System including the Canadarm2 robotic arm, the railcar and its tracks to be operable for the first time. Like previous truss elements, the P3/P4 was an open structure, but there was a major difference. It had the alpha rotary joint at its midpoint that could rotate 360° in either direction to enable its solar arrays to track the Sun as the Station circled the Earth.[59] This joint rotated around the Station's Y axis, perpendicular to the transverse truss's port-starboard line. If one could imagine sitting on the truss and looking down its length, the joint would enable the solar array masts to rotate one way and then the other, like a giant needle swinging on a compass.

The addition of the P3/P4 doubled the electrical power produced up to this point by the American sector on the Station to about 42kW. Not only was the complete Mobile Servicing System brought into action for the first time to help with the assembly task, but so was the original robotic arm on the side of Atlantis's payload bay. It was rather like a grand ballet in space – a mechanical *pas de trois*, with the two robotic arms working together to handle and pirouette a very large metallic ballerina: the P3/P4 Truss. First, Atlantis's robotic arm removed the P3/P4 from the payload bay and swung it over to the port side of the Station where it was grappled by Canadarm2, now mounted on the railcar, which had also moved to the port side on its tracks. With both robotic arms holding it in an embrace, the P3/P4 was left overnight. The next day, Atlantis's arm released the P3/P4, Canadarm2 rotated it and moved it out to the port end of the truss. Finally, once it was correctly aligned with the structure, the arm pulled it into place. Six spacewalks were needed on this mission. Once all the assembly tasks were accomplished, the solar arrays and thermal radiator on the P3/P4 piece were fully deployed in their final positions.

The next building block up was the P5 Truss on Mission 12A.1 (STS-116) with Discovery, launched on 9 December 2006. The P5 was a short spacer framework mounted on the end of the P3/P4 piece. It extended the transverse truss structure beyond the centreline of the P3/P4's solar arrays, enabling the attachment of the next truss building block outboard and clear of the P3/P4. Discovery arrived and docked in the usual manner at the Pressurized Mating Adapter at the forward end of the Destiny. As with the previous mission, it was a carefully planned exercise in mechanical choreography to extract the P5 piece from Discovery's payload bay with the Orbiter's robotic arm, hand it over to the Canadarm2 arm on its railcar and then move it along the rail track to the port side to mount it on the end of the P3/P4 piece. Two spacewalks supported this task.

Then the spacewalking astronauts turned their attention to the P6 piece. The P6 had been temporarily installed on top of the Z1 building block in November 2006 to provide interim electrical power to the Station before assembly of the full-length truss and the completion of all its sets of solar arrays.[60] Moving the P6 from its temporary to its permanent home would involve the retraction of both its solar arrays into their stowage boxes on each side of the central mast and this work needed doing first. The P6 array blankets had been fully extended in space for over six years, far longer than originally planned as a temporary measure, due to the delay in Station assembly following the loss of Columbia. The lightweight accordion-like solar blankets were designed to retract automatically into their boxes but the accordion strips failed to fold up and stow properly and extra astronaut spacewalks became necessary to provide 'hands-on' support during the folding process.[61] On 13 December, the astronaut repair crew spent six hours trying to fold up and retract the port array on the P6, without success. Guidewires were snagging. Repeated attempts the following day were also unsuccessful. On 16 December, two more spacewalks tackled the uncooperative P6 port arrays. Manipulating the grommets and guidewires, the astronaut repair crew shook the array while their colleagues inside the Station reeled it in, one bay at a time and achieved 65 per cent retraction.

S3/S4 Truss Both robotic arms went to work to install the S3/S4 Truss piece on the Station. The Shuttle Orbiter's robotic arm **01** removed the S3/S4 Truss **02** from the Orbiter cargo bay swung over the Shuttle's starboard side **A** where it was grappled by the Canadarm2 robotic arm **03** mounted on its railcar **04**. The next day, the Orbiter's arm released the S3/S4 Truss piece. Canadarm2 manipulated it and moved it under the existing transverse truss structure **B** towards aft **C** and then out towards the starboard end of the transverse truss **D** to attach it to the end of the existing truss structure **E**. The second solar array on the temporarily located P6 Truss **05** was folded back into its blanket box **F**. The giant solar arrays **6** on the S3/S4 Truss were swung into position **G** and fully extended fore and aft **H**. Eight spacewalks were required on this mission. Note: some elements omitted and the sequence simplified in this illustration for clarity. Image: author.

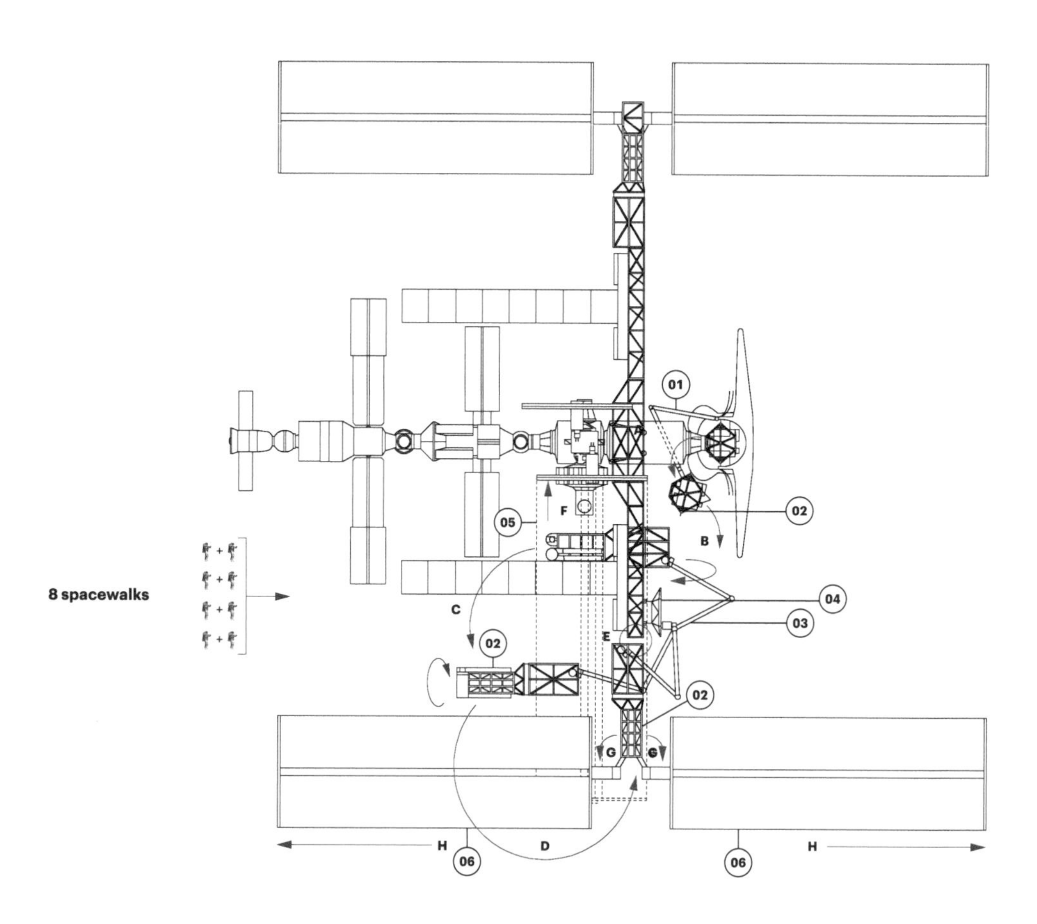

View from above

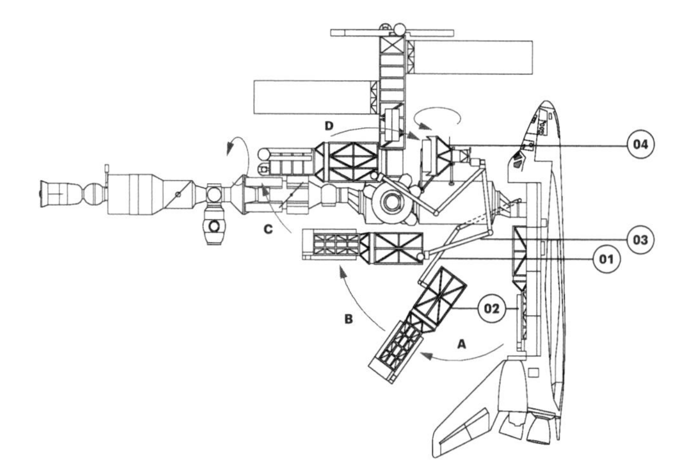

View from starboard side

On 18 December the repair crew tried again and finally completed the job. The starboard array on the P6 was still extended and would be dealt with on an upcoming mission. By the completion of this mission, astronaut Robert Curbeam, who was one of the repair crew, had set a new record for the most spacewalks on a single Shuttle mission – four.[62] Eight spacewalks – four each by two astronauts – were required to complete this arduous mission. Of these, the last two were unplanned and added during mission to deal with the solar array retraction problem.

In January 2007, the heads of all the space agencies involved in the Space Station met in Paris to discuss the progress of its assembly. They expressed confidence in the Space Shuttle's ability to complete the Station by 2010 but were concerned about several unresolved issues. First, a plan to increase the Station's crew size to six by mid-2009 would test the limits of the three docking ports; second, the Shuttle's retirement in 2010 would leave the Station without an experiment return capability until the Shuttle's successor Orion became operational; third, the NASA-Roscosmos agreement on Russian Soyuz and Progress flights would need updating.[63] In March there was some negative news from NASA. Administrator Michael Griffin informed a Senate panel that the Shuttle's successor Orion and its Ares launcher would not be ready and operational until 2015 due to a budget shortfall.[64] This would extend the period that America would be without a crew launch capability to five years from the Shuttle's planned retirement year in 2010. The same month, Space Station programme managers told a safety task force mandated by Congress that the Station would face a supplies shortfall of 5.5 tonnes before 2010, rising to 54 tonnes by 2016 if the Station had to rely on a mixture of Russian, European and Japanese launch vehicles for resupply.[65] The problem of how to keep the Station going after the Shuttle's retirement was beginning to cause concern.

On 28 March 2007, NASA staff appeared before a Senate hearing to discuss the 2010 Shuttle retirement impact on the issue of spare parts for the Station. NASA's original plan had been to bring spare parts down from the Station as cargo, repair and service them and then return them. NASA estimated that the total mass of Station hardware needing repair or exchange was 2,267kg.[66] The loss of the Shuttle would mean that the spare parts cargo capability would be dramatically reduced. No other spacecraft had anything like the Shuttle's return payload capability – neither the Russian Soyuz, nor the Shuttle's proposed successor Orion, nor, for example, the two commercial Station resupply vehicles under early development at the time by Space Exploration Technologies (SpaceX) and Rocketplane Kistler. To remedy this, NASA would order new pumps, processors and other parts on a regular basis and deliver them to the Station as necessary. Defective or worn-out parts on the Station would not be returned for repair, but stuffed into Russian Progress or European Automated Transfer Vehicle spacecraft with other waste materials and destroyed by re-entry burn-up, which was

S3/S4 Truss Space Shuttle Mission STS-117 delivered the S3/S4 piece of truss structure to the International Space Station in June 2007. Shown here, the S3/S4 Truss on the Station's starboard side was the mirror image of the P3/P4 Truss piece on the port side that had arrived on an earlier mission.

the normal way of disposing of these spacecraft. Of particular concern were replacements for the Station's control moment gyroscopes, vital for maintaining the Station's correct attitude on orbit and the cause of problems in the past – they had been replaced in 2005. NASA considered that they were too heavy and too delicate to deliver on other launch vehicles. Spares of these would need to be delivered to the Station while the Shuttle fleet was still operational.

The assembly timetable was affected again in early 2007. Mission 13A carrying the next piece of the truss structure had been due to launch in March, but thunderstorm hail damage to the foam insulation on the External Tank at the launchpad resulted in the need for repairs and delayed the launch. The spray-on insulation used on the tank was several centimetres thick and made of foam that had no impact resistance and was easy to damage. The fact that hail could damage it so easily was an indication of its inherent weakness as a material for spaceflight use. Mission 13A (STS-117) with Atlantis launched on 8 June 2007 carrying the S3/S4 Truss.[67] The S3/S4 was virtually identical to the P3/P4 piece but its mirror image. Assembly procedures using the two robotic arms were similar to those used for the P3/P4's attachment, but the railcar carrying the Canadarm moved the S3/S4 in the opposite direction to the starboard end. Deployment of the S3/S4 solar arrays would add another 21kW to the American Station sector's power capability. This mission also saw the retraction of the starboard solar array on the P6 piece. The retraction of the first array on the P6 had been a difficult procedure. This time, the astronauts were prepared and the task was carried out effectively.[68]

Shortly after the attachment of the S3/S4 piece and its solar arrays deployment, there was another emergency on board. A smoke alarm sounded, indicating one of the Russian modules as the source. The crew went to look and found nothing, but it became apparent that computers had malfunctioned. Those in question – made in Germany – were based on the Russian side of the Station. They operated the Station's attitude control system – a critical task – and performed other important functions. The evacuation of the Station was a possibility. The Orbiter Atlantis, docked at the Station, immediately took over attitude control while engineers on the ground and the crew on board tried to locate the problem. It appeared to be electronic noise generated by the solar arrays on the newly attached truss piece, which produced a current surge that caused the computers to shut down. The crew rigged some cabling to bypass the surge protection subsystem and the computers restarted. The incident was serious and the consequences potentially disastrous. Replacement of the problematic computers became a top priority.[69]

In July 2007, a House Science and Technology Subcommittee in Washington held a hearing on the Space Shuttle and the International Space Station. The Subcommittee reviewed the status of both programmes and

CSA ASC

esa
COLUMBUS

voiced concern about several issues. There was an ongoing labour dispute at Kennedy Space Center with United Space Alliance, the consortium operating the Space Shuttle, and machinists were on strike. The Subcommittee wondered about the impact of the strike on the safety of Shuttle operations. There was a question about the Station's ability to be a productive research facility at a time when NASA was cutting back on the Station's research budget due to continuing budget difficulties. There was a question about the future use of the Station after the Shuttle system retirement in 2010 and how crews would travel back and forth in the absence of an American launch system.[70] This last question was indeed a big one. The Space Shuttle's retirement was now irreversible, even if NASA had had a change of heart, which it had not. There was no confirmation that the Shuttle's successor Orion would be able to undertake missions to the Station, and it would not be ready until 2015, which was then eight years in the future. The only solution was for America to rely on Russia to provide the taxi service for the Station's crews with its Soyuz spacecraft and the delivery service for the Station's supplies with its Progress spacecraft, just as it had done in the aftermath of the Columbia disaster.

The Orbiter Endeavour lifted off on 8 August 2007 on Mission 13A.1 (STS-118) with the S5 Truss. The S5, like the P5 installed on an earlier mission, was a short spacer piece that was attached to the end of the S3/S4. Its purpose was the same: to extend the transverse truss beyond the centreline of the S3/S4's solar arrays, clearing the way for the addition of the next truss building block outboard. The S5's installation procedure was similar to that of the P5 but mirrored. Moving the short P5 and S5 pieces into position was not easy. Observation of the far ends of the transverse truss was restricted and the Canadarm2 robotic arm was operating at full extension. A crew member working at the control console inside the Station carefully inserted the S5 into the gap between the last pair of solar arrays with little room to spare for manoeuvre past the S3/S4 next to it. Two astronauts on spacewalks monitored the procedure. Canadarm2 left the S5 loosely attached to the S3/S4 in the 'soft-dock' position. Then two astronauts securely bolted the new piece of truss in place.[71] Eight spacewalks – four each by two astronauts – were needed to complete this mission.[72] Between this and the next mission, 10A, the Pressurized Mating Adapter on the port side of Unity was moved by Canadarm2 to the berthing port on the underside of Unity, facing nadir. This move was necessary to free up the port side berthing port on Unity for the attachment of the next node module to arrive.

The lack of an American launch capability to the Station after 2010 continued to occupy the minds of the Station's partners and in November, the European Space Agency announced it wanted to start work on a crewed version of its Automated Transfer Vehicle. The agency would propose the initiative to its government ministers in 2008.[73] Meanwhile, at the Guiana Space Centre, preparations were speeding up for the launch to the Station

Columbus Module Tethered by a safety cable, astronaut Rex Walheim carries out work on the exterior of the Columbus Laboratory Module after its berthing at the Station. Columbus was Europe's laboratory contribution to the International Space Station. This photograph is taken looking towards the starboard end of the transverse truss structure with the fully deployed solar arrays on the S3/S4 Truss visible. The projecting circular disc near Rex Walheim's boots is a grapple fixture for the end of the Canadarm2 robotic arm.

Columbus Module A cutaway view of the Columbus Module showing its construction. Forming the inside of the hull is a pressure-resisting skin made of waffle-shaped, curved panels mounted on a machined aluminium ring and longeron framework. Forming the hull's exterior is a grid of aluminium bumper panels to provide protection against micrometeoroid penetration. Sandwiched between the two are thermal insulation layers. The central square-section corridor is bounded on all four sides by experiment racks. These are supported off the module's hull by triangular Stand-Off frames behind which run air ducts, fluid lines, gas lines and cabling. All the racks can hinge inwards to provide crew access to the hull surfaces behind. Image: ESA.

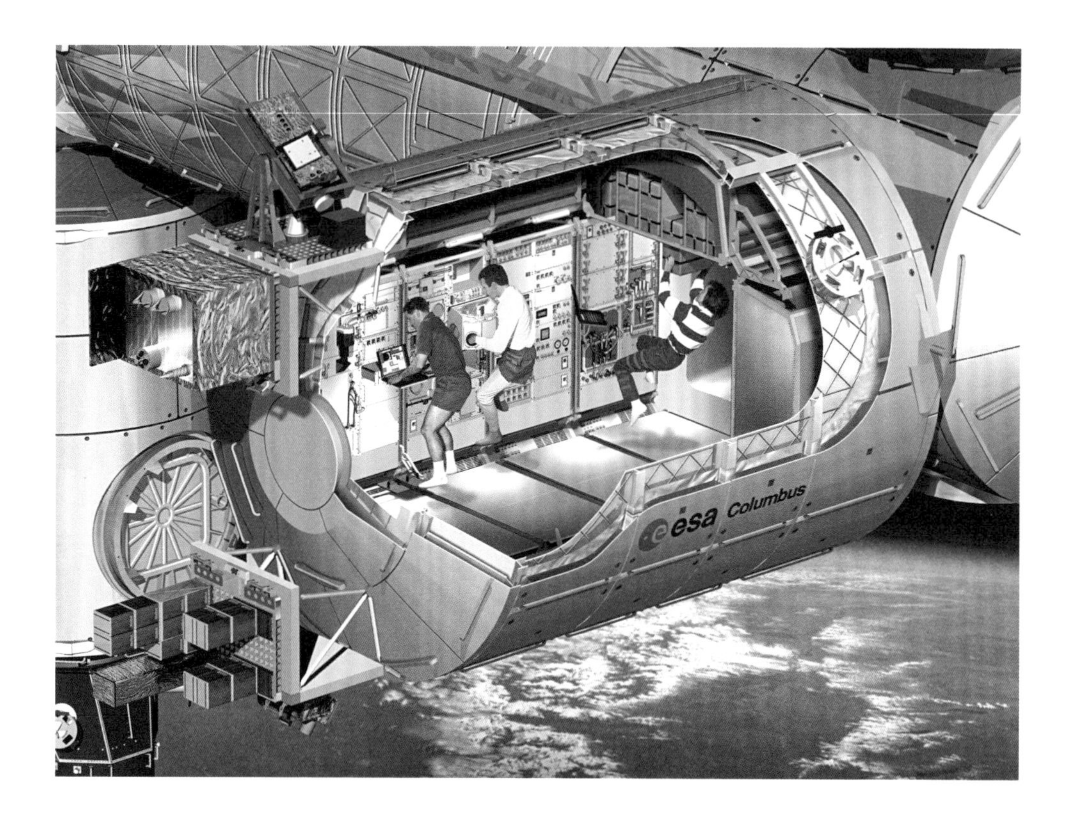

of the first Automated Transfer Vehicle named Jules Verne. The Jules Verne had originally been due to launch in 2004 but the European Space Agency had pushed the date back to 2008. The European vehicle was the most complex spacecraft Europe had ever designed and built and technical problems had caused a succession of delays.[74]

Mission 10A (STS-120) launched on 23 October 2007 with Node-2 named Harmony in Discovery's payload bay.[75] It was another assembly milestone designated as 'ISS US Core Complete'. The European Space Agency had supplied Harmony for the Station under a barter agreement with NASA. Harmony Node-2 was going to provide attachment points for the European and Japanese pressurised modules to be delivered on upcoming missions. Harmony was due for berthing at the forward end of Destiny. However, this was the same place where the Orbiters had been docking at the Station since assembly Mission 3A. Discovery would have to park Harmony somewhere on the Station and depart, leaving the forward berthing port free and enabling Canadarm2 to move Harmony to its final home. In the process, the forward Pressurized Mating Adapter would have to be detached and then reattached. Discovery arrived in the normal manner from forward of the Station's flight path and docked with the Pressurized Mating Adapter at the forward end of Destiny. A complicated assembly sequence of three steps then began. First, Canadarm2 'walked' itself off its railcar along the truss and latched on to Destiny's underside from where it grappled Harmony and lifted it out of the payload bay. It then swung Harmony round in an arc to port and mounted it at a vacant port on the port side of Unity Node-1. Second, after Discovery departed the Station, Canadarm2 removed the Pressurized Mating Adapter on the front of Destiny and, swinging it around to port in a similar arc, mounted it at the outward facing vacant port of Harmony. The third step involved Canadarm2 retrieving Harmony and returning it to the forward end of the Station to mount on the front end of Destiny.

This mission also saw the transfer of the P6 Truss from its temporary mounting on top of the Z1 Truss to its final home at the far end of the port side of the long transverse truss structure. This was another complicated manoeuvre involving both the Orbiter and Canadarm2 robotic arms. Both solar arrays on the P6 piece had been stowed back in their boxes on previous missions. Now, the two robotic arms passed the P6 between them in another carefully choreographed sequence ending with its attachment to the end of the P5 piece.[76] First, the P3/P4 piece on the port side rotated 90° to align its solar arrays on the zenith-nadir axis where they would be out of the way of the P6 piece on the move. Next, Canadarm2 removed the P6 from its mounting and swung it out forward of the transverse truss line to the side of the Orbiter. Then, Discovery's robotic arm took the P6 from Canadarm2 while the railcar moved Canadarm2 towards the port end of the truss. Finally, Canadarm2 swung the P6 around in an arc to the end of the P5 piece and attached it in place. This mission required another eight spacewalks.[77]

Laboratory work

European and Japanese additions, 2008

Until now, America and Russia had shared the limelight as the principal actors on the orbital stage where the Station's assembly was playing with Canada taking, almost literally, a walk-on part with its robotic arm. Now Europe was about to make a grand entrance with its Columbus Laboratory Module and Jules Verne, the first Automated Transfer Vehicle. Development and manufacturing of the Columbus Laboratory Module had begun in 1996 when the European Space Agency signed a Phase C/D contract with EADS Space Transportation (named DASA at the time). EADS would outfit and integrate Columbus and Alcatel Alenia Space in Turin, Italy would manufacture its pressurised shell. Work on the science experiments to go into Columbus's racks began in parallel. Preliminary Design Review took place at the end of 1997 and Critical Design Review at the end of 2000. In 1996, the target launch date had been set for 2002 but the extension of the Station's assembly timetable caused by the delay in the launch of

Automated Transfer Vehicle The Automated Transfer Vehicle named Jules Verne approaches the Zvezda Service Module at the aft end of the International Space Station in March 2008 during practice manoeuvres that led up to the spacecraft's first docking with the Station. Developed by the European Space Agency, the Automated Transfer Vehicle was designed to ferry 7.5t of cargo and consumables up to the Station about every seventeen months and assist with orbital stationkeeping by providing a periodic reboost capability.

the Russian Zvezda Service Module and the Orbiter Columbia disaster in 2003 added up to a total delay of six years until its eventual launch in early 2008.[78]

The importance of the Columbus Laboratory Module to Europe in general and Germany – its principal proponent – in particular was underscored by Chancellor Angela Merkel's attendance at the module's development completion ceremony at Bremen in April 2006. Germany had funded half Columbus's development costs. As the launch day approached, newspapers across Europe carried articles about the project and what it would achieve at the International Space Station. Europe was justifiably proud of Columbus. Reaching agreement on it among its participating nations – Belgium, Denmark, France, Germany, Italy, Netherlands, Norway, Spain, Sweden and Switzerland (remarkably, the United Kingdom was absent) – was as much of an achievement as developing and launching it. François Fillon, Prime Minister of France, perhaps spoke for them all when, in a letter to French astronaut Léopold Eyharts, who would travel up with Columbus, he referred to the module's launch as emblematic for Europe because it showed how European nations could build scientific and technological achievements when they joined forces.[79]

Mission 1E (STS-122) lifted off with Atlantis carrying the Columbus Laboratory Module on 7 February 2008.[80] Columbus was ready to expand the science activities already in progress in the American-built Destiny Laboratory Module with additional experiments in the life sciences, human physiology, biology, fluid physics and material sciences. Outside, at the outboard end of Columbus, there would be more experiments on the European Technology Exposure Facility.[81] Unlike some earlier assembly missions that involved a series of complex procedures using both the Orbiter and Canadarm2 robotic arms, the attachment of the Columbus building block was relatively straightforward. Canadarm2 retrieved Columbus directly from Atlantis's bay and swung it across to berth it on the starboard port of Harmony. Astronauts carried out final attachments over six spacewalks.[82]

Following hard on the heels of the successful launch of Columbus was the equally successful launch of ATV-1, the Jules Verne Automated Transfer Vehicle on an Ariane 5 rocket from the Guiana Space Centre on 9 March 2008. The Automated Transfer Vehicle was the most important of the three Station building blocks the European Space Agency was providing (the other two being the Columbus Laboratory Module and Harmony Node-2). Europe had built pressurised laboratories before with the Spacelab laboratory module system, but the Jules Verne was completely new. As an independently controlled spacecraft designed for delivering supplies to the Station, it incorporated a multitude of technologies and capabilities not previously developed in Europe. Work on it had begun in 1992 under a joint study between the European Space Agency and NASA, and the Council of the European Space Agency had given it the go-ahead in 1994 with formal approval a year later at ministerial level.[83] EADS Astrium Space Transportation was the prime contractor and headed a large industrial team. RSC Energia supplied the Russian docking system as the vehicle would be docking at the aft end of the Russian Zvezda Service Module. As was to be expected with a new spacecraft design, its development ran into problems. There were concerns during testing over the vehicle's thermal and mechanical reliability and these delayed its readiness until 2007.[84]

The Automated Transfer Vehicle was a multi-functional spacecraft with its own propulsion and navigation systems. It combined the fully automatic capabilities of an unmanned vehicle with human spacecraft standards and safety requirements and was powered by four bipropellant engines. Its basic structure was an aluminium alloy cylinder 1.3 metres long and 4.5 metres in diameter at its maximum girth, clad in micrometeoroid protection panels. X-shaped solar arrays extended from the cylinder. At the forward end was the Russian docking adapter. Inside was a pressurised dry cargo chamber and a shirtsleeve environment for crew access from the Station for unloading supplies through the docking adapter hatch. At the aft end of the dry chamber was a bulkhead. Behind that was an unpressurised wet cargo chamber containing propellant, water and air resupply tanks for the Station.[85]

Automated Transfer Vehicle A cutaway view of the Automated Transfer Vehicle shows its construction and interior arrangement. The vehicle had two compartments – one pressurised and the other unpressurised – separated by a conical bulkhead. The unpressurised compartment, on the right in this view, contained the vehicle's propellant, water and air supply tanks, main engines, electrical system, solar arrays, and some attitude control thrusters. The pressurised compartment, on the left in this view, provided a shirtsleeve environment to crews for unloading and transferring the supplies and consumables stored in racks. Like the Columbus Module, the hull had an internal pressurised skin, an outer protective bumper shield and a thermal insulation layer in between. Image: ESA.

Because the vehicle was untested in space, its launch, rendezvous and docking with the Station were handled with extreme caution. It was put through its paces well before any docking occurred. The Jules Verne's journey time to the Station lasted twenty-five days and involved an elaborate sequence of demonstrations and dress rehearsals to ensure that the spacecraft could manoeuvre itself safely in the Station's vicinity. After its launch and arrival on orbit, it spent about nine days reaching a 'parking' orbit 2,000km ahead of the Station where it successfully practised a collision avoidance procedure.[86] A day of demonstrations on 29 March brought the vehicle up to 39km behind and 5km below the Station where it practised an approach. This involved closing to 3.5km behind and 5km below and then performing an escape manoeuvre and backing away to 39km again. A second day of demonstrations on 31 March brought it to within 11 metres of the docking port on Zvezda using an innovative optical navigation system developed in Europe.[87] Again, it retired to a safe distance, this time 39km again, aft of the Station. On 3 April the controllers gave the final go-ahead and the Jules Verne docked with Zvezda. On 5 April, the Station crew opened the hatch and began transferring supplies.[88] The vehicle's rendezvous and docking at the Station had been flawless. Four other Automated Transfer Vehicle missions were to follow. With the success of the Columbus Laboratory Module and the Jules Verne Automated Transfer Vehicle, Europe had become a fully fledged International Space Station partner alongside America and Russia.

Expedition 17 Astronaut Greg Chamitoff and cosmonauts Sergei Volkov and Oleg Kononenko pose for a photograph inside the Automated Transfer Vehicle named Jules Verne after its arrival at the Station in April 2008. They are holding Jules Verne memorabilia brought up in the vehicle, including an early edition of his book *De la Terre à la Lune* first published in 1865.

After the success of the European missions, it was Japan's turn. The next building block lifted off for the Station on Mission 1J/A (STS-123) on 11 March 2008. In Endeavour's payload bay were the Japanese Kibo Experiment Logistics Module and the Canadian Dextre manipulator device for Canadarm2. Kibo means 'hope' in Japanese. The Kibo Experiment Logistics Module was the first of three building blocks that Japan was going to provide to the Space Station. The other two were a long laboratory module and an exterior experiment platform. The Kibo Experiment Logistics Module had a pressurised internal environment with storage for experiment payloads, experiment specimens and samples, spare parts and tools. The module's final home was to be on top of the Kibo Pressurised Module due to be delivered on the next mission. In the meantime, Canadarm2 parked it in a temporary position on the zenith port of Harmony.[89] Though only used for storage, it had a shirtsleeve environment for Station crews.[90] The Kibo Experiment Logistics Module was the only storage module dedicated to a particular pressurised laboratory on the Station.

Also delivered was the Dextre versatile manipulator designed for attachment to the end of the Canadarm2 robotic arm or to module exteriors. With the official title of the Special Purpose Dexterous Manipulator, Dextre was designed for manipulating small components on the Station's exterior that needed precise and careful handling. Controlled by astronauts inside the Station, it would be particularly useful at carrying out repair and maintenance duties. It had a tool kit comprising grippers, socket wrenches, special tools, lights and a video camera system. Each of Dextre's two limbs had a wrist, elbow and shoulder joint – each with roll, pitch and yaw movement. Both limbs had force-moment sensors and end effectors for repair, replacement and maintenance duties.[91] Built by MacDonald Dettwiler Space and Advanced Robotics Ltd of Brampton, Ontario, Dextre was the fourth and last piece of the Mobile Servicing System to be delivered to the Station.

The second Japanese building block set off on Mission 1J (STS-124) on Discovery on 31 May 2008. It was the Kibo Pressurised Module. This building block was to be the largest pressurised module on the Station in terms of internal volume, as well as the longest. As with the American and European modules, experiments were housed in racks that lined the interior, leaving a central corridor for crew access. Unique to the Kibo cluster were Japanese robotic arm capabilities and a control console for operating exterior Japanese experiments, as well as a scientific airlock that enabled transfer of experiments from inside to outside and vice versa.[92] The weight of the Kibo Pressurised Module exceeded the Shuttle's lift capacity so Discovery delivered it on this mission without some racks installed.[93] The crew installed the Kibo module in two operations using the Canadarm2. The robotic arm moved the module out of Discovery's cargo bay and mounted it on the port side of Harmony. Then later in the mission, the arm moved the Kibo Experiment Logistics Module from its temporary position on the top of Harmony to its final home on top of the Kibo Pressurised Module. Six spacewalks took place on this mission.[94]

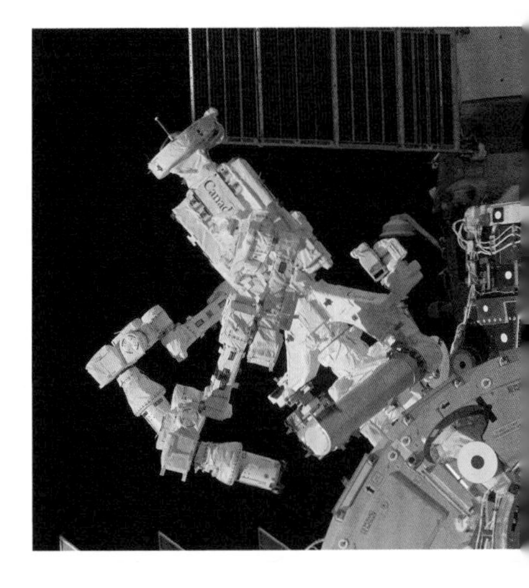

Dextre manipulator The Special Purpose Dexterous Manipulator named Dextre is a versatile robotic device designed to manipulate small components on the Station's exterior that need precise and careful handling. Attached to the end of the Canadarm2 robotic arm, it has two limbs that resemble short arms. Each limb has a wrist, elbow and shoulder joint, each with three degrees of freedom. The result is a tool that is exceptionally agile and capable of performing a wide range of tasks controlled by astronauts inside the Station. Dextre has its own set of tools, lights and a video camera system. The photograph shows Dextre after its delivery to the Station in March 2008 and initial attachment on the side of Harmony Node-2.

Shuttle's doom
Final Space Shuttle Mission Manifest, 2008

In July 2008, with the end of the Space Shuttle programme drawing closer, NASA released an 'International Space Station Transportation Plan' showing how it would maintain access for crew and supplies to the Station until the planned Ares-Orion rocket-spacecraft combination was ready in 2015.[95] This assumed that Ares-Orion mission applications still included visits to the Station and this was by no means certain. The first line on the first page of the plan's summary stated that the 'Shuttle Retirement in 2010 Impacts ISS Upmass/Return Capability'. This was a gross understatement. It should have said that the Shuttle's retirement would severely compromise the Station's ability to continue to exist – for nothing less was at stake – unless a reliable replacement was ready to take over. While the Ares-Orion vehicle combination might have the capability to ferry crews between the ground and the Station, there was nothing in the foreseeable future that would be able to replace the Shuttle Orbiter bays for the delivery of modules and

Kibo laboratory cluster The exterior photograph shows the triad of building blocks that together formed Japan's laboratory contribution to the International Space Station. Shuttle Orbiters delivered the three building blocks on three separate missions. The large module at centre right is the Kibo Pressurised Module, the longest laboratory module on the International Space Station. Mounted on top of it in the centre of this view is the Kibo Experiment Logistics Module, a short cylindrical module for storing experiment hardware, parts, specimens and tools. Projecting from the end of the Kibo Pressurised Module to the left is the Kibo Exposed Facility, an exterior experiment platform. Above the platform is the Kibo group's own robotic arm for manipulating experiments.

other large payloads. Only Russia with its Proton launch vehicle would be able to deliver module-sized payloads.

NASA's plan for continued access to the Station rested on two assumptions. First, the introduction of commercial initiatives entitled 'Commercial Orbital Transportation Services' would help to solve the supply delivery problem. The private sector, aided by NASA, would develop cargo vehicles to take over delivery duties. Secondly, purchase of more Soyuz missions from Roscosmos in Russia to ferry crews was possible.

However, it was by no means certain the American private launch companies selected by NASA would be able to produce reliable launch vehicles in time to take over the Shuttle's cargo duties. NASA planned to give the companies seed money to press on with their rocket development according to an aggressive timetable. The two chosen companies – SpaceX and Orbital Sciences Corporation – were to carry out Station docking demonstrations of their spacecraft in March and December 2010 respectively. These target dates were highly optimistic. NASA acknowledged that the Station would have to rely on European and Japanese cargo delivery vehicles until the newcomers were ready. Little mention was made in the plan of NASA's reliance on Russia for post-2010 crew flights to and from the Station, other than the need for relief from the Iran-North Korea-Syria Nonproliferation Act that prevented further purchases of Soyuz missions from Russia.[96] On the last page of the plan's summary was a photograph of the Station on orbit with the words 'onwards and upwards'. It was a plan that seemed to be full of elevated if not inflated predictions.

Also in July 2008, NASA published its final launch manifest that would usher in the end of the Space Shuttle era and the completion of the International Space Station. The manifest showed nine Shuttle missions to go (a tenth was the last Hubble Telescope servicing mission) and the Station crew increasing from three to six between May and July 2009. This would mean that two Soyuz spacecraft – each with three seats – would always need to be docked at the Station to act as lifeboats in case of emergency evacuation. A series of these spacecraft would occupy two docking ports permanently. A third docking port was needed by visiting Soyuz vehicles exchanging crews or Progress vehicles delivering supplies. Europe's Automated Transfer Vehicle would use a fourth docking port during its periodic visits.[97] Therefore, four Station ports would be put into operation, two pointing to nadir, one pointing to zenith and one pointing aft. The manifest showed the last Shuttle mission on 31 May 2010.[98] The same week, the Station's partners met in Paris to discuss its extended life. The major issue was the financial and technical measures needed to maintain Station operations until at least 2020. Russia, Japan, Europe and Canada had been concerned for some time about NASA's commitment to extending the Station's life after the Shuttle's retirement in 2010. The European Space Agency wanted to investigate expanding the number of partners to broaden

JAPAN
Canada

the Station's research value. America was sending mixed signals. NASA Administrator Michael Griffin said that America would support the Station until 2016 but a senior Republican advisor to the Senate Commerce, Science and Transportation Committee gave 2020 as the earliest Station termination date.[99] It was a matter for the new President to address after the November 2008 election.

The year 2015 marked the end of the International Space Station partners' international agreements and follow-on agreements would be necessary to continue operations. Whether the partners would be able to persuade their respective governments to invest more money in the Station was an open question. A European Space Agency representative at the July meeting was quoted as saying, rather ambitiously, that Europe would begin development of its own 'mini-station' based on Automated Transfer Vehicle pressurised modules in case the Station's extended lifespan fell through. There seemed to be some indecision at NASA. The agency had no firm plan to use Russian Progress vehicles for Station resupply after 2011 but neither did it have plans for switching to the privately developed commercial vehicles if they were ready by then.

Japan's Kibo module A view inside the Kibo Pressurised Module a few days after its delivery to the Station in June 2008. Astronaut Ron Garan in the foreground and from left to right in the background, astronauts Mike Fossum, Akihiko Hoshide, Mark Kelly and Ken Ham are at work on setting up experiments.

NASA's crew launch predicament after 2010 was eventually resolved, on paper at least, in September when President Bush relaxed the restrictions on purchasing Soyuz flights from Russia. Buried deep inside a 143-page Act encompassing several appropriations bills covering disaster relief, offshore drilling, security and defence was a single obscure sentence that read 'SEC. 125. Section 7(1)(B) of Public Law 106-178 (50 USC 1701 note) is amended by striking "January 1, 2012" and inserting "July 1, 2016."'[100] At a stroke, the White House had given NASA carte blanche to buy more Soyuz flights up to the arrival of the Ares-Orion combination, now planned for 2016 but still unconfirmed for Station visits. The case for the waiver was made in a Congressional Research Service report to Congress, which stated bluntly that after the Shuttle's termination, Soyuz was 'the only vehicle available after that date to transport astronauts to and from the ISS'. Despite Russia's intervention in Georgia, the report said that 'since 2005, Russia has stepped up cooperation with the United States and countries over Iran's nuclear program. President Bush has praised the Russian President for his leadership in offering a solution to the Iranian nuclear negotiations'.[101] NASA was now free to end the Space Shuttle programme.

On 14 November 2008, Shuttle Mission STS-126 with Endeavour took off on what NASA described as an 'Extreme Home Improvement' flight. It was a housekeeping and repair mission. Outside the Station, one of the solar array rotating joints on the starboard side had been suffering from the space equivalent of osteoarthritis. There was deterioration of the joint's ability to rotate freely, caused by small shavings that had worn away from surfaces inside the joint's rotating mechanism. This meant that the solar array could not track the Sun properly, resulting in reduced power generation.

Astronauts went on spacewalks to clean out the joint and lubricate it. Inside the Station, the expansion of the crew's size from three to six was due in a few months and improvements to accommodate the larger crews were necessary. Endeavour had brought up two 'bedrooms', another 'kitchen', an additional 'bathroom' and a waste water recycling system. A water recycler able to convert urine, sweat and other waste water into clean drinking and washing water was essential to reduce the amount of fresh water ferried up from the ground. The urine recycler, however, proved to be difficult to start up and the mission length was extended. Eventually it was declared a success and the plumbing team proudly held up a bottle of recycled water on NASA television bearing a label that read 'Yesterday's Coffee'.[102]

The International Space Station marked a historic day on 20 November 2008. It was the tenth anniversary of the launching of the Zarya Control Module, the first building block. After ten years on orbit, the Station was beginning to show some wear and tear. The chief cause was the impact of orbital debris. The Station orbits the Earth at a typical altitude of 400km. The actual altitude varies between about 330km and 400km. The density of the upper atmosphere expands outwards during times of high solar activity and has a drag effect on the Station, causing its altitude to drop. When this occurs, an altitude reboost is necessary. In this fluctuating altitude band, the Station passes through an orbital debris environment. Although it is relatively mild, the Station's size tends to increase the likelihood of impacts. There are basically two types of debris on orbit – natural objects from the Solar System or beyond such as micrometeoroids and materials of terrestrial origin such as fragments of old spacecraft. Particles smaller than 3mm in size are numerous but rarely cause harm. Those from 3mm to 10mm can result in significant damage, such as a small leak in a pressurised hull, which could affect the safety of the Station and its crew. Shield panels around the exterior of pressurised modules and pressure vessels are strong enough to withstand the impact of particles up to about 10mm in diameter travelling at a collision velocity of up to 10km per second. The debris shielding on the Station amounts to about 23 tonnes or approximately 5 per cent of its total mass. Anything larger than 10mm in size has the potential to cause serious or even catastrophic damage. Ground-based radar can spot and track anything larger than 100mm and sometimes pieces as small as 50mm in the Station's path and the Station can manoeuvre out of the way. Particles between 10mm and 100mm in size, therefore, continue to pose a potential threat against which the Station has little effective defence at the present time. Whether the Station's partners will upgrade the shielding to deal with particles in this range remains to be seen.[103]

There was evidence of debris impacts on the Station from crew observations on spacewalks, exterior photography and components returned to Earth. Damage had occurred to solar arrays, window panes, shielding panels, handrails, antennae and insulation. An impact on a thermal

Moving a crew compartment Astronauts Sandra Magnus and Greg Chamitoff move a bulky personal crew compartment around inside Harmony Node-2 on the Station. The compartment had arrived on Space Shuttle Mission STS-126 in November 2008. This was described by NASA as an 'Extreme Home Improvement' flight.

S6 Truss Both robotic arms were involved in installing the S6 Truss piece. The S6 was the last truss building block to be installed on the Station. As with previous truss pieces, the process of attaching the S6 to the Station involved several carefully planned steps. These were complicated on this occasion because the Orbiter's robotic arm was on the port side of the Station while the destination for the S6 was on the far starboard side. In **view from port side**, the sequence began with the Canadarm2 arm **01** moving the S6 building block **02** out of the Orbiter's payload bay **A** and handing it over to the Orbiter's robotic arm **03**. In **view from above**, Canadarm2 then moved in the starboard direction along the truss and retrieved the S6 from the Orbiter's arm **C**. Then the mobile railcar **04** with its arm firmly grasping the S6 moved towards the starboard end of the transverse truss **D**. The solar arrays **05** on the S4 Truss **06** rotated to their local vertical position **E** to provide an unobstructed path for the attachment of the S6. Then Canadarm2 swept it around **F** and brought the S6 towards its final home **G** and mounted it on the end of the S5 Truss **H**. Finally in position, the solar arrays on the S6 deployed outwards. Six spacewalks were required for this assembly sequence. Image: author.

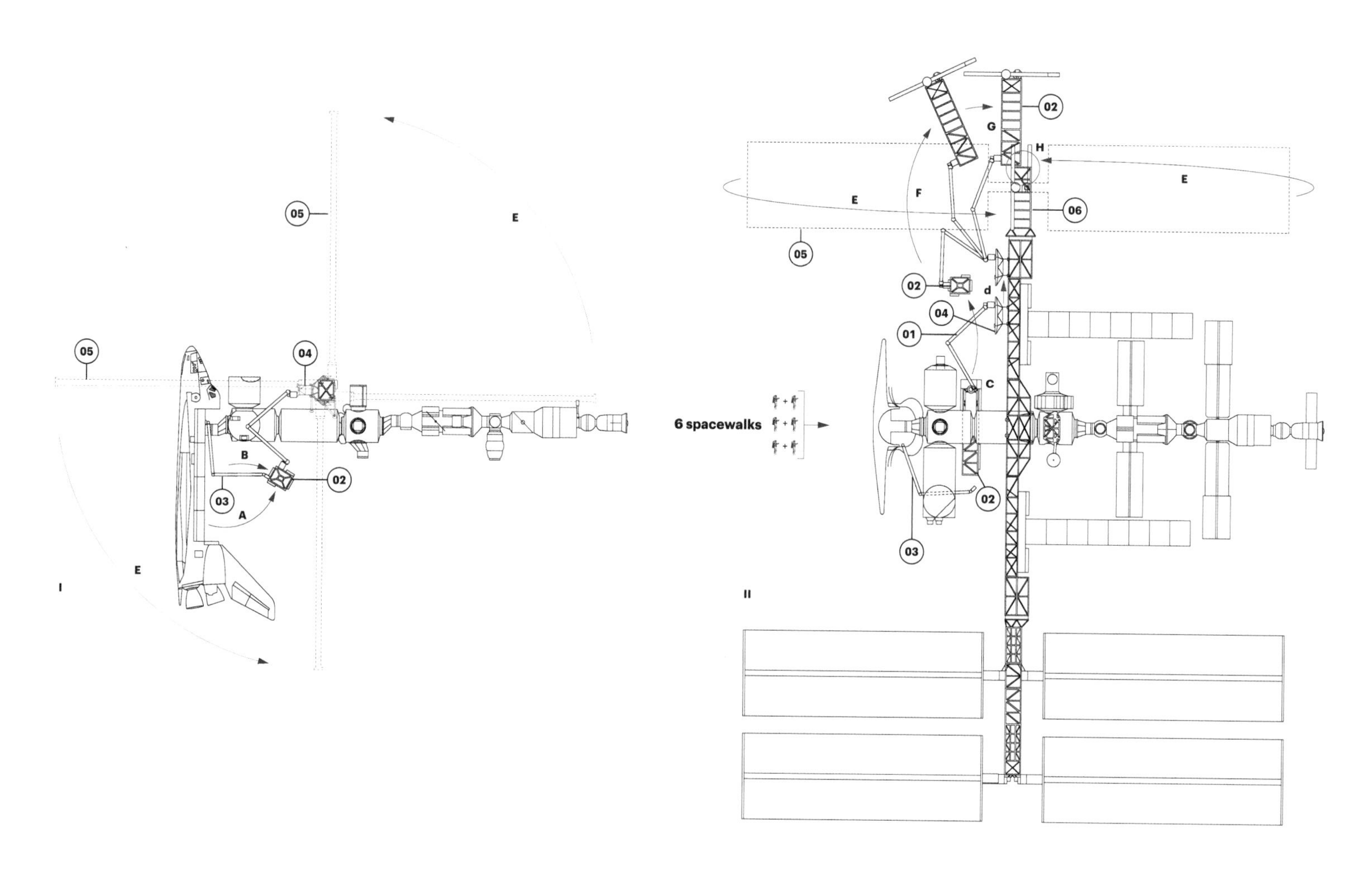

View from port side

View from above

blanket on the outside of Zarya had caused a tear over 60 × 30mm in size and the particle went on to penetrate the steel mesh, fibreglass and aluminium honeycomb layers beneath. Up to 2008, though potential threats were showing up on the radar about three times per month, analysis showed that the probability of strikes was extremely low and no action was taken. When an avoidance procedure was required, thrusters on a docked Progress or Automated Transfer Vehicle carried it out. There were seven collision avoidance manoeuvres during the first four and a half years of operation but only one during the next five and a half years. This improvement was due to better analysis of impact probabilities rather than a reduction in the amount of orbital debris. The Station itself has become a source of orbital debris. Painted surfaces have broken down and become brittle over time, resulting in the flaking off of paint particles. Sixty-five recorded releases of Station debris occurred in the first ten years of operation, some deliberate, others accidental. There had been deliberate jettisoning of towels, equipment covers and carriers, and an old spacesuit that floated away from the Station without a lifeline. Accidental losses included a camera, tools, a tool bag and a foot restraint. Despite these various events, it appeared that the Station was withstanding the orbital debris environment well and there was every reason to suppose that this reassuring state of affairs would continue.[104]

On 30 December 2008, NASA released a report describing the last moments of the crew of the Shuttle Orbiter Columbia.[105] The report was the continuation of investigation work begun shortly after the tragedy by the Columbia Accident Investigation Board. The introduction to the new report stated that it was 'the first comprehensive, publicly available accident investigation report addressing crew survival for a human spacecraft mishap ...' It was the first open report to explain in detail how the astronauts perished and it made some important recommendations for future crew safety. Among its recommendations were that crew survival measures should be considered early in the design process, crew suit and seat design should minimise injury, and post-disaster evidence and information management be improved. Its publication marked the end of a painstaking enquiry and brought closure to the history of Space Shuttle accidents.

The dangers of spaceflight were apparent in March when a 10mm sized particle of debris from an old satellite passed within three miles of the Station. The three members of the crew took the precaution of entering the Soyuz spacecraft for several minutes until the particle had zoomed by. Though this was not a new event – Station crews had sheltered in Soyuz on previous occasions – the international media, usually oblivious to the Station's progress, jumped on the story. 'Space Station has close call with space junk,' ran CNN; 'Space Station crew in near miss with space junk,' said National Public Radio; 'Space Junk Forces Astronauts To Evacuate Station' was the headline in the *Guardian* newspaper – an absurd exaggeration. The story quickly faded but left behind heightened concerns about the

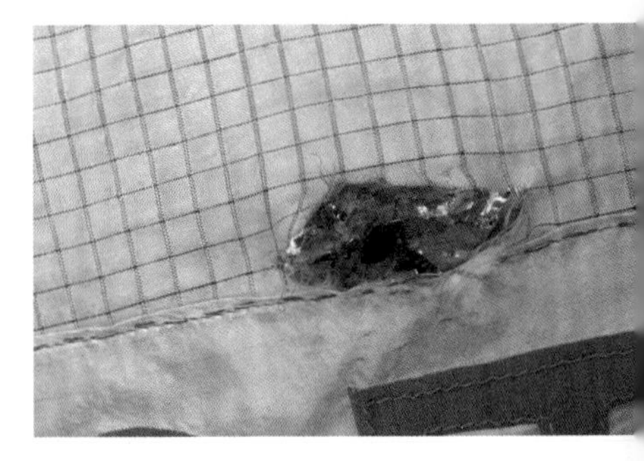

Micrometeoroid impact damage This photograph, taken by cosmonauts Fyodor Yurchikhin and Oleg Kotov on a spacewalk on 6 June 2007, shows micrometeoroid impact damage to multilayer insulation on the exterior of the Zarya module. Damage from orbital debris and micrometeoroids has been seen on solar arrays, window panes, shielding panels, handrails and antennae on the Station. Image: Fyodor Yurchikhin and Oleg Kotov.

accumulation of debris on orbit following the collision a few weeks earlier of a disused Russian satellite and a commercial communications satellite. At a relative speed of over 11km per second at impact, the two spacecraft had annihilated each other and created a vast new cloud of orbiting debris. According to an estimate by the European Space Agency, there were more than 600,000 objects larger than 1mm in size on orbit.[106] The possibility of further close calls was always present.

The Japanese Kibo external experiment platform The Japanese external experiment platform called the Kibo Exposed Facility arrived at the Station on Space Shuttle Mission STS-127 in July 2009. All the experiments on the platform are exposed to the space environment. Astronauts move experiments out onto the platform or bring them back inside through an airlock at the end of Japan's laboratory module, using a small robotic arm.

Subdued finale
Addition of the final elements, 2009-2011

On 15 March 2009, the Space Shuttle took off with the last major truss building block for the Station. Mission 15A (STS-119) with Discovery departed with the S6 Truss – the final piece for the truss's starboard side.[107] Attachment of the S6 piece would raise the Station's solar-powered electrical generation capability to full strength. Total available power came from a combination of American and Russian solar arrays. Together they would generate between 80kW and 100kW of power. Of this, 30kW was available for running scientific experiments while the rest was used for operational purposes. Moving the S6 from the docked Discovery to the starboard end of the truss required another carefully orchestrated operation of the robotic arms. It was particularly challenging this time because, with Discovery docked to the Station's forward end with its nose high and tail low, Discovery's robotic arm was on the port side of the truss – the opposite side from where the S6 needed to go. First, the Canadarm2 robotic arm on its railcar swung the S6 out of Discovery's payload bay, rotated it and passed it to Discovery's robotic arm. Then, while Discovery held the S6, Canadarm2 travelled along the tracks to the starboard side of the truss and grasped the S6 from Discovery's arm. Next, Canadarm2 swung the S6 round from port to starboard in a great sweep to point it out towards the starboard end. Then, to make way for Canadarm2 to ferry the S6 down to the starboard end of the transverse truss, the solar wings on the S3/S4 piece rotated vertically so that they were out of the way. Finally, Canadarm2 moved along the truss holding the S6 beneath the truss line, lifted it up to align with the end of the main truss and anchored it in place. The S3/S4 solar arrays then rotated back to the normal opening position. The transverse truss structure – all nine building blocks of it – was now complete. It had taken nine Shuttle missions to ferry up all the pieces.

S6 Truss The Canadarm2 robotic arm has the S6 Truss in its grasp during the installation of the truss piece at the Station in March 2009. Canadarm2 moved the S6 out of the Orbiter's payload bay and handed it over to the Orbiter's robotic arm that, in turn, handed it back to Canadarm2 for further manipulation, leading up to its installation at the far starboard end of the transverse truss.

The Station was now on full power at last, though the earlier problem with a rotating joint on the other S3/S4 solar arrays had yet to be fully resolved and it was still unable to track the Sun properly.[108] After reaching full power the Station also reached full crew status with the increase of its regular crew size from three to six. It was another important assembly milestone designated as 'Established Six Person Crew Capability'.

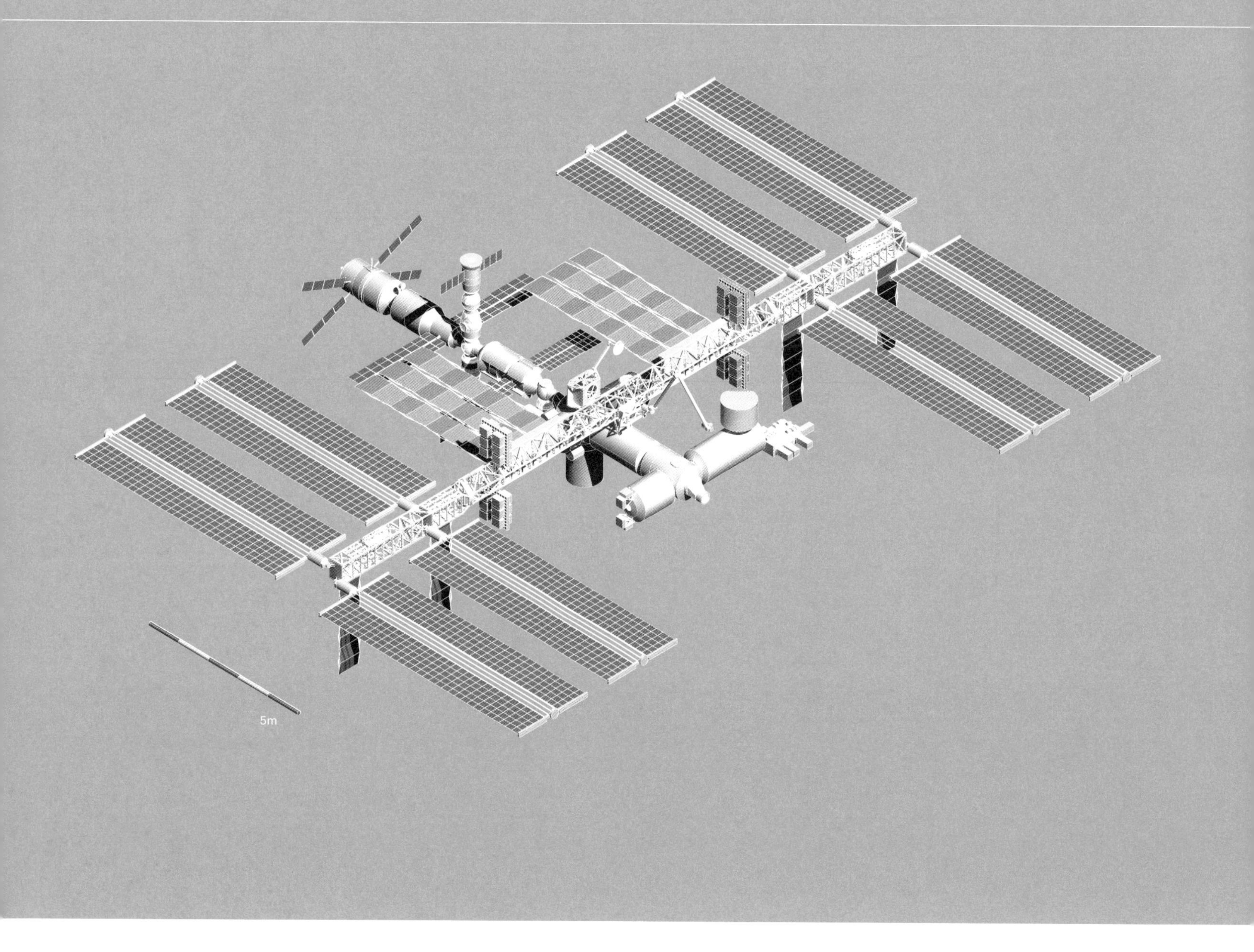

2011 International Space Station design – assembled and disassembled The major differences between the Station's 1994 configuration after Russia joined the team and the Station's 2011 configuration at its completion, shown here, were the reduction in the number of Russian building blocks and the omission of an American module for crew habitation. These illustrations show all the main building blocks of the 2011 Station configuration. They include the Soyuz TMA spacecraft, of which two are always docked at the Station as emergency return vehicles for the six-person crews. Also shown is the European Automated Transfer Vehicle for cargo delivery, of which there were five missions. Not shown are Japan's HTV Transfer Vehicles for cargo delivery, Russia's Progress cargo spacecraft, SpaceX's Dragon cargo spacecraft or Orbital Science's Cygnus cargo spacecraft. Image: author.

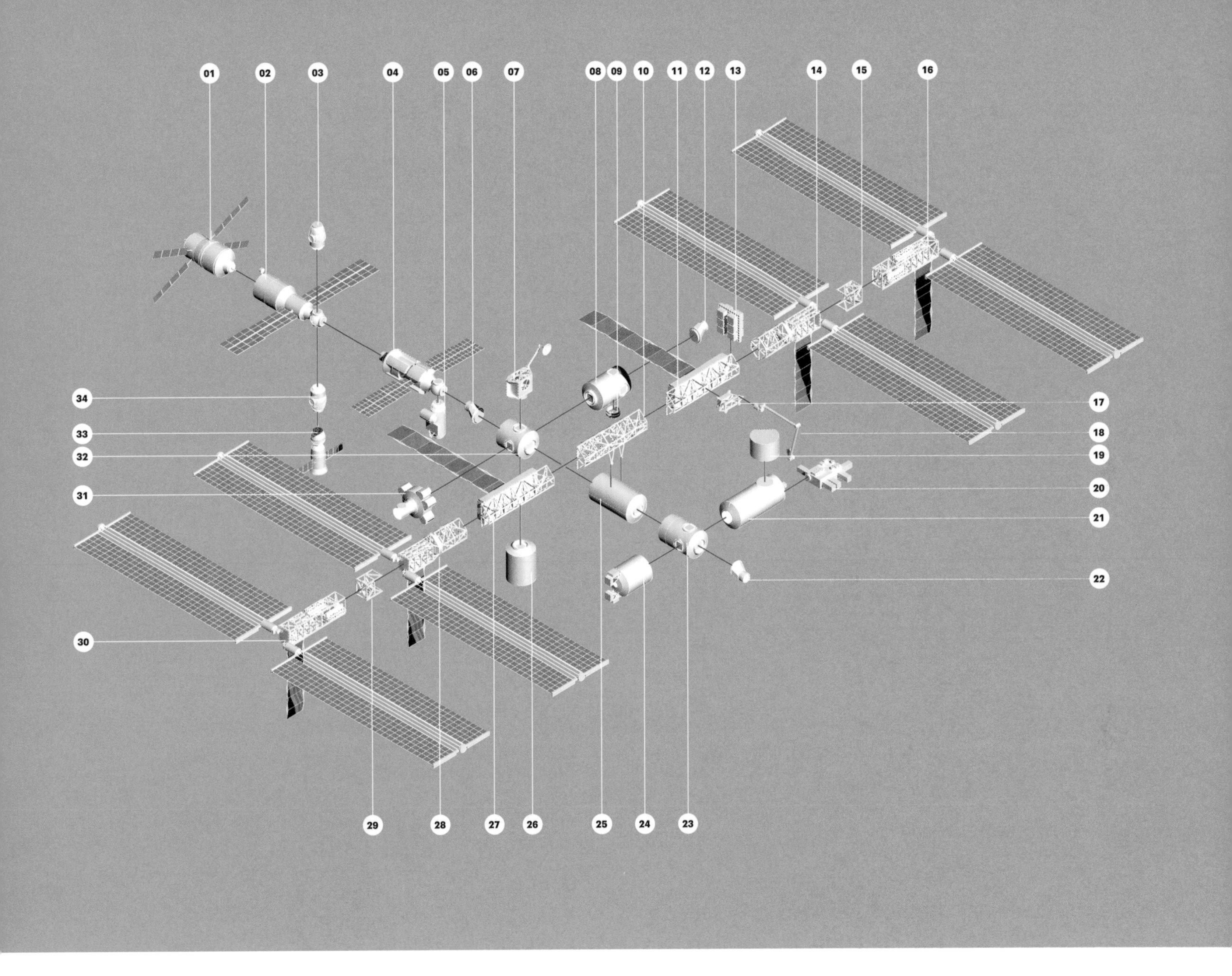

01 Automated Transfer Vehicle (5 missions)
02 Zvezda Service Module
03 Poisk Mini-Research Module
04 Zarya Control Module
05 Rassvet Mini-Research Module
06 Pressurized Mating Adapter-1
07 Z1 Truss
08 Tranquility Node-3
09 Cupola
10 S0 Truss
11 P1 Truss
12 Pressurized Mating Adapter-3
13 Express Logistics Carrier (4 places)
14 P3/P4 Truss
15 P5 Truss
16 P6 Truss
17 Mobile Base & Transporter
18 Canadarm2 robotic arm
19 Kibo Experiment Logistics Module
20 Kibo Exposed Facility
21 Kibo Pressurised Module
22 Pressurized Mating Adapter-2
23 Harmony Node-2
24 Columbus Laboratory Module
25 Destiny Laboratory Module
26 Leonardo Multipurpose Module
27 S1 Truss
28 S3/S4 Truss
29 S5 Truss
30 S6 Truss
31 Quest Joint Airlock
32 Unity Node-1
33 Soyuz TMA spacecraft (2 places)
34 Pirs Docking Compartment

On 29 May 2009, a Soyuz spacecraft delivered three new crew members to complement the three already based at the Station. For the first time in its ten-year history, the Station had a full crew permanently on board. NASA expected that science research would triple at the Station as a result, as assembly activities began to wind down, enabling the crew to focus more on the science.[109] On 11 June, the Station's population rose again to thirteen people when the Orbiter Endeavour arrived with a crew of seven. By this time, the Station's pressurised volume stood at about 760m^3, roughly equivalent to a three-bedroom home with a kitchen and two bathrooms.[110]

The final piece of Japan's suite of Station building blocks lifted off on Mission 2J/A (STS-127) with Endeavour on 15 July 2009. It was the Kibo Exposed Facility, an exterior experiment platform with the form of a rectilinear metal framework. It arrived with a separate payload carrier called the Experiment Logistics Module – Exposed Section, later returned to the ground on the same mission after the transfer of its payloads to the platform.[111] The Kibo Exposed Facility would provide a multipurpose attachment deck for a suite of scientific experiments that required exposure to space. Experiments covered Earth observation, the space environment, materials technology and communications and data transfer. The platform offered twelve experiment and payload attachment points on its exterior surfaces. The structural framework supplied power, thermal control and communications capabilities to each experiment. Many experiments were modular with a common shape, size and attachment interface, enabling them to be exchanged in future for new experiments. The platform attached to the Kibo Pressurised Module with a berthing mechanism that provided for the transfer of power, thermal control, data and communications functions. The Kibo suite also had its own robotic arm to handle the exterior experiments. After docking, Endeavour's robotic arm removed the platform from its payload bay and passed it to Canadarm2 for mounting on the port endcap of the Kibo Pressurised Module.

When it took over government in January 2009, the Obama Administration inherited the Constellation space plan from the Bush Administration. Constellation was an ambitious human exploration agenda that had been crafted in mid-decade by NASA Administrator Michael Griffin and senior NASA management.[112] It required a lot of expensive new hardware – the Ares 1 and the Ares 5 rockets, the Orion crew capsule and the Altair lunar lander – to achieve the aim of putting American astronauts back on the Moon. A massive budget increase for NASA stretching out over several decades combined with major savings from the termination of existing NASA programmes would be necessary to carry out the new plan. There would be no more Shuttle missions after 2010 when the International Space Station assembly would conclude and Shuttle expenditure would cease thereafter. Additionally, at a meeting with the Station's international partners in 2008, Michael Griffin had floated the idea of ending American involvement in the

Space Station altogether in 2016. Withdrawal would produce a significant saving. Yet the idea of America pulling out of the International Space Station so soon after its completion was absurd.

Thus it was that at the beginning of 2009, against a background of a crisis in the American economy, the Obama Administration decided to take a close look at NASA and its future plans. Michael Griffin left NASA early in the year and in May, John Holdren, Director of the Office of Science and Technology Policy at the White House, asked the NASA Acting Administrator Chris Scolese to arrange an independent review of its human spaceflight plans by a blue ribbon panel of outside experts and report back by August.[113] NASA formed the ten-person US Human Spaceflight Plans Committee, chaired by Norman Augustine, the widely respected former Chairman of Martin Marietta who yet again was asked to oversee an enquiry into NASA's affairs. The committee published its findings in a summary followed by a full report in September and October.[114, 115]

The report advocated a new approach in which the Constellation programme would be dropped and replaced by a 'flexible path' of sending robots and later crews to various destination points beyond Earth and boosting space science and technology in the process. About the Station, the committee found 'that the return on investment of ISS to both the United States and the international partners would be significantly enhanced by an extension of ISS life to 2020. It seems unwise to de-orbit the Station after 25 years of assembly and only five years of operational life. Not to extend its operation would significantly impair US ability to develop and lead future international spaceflight partnerships.' It seemed an obvious conclusion. Five principal alternatives were proposed for the human spaceflight programme of which two had variants, making a total of seven paths from which the Obama Administration could choose, each with a different price tag. Of these, two were baseline alternatives with constrained budgets that showed the Station de-orbited in 2016, while five showed its life extended to 2020. The committee also pointed out that 'the strong and tested working relationship among international partners is perhaps the most important outcome of the ISS'.

The next mission was Russian. Mission 5R with the Poisk Mini-Research Module launched on a Progress spacecraft on 10 November 2009. Poisk, which means 'search' in Russian, was built by RSC Energia. It was the virtual twin of the Pirs Docking Compartment that had arrived at the Station in September 2001. Like Pirs, Poisk provided a docking port for Russian Soyuz or Progress spacecraft.[116] Poisk also provided a small amount of pressurised volume for research and, like Pirs, could function as an airlock for spacewalks. The Progress spacecraft automatically docked Poisk to the zenith port of Zvezda and departed the Station afterwards.

At the beginning of February 2010, NASA released its proposed budget for 2011.[117] It showed an increase of $6 billion over five years from 2010,

Cupola Astronaut Ron Garan peers out of one of the Cupola's windows after its arrival and installation on the Station in February 2010. The protective shields on the outside of the window panes are all in their fully open positions.

confirmed that the Shuttle retirement year had moved to 2011, and showed the cancellation of the Constellation programme. Constellation was to be replaced by a new research and development plan aimed at improving a basket of space technologies that would ultimately lead to more efficient human space exploration missions. Most importantly for the Space Station, the budget provided $2 billion of extra funding up to 2014 and supported the extension of the Station's life until at least 2020. The stated goal was 'to fully utilise the Station's R&D capabilities to conduct scientific research, improve our capabilities for operating in space, and demonstrate new technologies developed through NASA's other programs'. Two days later, representatives of the Station's international partners met to reaffirm their commitment to research on the Station.[118] They noted expansively that: 'ISS continuation could bring great benefit to all humankind' and stressed the need 'to improve ISS utilization, productivity and operational efficiency by all possible means'. However, with the Station's dependence on the Shuttle as its delivery workhorse almost at an end, it was possible that operational efficiency could decline due to diminished cargo resupply capabilities. As to productivity, if this depended on flights to return products or experiments, it would grind to a halt as the Orbiters were still the only spacecraft able to return payloads to the ground.

Mission 20A (STS-130) with Endeavour lifted off on 8 February 2010 carrying Node-3 named Tranquility and the multi-pane windowed Cupola. Thales Alenia Space in Europe built both elements under contract to NASA. The Tranquility node was similar in design to the Harmony node delivered on an earlier mission and built by the same company. The late arrival of Tranquility enabled its builder to improve its design on that of the earlier node. It contained more sophisticated crew and life support equipment needed to support the larger six-person crews on the Station from February 2010 onwards. Included were an air revitalisation system, an oxygen generation system, a water recovery system with urine processing, waste storage racks and more physical exercise equipment.[119]

In architecture, the word 'cupola' is defined as a 'dome on a circular or polygonal base crowning a roof or turret'.[120] Whether it was the right word for the seven-paned window on the International Space Station is debatable. 'Observatory' might have been better, for its role was to provide panoramic views outside the Station. A workstation in the Cupola would provide the crew with direct observation and control of Canadarm2 for the first time.[121] The crew would be able to watch spacewalking activities and the approach and rendezvous of visiting spacecraft with their own eyes as well as relying on monitor screens. After Endeavour's arrival at the Station, Canadarm2 retrieved Tranquility with the Cupola mounted on its end port from the payload bay and swung it out and around to berth it on the port side of Unity. The robotic arm then detached the Cupola from Tranquility's end port and moved it to Tranquility's Earth nadir port.

Cupola Astronaut Nicholas Patrick carrying out a spacewalk after the arrival of the Cupola at the Station in February 2010. This photograph shows him removing the launch restraint bolts from the exterior shields of the Cupola's seven windows.

The last Space Shuttle mission This photograph of the interior of the Zvezda Service Module in July 2011 shows a group of astronauts and cosmonauts during Shuttle Mission STS-135. It was the final flight of the Space Shuttle before its retirement. In the photograph is the four-person crew of the Orbiter Atlantis – astronauts Chris Ferguson, Doug Hurley, Sandra Magnus and Rex Walheim – and the six-person Station Expedition 28 crew – astronauts Satoshi Furukawa, Ron Garan and Mike Fossum and cosmonauts Andrey Borisenko, Alexander Samokutyaev and Sergei Volkov. Taped to the top centre of the bulkhead wall behind the group is a black-and-white photograph of cosmonaut Yuri Gagarin, the first human in space.

There was welcome news in March 2010. The International Space Station won the Collier Trophy, aviation's top award, from the National Aeronautic Association.[122] The Station joined a long list of illustrious past winners including Glenn Curtiss for the development of the flying boat in 1912, Pan American Airways for the introduction of a trans-Pacific route in 1936 and Burt Rutan for SpaceShipOne. As the Station won the trophy, the heads of the space agencies of Canada, Europe, Japan, Russia and the United States met in Tokyo to review cooperation on the Station. They noted in a press release the new American commitment to the Station until 2020 and stressed the expanded opportunities for research made possible by the full crew and assembly completion.[123] They again underlined the need to pursue operational efficiency by all possible means. There was no mention of the Shuttle's approaching end or the consequences that would flow from it, only a comment at the end about 'assuring the most effective use of essential capabilities, such as space transportation for crew and cargo', for the life of the programme.

It was getting very near the end indeed for the Space Shuttle. Mission ULF4 (STS-132) with Atlantis launched on 16 May 2010 carrying the Russian Rassvet Mini-Research Module. It was the penultimate assembly mission. Rassvet, which means 'dawn' in Russian, was the only Russian module launched to the Station on a Shuttle Orbiter. Rassvet was fabricated from parts originally intended for the cancelled Russian Science Power Platform. It had a hull length of 6 metres and diameter of 2.35 metres. Designed for cargo storage and payload operations, it featured eight workstations for science experiments and educational research.[124] Mounted on Rassvet's exterior surface was a spare joint for the European Robotic Arm and outfitting equipment for a Russian multipurpose laboratory module named Nauka that was due for launch to the Station in 2014 on a Proton. Canadarm2 removed Rassvet from the payload bay and swung it around to berth it on to the nadir port of the spherical node on the forward end of Zarya where it would provide a fourth docking port for arriving Soyuz and Progress spacecraft.

The last building block mission to the Station was Mission ULF5 (STS-133) that launched on 24 February 2011 with Discovery, carrying the Leonardo Permanent Multipurpose Module, named after Leonardo da Vinci. Leonardo was one of three logistics modules that Italy had built and supplied, the others being Raffaello and Donatello. Leonardo – formerly known as the Multipurpose Logistics Module – had flown to the Station and returned from it before in its role as a space shipping container.[125] Upgrading Leonardo to enable it to function as a permanent part of the Station required improvements to its micrometeoroid and space debris shielding and changes to its internal systems and software. Canadarm2 removed Leonardo from Discovery's payload bay and berthed on the nadir port of Unity Node-1. It was the last space mission of the Orbiter Discovery.

With the delivery of Leonardo, the International Space Station had reached its Assembly Complete milestone. To all intents and purposes, the job was done. On 16 May 2011, Endeavour launched on Mission ULF6 (STS-134) with its payload bay packed with spare parts, equipment and supplies. It was the last space mission of the Orbiter Endeavour. Endeavour also had in its payload bay a very important experiment called the Alpha Magnetic Spectrometer. It was a state-of-the-art particle physics detector to help in the search for clues to the origin of dark matter and the existence of antimatter in the Universe.

Normally, the end of construction of a great project is celebrated in a 'topping out' ceremony of some kind. Sadly, the International Space Station was too inaccessible for such an event and the partners lost an opportunity to present the completion of the Station to the world at large. The Assembly Complete milestone passed without much attention.

The very last Space Shuttle flight to space was made by the Orbiter Atlantis. It was the 135th Space Shuttle flight since the programme began in 1981. Atlantis launched on 8 July 2011 on Mission ULF7 (STS-135) carrying the Raffaello Multipurpose Logistics Module delivering more parts, supplies and consumables. It was a fairly routine Shuttle flight. As it had done before with all the Shuttle missions to the Station, NASA issued a mission press kit.[126] On the front page the agency dedicated the final mission to 'the courageous men and women who have devoted their lives to the Space Shuttle Program and the pursuit of space exploration'.

In the introduction to the press kit, NASA reminded us of the Shuttle's achievements. 'For 30 years, the Space Shuttle has been the US human access to space. It has capabilities no other spacecraft can claim. No other spacecraft is likely to match those capabilities in this generation. It is the fastest winged vehicle ever to fly, with an orbital velocity of 17,500mph, ten times the speed of a high-powered rifle bullet. It is the only winged vehicle to reach orbit, and the only reusable space launch and landing vehicle. The shuttle can carry cargoes of substantial weight and dimensions. It has taken into space more than half the mass of all payloads launched by all nations since Sputnik in 1957 – 3,450,143 pounds (through STS-132) and counting as the final shuttle launch approaches. More singular still is the Shuttle's ability to return payloads from space. It has brought back from orbit more than 97 per cent of all mass returned to Earth, a total of 225,574 pounds (through STS-132) before the upcoming final flight. It has launched 802 crew members including those lost on Challenger and Columbia. Crew members returning on the shuttle numbered 789. Many crew members flew more than once. A total of 356 different individuals have flown aboard the shuttle (all through STS-132). It leaves a significant legacy.'[127] The Space Shuttle's legacy is the International Space Station, which would have been impossible to build and occupy without it.

Scott Kelly prepares For a spacewalk
Astronaut Scott Kelly has just put on a spacesuit inside the Quest airlock on the Station as he prepares to go outside for a spacewalk in October 2015. Floating just in front of him is a checklist.

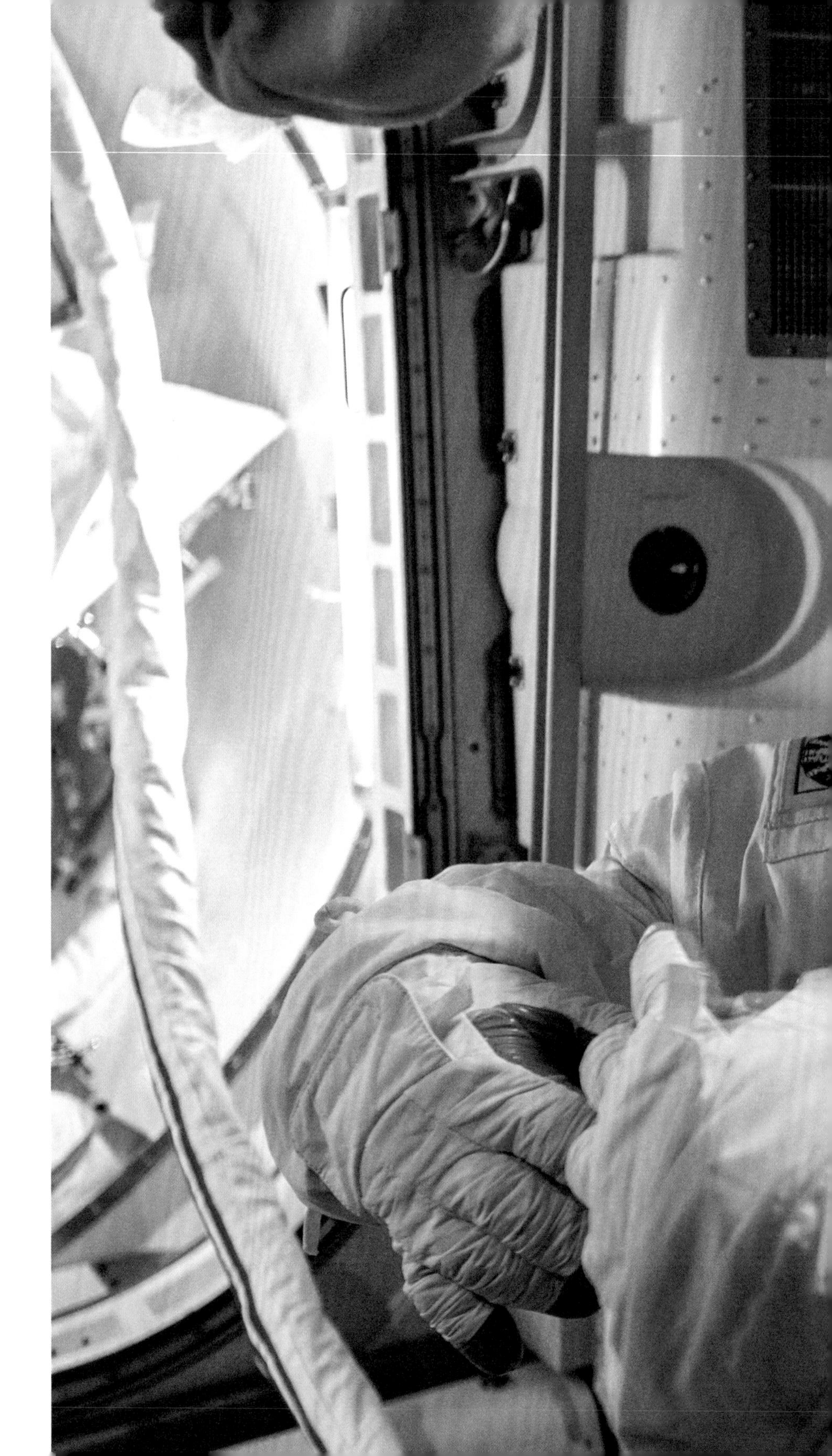

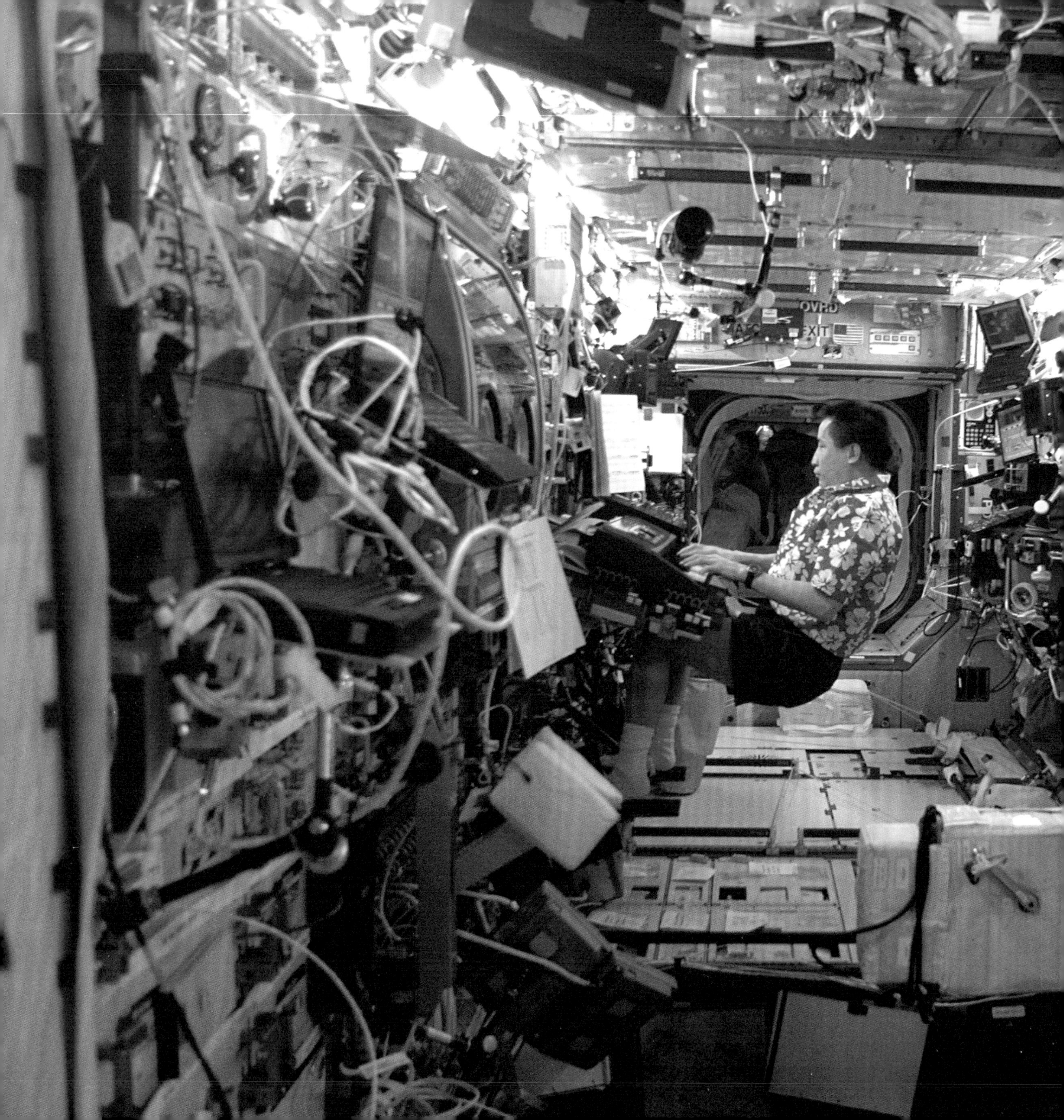
OVRD
EXIT

Leisure Astronauts and cosmonauts do not get much free time on the Station with their busy work schedules. When they go off duty, they pursue personal hobbies or pastimes or simply look out of the windows down at the Earth. Astronaut Edward Lu, science officer and flight engineer on Expedition 7 from April to October 2003, is playing a musical keyboard during his free time in the Destiny Module.

Cupola A view out from the interior of the Cupola after its installation at the Station in February 2010. The Cupola's windows look directly down upon the Earth's surface. Visible through the windows is the North Atlantic Ocean.

Epilogue

Science and technology payback

Research results and their benefits to Earth

On 2 March 2014, the science fiction movie *Gravity*[1] won seven Oscars, the most for any contending movie that year. It was a huge box office success. The International Space Station played a prominent role in the film's action, portrayed through exceptionally convincing special effects filmed in 3D. For NASA, the moment of Oscar glory offered a public relations opportunity and the agency moved swiftly to congratulate Alfonso Cuarón, the film's director, and everyone involved in its production. In the words of NASA's website: 'two astronauts find themselves adrift in space and struggling for survival after their spacecraft was destroyed by space debris. Although this scenario makes for gripping Hollywood entertainment, NASA actively works to protect its astronauts and vehicles from the dangers portrayed in the movie'.[2] The NASA website featured video clips of astronauts on the Station offering their congratulations, an astronaut interviewing one of the stars, and helpful explanations on the orbital debris issue, spacewalk training and other topics.

There was a link to another website page on the Station with a short public relations film entitled *Benefits for Humanity: In Their Own Words*[3] which featured five short stories on how technology developed or used on the International Space Station was helping to save lives and improve

Cupola Astronaut Tracy Caldwell Dyson, flight engineer on Expedition 24 from June to September 2010, looks through the Cupola's windows. This photograph was a self-portrait taken in natural lighting conditions. Image: NASA and Tracy Caldwell Dyson.

livelihoods on Earth. There was a woman in Canada who had successfully undergone highly complex surgery to remove a brain tumour using the surgical device called NeuroArm, an evolution of the Station's Canadarm robotic arm and its Dextre end manipulator. A man in New Mexico demonstrated the use of Station-derived water purification equipment to cleanse contaminated well and river water, providing drinking water for the local community. A family-run farm in Minnesota that grew a mixture of crops used a camera mounted on the Station to monitor in visible and infra-red light the progress of their seasonal crop biomass growth. A boy scout gathering in Illinois invited youngsters with science questions to telephone the Station and speak directly to an astronaut. In Brazil, a rural doctor used Station telemedicine techniques to save a woman's life by linking his ultrasound scanner to distant medical specialists who advised him on an emergency procedure.

Though the message was simple – the International Space Station was producing technology that was useful and valuable in many walks of life and in many places across the planet – it was unable to do justice to the much bigger research story that was unfolding on the Station. Within two years of its completion, the Station was generating a multitude of examples of the benefits it was bringing to Earth in the realms of human health, Earth observation, disaster response and global education.

In the human health field, the Canadian NeuroArm shown in the film was successfully treating dozens of patients, while the American water purification equipment was finding remote community applications worldwide. Japan was growing high quality protein crystals and studying their molecular structure to improve medical treatment of diseases. A European respiration analyser used by astronauts in space was helping asthma sufferers to monitor their breathing. A European study of accelerated osteoporosis in astronauts and the elderly was showing the aggravating effect of high salt intake. American researchers were developing vaccines to help fight the bacteria that cause food poisoning. An American educational programme was produced to train non-expert operators in the use of the telemedicine ultrasound scanners. Reduced immunity experienced by astronauts in space was leading to a new medical device to detect immunity changes to help those suffering from shingles. An American team was developing tiny capsules containing liquid drug solutions for improved delivery of cancer-fighting drugs to patients.[4]

In the Earth observation and disaster response field, a variety of cameras and imagers mounted on the Station was proving valuable. American photography of reef areas was providing island communities with accurate maps and helping them to manage changing reef ecosystems and halt their decline. An American agricultural camera was keeping track of local flooding and providing information to farmers in North Dakota. Using wavelength techniques, American researchers had mounted a camera on Japan's external experiment platform to record coastlines and coastal water depth, clarity, floating matter and seabed characteristics. In the same vein, Italian scientists were monitoring the ecological state of health of Venice's lagoon using high-resolution photographs taken by Station crews. Responding to a natural disaster, the Station's crew had taken helpful photographs of the flooding caused by the 2011 tsunami on the east coast of Japan. A Japanese sensing instrument was studying the recovery and stability of ozone in the Earth's stratosphere. Europe was planning to use the Station as a platform for studies of Earth's changing climate to supplement its fleet of Earth observation satellites. A Russian digital photography survey was providing information for the development of a system for worldwide warning and control of disasters from natural or human causes.[5]

The International Space Station is a perfect platform for Earth observation. As NASA has pointed out, the Station has a different orbit with several advantages over other types of remote sensing satellite. It orbits closer to the Earth and sees the planet at different times of day and according to a different timetable. It provides more observation time of forests and vegetation in temperate regions than is possible with observation satellites commonly placed in polar orbits. In particular, it has one big advantage over all Earth observation satellites. The Station's multi-windowed Cupola provides panoramic views of the Earth's surface and its crew can respond quickly if a need arises to photograph something urgently, such as a forest fire or a volcanic eruption. The profusion of mounting points on the Station's exterior offers economical instrument siting opportunities. Newly developed and improved instruments no longer need to be built into expensive satellites. A Station resupply mission can ferry them up and they can be bolted on to the Station's structure during spacewalks.

In a bid to boost the Station's usefulness in the Earth observation field, NASA announced a new initiative for Station-based Earth observation in September 2014. Mounted on the Station over the next several years will be six new Earth science instruments. The first of these is called ISS-RapidScat. It will monitor ocean winds as part of a broader Station research programme on weather prediction and hurricane tracking. A second instrument named the Cloud-Aerosol Transport System is a laser device that will measure clouds and the distribution of airborne pollution, mineral dust, smoke and other atmospheric particles. Next is the Stratospheric Aerosol and Gas Experiment III, which will measure aerosols, ozone, water vapour and other gases in the upper atmosphere to help with the study of the recovery of the ozone layer. Following these will be a Lightning Imaging Sensor to track lightning activity over tropical and mid-latitude regions. Another instrument, the Global Ecosystem Dynamics Investigation, will use lasers to study the structure and health of forest canopies in ecosystems ranging from the tropics to high northern latitudes. The sixth instrument,

Examples of experiments and instruments on the International Space Station

RapidScat is a replacement for a deactivated weather satellite called QuikScat. The RapidScat instrument monitors ocean wind speed and direction to provide essential measurements used in weather predictions, including the tracking of hurricanes. The Jet Propulsion Laboratory devised a Station-mounted replacement for the satellite as it was cheaper and faster to develop and launch. The photograph shows technicians at the Jet Propulsion Laboratory preparing RapidScat for shipment to Kennedy Space Center for launch to the Station on the SpaceX-4 commercial cargo vehicle in September 2014. The Canadarm2 robotic arm installed RapidScat on the exterior experiment framework on the Columbus Module.[6] Image: NASA/JPL-Caltech.

NeuroArm is a robotic device designed for neurosurgery using image guidance and magnetic resonance imaging technology. Conceived by a team at the University of Calgary and unveiled in 2007, its design drew on the robotics expertise of MacDonald Dettwiler and Associates who developed the Canadarm2 robotic arm. A surgeon at a workstation conducts surgery by controlling the NeuroArm in a similar manner to astronauts on the Station who operate the Canadarm2 and its Dextre end manipulator. The workstation environment replicates that of the theatre where surgery is taking place. NeuroArm is now under commercial development for worldwide medical applications.[7] Image: University of Calgary.

MISSE (Materials International Space Station Experiment) is a research programme managed by NASA Langley Research Center for testing materials samples in the space environment. Samples arrive in experiment containers, which are attached to the Station's exterior and opened to expose the contents. After a typical duration of one year, samples are returned to the ground for analysis. The MISSE programme has tested samples including polymers, coatings, composites, switches, sensors, mirrors, seeds, spores, bacteria and especially materials for the radiation shielding needed on long human missions. The photograph shows the two parts of an open MISSE container containing numerous samples and specimens.[8]

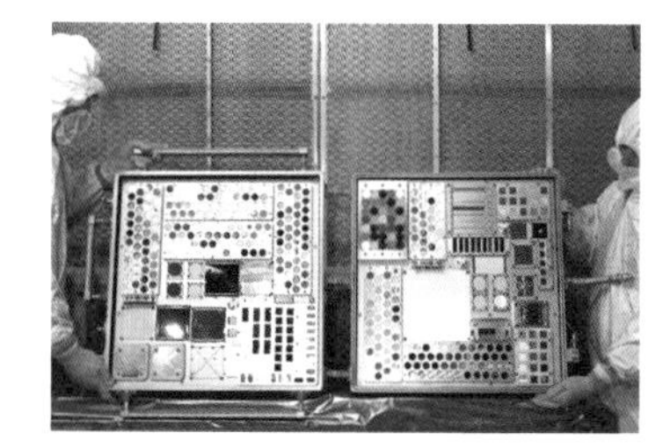

FOOT (Foot/Ground Reaction Forces During Space Flight) studies the physical loads on the lower bodies of the Station's crew and their muscle activity. The aim is to gain more knowledge of astronaut bone and muscle loss in space. The results will help to combat the effects of microgravity on long missions and benefit the treatment of osteoporosis on Earth. Developed by the Human Research Program at Johnson Space Center, FOOT requires astronauts to wear a special suit that measures muscle exertion, leg-joint movement and foot forces while using exercise equipment, as shown here, and then again on Earth. Research has found that exercise regimes on previous space missions were inadequate.[9]

Tomatosphere This Canadian educational experiment aimed to get young people interested in space science and horticultural technology. It involved sending 600,000 tomato seeds to the Station for exposure to the space environment and then returning them to Earth. The photograph shows Canadian astronaut Chris Hadfield holding a bag of seeds. Students in 13,000 schools across Canada planted the seeds to find out how well they grew. The experiment enabled students to engage in and contribute to science at their learning level. It provided them with experience of scientific research methodologies and perhaps inspired some of them to pursue studies and careers in the sciences.[10]

Robonaut is a humanoid robot. It has a torso with a rotating waist, dexterous arms, a head with two high image cameras for eyes and a back-mounted battery pack. Its job on the Station is to manipulate hardware, working in risky environments and responding safely to unexpected situations. Developed by NASA and General Motors, it is presently confined to the Station's interior but its greatest benefit may be to assist on spacewalks. The robot is controlled by a teleoperation system that mimics human physical motions. A crew member wears specialised gloves, a vest and a visor that provides a three-dimensional view through the robot's eyes. With human-like hands and arms, Robonaut is able to use the same tools that the crew uses.[11]

PESTO (Photosynthesis Experiment and System Testing and Operation) studied the photosynthetic response of plant tissues grown in microgravity. It found that the absence of gravity altered leaf development, plant cells and chloroplasts (cell structures that conduct photosynthesis) but was not harmful to the plants. The photograph shows dwarf wheat plants grown under high illumination and controlled carbon dioxide conditions through several generations. Developed by the University of Limerick and Dynamac Corporation, PESTO demonstrated the possibilities of growing plants in space that will be essential on long missions to reduce the need for on-board consumables and their resupply.[12]

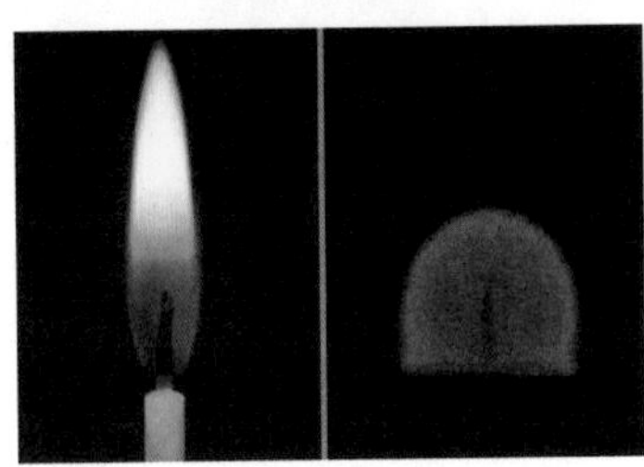

SAME A fire on board a spacecraft is one of the greatest potential hazards. SAME (Smoke and Aerosol Measurement Experiment) is a NASA experiment that measures properties such as particle size distribution of smoke generated by spacecraft fires. The processes of combustion in microgravity are different from those in Earth's gravity. The photograph illustrates this by showing a flame in one-g on the left and one in microgravity on the right. The aim of the experiment is to gain a better understanding of smoke and to develop improved means of smoke detection on future spacecraft. The SAME experiment examined smoke produced by burning Teflon, Kapton, cellulose and silicone rubber – all materials used on spacecraft.[15]

HDEV (High Definition Earth Viewing) is a set of four commercial high definition cameras mounted on the Station to take imagery of the Earth. Johnson Space Center developed the installation. HDEV tests the cameras' ability to operate when exposed for long durations to the orbital environment. One camera points forward into the Station's flight path, two point aft and one points towards nadir. The photograph shows a camera assembly. Video imagery is encoded into a compatible format with Ethernet for transmission to the ground. In this format, the video can be viewed from any computer connected to the internet. Much of HDEV's operation is carried out by student teams through an educational outreach programme.[13]

SPHERES (Synchronized Position Hold, Engage, Reorient, Experimental Satellites) are miniature satellites the size of bowling balls. Their purpose is to test a set of instructions that can be used by spacecraft to perform independent rendezvous and docking manoeuvres. The satellites are eighteen-sided polyhedrons with a diameter of 20cm and a weight of 3.5kg. Equipped with power, propulsion, computers and navigation equipment, they fly in formation inside the Station. They communicate with each other and with a laptop computer through a low-power 900MHz wireless link. SPHERES was developed by Massachusetts Institute of Technology and is supported by the US Department of Defense.[16]

NightPod The European Space Agency developed NightPod as a tracking instrument designed to assist cameras in taking improved photographs of the Earth, especially at night or in low light conditions. It compensates for the Station's movement in flight, enabling longer camera exposure times. The result is high-resolution images far superior to those from previous programmes. The instrument is particularly helpful at photographing city light pollution, vegetation fires, marine and road traffic, volcanic activity, urban pollution and fishing. It can be adapted to look out into space as well. The photograph shows the NightPod equipment with a camera set up in front of the central window in the Cupola on the Station.[14]

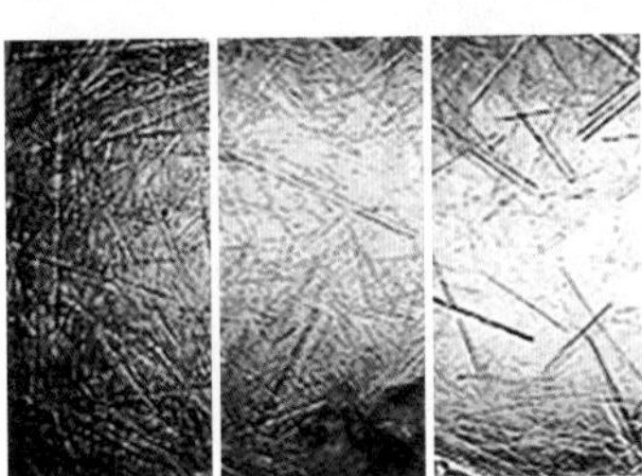

DCPCG Research has found that growing protein crystals in microgravity results in higher quality crystals than those grown on Earth. The aim of the DCPCG (Dynamically Controlled Protein Crystal Growth) experiment was to improve the protein crystallisation process by controlling the elements that influence crystal growth. DCPCG was led by the University of Alabama. It tested four types of protein. Researchers controlled the protein concentration and the diffusion process and determined the differences in vapour diffusion rates (the speed at which the liquid surrounding a protein solution evaporates, leaving behind a protein crystal). From left, the photograph shows fast, medium and slow evaporation rates.[17]

ELC4

the ECOsystem Spaceborne Thermal Radiometer Experiment, is a high-resolution thermal imaging spectrometer that will study water use and water stress in vegetation.[18]

The Station has also helped to save lives at sea. Ships on voyages are required to transmit signals of their position and identity through the Automatic Identification System. Until now, these signals were only received by stations along coastlines and the Earth's curvature blocked the signals when ships passed over the horizon. European Space Agency researchers in Norway wondered if a Station-based receiver would be able to track ships that went out of range and an astronaut mounted a test receiver on the Station's exterior to find out. On 25 January 2012, a Norwegian vessel named *Hallgrimur* capsized at sea. It was out of range of coastal receiving stations but the receiver on the International Space Station picked up *Hallgrimur*'s distress signal and relayed its position to a nearby ship that came to its aid. A lone survivor who had been in the water for several hours was rescued.[19]

In the global education field, the telephone conversations between young people and Station astronauts shown in NASA's film have taken place in hundreds of link-ups around the world. NASA used astronauts on the Station to inspire schoolchildren to get fit by sharing with them their physical training and exercise regimes. The Station's water recycling system served as an example to educate students about global water resources and conservation. In another photographic project, American college mathematics and science students controlled a camera on the Station to observe specific locations on Earth while utilising mapping and mathematical Station tracking techniques. A Russian educational initiative for schools ran four experiments. In the first, cosmonauts grew crystals out of solution to show the effects of weightlessness; then a VHF radio beacon on the Station enabled student radio enthusiasts to communicate with cosmonauts; relayed through direct downlinks to students on the ground was video coverage of life on board; and there was a public relations campaign of cosmonaut achievements on the Station intended for the general public.[20]

NASA was aware of the importance of underlining the Station's leading research role. After completion of the Station's assembly in 2010, the agency had begun to shift the emphasis away from its epic construction story to its future research use. NASA described it as positioning the International Space Station for the utilisation era and pointed out that, while the design, assembly and operations were remarkable human achievements in their own right, the opening up of the Station as a laboratory during the decade ahead would offer unprecedented opportunities for advancing research and development.[21] The Station's international partners were working together to maximise its value. Scientific, technological and industrial uses would result from new initiatives aimed at contributing to the future of space exploration and the missions of non-space organisations. The Station would support

Alpha Magnetic Spectrometer-2
Opposite: Astronauts Greg Chamitoff (left) and Andrew Feustel (right) seen at work after the installation of the Alpha Magnetic Spectrometer-2 on the Station. About the size of an airport shuttle bus, the spectrometer's mission is to record the passage of charged cosmic particles that pass through it, helping among other things to shed light on the origin of dark matter in the Universe. Developed and built by CERN in Switzerland, the Alpha Magnetic Spectrometer-2 has transformed the International Space Station into a world-class particle physics research centre. Above: The Alpha Magnetic Spectrometer-2 is one of the International Space Station's most important astrophysics experiments with sixty institutes from sixteen countries involved in using it as a research tool.

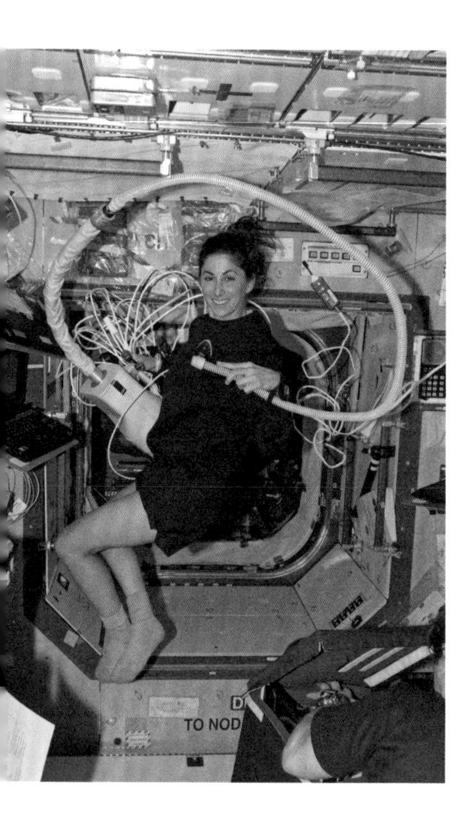

Housekeeping Astronaut Nicole Stott doing some housekeeping inside the Destiny laboratory with a space-rated vacuum cleaner during Expedition 20 in September 2009.

these initiatives. 'A new global economy in space is approaching the tipping point', claimed NASA.

By March 2014 the Station's research statistics were impressive. It had hosted 1,554 experiments since research had begun on it in 1999, of which 1,087 were permanent or completed; and 1,824 scientist investigators from eighty-two countries had participated in the research, which spanned disciplines of biology and biotechnology, Earth and space science, human research, physical science, technology development and educational and cultural activities. American, Japanese and Russian experiments tended to focus on biology and biotechnology, in Japan's case heavily. Canadian experiments tended to focus on human research. European research was fairly evenly balanced across all disciplines. Across the Station as a whole, the most experiments were in biology and biotechnology while the fewest were in Earth and space science. Research results had been published in 747 articles in scientific journals and 189 conference papers.[22]

There have been research results of great value. For example, adequate energy intake, vitamin D and strenuous physical exercise with muscles working against resisting loads are proved to be effective in reducing bone mineral loss in weightlessness – an issue of great concern for future human exploration missions. New candidate treatments for muscular dystrophy and prostate cancer have resulted from medical research on board. Universal equations on the behaviour of fluids in microgravity have derived from capillary flow experiments. Plant studies have confirmed that roots grow towards water and nutrients independently of gravity – a valuable discovery for the cultivation of food plants in microgravity.[23] Despite the impressive equipment and facilities available on the Station to scientists on the ground, no research results achieved so far can quite rank as a major discovery or scientific breakthrough. Since research began on the Station, the scientific payback has been gradual and cumulative. Though the Station's achievements show that it has justified its existence as a world-class research laboratory, it has yet to hit the scientific jackpot.

Externally, the Station has several multipurpose locations for experiments. Mounted along the transverse truss are four structural pallets called Express Logistics Carriers with electrical power and data connections to the Station's systems. They offer vantage points and attachment sites for space and Earth observation and for technology and exposure testing in the space environment. Japan's Kibo Exposed Facility has ten standard and three heavy sites for experiments, providing them with thermal cooling as well as electrical power and data connections. It hosts experiments in communications, space science, engineering, technology demonstration, materials processing and Earth observation. It also has a capability for launching small satellites. The European Columbus Module has an External Payload Facility comprising two structural frames mounted on its outer endcap. It offers a total of four sites for experiments, also with electrical power and data connections. It hosts experiments in solar observation, atomic clocks and atmospheric monitoring. Supporting and serving experiments on the exterior of the Russian modules is a mixture of multipurpose workstations, biaxial pointing platforms, handrail attachment clamps and magnetic mechanical anchors.[24]

Perhaps the most impressive experiment mounted outside the Station is the Alpha Magnetic Spectrometer-2. Begun in 1994 and developed by CERN in Switzerland, it is a collaborative research project between Finland, France, Germany, Italy, the Netherlands, Portugal, Spain, Switzerland, Turkey, China, Korea, Taiwan, Russia, Mexico and the United States.[25] The project is led by Professor Samuel Ting, a Nobel Laureate at the Massachusetts Institute of Technology. The Alpha Magnetic Spectrometer-2 arrived on the penultimate Space Shuttle flight, ULF6 (STS-134). It is a state-of-the-art particle physics detector devoted to the search for antimatter, dark matter and the origin of cosmic rays, as well as the study of various types of particles and their flux. At the heart of the spectrometer is a large cylindrical magnet made of NeodymiumIronBoron that creates a magnetic field to bend the trajectories of charged cosmic particles that travel through space. The particles are diverted to several instruments. The first of these is the Transition Radiation Detector that distinguishes between proton and electron particles. Next are two Time-of-Flight Detectors that time the passage of particles through the magnetic field. Next are the Silicon Trackers that monitor the curved paths that the particles take. Then there is the Tracker Alignment System that checks the performance of the trackers and the Anti-Coincidence Counter that specialises in detecting stray particles. The Ring Imaging Cherenkov Detector measures particle speed and the Electromagnetic Calorimeter measures particle energy. The Alpha Magnetic Spectrometer-2 gathers more than seven gigabits of data from these instruments per second and feeds it to 650 computers on board the Station. The computers process the data and then transmit it to Earth at about six megabits per second. Knowing where in space the spectrometer is looking as it gathers this colossal quantity of data is important and it has two Star Trackers and a Global Positioning capability to help with its pointing.[26]

After its first forty months in space, the Alpha Magnetic Spectrometer-2 had recorded 54 billion cosmic ray events, of which 41 billion had been analysed by participating universities worldwide. The spectrometer analysis of the numbers of electrons and positrons passing through it is greatly advancing the understanding of the origin of dark matter.[27] With the arrival of the internationally developed spectrometer, the International Space Station was transformed into one of the world's leading astrophysics and particle research centres.

At a seminar at the United Nations in 2012, NASA summarised America's own research activities on the Station up to that point. They followed a dual approach. On the one hand, NASA's own utilisation of the Station was aimed

Training Before beginning their missions on the International Space Station, astronauts and cosmonauts are put through intensive training programmes on the ground that can last for years. Here astronaut Samantha Cristoforetti is seen training before her mission as flight engineer on Expedition 42/43 that began in November 2014. Assisted by an instructor, she is practising in a replica of the Station's Cupola in the Avionics Systems Laboratory at Johnson Space Center where background video projection simulates the Station's exterior elements at life size.

at advancing space exploration capabilities. It covered astronaut health and space exploration countermeasures, testing of new space technologies, validation of operational procedures for long-duration missions and research on physical and life sciences in weightlessness. On the other hand, the Station had now become a National Laboratory, open for business with other government agencies, commercial companies and non-profit organisations. There were new initiatives for barter deals with the Canadian, European and Japanese partners.[28] NASA wanted to open the Station up to a wider community of global users and make it easier for them to access its facilities. It remains to be seen, however, whether the agency's proactive attitude will reach out effectively to a new family of experimenters around the world.

Research Station crews follow highly planned and detailed daily timetables with research and other tasks defined in increments of time that can be as short as five minutes. Much of the working day is spent operating and tending experiments and instruments in the Station's laboratories. Top: astronaut Shannon Walker, flight engineer on Expedition 24 from June to September 2010, is working on the SAME experiment (Smoke Aerosol Measurement Experiment) inside the glovebox in the Columbus Module. Bottom: astronaut William McArthur, commander and science officer on Expedition 12, from October 2005 to April 2006, is setting up the calibration arm on SLAMMD (Space Linear Acceleration Mass Measurement Device) that is attached to a rack in the Destiny Module.

Some rooms with a view
Living and working on board

By March 2014, thirty-nine crew expeditions had taken place on the Space Station with a total crew time of 19,623 hours. Flight plan timetables are transmitted to the Station listing every activity the six crew members of each expedition are due to carry out each day. They are full of activities and tasks intricately planned in increments as short as five minutes. The following example of a weekly timetable is typical:

On Monday the crew wakes at 06:00. For the first five to ten minutes, each member carries out an equipment inspection or test or monitors an experiment such as overnight circadian rhythms.[29] From 06:10 to 06:45, four members take care of personal hygiene needs while urine, blood and saliva samples are collected by the other two. The crew commander carries out a morning inspection at 06.15 and deals with personal hygiene needs afterwards. Most crew members have breakfast between 06.35 and 07:30 but not necessarily exactly at the same time as some are working on experiments. From 07:30 to 07:55 there is a daily planning conference that takes place with an audiovisual link to the ground.

After the planning conference, morning work starts in earnest and continues until lunchtime. The commander takes physical exercise, assists in blood pressure recording, acts as a subject for ocular tests, carries out cabin atmosphere condensate removal, closes a window shutter, reviews the rendezvous and berthing procedures for a visiting spacecraft and conducts an experiment debrief. The first crew member (each has a number) engages in photography, prepares experiments, participates in a conference with a ground radio station, sets up a camcorder, takes physical exercise, works on experiments and participates in a television conference with schoolchildren. The second upgrades software, operates experiments, fills out a log sheet, takes physical exercise and participates in the school teleconference. The third deals with blood and urine samples, sets up and carries out

ocular tests, answers a vision questionnaire and works on experiments. The fourth participates in the ground radio station conference, carries out hardware maintenance, takes physical exercise, downlinks exercise data, closes window shutters and participates in the school teleconference. The fifth sets up video equipment, takes physical exercise, prepares experiments, reviews the rendezvous and berthing procedures for a visiting spacecraft with the commander and takes part in a psychological evaluation programme.

The lunch period begins at 12:40 and continues to 13:55 but again not everyone takes it exactly at the same time. After lunch, the commander is involved for two hours with an experiment, takes care of an environmental control mechanical task, prepares for evening work, sets up a video-recorder and takes physical exercise. The first crew member follows a Station orientation programme, discusses an experiment with the ground, prepares for evening work, takes physical exercise and follows a Station adaptation and orientation programme (he or she had arrived just two weeks earlier on a Soyuz flight and was still a 'rookie'). The second (another recent arrival) works on an experiment, adjusts optical telescope settings, cleans an air circulation screen, follows the Station adaptation and orientation programme, helps set up the video-recorder, takes physical exercise and prepares for evening work. The third stows blood and urine collection

Exercise Frequent strenuous physical exercise is vitally important for astronauts and cosmonauts to maintain their health and fitness in the weightlessness of space. In the absence of gravity there is a need to stress bones and muscles and increase cardiovascular performance in much the same way as on Earth except that longer periods of exercise are needed. To the right, astronaut Richard Mastracchio, flight engineer on Expedition 38 from November 2013 to March 2014, exercises on a treadmill. He is restrained by a bungee harness. To the left, he is exercising on a piece of equipment that uses piston-driven vacuum cylinders with adjustable resistance to simulate the lifting of different weights. There are six different types of exercise equipment on the Station.

5
3

equipment, helps set up the video-recorder, takes physical exercise, opens a window shutter and prepares for evening work. The fourth prepares to photograph the undocking of a Progress spacecraft from a Russian docking module. This is the main event of the day. He or she then carries out a Station control operation, discusses an experiment with the ground, prepares for evening work and takes physical exercise. The fifth reviews a hardware deployment plan on the Columbus Module, writes a journal entry, helps with an experiment, works on the communications system, downloads camera images and stows experiment equipment. From 19:05 to 19:30 there is another daily planning conference. Evening dinner and off-duty time for the crew begins at 19:30 and continues until 21:30. Crew members are free to eat and relax during this two-hour period. After this, there is equipment to stow. Then each person retires to his or her private compartment to sleep.

On Tuesday the crew wakes at 06:00 and mostly engages in experiment and payload operations during the day.[30] There are Station housekeeping and maintenance duties to carry out. Three crew members take part in a television programme and three have private medical conferences with the ground.

On Wednesday the crew wakes at 07:30.[31] The main event of the day is the docking of a Progress resupply spacecraft. Two crew members are involved with this for over two hours. The crew goes to bed at 23:00.

On Thursday the crew again wakes at 07:30.[32] The main event is the opening of the hatch to Progress and the unloading of its cargo. Two crew members first check for air leaks between Progress and its docking module and then open the hatch and test the air quality inside the spacecraft. Then, one of them moves the docking mechanism out of the way and two others enter Progress to unstow priority cargo and move it into the Station. They continue to unstow Progress for a few hours. The crew retires at 21:30.

Friday is an off-duty day for the crew but they wake at 06:00.[33] At 09:40 there is a full crew televised conversation with the Russian President via a Russian television station. This continues until 11:00 and includes back-up time in case of live link-up problems. This is followed by participation in a space-related competition by three crew members on the same television channel. After lunch, three crew members take physical exercise while the other three have the afternoon off. Bedtime for all is at 21:30.

On Saturday the crew wakes at 06:00.[34] The main event of the day is a reboost of the Station's orbit using the thrusters on the docked Progress spacecraft. This is carried out automatically under the direction of the ground. After breakfast, three of the crew members deal with weekly housekeeping duties which last from 07:30 to 10:30. Three have private family discussions at different times during the day and in the afternoon all crew members take physical exercise.

Sunday is an off-duty day.[35] The crew wakes at 06:00. Much of the morning is taken up with more housekeeping duties by three crew members

Spacewalks Above: Russian cosmonauts Gennady Padalka, commander of Expedition 32 from July to September 2012, and flight engineer Yuri Malenchenko, carry out a joint spacewalk that lasted just under six hours. Its purpose was to transfer equipment from the Pirs Docking Compartment to the Zarya module to prepare Pirs for its eventual replacement by a new Russian multipurpose laboratory module. The cosmonauts also installed micrometeoroid debris shields on the exterior of the Zvezda module and deployed a small science satellite. Opposite: Astronaut flight engineers Chris Cassidy and Tom Marshburn on Expedition 35 from March to May 2013 inspect and replace a pump controller box that was leaking ammonia coolant on the Station's truss structure. This joint spacewalk lasted over five hours.

while the others take physical exercise. The first three crew members have private family discussions, then the others have private psychological conferences with the ground. In the afternoon there is more physical exercise, some maintenance work, some preparation for the next day's work and another daily planning conference. Dinner and leisure time begin at 19.15 and continue until 21:30 when the crew goes to bed.

Supporting daily life on the Station is a lifeline of regular contact between the Station's crew and its controllers on the ground. There are dual mission control and programme management centres for the International Space Station, one in America and the other in Russia. In Russia, mission control and management are split between the Moscow Mission Control Centre in Korolev on the outskirts of Moscow and the International Space Station Programme Management at Roscosmos Headquarters in Moscow. In America, the International Space Station Mission Control Center and Program Management are both based at Johnson Space Center in Houston. NASA's Ames Research Center, Glenn Research Center, Marshall Space Flight Center are also involved in Station operations in various ways.

International Space Station Mission Control Centers A view of the International Space Station Flight Control Room in Moscow Mission Control Center at Roscosmos at Korolev, Russia. Korolev is situated on the outskirts of Moscow. Image: Bill Ingalls.

An elaborate radio and satellite communications network performs several vital tasks to keep the Station connected with the centres on the ground. It enables the crew to talk to both the ground and to arriving vehicles; it allows the ground to monitor, maintain and control Station systems and payloads; and it feeds experiment information to scientists. The two-way audio and video communications system links crew members with arriving vehicles such as Soyuz spacecraft and with other crew members inside the Station or outside on spacewalks. The same system connects the crew with Station control teams at the two mission control centres and with payload scientists via other receiving stations. The communications system permits audio, video and electronic data to travel backwards and forwards between the Station and the ground on a daily basis.

Enabling this communications lifeline is a transmission network that must operate continuously. The two mission control centres are at different geographical points on the Earth's surface. They maintain constant contact with the Station that encircles the Earth about every ninety minutes and therefore passes out of sight of both centres for a portion of each orbit. To overcome this interruption, a fleet of geosynchronous satellites called the Tracking and Data Relay Satellites acts as a go-between and relays transmissions between the Station and the ground regardless of where the Station is overhead. These satellites operate in S and Ku bands that are commonly used for satellite communications. The Station communicates with arriving spacecraft and astronauts and cosmonauts on spacewalks on VHF band.

There are several other control centres around the world that have contact with the Station. Japan uses its own relay satellite to connect its modules with its control centre in Tsukuba. Europe has an International Space Station Programme Management Centre at Noordwijk in the Netherlands and Canada has one in Quebec. In Europe there is a control centre for the Automated Transfer Vehicle at Toulouse, France and another one for the Columbus Module at Oberpfaffenhofen, Germany.[36] Beyond these, there are numerous scientists and technicians based at research centres, universities and other establishments worldwide who work on Station experiments.

One of the most important activities the crews perform on the Station from time to time is to operate the Canadarm2 robotic arm and its mobile railcar system. It is impossible to build a skyscraper without a crane and it would have been impossible to build the Station without this highly capable tool. With the Station complete, it is as important as ever. There are two sites on the Station where there are identical workstations for controlling the arm and its railcar – the Destiny Module and the Cupola. The sites are called the Robotic Workstations. They can be used simultaneously – one for control and the other for monitoring – and their roles are interchangeable. The Cupola, with its superb views, is the best place for working the arm to grapple and berth an arriving resupply spacecraft or to move about an astronaut in a spacesuit perched on its end for maintenance or repair work.

The arm and railcar can be controlled by the Station's crew, by flight controllers at the International Space Station Mission Control Center in Houston or by the Operations and Engineering Center at the Canadian Space Agency near Montreal. Each workstation has everything an astronaut needs to drive the railcar and work the arm, the dexterous manipulator at the arm's end and the arm's video cameras. A display and control panel contains a bank of switches, buttons and indicators for activating the various features of the system. Three liquid crystal display monitors show multiple video views. The monitors have the ability to overlay static and dynamic telemetry data on to the live video to provide additional information to an astronaut watching the arm's motion. There are two hand controllers. The first of these called the Translation Hand Controller has three degrees of freedom and enables its operator to control the robotic arm in side-to-side, forward, backward, and up-and-down directions of motion. The second is called the Rotation Hand Controller and also has three degrees of freedom. It enables control of the arm in roll, pitch and yaw rotations. The left hand works the Translation Hand Controller and the right hand works the Rotation Hand Controller. Selection of the rates for controlling translation and rotation depends on the size and mass of the object and its position relative to the surrounding structure. At a comfortable distance from the structure, an astronaut can choose coarse rates and can then switch to fine rates as the distance reduces. There are ten different settings for each type of rate.

The robotic arm and end manipulator Dextre can manoeuvre at very small rates and can position objects with exceptional precision, down to fractions

International Space Station Mission Control Centers A view of the International Space Station Flight Control Room in Johnson Space Center adjoining Clear Lake City, south of Houston, in Texas.

of a degree or millimetre per second. Only Station crews can use the hand controllers – they cannot be worked by flight staff on the ground. There are two major modes for controlling the arm's multiple joints. The first is the predetermined movement of individual joint parts and the second is movement related to a point of reference on the arm itself or on the object it is holding. The reference point follows the controller's movement. The actions of human joints responding to the brain help to explain the difference between the two modes. Consider, for example, a point of reference as a key to be inserted into a lock. In the first mode, the brain can tell an arm to move in a joint-by-joint sequence – shoulder, elbow, wrist and finger – to insert the key. In the second, the brain can simultaneously tell all the arm's joints to work together in a coordinated manner to perform the same task.[37] The process of learning to operate the robotic arm and its railcar involves many months of training for both astronauts and ground controllers. Training takes place in simulators at facilities at Johnson Space Center and at the Canadian Space Agency and covers both the robotic system and how it interacts with other Station systems.[38]

Canadarm2 workstations At the robotics workstation in the Cupola, astronaut Karen Nyberg, flight engineer on Expedition 36, from May to September 2013, is practising the use the Canadarm2 robotic arm in preparation for the grappling and berthing of Japan's H2 Transfer Vehicle and then plugging it on to the Earth-facing port of the Harmony node. There is a similar workstation in the Destiny Module and they can be used simultaneously to carry out an operation.

In May 2014, the Canadarm2 became the first robotic device to repair itself in space. The Dextre mechanical manipulator on the end of the arm replaced a camera on one of the arm's limbs and another camera on the arm's mobile railcar.[39] This event marked a major step towards the day that robots will be able to repair, refuel and reposition satellites on Earth orbit. Canadarm2's amazing success has made one thing clear. Whatever succeeds the International Space Station as the next human space exploration destination – it looks increasingly like Mars – should have crew-controlled robotic devices to help construction, improve safety and perform a host of other tasks. From now on, it will be a mistake not to employ this extraordinarily effective technology wherever it can be helpful in the space exploration field.

The Station's assembly job done, one of the most valuable tasks performed by the robotic arm system today is to help berth arriving spacecraft. Now that the Space Shuttle is no longer flying, the duty of delivering fresh supplies to the Station has been partly taken over by two privately developed vehicles named Dragon and Cygnus. These spacecraft are new additions to the Station's fleet of delivery vehicles. Both were developed using private funds under a NASA commercial initiative.

SpaceX (Space Exploration Technologies Corporation) operates Dragon. It is launched from Kennedy Space Center on a new rocket also built by SpaceX called the Falcon 9. Dragon comprises a conical capsule that carries 3.3 tonnes of pressurised cargo in $6.8m^3$ of volume and a cylindrical trunk that carries the same mass of unpressurised cargo in a volume of $14m^3$. The capsule returns to Earth with experiments and other materials for an ocean splashdown and recovery while the trunk is jettisoned and destroyed during re-entry.

The Dragon spacecraft Crew members on Expedition 34, from November 2012 to March 2013, took this photograph of the automated Dragon spacecraft after its arrival at the Station on 3 March 2013. The photograph shows it grappled by the Canadarm2 robotic arm that brought it in to berth for a three-week stay at the Station. Dragon was developed as a commercial space venture by SpaceX of Hawthorne, California. By November 2014, the Dragon spacecraft had completed four successful round trips to resupply the Station.

Orbital Sciences Corporation operates Cygnus. By contrast to the Dragon, the Cygnus spacecraft is a cylindrical module launched on a Taurus II rocket, also built by Orbital Sciences Corporation. It can deliver 2 tonnes of pressurised cargo only and is not recoverable. It is loaded up with trash and waste before it departs the Station and is destroyed on reentry.[40]

The robotic arm system is used to receive Dragon. Flight controllers on the ground perform a checkout of the entire system in advance and position the railcar, robotic arm base, robotic arm and dexterous manipulator in the required configuration to await the arriving spacecraft. They designate the robotic workstation in the Cupola as the principal workstation with the other one in the Destiny Module ready as a back-up to take over in an emergency. Ground controllers also get the video cameras ready on the arm and the Station and verify the video transmissions.

Dragon, Cygnus and the Japanese HTV vehicle approach the Station from below because the laser hardware used to fix their distance from the Station is mounted on the underside of the Japanese laboratory module. Dragon approaches the Station under the control of its digital autopilot that commands its thruster jets to keep it within a prescribed position and attitude range called a deadband. When it has reached this state just clear of the Station, the crew commands it to switch to free drift mode and the spacecraft comes to a halt. An astronaut at the workstation in the Cupola very carefully manoeuvres the Canadarm2 robotic arm with its small end manipulator to close over a grapple shaft on the Dragon using both types of hand controller. The astronaut commands the manipulator to lock on to the shaft and capture the spacecraft. Ground controllers then take over and berth Dragon on the Station's Harmony node.

Which way forward?
The future of the International Space Station

The International Space Station will carry on operations until 2024 and possibly even beyond that. In whole or in part, it will continue its job as a leading research facility and pursue major scientific discoveries. But scientific research on its own is not enough if the Station is to encompass the original objectives NASA laid down for it in 1984. It is important to keep returning to those early visions for the Station as a reminder of its true potential for there is much that it can yet achieve beyond its present role.

August 2013 saw the publication of a *Global Exploration Roadmap* for a new exploration strategy formed by a group of leading space agencies around the world. It could give the Station a new lease on life.[41] It begins with the International Space Station as an overture and ends after 2030 with human missions to Mars as a grand finale. In between, the strategy is flexible because a preferred path does not yet exist.

The *Global Exploration Roadmap* report is initially vague on the International Space Station, describing its role as carrying out 'general research and exploration preparatory activities'. Later, it states that the Station is 'an excellent platform on which to prepare for future exploration missions. New activities in the areas of exploration technologies, human health research and operations simulations have been identified, and many have begun operation on board. Critical capabilities used to support the International Space Station will be advanced toward exploration requirements, such as lower mass, lower power and high reliability.'
In an admission that Station activities are not running at full throttle, the strategy will be 'fully utilizing' the Station. The report notes that 'advanced habitation capabilities beyond those currently in use on the International Space Station will be necessary to enable future deep space missions'.

The end of the report is specific on the areas of investigation that the Station could pursue. There will be more research on crew health and performance in weightlessness and countermeasures – including improved nutrition – to help to neutralise cardiovascular, musculoskeletal, neurological and behavioural changes. There will be a one-year tour on board for an astronaut and a cosmonaut to study the adverse effects of long-term weightlessness in the light of improved biomedical techniques to deal with bone and muscle loss. An investigation into greater crew control over space missions with reduced ground contact will simulate the communications delays of up to twenty minutes each way on Mars flights. The Station will become a testbed for improved life support systems and hardware that must do a more effective job of recycling all the water and oxygen on board as little or no resupply from Earth will be possible on a deep space mission. Also tested will be improved solar arrays and fuel cells that use new technology to boost their power output. The Station will continue to demonstrate the success of using robotic and telerobotic systems to carry out construction and maintenance tasks. There will be new and better spacesuits for the crews. There will be an evaluation of new spacecraft navigation, docking and berthing systems with more emphasis on automation, and new research on the technology of entry and descent into Earth's atmosphere. Finally, an inflatable module prototype due for attachment to the Station from 2015 onwards will investigate the effects of the harsh ultra-violet, thermal and radiation environment on the flexible materials of which such modules are made.[42] Yet many of these new research roles for the Station envisioned in the *Global Exploration Roadmap* are really a continuation of the research scope already in progress – a multitude of investigations into realms of science and engineering that are moving them forwards cautiously and incrementally. Still missing is a drive to stretch the Station's capabilities and exploit its full potential.

At a press conference in April 2014, NASA Administrator Charles Bolden made it clear that Mars is the next favoured exploration destination and outlined the steps needed to reach that objective. The agency had answers ready to the perpetual question 'why human space exploration?' Scientific and human exploration and pioneering are hallmarks of advanced civilizations. They broaden human experience, expand knowledge, drive innovation and spur commerce. Space exploration boosts humanity's motivations by igniting imaginations, unlocking discoveries and creating a better future. Robots can investigate remote destinations to begin with, followed by humans who can add speed, intuition and efficiency to the exploration process. Last but not least, human space exploration is good for national prestige and international cooperation.[43] NASA stated that a crewed Mars mission will take two to three years to complete and that the International Space Station will have several important roles to play. The agency echoed most of the objectives outlined in the 2013 *Global Exploration Roadmap*. The Station will improve knowledge about the long-term medical, psychological and biomedical effects of weightlessness on humans.

The Cygnus spacecraft This photograph shows the Canadarm2 robotic arm about to release the automated Cygnus spacecraft after its three-week stay berthed to Harmony at the Station in October 2013. Cygnus was developed by Orbital Sciences Corporation of Dulles, Virginia as a commercial space venture. By November 2014, Cygnus had carried out two successful resupply missions to the Station. Unlike the Dragon spacecraft which is recoverable, Cygnus is destroyed on atmospheric re-entry.

NASA announced that in 2015 one astronaut and one cosmonaut would begin a full-year tour of duty on the Station. Prior to that, four Russian cosmonauts had spent a year or more of continuous time in weightlessness on consecutive tours on the Mir Space Station. Now an American astronaut would join that exclusive club. During the course of the year, there would be multiple joint investigations and data sharing by America and Russia with the focus on seeking more knowledge for the long Mars mission.[44] There is more important medical research to be done that ought to involve all the Station's partners. In one example, twenty-one American astronauts who have spent time on board the Station have developed changes to their vision. They have developed a shift towards farsightedness (hyperopia), diminished vision within the visual field (scotoma), distension of the optic nerve, changes to the eyeball's form and fluid build-up around the optic nerve (oedema). The indications are that increased intracranial pressure among astronauts may be the cause.[45] This is a recent hypothesis and obviously there will be a need for countermeasures if the medical research findings bear it out.

The Station can begin to demonstrate the types of advanced life support technologies that will be essential on a mission to Mars with a round-trip time of between one and three years. As NASA puts it, 'As we leave Earth, we still need air to breathe, water to drink and food to eat. Sustaining people in space requires managing all of their "ins and outs".'[46] While environmental control and life support systems technology is now highly developed for applications such as the Station, long exploration missions will need much better ways of handling oxygen, water and waste recycling. The recycling of oxygen from carbon dioxide – already well-understood in the chemical engineering field – must achieve a high level of efficiency with the small amount of make-up oxygen needed supplied by water electrolysis. The same target applies to supplying clean water by highly efficient grey water and urine recycling and the recovery of humidity condensate. Solid waste –

Human Mission to Mars At a press conference in April 2014, NASA Administrator Charles Bolden and his management team confirmed that Mars is the preferred destination for the next big human exploration step into the Solar System. This panoramic view of 'Husband Hill' on Mars was take by the exploration rover Spirit in November 2005. Image: NASA/JPL-Caltech/Cornell.

whether packaging or human waste – needs to be biodegraded into something useful, such as soil for plant growth. According to NASA, the technology needed for raising the efficiency of recovering oxygen from carbon dioxide is at a low level of readiness for testing on the Station. It is among NASA's highest research priorities. Too much crew time is being spent maintaining the Station's troublesome Carbon Dioxide Removal Assembly that takes carbon dioxide out of the air for conversion back into oxygen. NASA is looking at several options for improving its reliability. The Sabatier system on board that converts carbon dioxide back to oxygen (by combining its carbon with hydrogen to produce methane) is only operating at 50 per cent efficiency. As NASA says, 'Although promising technologies exist, the key challenge is to enable greater recovery of (oxygen) without the need for prohibitive amounts of complex equipment or expendables to accomplish the job.'[47]

The ability of plants to convert carbon dioxide to oxygen through natural photosynthesis in an encapsulated environment and produce edible cereals or vegetables in the process is a subject that spacefaring nations have studied for decades. America and Russia have accumulated a lot of research experience in this field. Yet, the technology has barely progressed beyond ground laboratory experiments. The plant experiments on the Station come from the life sciences domain and focus mostly on plant behaviour and biology in weightlessness. In an article in January 2014, *National Geographic* gave five reasons why NASA needs greenhouses on space missions: they can produce fresh food instead of the processed and packaged food that crews eat now; they can have therapeutic value by helping to relieve stress; they can improve the visual appearance of a clinical spacecraft interior; they can regenerate and revitalised the cabin atmosphere; and they can help to mark the passage of time.[48] According to NASA, the problem with installing a greenhouse chamber on the Station is that it requires too much room. The quantity of vegetation needed to clean the air usefully and provide sufficient food is too great for the Station's limited volume in the pressurised modules. Though plant experiments may well increase on the Station, they will not be able to replace the life support system. Greenhouses will have to wait for a Mars base.

The International Space Station will soon break its bonds with Earth in another way. In future, there will be no need to rely on national launch programmes to transport crews to and from the Station. Private space companies are stepping into that role – at least as far as America is concerned – under a NASA initiative called the Commercial Crew Program. Following its highly successful campaign to delegate Station resupply flights to SpaceX and Orbital Sciences, NASA awarded competitive contracts called Space Act Agreements – basically cost-shared agreements in this case – to private ventures to help them to develop and operate a new generation of vehicles to deliver crews to the Station and return them to Earth. In the first round in 2010, NASA chose five companies to develop design ideas. In the second round in 2011 – independent of the first – the agency chose four companies of which three were from the first round: Blue Origin, the Boeing Company, Sierra Nevada Corporation and SpaceX. NASA was looking for vehicles to deliver four astronauts to the Station and return them at least twice a year. Crew safety in an emergency during launch and ascent was of paramount importance. The vehicles would provide a twenty-four-hour safe haven in an emergency in space and would be able to stay docked at the Station for over 200 days. The companies were free to develop their designs while meeting NASA standards and milestones and the agency would provide technical expertise wherever needed.[49]

The Boeing CST-100 crew capsule Boeing in Houston is developing this design for a crew capsule under NASA's Commercial Crew service initiative. In August 2014, NASA selected Boeing and SpaceX to proceed with full development, fabrication and introduction of their crewed vehicles and services to deliver astronauts to the International Space Station and return them to Earth. Designated as the CST-100, the capsule will be capable of transporting a crew of up to seven or a mixture of crew and cargo. The first uncrewed test flight to the Station is due in 2017.

By April 2014 and the third round, Blue Origin had pressed on without waiting for further funding and developed a biconic capsule for launching on either an existing Atlas V rocket or Blue Origin's own reusable rocket that it was developing. Meanwhile, NASA had proceeded with third round agreements with Boeing, Sierra Nevada and SpaceX. Boeing was developing a capsule called the CST-100 using the Atlas V as the launcher. Sierra Nevada had partnered with Lockheed and was working on a lifting body vehicle with winglets called Dream Chaser, also for launch on the Atlas V. SpaceX had based its vehicle on its own Dragon resupply spacecraft and Falcon 9 launcher. NASA's plan was to award contracts in 2014 for the full development and flight certification of the crewed vehicles to provide transportation for astronauts to and from the Station to be ready no later than 2017 which in NASA's own words was 'as soon as possible'. The agency recognised that the biggest threat to the accelerated development timetable was its own ability to handle the rapid design pace. The fast development of space missions was perhaps something NASA had not been accustomed to since President John F Kennedy's commitment in 1961 to send Americans to the Moon by the end of that decade. The agency placed its trust in the competitive process to unlock the best ideas and achieve a productive outcome. NASA likened the Commercial Crew Program to that of the Space Shuttle's early development in the late-1960s. Then, the agency had held a competition between five aerospace companies for the new vehicle, out of which one emerged as the successful contractor.[50]

The importance of the International Space Station and the shift of its crew and resupply duties to the private sector was evident in June 2014 when the National Research Council of the National Academies published its findings of a study on the future of NASA and its human spaceflight programme. They called it *Pathways to Exploration – Rationales and Approaches for a US Program of Human Space Exploration*.[51] The NASA Authorization Act of 2010 had instructed the agency to commission the National Academies to carry out the study and to investigate 'the goals, core capabilities and direction of human spaceflight'. The following passage is the first paragraph of the report's summary:

'The United States has publicly funded its human spaceflight program on a continuous basis for more than a half-century. Today the United States is the major partner in a massive orbital facility – the International Space Station – that is a model for how US leadership can engage nations through soft power and is becoming the focal point for the first tentative steps in commercial cargo and crewed orbital space flights. And yet, a national consensus on the long-term future of human space flight beyond our commitment to [the Station] remains elusive.'

The National Research Council had made a telling point about the lack of a proper vision. There had been no focused plan in the 2013 *Global Exploration Roadmap* initiative, nor in NASA's April 2014 press conference about space exploration. NASA did what the White House told it to do. In 1962, President Kennedy told it to land on the Moon. NASA did. In 1984, President Reagan told it to build a Space Station in Earth orbit. NASA did that, too. Now the White House was silent on committing itself to the next big move in space.

In September 2014, the Commercial Crew Program took a step forward when NASA announced the winners of the Commercial Crew Program competition. They were Boeing and SpaceX. Making the announcement at Kennedy Space Center, NASA Administrator Charles Bolden explained the reasoning. 'From day one, the Obama Administration made clear that the greatest nation on Earth should not be dependent on other nations to get into space. Thanks to the leadership of President Obama, the hard work of our NASA and industry teams, and support from Congress, today we are one step closer to launching our astronauts from US soil on American spacecraft and ending the nation's sole reliance on Russia by 2017. Turning over low-Earth orbit transportation to private industry will also allow NASA to focus on an even more ambitious mission – sending humans to Mars.'

Boeing would get $4.2 billion and SpaceX would get $2.6 billion to finish the job of developing and flying their spacecraft. The contracts provided

The Boeing CST-100 crew capsule
Top: A mock-up of the interior of the CST-100 crew capsule. Two body restraints (seats) are horizontally positioned with head to the left and feet to the right. The restraints provide support for astronauts during launch and ascent when the capsule is subject to increased g-forces. This mock-up is being used to evaluate crew safety, interfaces, communications, manoeuvrability and ergonomics. Bottom: Boeing's structural test prototype of its CST-100 crew capsule.

for at least one crewed flight test per company with at least one American astronaut on board to verify that the rocket and spacecraft would perform properly through launch, ascent, rendezvous and docking with the Space Station. After the test phases, NASA would certify the vehicles and Boeing and SpaceX would conduct between two and six crewed missions to the Station with the spacecraft serving as lifeboats when there. Both companies would own and operate their vehicles and be able to sell flights to customers beyond NASA with the aim of reducing costs for all. The target date for test flights was 2017.[52]

Also in September 2014, NASA addressed the question of the International Space Station's life cycle. The Station had completed fifteen years of operation by November 2013 and it had reached the end of its lifespan as originally designed and tested during its development years. Two months later, NASA announced its intention to extend Station operations to 2024 while a further extension until 2028 was also under review. Its overall lifespan would certainly extend to twenty-six years with the aim of drawing as much research value as possible from the huge investment that America had made. According to a new analysis by NASA's Office of Inspector General, that investment now stood at $74.4 billion. It covered the Station's development, operations, research and transportation up to 2013. $30.7 billion of it was for thirty-seven Space Shuttle assembly and operations flights that the General Accounting Office had previously costed at $759 million each. The Office of Inspector General estimated that the Station's maintenance and operations would cost $3-$4 billion per year going forward until 2024. This would raise the Station's total life-cycle costs (excluding decommissioning) to America up to 2024 to $105-$115 billion.[53]

Appendix 4 shows the $74.4 billion figure in the context of the Station's changing development and fabrication costs from its go-ahead in 1984 up to the start of its assembly on orbit.

The Office of Inspector General also commented on the implications of extending the Station's life until 2024. In its investigation, the Office found that, though no major obstacles existed, there were several areas of concern that affected the Station's future use. To begin with, there was the degradation of the Station's giant solar arrays, which was proving to be faster than expected. This would reduce their electrical power generation output. The Station might need some extra solar arrays. There had also been sudden failures of various important hardware items that had needed unplanned spacewalks to repair. Delivering large replacement parts to the Station such as solar arrays or radiators was now a problem as no space vehicles with a large cargo capacity had existed since the Shuttle's retirement. Anticipated lack of future space parts was also an issue. There was particular concern about replacement pumps for the thermal control system because they had already failed twice. Another worry was the batteries that provided the Station with electricity during the night portion of each orbit. These would need replacing and would rely on the Japanese H-II Transfer Vehicle to deliver them, a space vehicle that Japan was reluctant to keep in service beyond 2020.

The Inspector General's report also stated that NASA's cost projections for future Station operations appeared to be optimistic. Though the agency had estimated the annual budget for this at $3-$4 billion per year, the Inspector General's office thought it would be higher. NASA estimated it would be paying the Russians to fly American astronauts to the Station on Russia's Soyuz spacecraft at over $70 million a seat from 2016 onwards, but the agency anticipated higher seat costs with the privately developed crew vehicles of Boeing and SpaceX when they entered service in 2017. This led to another concern. With the Space Shuttle gone, Russia's Progress spacecraft and Europe's Automated Transfer Vehicle and their propulsion systems had been responsible for periodically reboosting the Station's orbit. In July 2014, the European vehicle had carried out its last mission. The Station's older Zarya module also had a propulsion system but its engines were no longer usable. A question mark, therefore, now hung over the future means of maintaining the Station's orbital altitude reliably.

NASA's international partners had not yet committed themselves to supporting Station operations beyond 2020. If they decided not to do so, the remaining partners would certainly face higher costs. Though Russia had not highlighted major hardware lifetime extension issues and Japan and Europe had cleared all their hardware until 2028 with adequate spare parts if necessary, there were no guarantees that any of them would remain involved beyond 2020.

There was also the matter of the Station's scientific research. The Inspector General's inquiry had found that crew weekly time spent on experiments, the number of different experiments and the use of allocated volume to them had all continued to grow in recent years. This was good news. The Station's focus on biomedical research and the study of the health of humans in weightlessness had highlighted twenty-three types of risk associated with long-duration spaceflight that required lengthy study over several years. They covered decompression sickness, reduced muscle mass, fatigue-induced mistakes and cardiac rhythm problems. However, the full evaluation of twelve of these would require use of the Station until 2024 and it would only be possible to complete five if the Station ceased operations in 2020. There remained eleven risks that would need study beyond 2024. These covered human-computer interaction, inadequate food systems, errors due to training deficiencies, launch and landing injuries to crews, early osteoporosis, altered immune response, bone fracture, medication side effects, vestibular/sensorimotor problems, behavioural disorders and in-flight medical limitations. There were also two areas that Station research would be unable to cover adequately: the functional fusion of human with automated and robotic capabilities and the long-term effects of radiation on human health.[54] There was a lot of extremely important research for the Station still to do in the years ahead.

In its report, however, NASA's Office of Inspector General did not mention past problems with the Station's attitude control system (not the same as the Station's orbital reboost system) in its list of mechanical and electrical failures and other hazards. This system is vital to the Station's orientation as it orbits the Earth. The job of controlling the Station's attitude goes to the four control moment gyroscopes. About the size of golf carts and shaped like thick circular discs, they are large mechanisms comprising rotors spinning at high speed to generate angular momentum. Mounting the rotors on dual gimbals like a traditional ship's compass allows them to tilt in different planes. As the rotors tilt, they exchange some of their momentum with the Station through gyroscopic torque, causing the Station to move. Each gyroscope weighs about 272kg and spins at a constant speed of 6,600 revolutions per minute. The Station needs at least two of them working together to control its attitude as it orbits the Earth. Four of them arrived at the Station on the Z-1 building block on STS-92 in 2000 and were activated in 2001. A year later, one of them had failed, leaving three to handle the Station's attitude until the arrival of a replacement on a Shuttle mission in August 2005. Then another one was shut down in October 2006, leaving the Station operating with three again until a Shuttle mission delivered a replacement in August 2007.[55] Further failures of the gyroscopes would be a serious setback due to the difficulty of delivering such heavy and sensitive replacements on the launch vehicles currently available.

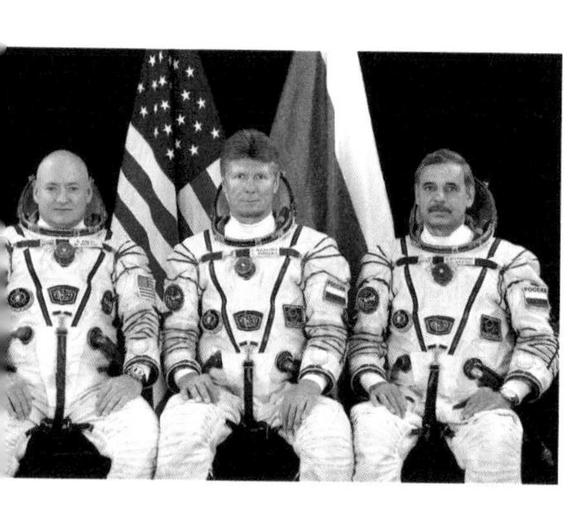

The One-Year Mission & Twins Study Astronaut Scott Kelly (left) and cosmonauts Gennady Padalka (centre) and Mikhail Kornienko (right) were the prime crew members on Expedition 43 to the Station, which launched in March 2015. Scott Kelly and Mikhail Kornienko were participating in the 'One-Year Mission.' Scott Kelly was also participating in the 'Twins Study' with his twin brother, retired astronaut Mark Kelly, who remained on the ground. The research focus of both the One-Year Mission and the Twins Study was on the effects on the human body of spending one year of weightlessness. Image: Roscosmos/GCTC.

It is not often that a popular weekly newspaper devotes a major feature to spaceflight, but in its issue on 29 December 2014, *Time* magazine ran a cover story on astronaut Scott Kelly and the one-year tour he was soon to begin on the International Space Station.[56] Kelly would spend one year on the Station with cosmonaut Mikhail Kornienko on a mission set for launch on a Soyuz spacecraft from Baikonur Cosmodrome in March 2015 with cosmonaut Gennady Padalka joining them on the flight up. What made this particular mission unique was that Scott Kelly had a twin brother Mark, who was a retired astronaut, and who would participate in the mission's research. With one twin in space, the other on the ground, both with the same age and genetics, and both with very similar health and physiology, NASA would take advantage of an exceptional opportunity to study and test the physiological and behavioural effects of weightlessness on Scott Kelly over the year and compare them with those of Mark Kelly who would stay in normal gravity but undergo the same tests. NASA called it the 'One-Year Mission & Twins Study'.[57] There were two parts to it. The One-Year Mission with Scott Kelly and Mikhail Kornienko would span seven major areas of research. These would cover the human performance of tasks, behavioural and psychological health, ocular health and visual impairment, metabolism and immunology, exercise and physical health, microbiology and human factors. The Twins Study with Scott and Mark Kelly would involve ten more investigations. These would examine the effects of weightlessness on major organs such as the heart, muscles and brain, and on their perception, reasoning and alertness. They would compare the twins' dietary differences and digestive system microbiology. They would look in detail at such things as the twins' genetics, biochemical profiles and their reactions to influenza vaccinations.

Gennady Padalka, Scott Kelly and Mikhail Kornienko arrived at the Station on 27 March 2015 after a perfect launch from Baikonur. Scott Kelly commented that 'one of the differences here is that we're doing it as an international partnership, and if we're going to go beyond low-Earth orbit again, perhaps to Mars, because of the cost and the complexity it will most likely be an international mission, so we see this as a stepping stone to that'. He had used the word 'international' twice in the same sentence.

Human exploration of Mars is going to be difficult and challenging. It will only become possible through another shared vision and commitment by the world's spacefaring nations of the type that made the International Space Station possible. No single nation will be able to do it alone.

In October 2015, NASA published a document called 'Journey to Mars – Pioneering Next Steps in Space Exploration'[58] that summarised the agency's current thinking about the exploration of the planet. Unlike the stack of complicated exploration plans published since the 1980s that amounted to little more than stuff that dreams are made of, 'Journey to Mars' was different. It was cautious and pragmatic. The path forward had three clear stages. First was 'Earth Reliant' with its focus on the International Space Station for long-duration research and commercial spacecraft for crew and cargo transportation. Second was 'Proving Ground' at lunar distance with its aim to test out the technologies and operations needed to show that crewed missions to Mars were feasible and safe. Third was 'Earth Independent' with human and robotic missions to the planet itself that would break the bonds with Earth because, so far away, they would have to survive without support. The message of 'Journey to Mars' was that the further away from Earth that we travel as a species, the more independent and resourceful we must be. Playing a prominent role in this new vision was the International Space Station. NASA pointed out that the Station was the 'only microgravity platform for the long-term testing of new life support and crew health systems, advanced habitat modules, and other technologies needed to decrease reliance on Earth'. In one sentence, NASA had spelled out the vital importance of the Station to our next leap across the Solar System.

The Twins Study Astronaut Scott Kelly (left) and his twin brother, retired astronaut Mark Kelly (right), participated in the 'Twins Study' mission, which began in March 2015. Scott Kelly embarked on a year in weightlessness on the International Space Station while Mark stayed on the ground. Both would be subject to a battery of medical tests which, due to the twins' identical genetic and very similar physical characteristics, would shed light on the effects of long-term weightlessness and how it differs from gravity.

Canada

The Special Purpose Dexterous Manipulator
The mechanical manipulator attached to the end of the Canadarm2 robotic arm is called Dextre. It is vital for Station construction and maintenance activities. With advanced stabilisation and precise handling capabilities, Dextre can remove and replace very small components. It has lights, four cameras, a stowage platform, and can operate three robotic tools. Built by MDA Space Systems in Ontario, Canada, Dextre has two arms and is an evolution of the technology used in the Canadarm2. Guiding Dextre's high accuracy and gentle touch is its precise sensing of forces and torques in its grip. Dextre can pivot at its waist while its shoulders support two identical arms of seven joints each. This results in considerable freedom of movement. Mounting Dextre is possible either on the end of Canadarm2 or on the Mobile Base System (the railcar that runs along the transverse truss).

Truss structure This photograph of the Station's P6 Truss was taken during a spacewalk on 6 November 2015. It shows the complexity of the cables and fluid lines inside the structural frame. Cable in-line connections, such as those in the foreground, are designed to connect or disconnect easily by a gloved spacesuit hand during a spacewalk.

ELC1

Goodbye from the Space Shuttle The International Space Station seen from the Orbiter Atlantis after it had departed the Station on 19 July 2011. It was the final Space Shuttle mission to the Station and this was one of the very last photographs taken of the Station from the Space Shuttle.

Appendices

Appendix 1 System engineering flow diagram for a space project

Pre-Phase A
Advanced Studies

Mission Feasibility

- Define mission objectives and top level functional and performance requirements
- Ensure mission technical and programmatic feasibility
- Confirm customer's mission need

Phase A
Preliminary Analysis

Mission Definition

- Establish validated (segment level) requirements which meet mission objectives
- Establish architectural and top level operations concept
- Identify technology risks and mitigation plan
- Refine programmatic resource need estimates

Phase B
Definition

System Definition

- Establish validated requirements for end items (full functional baseline)
- Complete 'architecture' design
- Mitigate technical risk: critical technology, long lead items
- Set firm estimates of programmatic and technical resources
- Mitigate programmatic risks
- Show system can be built to cost, schedule, performance constraints

Preliminary Design

- Establish a design solution that fully meets mission needs
- Complete test and verification plan
- Establish design dependent requirements and interfaces
- Complete 'implementation' level of design

Phase C
Design

Final Design

- Establish complete, validated detailed design
- Complete design speciality audits
- Establish manufacturing processes
- and controls
- Finalise and integrate interfaces

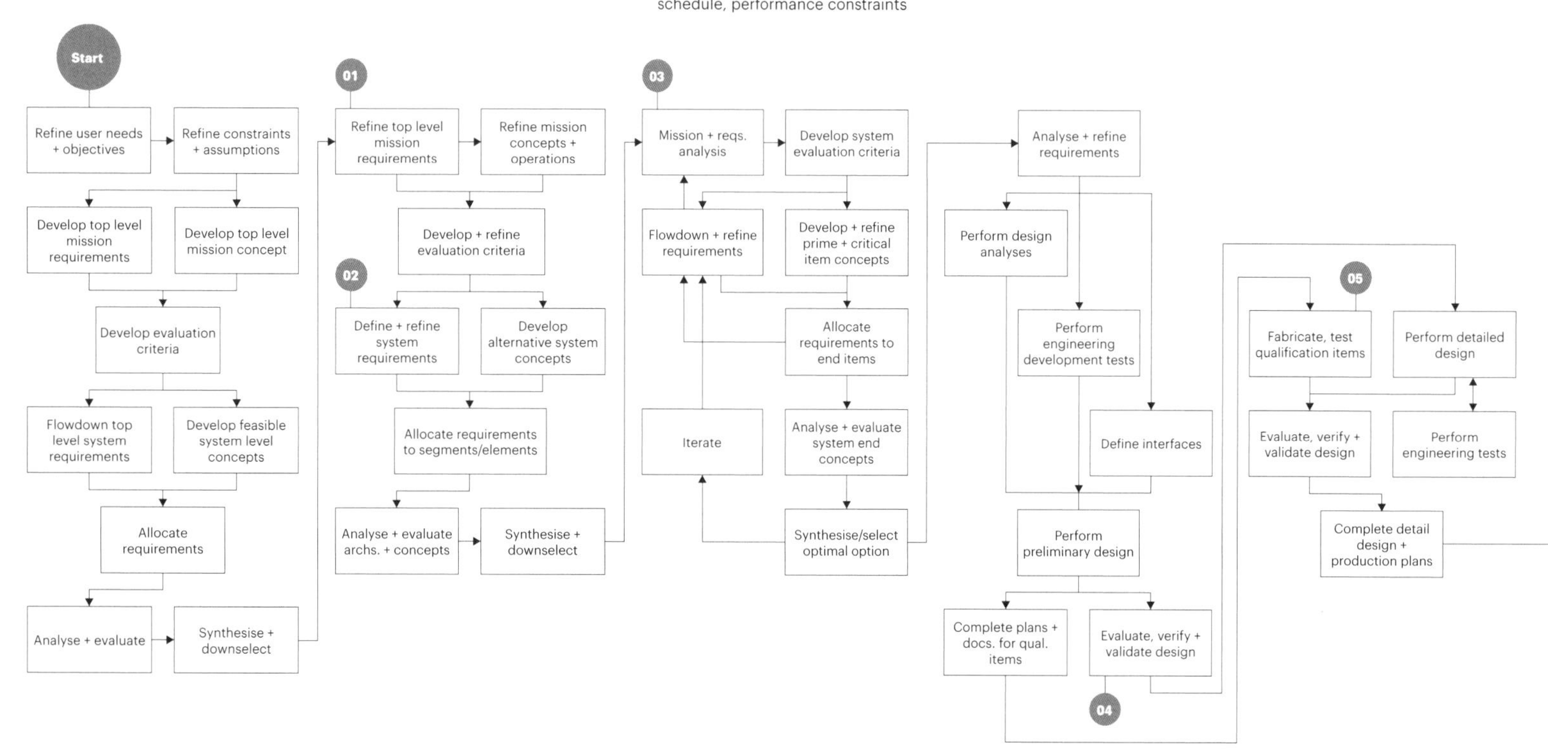

The development, fabrication and launch of a space vehicle or space station (in whole or in part) utilises 'systems engineering'. NASA defines this as a 'methodical, disciplined approach for the design, realization, technical management, operations and retirement of a system.' The 'system' can include people, hardware, software, facilities, supplies, policies and documents – everything that has a role in achieving the goal. In NASA's words, it is: 'a way of looking at the "big picture" when making technical decisions'. This diagram shows a typical flow sequence for a space project. It begins with outline design concepts and mission feasibility in pre-Phase A, Advanced Studies and ends with hardware disposal after mission completion in Phase E, Operations. Intermediate phases cover preliminary and detailed design, fabrication, testing, qualification, integration, transportation, pre-launch processing, launch and mission operations. Acting as milestones in the flow sequence are a series of major reviews that focus on, evaluate and approve the project's progress throughout the Phases.

Sources: (i) National Aeronautics and Space Administration, "NASA Systems Engineering Handbook," SP-6105, NASA Headquarters, June 1995; (ii) National Aeronautics and Space Administration, "NASA Systems Engineering Handbook," SP-2007-6105, Rev. 1, NASA Headquarters, December 2007. Image: adapted from NASA charts and diagrams by the author.

01 Mission Concept Review
02 System Requirements Review
03 System Definition Review
04 Preliminary Design Review
05 Critical Design Review
06 System Integration Review
07 Test Readiness Review
08 System Acceptance Review
09 Flight Readiness Review
10 Post-Launch Assessment Review
11 Critical Readiness Review

Phase D
Development

Fabrication & Integration

- Produce items that conform to specifications and acceptance criteria
- Assemble and integrate the system
- Verify and validate the system
- Develop the capability to use the system to perform a mission
- Prepare facilities for production, maintenance and operation

Preparation For Deployment

- Configure system for launch and deployment
- Establish readiness to launch and deploy

Deployment & Verification

- Launch and deploy system
- Establish operational envelope of system
- Establish system logistics

Phase E
Operations

Mission Operations

- Carry out space mission
- Sustain mission system
- Improve/augment mission system

Disposal

- Decommission and dispose of mission system items

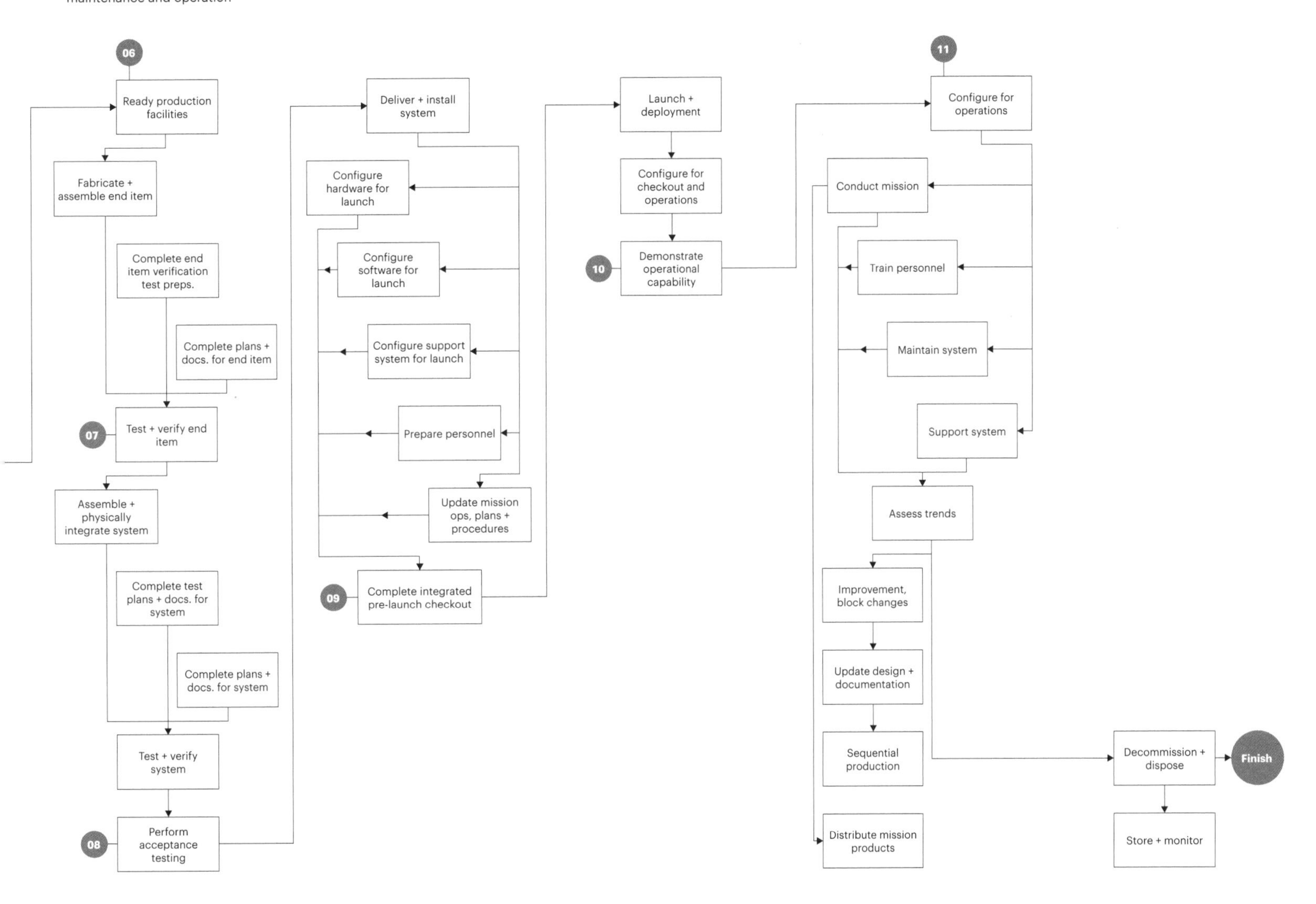

Appendix 2
Evolution of crewed space vehicle pressurised volumes

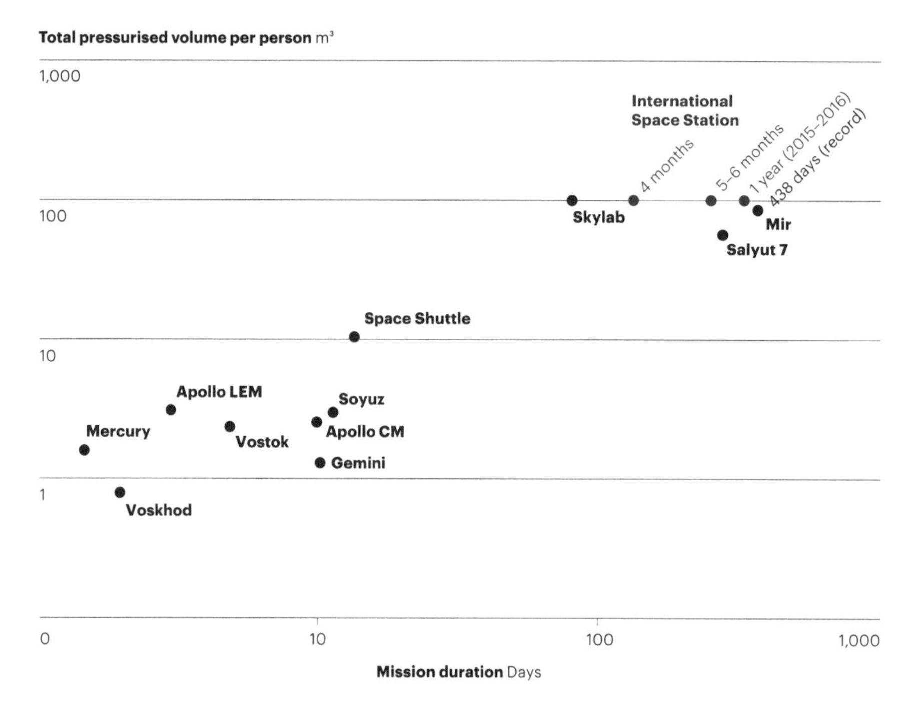

This graph shows the relationship between pressurised volumes inside space vehicles, outposts or stations and the length of their missions. Volumes have increased steadily with mission lengths. The Soviet Voskhod capsules of 1964 and 1965 had just under 1m³ of volume for their one cosmonaut on missions that lasted about one day each. The Skylab orbital outpost of the early 1970s provided its three-person crews with about 100m³ per astronaut on three missions that lasted twenty-eight, fifty-nine and eighty-four days respectively. NASA estimates that the International Space Station has 916m³ of pressurised volume and 388m³ of habitable volume after the deduction of room occupied by racks and other equipment. Each member of the six-person crew therefore has 153m³ of pressurised volume and 65m³ of habitable volume. Though the first International Space Station crew – Expedition 1 – spent nearly five months on board, early Expedition tours generally lasted around four months and increased to between five and six months by 2014. A one-year mission by astronaut Scott Kelly and cosmonaut Mikhail Kornienko began in March 2015. The record for the longest space mission is presently held by cosmonaut Valeri Polyakov who spent 438 days on the Mir Space Station.

Source: NASA, 'Reference Mission Version 3.0. Addendum to the Human Exploration of Mars: the Reference Mission of the NASA Mars Exploration Study Team', EX13-98-036, Exploration Office, Advanced Development Office, Lyndon B Johnson Space Center, June 1998. Image: adapted from a NASA graph with additions by the author.

Appendix 3

Fluctuating crew sizes during Space Station design and development phases

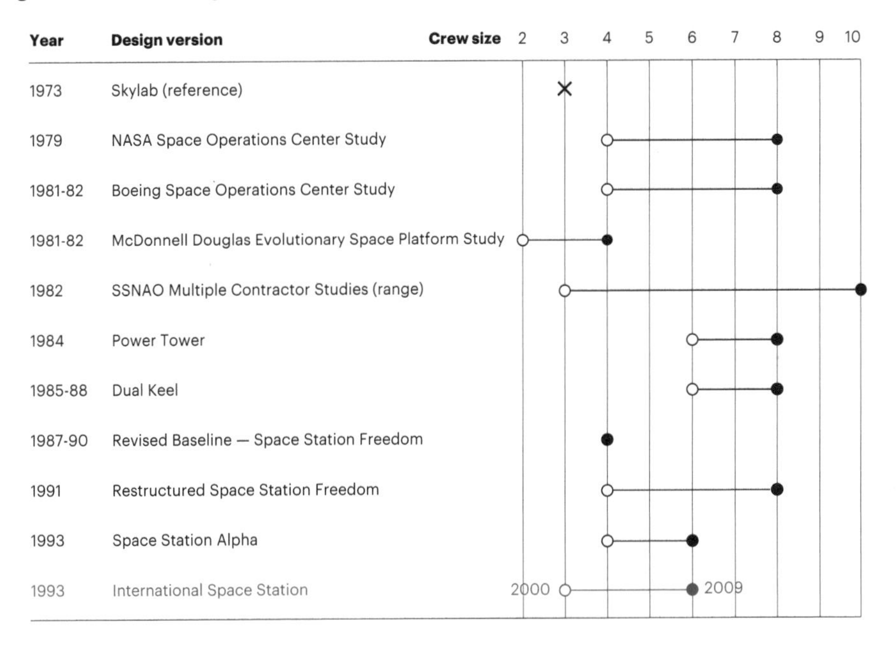

This chart shows changes to the anticipated sizes of Space Station crews during the Station's design and development cycle. Early concept studies by NASA and its aerospace contractors up to 1984 assumed a wide range of crew sizes before the Station's requirements were fully defined. The Power Tower and Dual Keel designs of the mid-1980s were based on initial six-person crews and final eight-person crews. From then on, up to 1993, initial crews were reduced to four as the Station's budget difficulties resulted in reduced intermediate Station facilities capable of supporting permanent crews. International Space Station initial crews were then reduced to three as this was the maximum number that a docked Soyuz spacecraft could seat, once that vehicle became the Station's means of emergency evacuation. With two Soyuz always docked at the Station after 2009, the crew size increased to six.

Sources: multiple NASA and contractor reports, studies and descriptions referenced in the main text.
Image: author.

Appendix 4

Evolution of development and fabrication costs for the American sector of the Space Station

This graph shows the rising design, development and fabrication costs in current (then) year dollars of the Space Station up to and including the International Space Station. NASA envisaged the hardware cost of the Station at the time of its go-ahead in 1984 at $8 billion. This was little more than an educated guess as the Station was still at the concept stage. As the Station's design, engineering, launch and operations became defined throughout the late-1980s, the costs of the American sector began to exceed budget allocations, sometimes substantially. At the time that Russia joined the project as a partner in 1993, the development and fabrication costs for the American sector stood at $19.4 billon in current year dollars. NASA claimed that bringing Russia into the project would reduce that figure to $17.4 billion, a claim disputed by the Government Accountability Office.

Total project costs that included transportation and operations were difficult to determine reliably during the Station's design and development. They depended on fluctuating factors such as the cost and number of Space Shuttle assembly and resupply missions, and the cost of the Station's operations up to and beyond its completion to some unspecified date. There was also a lack of cost information on the contributions of America's international partners. For this reason, changing total project costs do not appear on this graph. However, NASA's Office of Inspector General estimated in 2014 that the International Space Station's total project costs to the United States covering design, development, fabrication, launch, assembly and operations up to 2013 came to $74.4 billion in current year dollars. The graph includes this figure as a benchmark. The Office of Inspector General estimates that the Station's maintenance and operations will cost $3 billion to $4 billion per year going forward until 2024, the year beyond which there are no further funding commitments at the time of writing. On this basis, the International Space Station's total life-cycle costs (excluding decommissioning and disposal) up to 2024 will be in the range of $105 billion to $115 billion to the United States.

Sources: (i) United States Government Accountability Office, 'Space Station. NASA's Search for Design, Cost, and Schedule Stability Guidelines', Report to the Chairman, Committee on Science, Space, and Technology, House of Representatives, GAO/NSIAD-91-125, March 1991; (ii) Space Station Redesign Team, 'Final Report to the Advisory Committee on the Redesign of the Space Station', National Aeronautics and Space Administration, June 1993; (iii) United States Government Accountability Office, 'Space Station. Impact of the Expanded Russian Role on Funding and Research', Report to the Ranking Minority Member, Subcommittee on Oversight of Government Management, Committee on Governmental Affairs, US Senate, GAO/NSIAD-94-220, June 1994; (iv) Smith, MS, 'NASA's Space Station Program: Evolution of its Rationale and Expected Uses', Testimony before the Subcommittee on Science and Space, Committee on Commerce, Science and Transportation, US Senate, 20 April 2005; (v) Office of Audits, Office of Inspector General, 'Extending the Operational Life of the International Space Station Until 2024', NASA, IG-14-031 (A-13-021-00), 18 September 2014. Image: author.

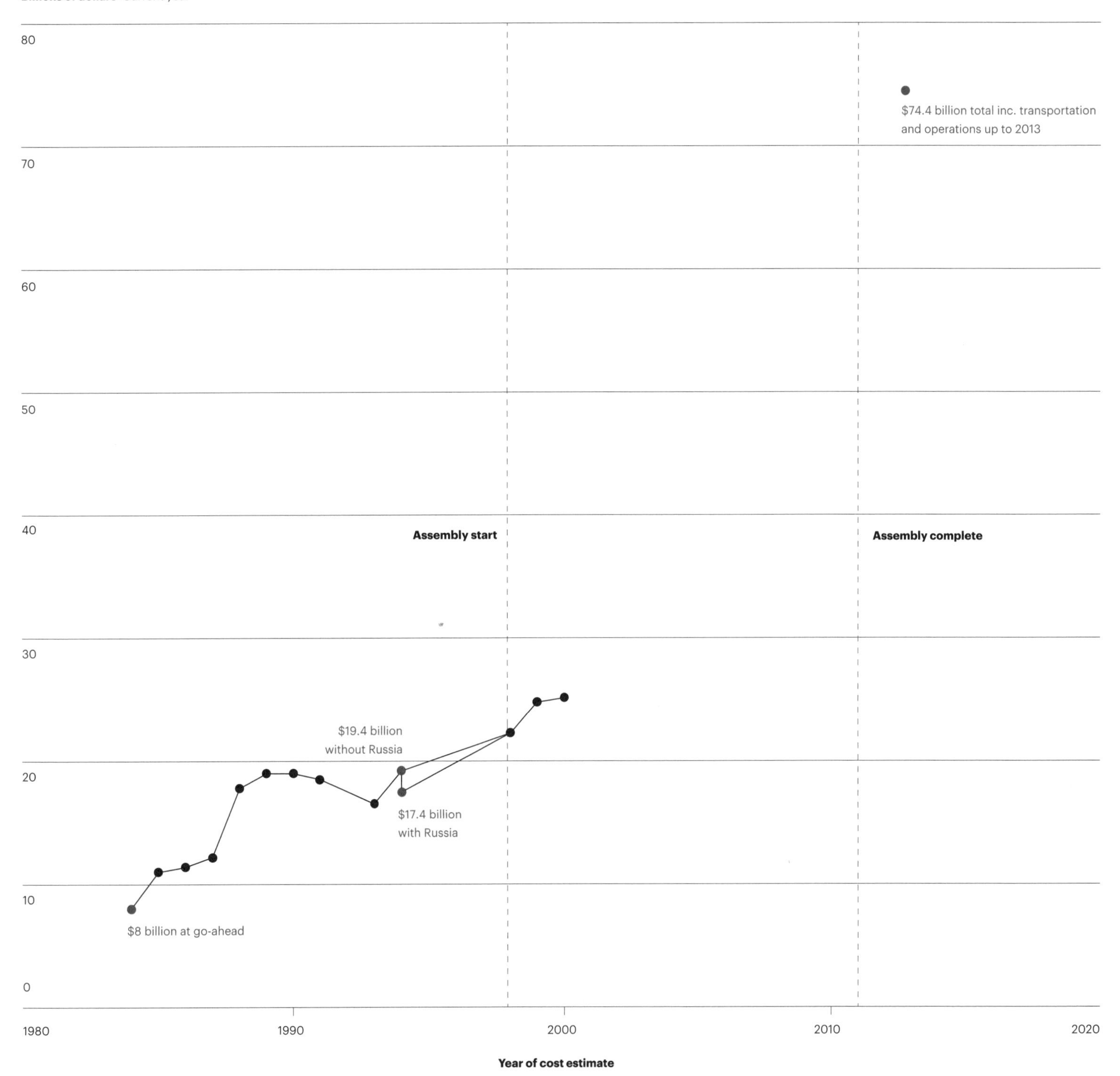
Billions of dollars Current year
80
70
60
50
40
30
20
10
0
$74.4 billion total inc. transportation
and operations up to 2013
Assembly start
Assembly complete
$19.4 billion
without Russia
$17.4 billion
with Russia
$8 billion at go-ahead
1980
1990
2000
2010
2020
Year of cost estimate

Glossary

Accelerometer Instrument for measuring acceleration.
Actuator Mechanism for activating physical motion, usually over a limited range.
Aft Towards the back of a ship, aircraft or spacecraft.
Age-hardening Time required for a metal or alloy to harden after heat treatment.
Airlock Compartment enabling passage between environments of different ambient pressure.
Androgynous Separate mating mechanisms of a docking system that are physically identical and compatible.
Annealing Heat treating metals and alloys to alter their physical properties.
Anthropometrics Scientific evaluation of the dimensions and measurement of the human body.
Antimatter Matter composed of antiparticles.
Antiparticle Subatomic particle with opposite electric charge to a given particle but with the same mass.
Antiproton Proton antiparticle with a negative charge.
Artificial gravity Gravity provided by artificial means such as a rotating space station.
Astrophysics Physics of the observed and sensed universe.
Attitude control Means of control and adjustment of a spacecraft's orientation with regard to its flight path, often using thruster jets.
Avionics Electronic systems used in a spacecraft or aircraft.

Berthing Mechanical linking of a spacecraft or payload to another spacecraft or space station using a manipulating device such as a robotic arm.
Biconic Composed of two joined cones, usually truncated and with different geometries.
Bifurcated Partly split into two halves with a gap between, forked.
Bioassay Measurement of the effects of a substance on living tissue or cells.
Biodegradation Chemical breakdown of materials by biological means.
Bipropellant engine Rocket engine that uses two types of propellant, such as liquid oxygen and liquid hydrogen.
Bone demineralisation Loss of calcium in bones leading to decrease in bone mass and density and increased fracture risk.
Boson Type of subatomic particle.
Building block Large prefabricated component of a space station, such as a module or piece of structure, delivered to orbit by a space vehicle.
Bulkhead Major transverse structural component in a pressurised hull or fuselage.
Bumper panels Panels that sheath the exterior of spacecraft or a space module for shield protection against micrometeoroid or debris penetration.
Bus Universal payload carrier or subsystems spine.
Busbar Major electrical conductor linking power sources such as solar arrays or batteries to a power distribution system.

Cantilever Structural member fixed at one end with the other unsupported end projecting outwards.
Centre of mass Point through which all the masses of a solid object act without inducing motion in the object.
Centrifugal force Outward force that acts on a body moving around a central point.
Centrifuge Device for spinning experiments, materials and organisms (including humans) about a central axis.
Clean room Laboratory or workroom with a mechanical air supply from which airborne organisms, pollutants and particles have been removed.
Close-out Act of sealing and readying a space vehicle before launch, often performed by a ground crew as the final task.
Coldplate Metal plate with internal tubing carrying a coolant to remove heat from experiments or equipment mounted on it.
Control moment gyroscope Mechanical disc or rotor spinning at high speed on gimbals that tilt in order to impart changing angular momentum to a spacecraft's attitude.
Crossbar Horizontal bar supported at its midpoint or at both ends.
Cryogenic Operatiing at very low temperatures.
Cupola Projecting dome, turret or observatory.
Cyclotron Device for accelerating charged particles in an electric field while following a circular path in a magnetic field.

Dark matter Hypothetical, non-visible matter that accounts for the majority of the matter in the Universe.
Deployable Capable of being mechanically extended.
Dipole Twin oppositely and equally charged or magnetized poles placed some distance apart.
Docking Mechanical linking of a spacecraft to another spacecraft or space station under automatic or crewed command and control.
Double-fold Foldable or unfoldable in one direction and then another direction.

Electrolysis Chemical decomposition of a substance caused by an electric current, such as water into oxygen and hydrogen.
Electromagnetism Force field that combines electric and magnetic force fields.
Electron Subatomic particle with a negative electrical charge.
Electrophoresis Separation of different substances under the influence of an electric field.
Endcap Part of a cylindrical pressurised module that closes off its end, often with a conical shape and central hatch.
Environmental control and life support system Physical means of providing environment to sustain human life inside a spacecraft or space station.
Erectable Capable of being assembled manually or by a mechanical device or a combination of both.
Ergonomics Scientific evaluation of the efficiency of humans in the working environment, including human-machine interfaces.
Extrusion Extension of a fixed geometrical shape or formed mass of material in one direction.

Fairing Streamlined outer enclosure of a rocket stage, such as a payload compartment.
Flight path Direction or course that an aircraft or spacecraft is travelling in or on.
Flightworthiness State of readiness for flight.
Free-flyer Space station or platform that orbits autonomously in formation with or separate from another space station or platform.

Galley Kitchen on a ship, yacht or space station.
Geostationary orbit Equatorial orbit on which an object constantly remains above the same point on the Earth's surface as the Earth revolves.
Gimbal Mountings using pivots or bearings on two or three intersecting axes to enable a gyroscopic device to spin freely.
Glovebox Enclosed chamber with sealed gloves on one side for handling hazardous material or specimens.
Gluon Type of subatomic particle.
Grapple Grasp and hold on to using a mechanical device, such as a robotic arm.
Gravity gradient Small and quantifiable variation in a gravitational force as one moves away from the centre of mass of a spacecraft.

Hadron Type of subatomic particle.
Hardpoint Major load-bearing anchoring or mounting point in an aircraft or spacecraft structure.
Harnessing Strapping of cables or fluid lines, sometimes in bundles, and their attachment to structure through which they pass.
Heat exchanger Device for the transfer of heat between different media (e.g. liquid to gas).
Hematology Study and treatment of disorders related to the blood.
Human factors Characteristics of humans that influence the design of mechanisms and systems which they control or operate.
Humanoid Having the characteristics of or resembling a human being.
Hydrogenation Combine or charge with hydrogen.
Hyperbaric Chamber atmospheric pressure or oxygen ratio higher than Earth's atmosphere.
Hypotenuse Diagonal side of a right-angled triangle opposite the right angle.
Immunology Study of the immune system in organisms.
In-situ resource utilization Use of resources at their original location.
Inertia Resistance of an object to change of its motion, direction or attitude when acted on by an external force.
Infrastructure Physical foundation of a nation, city or community comprising roads, railways, harbours, bridges, water supply, drainage systems, telecommunications systems, etc.
Integration Assembly of components and systems of spacecraft, space vehicles, satellites and other space elements before launch.
Ion Atom that has lost or gained one or more electrons.
Ionization Conversion of atoms into ions.
Ionizing radiation Radiation having energy capable of ionizing particles in a material that it penetrates.

Jig Rigid framework for the precise assembly of a component or part.

Lateral Running across from side to side, normally at right angles to the longitudinal direction.
Life sciences Branch of the sciences dealing with the study of living organisms, both flora and fauna.
Life support system (i) Physical means of providing environment to sustain human life inside a spacecraft or space station; (ii) Shortened version of Environmental Control & Life Support System.
Logistics Organization of the transportation, staging and delivery of payloads, equipment and supplies.
Longeron Longitudinal structural member.
Longitudinal Running lengthwise or in the longer dimension.
Low Earth orbit Orbit around the Earth with an altitude ranging from about 160km to 2,000km.

Mandrel Rotating shaft to which equipment, tools or workpieces are attached.
Manipulator Mechanical device for handling small components, experiments or parts, usually mounted on a robotic arm.

Megastructure Complex built structure of very large proportions and often using many small components.
Metabolism Chemical processes that help to maintain life in an organism such as the human body.
Microbiology Study of microscopic organisms.
Microgravity Extremely weak gravity that is approaching zero gravity or weightlessness.
Micrometeoroid Very small particle of rock broken off from a larger rock and traveling in space at very high velocity.
Module (i) Space station building block comprising a pressurized, habitable hull; (ii) a spacecraft, space vehicle or part thereof.
Muon Type of subatomic particle.

Nadir Downwards, towards the centre of the Earth.
Navigation Method of determining the position and course of an aircraft, spacecraft, ship, satellite or vehicle.
Neurosurgery Surgery performed on the human nervous system and the brain.
Neutral buoyancy State of an object or person ballasted to neither sink nor swim when submerged in water.
Neutron Subatomic particle with no electrical charge.
Node (i) Point of convergence and intersection of different structural elements; (ii) a small pressurized module to which larger modules are berthed or docked.

Orbital plane Plane defined and circumscribed by the orbit of a body around another body.
Orbital transfer vehicle Spacecraft for transporting payloads between different Earth orbits or different places on the same orbit.
Orthogonal Based on right angles.
Outgassing Emitting a gas or vapour.

Pallet Standardized container or platform for transporting equipment, experiments or supplies in a payload bay.
Pantograph Articulated framework capable of extending outwards in one direction and contracting inwards in the opposite direction.
Payload bay Compartment for transporting cargo in a spacecraft or aircraft.
Photosynthesis Process in plants and small organisms that uses energy from light to convert carbon dioxide and water into oxygen.
Pitch Rotation about a lateral axis (e.g. from port to starboard) of an aircraft, spacecraft or ship; 'Y' axis.
Platform Deck, structure or similar framework used for mounting equipment and carrying out operations.
Port (i) Left direction or side; (ii) Base, destination or home for a ship, aircraft or spacecraft.
Positron Type of subatomic particle.
Pre-breathing Respiration in pure oxygen to purge the bloodstream of nitrogen before a spacewalk.
Propulsion Pushing or driving forward an object such as a spacecraft.
Proton Subatomic particle with a positive electrical charge.

Quadrupole Two equal dipoles configured together as a unit.
Quark Type of subatomic particle.

Rack Structural framework or compartment for accommodating crew amenities, experiments or equipment inside a module.
Reboost Propulsive power needed to increase the orbital altitude of a spacecraft or space station.
Remote sensing Surveillance of the Earth or an astronomical body such as a planet or a moon at a distance by a spacecraft or observatory.
Reverse osmosis Process for purifying water by forcing it through a semipermeable membrane to remove particles.
Revitalization Regeneration of ambient atmosphere in a spacecraft space station by removal of carbon dioxide, temperature and humidity.
Robotic arm Limbed and jointed mechanical device for handling and manipulating objects such as modules, experiments or spare parts and remotely operated from a workstation.
Roll Rotation about a longitudinal axis (e.g. from forward to aft) of an aircraft, spacecraft or ship; 'X' axis.

Sabatier process Chemical reaction of carbon dioxide with hydrogen to produce water and methane.
Safe haven Refuge in a spacecraft or space station for crew members to occupy in an emergency.
Satellite (i) Fabricated device orbiting a planet or moon for purposes of telecommunications, navigation, observation or sensing; (ii) Natural astronomical object such as a moon orbiting around a planet.
Shirtsleeve environment Spacecraft or module with an interior environment fit for a crew to live and work in wearing normal clothing.
Single-fold Foldable or unfoldable in one direction.
Skin-stringer panels Structural combination of the skin of an aircraft fuselage or spacecraft hull with stiffeners such as thin longerons to form panels.
Sleep restraint Bed for use in weightlessness.
Solar array Folded or rolled solar cell assembly that deploys on a telescopic mast or similar articulated structure.
Solar dynamic collector Device for generating power from the sun using a reflecting dish to focus sunlight on an electrical generator.
Space station Occupied base, port, habitat, laboratory or combination thereof orbiting a planet or moon or located in deep space.
Spaceplane Spacecraft having the features of a winged vehicle with lift and control abilities in an atmosphere.
Spacewalk (i) Astronaut or cosmonaut wearing a spacesuit in the act of moving about or carrying out a task in space; (ii) Also known as Extra-Vehicular Activity (EVA).
Spin-stabilized Possessing gyroscopic attitude stability due to rotation.
Staging (i) Separation of one rocket stage or part from another; (ii) Operational activity area providing temporary occupancy of crews, supplies or spacecraft during passage from one place to another.
Stand-off Framework that acts as a structural intermediary between a pressurized hull and internal racks and compartments.
Starboard Right direction or side.
Stationkeeping Ability of a space station, space vehicle or satellite to maintain a prescribed orbit.
Strain gauge Instrument to measure the structural strain on an object.
Strongback Structural member that provides rigidity in bending and torsion.
Strut Long and often thin structural component that works in compression or tension.
Sun synchronous orbit Orbit that follows the sun's path so that that the observed sunlight angle on the Earth's surface is consistent.
Superconductor Material with zero electrical resistance.
Surveillance Close or distant methodical observation of an object or person using a variety of sensing devices.
Synchrotron Cyclic particle accelerator in which the guiding magnetic field is synchronized to the particle beam and increased with time to give increasing kinetic energy to the particles

Telecommunications Communications over distance by radio, television, telephone, telegraph or cable.
Telemedicine Application of information and communication technology to provide medical care to remote locations.
Teleoperation Operation of a device or mechanism from a distance.
Tetrahedral Having four sides.
Thermal radiator Radiator panel for discarding heat to space generated from inside a spacecraft or space station.
Torque Force imparting rotational motion to an object.
Translation Movement from one position to another, usually in a straight line.
Transom Horizontal structural member supported at each end.
Trapezoid Figure with four sides of which only two are parallel.
Treadmill Device used for physical exercise based on walking or running on a moving belt.
Trunnion Projecting hardpoint on the side of a pressurized module or other payload to restrain it firmly in a payload bay during flight.
Truss Open structural beam or framework fabricated from multiple smaller components such as struts.

Umbilical Flexible utilities connection between two space modules or spacecraft or between a launch vehicle and a launchpad infrastructure.

Vectorcardiology Specialized branch of cardiology.
Velocity vector The direction of flight in which a spacecraft or aircraft is travelling at speed.
Vernier thruster Small engine on a space vehicle that provides fine adjustment of velocity and attitude.
Vestibular Concerning sense of balance, spatial perception and sensation.

Wardroom Communal area or room on board a spacecraft or space station for crew meals, meetings and leisure use.
Weightlessness State of a person or object not subject to gravitational attraction that causes weight.

X-ray crystallography Technique for identifying the atomic and molecular structure of a crystal.

Yaw Rotation about a vertical axis (e.g. from keel to mast) of an aircraft, spacecraft or ship; 'Z' axis.

Zenith Upwards, towards the sky or cosmos on a line from the centre of the Earth.

References

Introduction

1 Hale, EE, *The Brick Moon*, The Atlantic Monthly, 1870, Project Gutenberg, 1999.
2 von Braun, W, 'Crossing the Last Frontier', *Collier's Weekly*, 22 March 1952.
3 Twentieth Century Fox, *On the Threshold of Space*, feature film, released April 1956, directed by Robert Webb, produced by William Bloom, starred Guy Madison, Virginia Leith and John Hodiak.
4 Metro-Goldwyn-Mayer, *2001: A Space Odyssey*, released April 1968, feature film, written, directed and produced by Stanley Kubrick, inspired by a short story called *The Sentinel* by Arthur C Clarke, starred Keir Dullea, Gary Lockwood, William Sylvester and Douglas Rain.
5 Howell, E, 'Skylab: first US Space Station', Space.com, 1 February 2013, website, available at: http://www.space.com/19607-skylab.html
6 NASA, 'Apollo-Soyuz. A pioneering partnership', 18 October 2013, website, available at: http://www.nasa.gov/mission_pages/apollo-soyuz/
7 *Oxford English Dictionary*, 2nd eD, 1989, Clarendon Press, Oxford.
8 NASA, 'Reference guide to the International Space Station. Assembly Complete edition', NP-2010-09-682-HQ, Washington DC, 2010.
9 Beggs, JB, Testimony before the US Congress, 27 March 1984, House Committee on Appropriations, Subcommittee on HUD – Independent Agencies, Department of Housing and Urban Development, Independent Agencies Appropriations for 1985, Part 6, NASA.
10 *Oxford English Dictionary*
11 *Encylopaedia Britannica*, 2014, online subscription version, Encylopaedia Britannica Inc.
12 Organisation Européene pour la Recherche Nucléaire, 'CERN faq. LHC the guide', CERN-Brochure-2009-003, Communication Group, February 2009.
13 Organisation Européene pour la Recherche Nucléaire, 'The history of CERN. CERN timelines', website, available at: http://timeline.web.cern.ch/timelines/The-history-of-CERN
14 Ibid.
15 Organisation Européene pour la Recherche Nucléaire.
16 Space Architecture Workshop, first International Space Architecture Symposium, Houston, Texas, 2002.
17 Metro-Goldwyn-Mayer, *Forbidden Planet*, released 1956, feature film, directed by Fred Wilcox, produced by Nicholas Nayfack, starred Walter Pidgeon, Anne Francis and Leslie Nielsen.
18 Warner Brothers, *Outland*, released 1981, feature film, directed by Peter Hyams, produced by Richard Roth and Stanley O'Toole, starred Sean Connery, Peter Boyle and Frances Sternhagen.
19 Twentieth Century Fox, *Prometheus*, released 2012, feature film, directed by Ridley Scott, produced by Ridley Scott,David Giler and Walter Hill, starred Noomi Rapace, Michael Fassbender and Guy Pearce.
20 Metro-Goldwyn-Mayer, *2001: A Space Odyssey*.
21 O'Neill, GK, *The High Frontier. Human Colonies in Space*, Jonathan Cape, London, 1976, pp.61-97.
22 NASA, 'Space Settlements. A Design Study', NASA SP-413, Richard Johnson and Charles Holbrow, edS, Washington DC, 1977.
23 Kennedy, KK, 'Vernacular of Space Architecture', *Out of this World, The New Field of Space Architecture*, Chapter 2, Library of Flight Series, Howe, AS, and Sherwood, B, edS, AIAA, 2009, pp.7-21.
24 Bigelow Aerospace, 'Expanding Humanity's Future in Space', 16 January 2013, website, available at: http://www.bigelowaerospace.com/beam_media_brief.php
25 *Encylopaedia Britannica*, 2014, online subscription version, Encylopaedia Britannica Inc.
26 Morgan, HM, *Vitruvius. The Ten Books on Architecture*, Book 1, Chapter 3, Harvard University Press, Cambridge, MasS, 1914, p.17.

Diversity and vision, 1979-1983

1 NASA, 'Launching NASA, a Brief History of NASA', NASA History Program Office, website, available at: http://history.nasa.gov/factsheet.htm
2 Compton, DW, and Benson, CD, *Living and Working in Space. A History of Skylab*, NASA SP-4208, NASA History Series, Scientific and Technical Information Branch, NASA, 1983, p.22-55.
3 Ibid, pp.104-125.
4 Ibid, pp.200-211.
5 Belew, LF, and Stuhlinger, E, *Skylab, A Guidebook*, George C Marshall Space Flight Center, NASA, 1973, pp.12-13.
6 Ibid, pp.62-80.
7 Turnhill, R, *Jane's Spaceflight Directory*, Jane's Publishing Company Ltd, London, 1984, pp.93-95.
8 Compton, DW, and Benson, CD, *Living and Working in Space. A History of Skylab*, pp.367-372.
9 NASA, *Biomedical Results from Skylab*, NASA SP-377, Johnston, RE, and Dietlein, LF, edS, Lyndon B Johnson Space Center, 1977, p.8.
10 Ibid, p.417.
11 Woodcock, GR, *Space Stations and Platforms*, Orbit Book Company, Florida, 1986, p.5.
12 Lord, DR, *Spacelab. An International Success Story*, NASA SP-487, Scientific and Technical Information Division, NASA, Washington DC, 1987, pp.5-19.
13 Krige, J, *Fifty Years of European Cooperation in Space*, Beauchesne Editeur, Paris, 2014, pp.238-239.
14 Lord, DR, *Spacelab. An International Success Story*, pp.51-53.
15 Krige, J, *Fifty Years of European Cooperation in Space*, pp.241-242.
16 NASA, 'Space Shuttle. Spacelab Payloads on Shuttle Flights', website, available at: http://www.nasa.gov/mission_pages/shuttle/launch/spacelab_shuttle.html.
17 Covington, C, et al., 'Space Operations Center, a Concept Analysis', JSC-16277, NASA Johnson Space Center, November 1979.
18 Logsdon, JM, 'Together in Orbit. The Origins of International Participation in the Space Station,' Monographs in Aerospace History #11, NASA History Division, NASA Headquarters, Washington DC, November 1998.
19 Logsdon, JM, and Butler, G, 'Space Station and Space Platform Concepts: A Historical Review', *Space Stations and Space Platforms – Concepts, Design, Infrastructure, and Uses*, Bekey, I, and Herman, D, edS, Progress in Astronautics and Aeronautics, Volume 99, AIAA, Washington DC, 1985.
20 McDonnell Douglas Astronautics Company, 'Evolutionary Space Platform Concept Study. Volume I – Executive Summary', MDC H0072, DPD-610, DR-4, NASA-CR-170829, Huntington Beach, May 1982.
21 McDonnell Douglas Astronautics Company, 'Evolutionary Space Platform Concept Study. Volume II – Technical Report. Part B – Manned Space Platform Concepts', MDC H0072, DPD-610, DR-4, NASA-CR-170829, Huntington Beach, May 1982.
22 Ibid.
23 Boeing Aerospace Company, 'Space Operations Center. System Analysis Study Extension. Final Report, Volume I, Executive Summary', D180-26785-1, NASA-CR-167555, Seattle, January 1982.
24 Boeing Aerospace Company, 'Space Operations Center. System Analysis. Final Report, Volume III, Book 1 of 2, SOC System Definition Report', D180-26495-3, Rev. A, NASA-CR-167559, Seattle, January 1982.
25 Turnhill, R, *Janes Spaceflight Directory*, p.110.
26 Pournelle, JE, *Space: The Crucial Frontier*, Citizens Advisory Council on National Space Policy, L5 Society, Tucson, Spring 1981.
27 Logsdon, JM, 'Together in Orbit. The Origins of International Participation in the Space Station,'
28 Ibid.
29 NASA, *Space Station Technology 1983*, Proceedings of the Space Station Technology Workshop, Williamsburg, Virginia, 28-31 March 1983, NASA Conference Publication 2293, NASA Scientific and Technical Information Branch, 1984.
30 Boeing Aerospace Company, 'Space Station Needs, Attributes and Architectural Options Study. Final Report. Volume 1. Executive Summary', D180-27477-7-1, Contract NASW 3680, April 1983.
31 Boeing Aerospace Company, *Space Station Needs, Attributes and Architectural Options Study*. Final Report. Volume 4. Architecture Options, Subsystems, Technology and Programmatics', D180-27477-7-3, Contract NASW 3680, April 1983.

32 Rockwell International, *'Space Station Needs, Attributes, and Architectural Options Study*. Final Executive Summary Report, SSD 83-0037, Contract NASW 3683, April 1983.
33 Rockwell International, *Space Station Needs, Attributes, and Architectural Options Study*. Volume II. Program Options, Architecture and Technology, SSD 83-0032-1, Contract NASW 3683, April 1983.
34 Lockheed Missiles & Space Company., *Space Station Needs, Attributes, and Architectural Options*. Final Presentation. Volume 1 – Executive Summary. LMSC-D889718. NASA-CR-172792. Contract NAS3684. Sunnyvale. April 1983.
35 Lockheed Missiles & Space Company, *Space Station Needs, Attributes, and Architectural Options*. Final Study Report. Study Summary, LMSC-D889718, Contract NAS3684, Sunnyvale, April 1983.
36 McDonnell Douglas Astronautics Company, *Space Station Needs, Attributes, and Architectural Options*. Architectural Options and Selection, MDC-H0537, Contract NASW-3687, Huntington Beach, April 1983.
37 McDonnell Douglas Astronautics Company, *Space Station Needs, Attributes, and Architectural Options*. Mission Requirements. MDC-H0533, Contract NASW-3687, Huntington Beach, April 1983.
38 General Dynamics, Convair Division, *A Study of Space Station Needs, Attributes, and Architectural Options*. Final Report, Volume 1 – Executive Summary, GDC-ASP-83-001, Contract NASW-3682, San Diego, April 1983.
39 General Dynamics, Convair Division, *A Study of Space Station Needs, Attributes, and Architectural Options*. Final Briefing, Contract NASW-3682, San Diego, April 1983.
40 Martin Marietta Aerospace, *Space Station Needs, Attributes, and Architectural Options Study – Final Report*. Mission Implementation Concepts, SOC-SE-03-01, Contract NASW-3686, Denver, April 1983.
41 Martin Marietta Aerospace, *Space Station Needs, Attributes, and Architectural Options Study – Final Report*. Executive Summary, SOC-SE-03-01, Contract NASW-3686, Denver, April 1983.
42 TRW, *Space Station Needs, Attributes, and Architectural Options Study – Final Review Executive Summary Briefing*, Contract NASW-3681, April 1983.
43 Gibbons, JH, *Civilian Space Stations and The US Future in Space*, Office of Technology Assessment, Congress of the United States, OTA-STI-241, November 1984, pp.79-82.
44 Turnhill, R, *Janes Spaceflight Directory*, pp.232-233.
45 Lord, DR, *Spacelab. An International Success Story*, pp.182-185.
46 Beggs, JB, 'Keynote Address', AIAA/NASA Symposium on the Space Station, Arlington, Virginia, 18 July 1983. Pub. in *Space Station: Policy, Planning and Utilization*, AIAA, New York, October 1983.
47 Culbertson, PE, 'Overview', AIAA/NASA Symposium on the Space Station, Arlington, Virginia, 18 July 1983. Pub. in *Space Station: Policy, Planning and Utilization*, AIAA, New York, October 1983.
48 Mark, H., 'Luncheon Speech', AIAA/NASA Symposium on the Space Station, Arlington, Virginia, 18 July 1983. Pub. in *Space Station: Policy, Planning and Utilization*, AIAA, New York, October 1983.
49 Powell, L. E., 'Architectural Options and Development Issues', AIAA/NASA Symposium on the Space Station, Arlington, Virginia, 18 July 1983. Pub. in *Space Station: Policy, Planning and Utilization*, AIAA, New York, October 1983.
50 NASA, 'Conceptual Design and Evaluation of Selected Space Station Concepts. Executive Summary', JSC-19521, Lyndon B Johnson Space Center, Houston, December 1983.
51 NASA, 'Conceptual Design and Evaluation of Selected Space Station Concepts. Volume I', JSC-19521, Lyndon B Johnson Space Center, Houston, December 1983.
52 Lyndon B Johnson Space Center, 'Conceptual Design and Evaluation of Selected Space Station Concepts. Volume II', JSC-19521, NASA, Houston, December 1983.
53 Ibid.
54 NASA, 'STS-9', Mission Archives, website, available at: http://www.nasa.gov/mission_pages/shuttle/shuttlemissions/archives/sts-9.html

Ambition and grandeur, 1984-1988

1 Logsdon, JM, 'Together in Orbit. The Origins of International Participation in the Space Station.' Monographs in Aerospace History #11, NASA History Division, NASA Headquarters, November 1998.
2 Woodcock, GR, *Space Stations and Platforms*, Orbit Book Company, Florida, 1986, p.8.
3 Reagan, RW, President, Address Before a Joint Session of the Congress on the State of the Union, 25 January 1984, House Chamber of the Capitol, Woolley, J, and Peters, G, The American Presidency Project, Santa Barbara, California.
4 Krige, J, *Fifty Years of European Cooperation in Space*, Beauchesne Editeur, Paris, 2014, pp.324-325.
5 Beggs, JB, Testimony before the US Congress, 27 March 1984, House Committee on Appropriations, Subcommittee on HUD – Independent Agencies, Department of Housing and Urban Development, Independent Agencies Appropriations for 1985, Part 6, NASA.
6 Smith, M, NASA's Space Station Program: Evolution of its Rationale and Expected Uses, 20 April 2005, Testimony before the Subcommittee on Science and Space, Committee on Commerce, Science and Transportation, United States Senate.
7 Logsdon, JM, 'Together in Orbit. The Origins of International Participation in the Space Station'.
8 Krige, J, *Fifty Years of European Cooperation in Space*, pp.325-327.
9 Kitmacher, GH, Design of the Space Station Habitable Modules, IAC-02-IAA.8.2.04, 53rd International Astronautical Congress, The World Space Congress, Houston, Texas, 16 October 2002.
10 NASA, 'Space Station Reference Configuration Description', NASA-JSC-19989, Systems Engineering and Integration, Space Station Program Office, Lyndon B Johnson Space Center, August 1984.
11 Woodcock, GR, *Space Stations and Platforms*, pp.170-172.
12 Mikulas, MM, Croomes SD, et al., 'Space Station Truss Structures and Construction Considerations', NASA Technical Memorandum 86338, NASA, Langley Research Center, January 1985.
13 NASA, 'Space Station Reference Configuration Description'.
14 NASA, 'NASA Systems Engineering Handbook', SP-6105, NASA Headquarters, June 1995.
15 NASA, 'NASA Facts, A History of US Space Stations', IS-1997-06-ISS009-JSC, Lyndon B Johnson Space Center, June 1997.
16 Gibbons, JH, 'Civilian Space Stations and the US Future in Space', Office of Technology Assessment, Congress of the United States, November 1984.
17 Kitmacher, GH, Design of the Space Station Habitable Modules.
18 NASA, 'NASA Systems Engineering Handbook'.
19 Kitmacher, GH, Design of the Space Station Habitable Modules.
20 Bertrand, R, Messerschmid, E, and Thomas D, 'Designing and Sizing Space Elements', *Human Spaceflight, Mission Analysis and Design*, edited by WJ Larson and LK Pranke, Space Technology Series, McGraw-Hill, 1999, p.336.
21 McCutchen, DK, Berka, R, Mikulas, MM, et al., Presentations at the Space Station Systems Integration Board, Lyndon B Johnson Space Center, 9 October 1985.
22 Mikulas, MM, Wright, AS, et al, 'Deployable/Erectable Trade Study for Space Station Truss Structures', NASA Technical Memorandum 87573, NASA, Langley Research Center, July 1985.
23 Dorsey, JT, 'Structural Performance of Space Station Trusses with Missing Members', NASA Technical Memorandum 87715, NASA, Langley Research Center, May 1986.
24 McCutchen, DK, Berka, R, Mikulas, MM, et al., Presentations at the Space Station Systems Integration Board.
25 NASA, 'Space Shuttle Mission STS-61B. Press Kit', Release No 85-153, November 1985.
26 Card, MF, et al., 'Construction and Control of Large Space Structures', NASA Technical Memorandum 87689, NASA, Langley Research Center, February 1986.
27 Mikulas, MM, Bush, HG, 'Design, Construction and Utilization of a Space Station Assembled from 5-Meter Erectable Struts', NASA Technical Memorandum 89043, NASA, Langley Research Center, October 1986.
28 Rogers, WP, et al., 'Report of the Presidential Commission on the Space Shuttle Challenger Accident, Volume I', Washington DC, 6 June 1986.
29 Reagan, RW, President, Address to the Nation on the Explosion of the Space Shuttle Challenger. 28 January 1986, The

American Presidency Project, University of California, Santa Barbara, 1999.

30 Sanger, DE, 'Cold And Vibration In Rocket Studied; Questions About Solid-fuel Booster Focus On Temperature And Changes In Pressure', *The New York Times*, 5 February 1986.

31 Feynman, R. P, *What Do You Care What Other People Think?*, WW Norton & Company, New York, 1988, pp.139-153.

32 NASA, 'STS-135', Shuttle Press Kit, July 2011.

33 United Press International, 'Astronaut Assails Space Station Plan', 19 July 1986.

34 Associated Press, 'NASA Teams to Take New Look at Space Station Plans', 20 August 1986.

35 Paine, TO, et al., 'Pioneering the Space Frontier', The Report of the National Commission on Space, Bantam Books, New York, May 1986.

36 NASA, 'Space Station Plan', NASA photo 86-H-324, 1986.

37 NASA, 'NASA Systems Engineering Handbook'.

38 Boeing Aerospace Company, 'Space Station Final Study Report. Volume II Study Results', NAS8-36526 Data Requirements 15, D483-50115-2, 18 January 1987.

39 Rockwell International, 'Space Station Evolutionary Growth Plan. Work Package 2, Definition and Preliminary Design Phase', NAS9-17365, DR-18, SSS 85-0209, 20 December 1985.

40 Ferebee, MJ, and Powers, RB, 'Optimization of Payload Mass Placement In a Dual Keel Space Station', NASA Technical Memorandum 89051, NASA, Langley Research Center, March 1987.

41 Space Station Program Office, 'Architectural Control Document, Extravehicular Activity System', JSC 30256, Lyndon B Johnson Space Center, NASA, !986.

42 Boeing Aerospace Company, 'Final Study Report. Volume I Executive Summary', NAS8-36526 Data Requirements 15, D483-50115-1, 18 January 1987.

43 Smith, M, NASA's Space Station Program: Evolution of its Rationale and Expected Uses, 20 April 2005.

44 NASA, 'Space Station Program Response to the Fiscal Year 1988 and 1989 Revised Budgets', Submitted to the Committee on Appropriations, US House of Representatives and the Committee on Appropriations, US Senate, NASA, April 1988.

45 United States General Accounting Office, 'Space Station. NASA's Search for Design, Cost, and Schedule Stability Guidelines', Report to the Chairman, Committee on Science, Space, and Technology, House of Representatives, GAO/NSIAD-91-125, March 1991.

46 NASA, 'Space Station Program Response to the Fiscal Year 1988 and 1989 Revised Budgets'.

47 GE Astro Space, 'Space Station Freedom – WP3. Attached Payload Accommodations Equipment User Handbook', Revision A, NASA-CR-195717, NAS5-32000, GE Astro-Space Division, March 1989.

48 NASA, 'Space Station Program Response to the Fiscal Year 1988 and 1989 Revised Budgets'.

49 Ride, SK, 'Leadership and America's Future in Space', A Report to the Administrator, NASA, Washington, DC, August 1987.

50 NASA, 'Space Station Program Response to the Fiscal Year 1988 and 1989 Revised Budgets'.

51 Krige, J, *Fifty Years of European Cooperation in Space*, pp.327-330.

52 Logsdon, JM, 'Together in Orbit. The Origins of International Participation in the Space Station'.

53 NASA, 'Space Shuttle Mission Archives', STS-26, NASA Human Spaceflight Gallery, available at http://spaceflight.nasa.gov/gallery/

54 Kotulak, R, et al., 'Americans Return to Space', *Chicago Tribune*, 30 September 1988.

55 National Research Council, 'Space Station Engineering Design Issues. Report of a Workshop', Workshop Committee on Space Station, Engineering Design Issues, Aeronautics and Space Engineering Board, Commission on Engineering and Technical Systems, National Research Council, National Academy Press, 1989.

56 Stepaniak, P, et al., 'Considerations for Medical Transport From the Space Station via an Assured Crew Return Vehicle', NASA TM-2001-210198, NASA, Lyndon B Johnson Space Center, July 2001.

57 Griffin, B, 'Design Guide. The Influence of Zero-G and Acceleration on the Human Factors of Spacecraft Design', NASA, Lyndon B Johnson Space Center, August 1978.

58 National Academy of Sciences, *Human Factors in Long-Duration Spaceflight*, Space Science Board, National Academy of Sciences, National Research Council, Washington DC, 1972, pp.24-25.

59 Connors, MM, Harrison AA, and Akins, FR, *Living Aloft. Human Requirements for Extended Spaceflight*, NASA SP-483, NASA, Ames Research Center, 1985, p.308.

60 Stuster, JW, 'Space Station Habitability Recommendations Based on a Systematic Comparative Analysis of Analogous Conditions', NASA-CR-3943, Contract NAS2-11690, NASA, Ames Research Center, 1986.

61 Cohen, MM, 'Testing the Celentano Curve: An Empirical Survey of Predictions for Human Spacecraft Pressurized Volume', 2008-01-2027, SAE Technical Paper Series, 38th International Conference on Environmental Systems, San Francisco, June 2008.

62 NASA, 'Man-Systems Integration Standards', NASA-STD-3000, Volume 1, Revision B, Lyndon B Johnson Space Center, July 1995.

63 Sherwood, B, and Capps, S, 'Habitats for Long-Duration Missions', *Out of this World. The New Field of Space Architecture*. Chapter 11, Library of Flight Series, Howe, AS, and Sherwood, B, edS, AIAA, 2009, pp.121-311.

64 NASA, 'Reference Mission Version 3.0. Addendum to the Human Exploration of Mars: the Reference Mission of the NASA Mars Exploration Study Team', EX13-98-036, Exploration Office, Advanced Development Office, Lyndon B Johnson Space Center, June 1998.

65 Shapland, D, and Rycroft, M, *Spacelab. Research in Earth Orbit*, Cambridge University Press, 1984, pp.31-36.

66 Rockwell International and Grumman Corporation, 'NASA Hab Architecture Evaluation Team Review', Bethpage, NY, 20 November 1985.

67 Jones, R, 'Performance of the International Space Station Interior', *Out of this World. The New Field of Space Architecture*, Chapter 4, Library of Flight series, Howe, AS, and Sherwood, B, edS, AIAA, 2009, pp.31-43.

68 McDonnell Douglas Corporation, 'Space Station Definition and Preliminary Design. Work Package No 2. Preliminary Design and Analysis Document', Book 15, Habitation Module Outfitting (Section 4.11), MDC H2028, Contract NAS9-17367, Huntington Beach, December 1985.

69 Hadfield, C, *An Astronaut's Guide to Life on Earth*, Little, Brown and Company, New York, 2013, pp.198-199.

70 Nixon, D, Miller, C, and Fauquet, R, 'Space Station Wardroom Habitability and Equipment Study', Space Projects Group, Southern California Institute of Architecture, NASA CR 4246, December 1989.

71 National Space Development Agency of Japan, 'Reference Configuration of the Japanese Experiment Module for Phase B Study', SU-284017A, Revision A, December 1985.

72 NASA, 'Space Station Program Definition and Requirements. Section 3: Systems Requirements', Space Station Projects Office, Marshall Space Flight Center, WP-01, SS-SRD-0001, 10 November 1986.

73 Ibid.

74 Hager, RW, 'Selected Advanced Technology Studies for the US Space Station', IAF-87-79, 38th International Congress of the International Astronautical Federation, October 1987, Brighton, UK.

75 Boeing Aerospace Company, 'Space Station Final Study Report. Volume II Study Results', NAS8-36526, Data Requirements 15, D483-50115-2, 18 January 1987.

76 Belew, LF, and Stuhlinger, E, *Skylab, A Guidebook*, George C Marshall Space Flight Center, NASA, 1973, p.63.

Crisis and resolution, 1989-1993

1 United States General Accounting Office, 'Space Station. NASA's Search for Design, Cost, and Schedule Stability Guidelines', Report to the Chairman, Committee on Science, Space and Technology, US House of Representatives, GAO/NSIAD-91-125, March 1991.
2 Smith, MS, 'NASA's Space Station Program: Evolution of its Rationale and Expected Uses', Testimony before the Subcommittee on Science and Space, Committee on Commerce, Science and Transportation, US Senate, 20 April 2005.
3 Harland, DM, and Catchpole, JE, *Creating the International Space Station*, Springer-Praxis Publishing, UK, 2002, p.121.
4 Ibid, pp.122-123.
5 Ibid, p.123.
6 Bush, GHW, President, 'Executive Order 12675 – Establishing the National Space Council', The American Presidency Project Online, produced by G Peters and JT Wooley, available at www.presidency.ucsb.edu/ws/?pid=60450.
7 Bush, GHW, President, 'Remarks on the 20th Anniversary of the Apollo 11 Moon Landing', Public Papers, George Bush Presidential Library and Museum.
8 NASA, 'Report of the 90-Day Study on Human Exploration of the Moon and Mars', NASA-TM-102999, November 1989.
9 The White House, Office of the Press Secretary, 'Fact Sheet: US National Space Policy', 16 November 1989, and 'National Space Policy' (attached document), 2 November 1989.
10 NASA, 'Report of the 90-Day Study on Human Exploration of the Moon and Mars', NASA-TM-102999, November 1989.
11 Dick, S, NASA Chief Historian, 'Summary of Space Exploration Initiative', undated, available at http://history.nasa.gov/seisummary.htm
12 United States General Accounting Office, 'Space Station. NASA's Search for Design, Cost, and Schedule Stability Guidelines', Report to the Chairman, Committee on Science, Space and Technology, US House of Representatives, GAO/NSIAD-91-125, March 1991.
13 Harland, DM, and Catchpole, JE, *Creating the International Space Station*, p. 123.
14 Augustine, NR, et al., 'Report of the Advisory Committee On the Future of the US Space Program', Advisory Committee On the Future of the US Space Program, Washington DC, December 1990.
15 Ibid.
16 Augustine, NR, and Wilkening, LL, Testimony Before the Space Science and Applications Subcommittee of the Committee on Science, Space and Technology, US House of Representatives, Washington DC, 3 January 1991.
17 Moorhead, RW, 'Space Station Freedom – A Restructured Program', NASA, MS, 11 January 1991.
18 Ibid.
19 Smith, MS, 'NASA's Space Station Program: Evolution of its Rationale and Expected Uses'.
20 Riel, FD, and Markus, G, 'Pre-Integrated Truss Assemblies for Space Station Freedom', McDonnell Douglas Space Systems Company, Huntington Beach, IAF-91-071, 42nd Congress of the International Astronautical Federation, 5 October, Montreal, Canada.
21 NASA, 'NASA Systems Engineering Handbook', SP-6105, NASA Headquarters, Washington DC, June 1995.
22 Riel, FD, and Markus, G, 'Pre-Integrated Truss Assemblies for Space Station Freedom'.
23 Ibid.
24 McDonnell Douglas Space Systems Company, 'On-Orbit Stage Configurations', IF03498, Set of drawings, Huntington Beach, May 1991.
25 Ibid.
26 Boutros, R, and Olson, RL, 'Post-Restructure US Laboratory and Habitation Module Configuration for Space Station Freedom', Boeing Defense & Space Group, Missiles and Space Division, Huntsville, Paper 911594, 21st International Conference on Environmental Systems, San Francisco, 1991.
27 Levesque, RJ, and Lauger, JB, 'Space Station Freedom Resource Node Status: First Quarter 1991', McDonnell Douglas Space Systems Company, Space Station Division, Huntington Beach, Paper 911595, 21st International Conference on Environmental Systems, San Francisco, 1991.
28 Kohrs, RH, Edwards, A, Craig, J, et al., 'ACRV. Project Status/Introduction/Overview', NASA, Washington DC, 1991.
29 NASA, 'Tier 1 Environmental Impact Statement for Space Station Freedom. Final Report', NASA-TM-109734, Washington DC, March 1991.
30 Harland, DM, and Catchpole, JE, *Creating the International Space Station*, p.129.
31 Synthesis Group, 'America at the Threshold. America's Space Exploration Initiative', Washington DC, 3 May 1991.
32 Space Studies Board, 'Annual Report –1991', Commission on Physical Sciences, Mathematics and Applications, National Research Council, Washington DC, 1991.
33 Daniell, RG, and Beck, JR, 'On the Design and Development of the Space Station Remote Manipulator System', IAF-91-074, 42nd Congress of the International Astronautical Federation, 5 October, Montreal, Canada.
34 Office of Inspector General, 'Faster, Better, Cheaper: Policy, Strategic Planning and Human Resource Alignment', Audit Report, IG-01-009, NASA, 13 March 2001.
35 NASA, STS-49 Press Kit, 'Satellite Rescue, Spacewalks Mark Endeavour's First Flight', Release 92-48, NASA Headquarters, Washington DC, May 1992.
36 NASA, Space Station Freedom Utilization Conference, NASA-TM-108742, Von Braun Civic Center, Huntsville, 3 August 1992, Office of Space Science and Applications, NASA Headquarters, Washington DC.
37 Goldin, D, 'Keynote Address – Remarks', Space Station Freedom Utilization Conference, NASA-TM-108742, Von Braun Civic Center, Huntsville, 3 August 1992, Office of Space Science and Applications, NASA Headquarters, Washington DC.
38 Chambers, LP, 'Life Sciences Utilization of Space Station Freedom', Space Station Freedom Utilization Conference, NASA-TM-108742, Von Braun Civic Center, Huntsville, 3 August 1992, Office of Space Science and Applications, NASA Headquarters, Washington DC.
39 Bayuzick, RJ, 'Research Objectives, Opportunities and Facilities for Microgravity Science', NASA-TM-108742, Von Braun Civic Center, Huntsville, 3 August 1992, Office of Space Science and Applications, NASA Headquarters, Washington DC
40 Rose, MF, 'Space Station Freedom as an Engineering Experiment Station: an Overview', NASA-TM-108742, Von Braun Civic Center, Huntsville, 3 August 1992, Office of Space Science and Applications, NASA Headquarters, Washington DC.
41 Scheib, J, 'WP-2 Attached Payload Accommodations', Payload Integration Meeting, User Engineering Integration, McDonnell Douglas Space Systems Company, Space Station Division, Huntington Beach, June 1992.
42 Aldrin, B, 'Fiscally Responsible Alternatives to the Space Station Freedom in the Post-Cold War Era', Laguna Beach, August 1992.
43 Station Redesign Support Team, 'Space Station Redesign Option A. Modular Buildup Concept', NASA TM-108415, NASA, Marshall Space Flight Center, 10 June 1993.
44 United States General Accounting Office, 'Space Station. Impact of the Expanded Russian Role on Funding and Research', Report to the Ranking Minority Member, Subcommittee on Oversight of Government Management, Committee on Governmental Affairs, US Senate, GAO/NSIAD-94-220, June 1994.
45 Goldin, D, 'Redesign Process', Letter to: Officials-in-Charge of Headquarters Offices; Directors, NASA Field Installations; Director, Jet Propulsion Laboratory, Office of the Administrator, NASA Headquarters, Washington DC, 9 March 1993.
46 Space Station Redesign Team, 'Final Report to the Advisory Committee on the Redesign of the Space Station', NASA, June 1993.
47 Ibid.
48 Ibid.
49 NASA, 'Super Lightweight External Tank', NASA Facts, Pub 8-40341, Marshall Space Flight Center, Alabama, April 2005.
50 Space Station Redesign Team, 'Final Report to the Advisory Committee on the Redesign of the Space Station', NASA, June 1993.
51 Vest, CM, et al., 'Final Report to the President', Advisory Committee on the Redesign of the Space Station, 10 June 1993.
52 Harland, DM, and Catchpole, JE, *Creating the International Space Station*, p.168.
53 US House of Representatives, NASA Space Station Freedom Appropriations, 23 June 1993, C-Span Video Library, available at www.c-spanvideo.org/program/43505-1
54 Eisenhower, S, *Partners in Space. Russian Cooperation after the Cold War*, The Eisenhower Institute, Washington DC, 2004, p.38.
55 *Encyclopedia Astronautica*, *Mir-2*, available at www.astronautix.com/craft/mir2.htm
56 Eisenhower, S, *Partners in Space. Russian Cooperation after the Cold War*, The Eisenhower Institute, Washington DC, 2004, pp.54-55.
57 Ibid, p.57.

58 Thirkettle, A, former ESA Project Manager for the Columbus module, Interview, May 2013.
59 Ibid.
60 Logsdon, JM, and Millar, JR, 'US-Russian Cooperation in Human Space Flight. Assessing the Impacts', Space Policy Institute and Institute for European, Russian and Eurasian Studies, Elliott School of International Affairs, The George Washington University, Washington DC, February 2001.
61 United States General Accounting Office, 'Space Station. Impact of the Expanded Russian Role on Funding and Research', Report to the Ranking Minority Member, Subcommittee on Oversight of Government Management, Committee on Governmental Affairs, US Senate, GAO/ NSIAD-94-220, June 1994.
62 Smith, MS, 'NASA's Space Station Program: Evolution of its Rationale and Expected Uses'.
63 Krige, J, *Fifty Years of European Cooperation in Space*, Beauchesne Editeur, Paris, 2014, p.328.
64 Ezell, EC, and Ezell, LN, *The Partnership. A History of the Apollo-Soyuz Test Project*, NASA SP-4209, The NASA History Series, Scientific and Technical Information Office, NASA, Washington DC, 1978, p.98.
65 Ibid, pp.190-193.
66 Ibid, pp.328-329.
67 NASA, 'History, Shuttle-Mir Background, Negotiations and Joint Planning', available at http://spaceflight.nasa.gov/history/ shuttle-mir/history/h-b-negotiations.htm
68 NASA and the Russian Space Agency, 'Implementing Agreement between the NASA of the United States of America and the Russian Space Agency of the Russian Federation on Human Spaceflight Cooperation', Moscow, 5 October 1992.
69 Morgan, C, *Shuttle-Mir, The United States and Russia Share History's Highest Stage*, NASA, NASA SP-2001-4225, The NASA History Series, Lyndon B Johnson Space Center, 2001, pp.32-35.
70 Ibid, pp.157-160.

Industry and preparation, 1994-1998

1 Waterman, J, former Operations Manager for Station Module Fabrication, The Boeing Company, Interview, 9 January 2007.
2 Ibid.
3 Eisenhower, S, *Partners in Space. Russian Cooperation after the Cold War*, The Eisenhower Institute, Washington DC, 2004, p.78.
4 NASA, 'Super Lightweight External Tank. NASA Facts', FS-2005-04-025-MSFC, Marshall Space Flight Center, April 2005.
5 Portree, DS, 'Mir Hardware Heritage', NASA RP 1357, NASA, Lyndon B Johnson Space Center, March 1995.
6 Ibid.
7 Morgan, C, *Shuttle-Mir, The United States and Russia Share History's Highest Stage*, NASA, p.168.
8 Portree, DS, 'Mir Hardware Heritage'.
9 Jorgensen, CA (editor), 'International Space Station Evolution Data Book. Volume 1. Baseline Design', Revision A, NASA/SP-2000-6109/VOL1/REV1. FDC/ NYMA,NASA, Hampton, October 2000.
10 Ibid.
11 NASA, 'International Space Station Fact Book', July 1999.
12 Jorgensen, CA (editor), 'International Space Station Evolution Data Book. Volume 1. Baseline Design'.
13 NASA, 'International Space Station Fact Book'.
14 Jorgensen, CA (editor), 'International Space Station Evolution Data Book. Volume 1. Baseline Design'.
15 The Committee on Science, House of Representatives, 'International Space Station Authorization Act of 1995', 104th Congress, 1st Session, Report 104-210, 28 July 1995.
16 United States General Accounting Office, 'Space Station. Impact of the Expanded Russian Role on Funding and Research', Report to the Ranking Minority Member, Subcommittee on Oversight of Government Management, Committee on Governmental Affairs, US Senate, GAO/ NSIAD-94-220, June 1994.
17 The Committee on Science, House of Representatives, 'International Space Station Authorization Act of 1995', 104th Congress, 1st Session, Report 104-210, 28 July 1995.
18 Ernst, S, former Project Manager for Z1 and P6 Assembly and Test, The Boeing Company, Interview, 13 January 2007.
19 The Boeing Company, 'Space Exploration. P3 and P4 to expand station capabilities, providing a third and fourth array', IDS Business Support, Communications and Community Affairs, The Boeing Company, St Louis. July 2006.
20 Ernst, S, former Project Manager for Z1 and P6 Assembly and Test, The Boeing Company, Interview, 13 January 2007.
21 Waterman, J, former Operations Manager for Station Module Fabrication, The Boeing Company, Interview, 9 January 2007.
22 NASA, 'Micrometeoroid and Orbital Debris (MMOD) Protection', International Space Station Interactive Reference Guide, Undated.
23 Waterman, J, former Operations Manager for Station Module Fabrication, The Boeing Company, Interview, 9 January 2007.
24 Ibid.
25 Compton, DW, and Benson, CD, *Living and Working in Space. A History of Skylab*, NASA SP-4208, The NASA History Series, NASA, Scientific and Technical Information Branch, 1983.
26 Curell, P, 'Lab Window Overview', The Boeing Company, October 2001.
27 Krige, J, *Fifty Years of European Cooperation in Space*, Beauchesne Editeur, Paris, 2014, p.331.
28 Eisenhower, S, *Partners in Space. Russian Cooperation after the Cold War*, pp.81-82.
29 Goldin, D, and Koptev, Y, 'Protocol, including terms, conditions and assumptions, summary balance of contributions and obligations to International Space Station (ISS) and resulting rights of NASA and RSA to ISS utilization accommodations and resources, and flight opportunities', 11 June 1996.
30 *Science*, 'NASA to buy Research Time to Bail Out Russian Agency', News This Week, Vol 282, 9 October 1998.
31 Eisenhower, S, *Partners in Space. Russian Cooperation after the Cold War*, p.89.
32 NASA, 'NASA Facts. International Space Station Environmental Control and Life Support System', FS-2002-05-85-MSFC, Marshall Space Flight Center, Alabama, May 2002.
33 Wieland, PO, 'Living Together in Space. The Design and Operation of the Life Support Systems on the International Space Station', NASA TM-98-206956/ VOL1, National Aeronautics and Space Administration, Marshall Space Flight Center, Alabama, January 1998.
34 Ibid.
35 NASA, 'NASA Facts. International Space Station Environmental Control and Life Support System', FS-2002-05-85-MSFC, Marshall Space Flight Center, Alabama, May 2002.
36 Jorgensen, CA (editor), 'International Space Station Evolution Data Book. Volume 1. Baseline Design'.
37 NASA, 'Reference Guide to the International Space Station. Assembly Complete Edition', November 2010.
38 Ibid.
39 NASA, 'Tier 1 Environmental Impact Statement for Space Station Freedom', Final Report, NASA-TM-109734, Washington DC, March 1991.
40 NASA, 'Final Tier 2 Environmental Impact Statement for the International Space Station', NASA-TM-111720, Space Station Program Office, Office of Space Flight, Washington DC, May 1996.
41 Ibid.
42 Kennedy, KJ, 'TransHab project', *Out of This World. The New Field of Space Architecture*, Chapter 4, Library of Flight series, AIAA, 2009, pp.81-88.
43 Ibid.
44 Krige, J, *Fifty Years of European Cooperation in Space*, p.332.
45 NASA and the Russian Space Agency, Memorandum of Understanding between the NASA of the United States of America and the Russian Space Agency Concerning Cooperation on the Civil International Space Station, Daniel Goldin and Yuri Koptev signees, Washington DC, 29 January 1998.
46 Cost Assessment and Validation Task Force. 'Report of the Cost Assessment and Validation Task Force on the International Space Station', NASA Advisory Council, 21 April 1998.
47 Ibid.
48 Li, A., Associate Director, Defense Acquisition Issues, National Security and International Affairs Division, United States General Accounting Office, 'Space Station. US Life-Cycle Funding Requirements', Testimony before the Committee on Science, House of Representatives, Washington DC, 24 June 1998.
49 United States Congress, 'Commercial Space Act of 1998', 105th Congress, 28 October 1998, [H.R. 1702], Public Law 105-303.
50 NASA, 'Commercial Development Plan for the International Space Station', Final

Draft, Washington DC, 16 November 1998.
51 Reichhardt, T, et al., 'Tension and relaxation in space-station science', *Nature*, Vol. 391, 19 February 1998.
52 Isakowitz, JS, Hopkins, JB, and Hopkins, JP, *International Reference Guide to Space Launch Systems*, AIAA, July 2004, pp.568-569.
53 Federation of American Scientists Space Policy Project, 'Baikonur Cosmodrome', available at: www.fas.org/spp/guide/russia/facility/baikonur/htm
54 NASA, International Space Station Assembly website, available at: www.nasa.gov/mission_pages/station/structure/elements/baikonur.html
55 Turnhill, R, *Jane's Spaceflight Directory*, Jane's Publishing Company Ltd, London, 1984, p.287.
56 Ibid, pp.284.
57 Isakowitz, JS, Hopkins, JB, and Hopkins, JP, *International Reference Guide to Space Launch Systems*, pp.570-573.
58 Turnhill, R, *Janes Spaceflight Directory*, 1984, p.284.
59 Ibid, 281-282.
60 European Space Agency, 'Europe's Spaceport', website, available at: www.esa.int/Our_Activities/Launchers/Europe_s_Spaceport/Europe_s_Spaceport2.
61 European Space Agency, 'CNES at Europe's Spaceport; Arianespace at Europe's Spaceport; ESA at Europe's Spaceport', website, available at: http://www.esa.int/Our_Activities/Launchers/Europe_s_Spaceport/Europe_s_Spaceport
62 Japanese Aerospace Exploration Agency, 'Tanegashima Space Center', website, available at www.jaxa.jp/about/centers/tnsc/index_e.html
63 Khrunichev State Research and Production Space Center, Moscow, 'Proton Launch System Mission Planner's Guide', LKEB-9812-1990, Revision 6, December 2004, International Launch Systems, McLean, Virginia.
64 Ibid.
65 Isakowitz, JS, Hopkins, JB, and Hopkins, JP, *International Reference Guide to Space Launch Systems*, pp.301-327.
66 Hall, RD, and Shayler, DJ, *Soyuz – A Universal Spacecraft*, Springer Praxis Publishing, UK, 2003, p.85.
67 TsSKB-Progress State Rocket and Space Scientific Production Center, Samara, Russia, 'Soyuz User's Manual', ST-GTD-SUM-01, Issue 3, Revision 0, Starsem, Evry, France, April 2001.
68 Ibid.
69 Turnhill, R, *Jane's Spaceflight Directory*, pp.167-197.
70 NASA, 'Shuttle Reference Manual', Space Transportation System, Background and Status, Washington DC, 1988.
71 Ibid.
72 NASA, Space Shuttle Launch Archives, Kennedy Space Center.
73 Report of the Presidential Commission on the Space Shuttle Challenger Accident, Washington DC, 6 June 1986.
74 Columbia Accident Investigation Board, Report Volume 1, August 2003.
75 NASA, 'Shuttle Reference Manual', Space Transportation System, Space Shuttle Requirements, Washington DC,1988.
76 Turnhill, R, *Jane's Spaceflight Directory*, 1984, p.214.
77 European Space Agency, 'Convention for the establishment of a European Space Agency & ESA Council Rules of Procedure', ESA SP-1271(E), March 2003.
78 Arianespace, 'Ariane 5 User's Manual', Issue 4, Revision 0, Evry-Courcouronnes, France, November 2004.
79 Ibid.
80 Japanese Aerospace Exploration Agency, 'H-IIB Launch Vehicle', website, available at: www.jaxa.jp/projects/rockets/h2b/index_e.html
81 NASA, 'NASA Super Guppy', website, available at: http://jsc-aircraft-ops.jsc.nasa.gov/guppy/
82 Airbus Industrie, 'Beluga, The Unmatched Airlifter for Outsized Cargo', website, available at: http://www.airbus.com/aircraftfamilies/freighter/beluga/
83 NASA, 'Space Station Processing Facility', FS-2005-06-020-KSC, NASA Facts, John F Kennedy Space Center, December 2006.
84 NASA, 'Space Station Processing Facility Processing and Support Capabilities', K-STSM-14.1.16-BASIC-SSPF, John F. Kennedy Space Center, April 1995.
85 NASA, 'Canister Rotation Facility', IS-2004-09-014-KSC, NASA Facts, John F Kennedy Space Center, 2006.

Endurance and achievement, 1999-2011

1 NASA, 'Zarya', Shuttle Press Kit, International Space Station, November 1998.
2 NASA, 'Summary Flight Plan. Zarya Orbital Events Summary', Shuttle Press Kit, International Space Station, November 1998.
3 NASA, 'STS-88. Payloads. Unity Connecting Module', Shuttle Press Kit, International Space Station, December 1998.
4 Lyndon B Johnson Space Center, 'Unity Connecting Module. Cornerstone for a Home in Orbit', NASA Facts, January 1999.
5 NASA, 'STS-88', Mission Archives, website, available at: http://www.nasa.gov/mission_pages/shuttle/shuttlemissions/archives/sts-88.html
6 Smith, MS, NASA's International Space Station Program, 25 February 1999, Testimony Before the Subcommittee on Space and Astronautics, Committee on Science, United States House of Representatives, Washington DC.
7 Lafleur, C, 'NASA International Space Station, ISS Assembly Sequence Revision E (June 1999)', Agence Science Presse, 2000.
8 NASA Advisory Council, 'Report of the Cost Assessment and Validation Task Force on the International Space Station', 21 April 1998.
9 United States Congress, Congressional Bill No HR 2684, 8 September 1999, Official Title of Legislation: H. Amdt. 423 to HR 2684: To eliminate funding from the International Space Station program.
10 Eisenhower, S, *Partners in Space. Russian Cooperation after the Cold War*, The Eisenhower Institute, Washington DC, 2004, p.87.
11 United States Congress, Iran Nonproliferation Act of 2000, 106th Congress, Public Law 106-178, 14 March 2000.
12 United States General Accounting Office, Letter from Allen Li to the Hon F. James Sensenbrenner Jr, Chairman, Committee on Science, House of Representatives, Subject: Space Station: Russian-Built Zarya and Service Module Compliance With Safety Requirements, 28 April 2000.
13 NASA, SM Free Flight and Docking Overview, SFOC-FL2212.
14 Lyndon B Johnson Space Center, 'The Service Module: A Cornerstone of Russian International Space Station Modules', NASA Facts, January 1999.
15 Nasawatch, 'Assembly Sequence Revision F Overview', 14 June 2000, Nasawatch, website, available at: www.nasawatch.com
16 NASA, 'STS-92', Shuttle Press Kit, October 2000.
17 NASA, 'Expedition One Crew Brings the Station to Life', Expedition 1 Press Kit, 25 October 2000.
18 The *New York Times*, 'Opening the Space Station', Archives, 1 November 2000.
19 Hall, RD, and Shayler, DJ, *Soyuz – A Universal Spacecraft*, Springer/Praxis Publishing, 2003, pp.377-398.
20 NASA, 'STS-97', Shuttle Press Kit, December 2000.
21 Ibid.
22 NASA, 'STS-98', Shuttle Press Kit, February 2001.
23 The Boeing Company, 'International Space Station. Common Berthing Mechanism', Boeing, website, available at: http://www.boeing.com/boeing/defensespace/space/spacestation/components/common_berthing_mechanism.page
24 NASA, 'International Space Station Familiarization', TD9702A, Mission Operations Directorate, Space Flight Training Division, Lyndon B Johnson Space Center, 3 July 1998.
25 McLaughlin, RJ, and Warr, WH, 'The Common Berthing Mechanism for International Space Station', 2001-01-2435, The Society of Automotive Engineers, 2001.
26 NASA, 'STS-100', Shuttle Press Kit, May 2001.
27 NASA, 'Space Station Assembly. Canadarm2', International Space Station, website, available at: http://www.nasa.gov/mission_pages/station/structure/elements/mss.html
28 Canadian Space Agency, 'Mobile Servicing System', website, available at: http://www.asc-csa.gc.ca/eng/iss/mobile-base/overview.asp
29 Hadfield, C, *An Astronaut's Guide to Life on Earth*, Little, Brown and Company, 2013. p.192.
30 NASA, 'STS-104', Shuttle Press Kit, July 2001.
31 NASA, 'Space Shuttle Consolidated Launch Manifest', 29 May 2001.
32 Zak, A, 'Spacecraft: Manned: ISS: Russian Segment: Docking Compartment', Russianspaceweb, website, available at: www.russianspaceweb.com
33 Young, AT, Chairman, 'Report by the

International Space Station (ISS) Management and Cost Evaluation (IMCE) Task Force to the NASA Advisory Council', 1 November 2001.
34 NASA, 'STS-110', Shuttle Press Kit, April 2002.
35 Ibid.
36 MacDonald Dettwiler Space and Advanced Robotics Ltd, 'Mobile Servicing System', MSS Backgrounder, July 2011.
37 United States General Accounting Office, 'Space Station: Actions Under Way to Manage Cost, but Significant Challenges Remain', July 2002.
38 American Institute of Physics, 'Task Force Questions Scientific Value of Scaled-Back Space Station', *American Institute of Physics Bulletin*, Number 86, 25 July 2002.
39 NASA, 'Space Shuttle Consolidated Launch Manifest', October 2002.
40 NASA, 'STS-112', Shuttle Press Kit, October 2002.
41 NASA, 'International Space Station Familiarization', TD9702A, Mission Operations Directorate, Space Flight Training Division, Lyndon B Johnson Space Center, 3 July 1998.
42 Favero, JP, 'International Space Station Mechanism Contact Simulation', The Boeing Company, *Aircraft Engineering and Aerospace Technology*, Volume 77, Number 1, 2005.
43 NASA, 'Quick-Connect Nut', NASA Image Exchange (NIX), website, NASA Marshall Space Flight Center, NASA STI (Scientific and Technical Information) Program, 20 January 2004, available at: http://nix.nasa.gov/?N=125&Ntk=All&Ntt=Quick%20Connect%20Nut&Ntx=mode%20matchallpartial
44 NASA, ' Public Safety: Secure Future', *Spinoff 1998*, Office of Aeronautics and Space Technology, Commercial Programs Division, Washington DC, 1998.
45 NASA, 'STS-113', Shuttle Press Kit, November 2002.
46 IMAX Corporation and Lockheed Martin Corporation in cooperation with the NASA, *SPACE STATION*, Narrated by Tom Cruise, produced and directed by Toni Myers and filmed in IMAX, 2002.
47 Berger, B, 'Loss of Columbia Will Test Iran Non-Proliferation Act', Space News, Business Report, 18 February 2003.
48 Columbia Accident Investigation Board, 'Report Volume 1', August 2003.
49 Columbia Accident Investigation Board, 'Report Volume 1', August 2003.
50 Covey, RO, Task Group Co-Chairman, 'Interim Report. Return to Flight Task Group', 20 January 2004 (and subsequent dates).
51 NASA, 'Implementation Plan for Space Shuttle Return to Flight and Beyond', October 2003 (and subsequent dates).
52 Aldridge, JR, EC, Chairman, *A Journey to Inspire, Innovate and Discover*, Report of the President's Commission on Implementation of United States Space Exploration Policy, June 2004.
53 Office of Science and Technology Policy, Executive Office of the President, 'US Space Transportation Policy', 6 January 2005.
54 NASA, 'STS-114 and STS-114 Return to Flight', Space Shuttle, Mission Archives, Undated.
55 Smith, MS, and Squassoni, S, 'The International Space Station and the Iran Nonproliferation Act (INA): The Bush Administration's Proposed INA Amendment', CRS Report for Congress, Order Code RS22270, 14 November 2005.
56 NASA, 'Shuttle/Station Configuration Options Team (S/SCOT) study', October 2005.
57 NASA, 'Space Shuttle Program. Flight Assignment Working Group Manifest', March 2006 (and subsequent dates).
58 NASA, 'STS-115', Shuttle Press Kit, September 2006.
59 Ibid.
60 NASA, 'STS-116', Shuttle Press Kit, December 2006.
61 NASA, live television coverage of the STS-116 mission, December 2006.
62 NASA, 'STS-116', Space Shuttle, Mission Archives, Undated.
63 de Selding, PB, 'Space Station Partners Voice Optimism but Note Challenges', *Space News*, 29 January 2007.
64 Berger, B, 'Griffin: Budget Shortfall Pushes Orion, Ares back to 2015', *Space News*, 5 March 2007.
65 Morring, F., 'Safe in Space', *Aviation Week & Space Technology*, 5 March 2007.
66 Berger, B, 'As Shuttle Era Ends, NASA Rethinks ISS Supply Chain', *Space News*, 9 April 2007.
67 NASA, 'STS-117', Shuttle Press Kit, June 2007.
68 NASA, live television coverage of the STS-117 mission, June 2007.
69 Morring F, and Taverna, M, 'Forging Ahead', *Aviation Week & Space Technology*, 25 June 2007.
70 SpaceRef.com, 'House Science and Technology Committee's Subcommittee Examines Challenges Facing Space Shuttle and International Space Station Programs', Press release, 24 July 2007, available at: www.spaceref.com.
71 NASA, 'STS-118', Shuttle Press Kit, August 2007.
72 NASA, 'STS-118', NASA Facts.
73 de Selding, P, 'Eyeing International Trend, ESA To Ask Members for Huge Investment', *Space News*, 12 November 2007.
74 Barrie, D, 'Schedule Squeeze', Aviation Week & Space Technology, 19 November 2007.
75 NASA, 'STS-120', Shuttle Press Kit, October 2007.
76 Ibid.
77 NASA, 'STS-120', NASA Facts.
78 European Space Agency, 'Columbus Begins Voyage of Discovery', ESA bulletin 127, August 2006.
79 Fillon, F, Prime Minister of France, Letter to Léopold Eyharts, 9 February 2008.
80 NASA, 'STS-122', Shuttle Press Kit, February 2008.
81 European Space Agency, 'Europe's Columbus Laboratory leaves Earth', Press Release No 7-2008.
82 NASA, 'STS-122', NASA Facts.
83 David, L, 'ATV's on-orbit triumph', *Aerospace America*, October 2008
84 Klesius, M, 'ATV: Unmanned, but everyone's on board', *Aerospace America*, June 2007.
85 European Space Agency, 'Jules Verne' Automated Transfer Vehicle (ATV)', Mission Information Kit, February 2008.
86 European Space Agency, 'Jules Verne ATV – roadmap of campaign until docking', Mission Statement, 22 July 2008.
87 European Space Agency, 'Impressive dress-rehearsal for Jules Verne ATV', Mission Statement, 31 March 2008.
88 European Space Agency, 'Space Station crew enters Jules Verne ATV', Mission Statement, 5 April 2008.
89 NASA, 'STS-123', Shuttle Press Kit, March 2008.
90 Japanese Aerospace Exploration Agency, 'Kibo Handbook', Human Space Systems and Utilization Group, JAXA, September 2007.
91 MacDonald Dettwiler Space and Advanced Robotic, 'Mobile Servicing System', MSS Backgrounder, July 2011.
92 Japanese Aerospace Exploration Agency, 'Kibo Handbook', Human Space Systems and Utilization Group, JAXA, September 2007.
93 NASA, 'STS-124', Shuttle Press Kit, May 2008.
94 NASA, 'STS-124', Mission Summary, May 2008.
95 Lueders, K, 'ISS Space Transportation Plan', Space Station Control Board, NASA, July 2008.
96 Ibid.
97 Morring, F, 'Finish Line in Sight', *Aviation Week & Space Technology*, 14 July 2008.
98 NASA, 'Launch Schedule. Consolidated Launch Manifest, Space Shuttle Flights and ISS Assembly Sequence', 7 July 2008.
99 de Selding, PB, 'Stakeholders Meet to Ensure Longevity of the Space Station', *Space News*, 14 July 2008.
100 Congress of the United States, 'Consolidated Security, Disaster Assistance, and Continuing Appropriations Act, 2009', HR 2638, January 2008.
101 Behrens, C, Specialist in Energy Policy, Resources, Science and Industry Division, and Nikitin, M, Analyst in Nonproliferation, Foreign Affairs, Defense and Trade Division, 'CRS Report for Congress. Extending NASA's Exemption from the Iran, North Korea, and Syria Nonproliferation Act', Order Code RL34477, 1 October 2008.
102 Malik, T, 'ISS Modifications Set the Stage for a Six-Person Crew', Space News, 1 December 2008.
103 Johnson, J, and Klinkrad, H, 'The International Space Station and the Space Debris Environment: 10 years on', Orbital Debris Office, Johnson Space Center and Space Debris Office, ESA Space Operations Center, 2008.
104 Ibid.
105 NASA, 'Columbia Crew Survival Investigation Report', NASA/SP-2008-565, December 2008.
106 European Space Agency, 'ESA Meteoroid and Space Debris Terrestrial Environment Reference, the MASTER-2005 model', available at: www.master-model.de
107 NASA, 'STS-119', Shuttle Press Kit, March 2009.
108 Morring, F., 'Plugged In', *Aviation Week & Space Technology*, 23 March 2009.
109 Daily Launch Newsletter, 'Soyuz Docks At ISS, Doubling Crew To Six', AIAA, 1 June 2009.
110 Reuters, 'Grand experiment: 13 people on space station', 11 June 2009.
111 NASA, 'STS-127', Shuttle Press Kit, July 2009.
112 Conolly, JF, 'Constellation Program

Overview', Constellation Program Office, NASA, October 2006.
113 Holdren, J, Director, Office of Science and Technology Policy and Assistant to the President for Science and Technology, letter to Chris Scolese, NASA Acting Administrator, 7 May 2009.
114 US Human Space Flight Plans Committee, Summary Report, Washington DC, September 2009.
115 US Human Space Flight Plans Committee, 'Seeking a Human Spaceflight Program Worthy of a Great Nation', Washington DC, October 2009.
116 Massachusetts Institute of Technology, 'Technology Review', 10 November 2009.
117 NASA, 'Highlights of NASA's FY 2011 Budget', Washington DC, 1 February 2010.
118 Multilateral Coordination Board, 'MCB Joint Statement Representing Common Views on the Future of the ISS', 3 February 2010.
119 NASA, 'STS-130', Shuttle Press Kit, February 2010.
120 Pevsner, N, et al., *A Dictionary of Architecture*, The Overlook Press, 1976.
121 Morring, F., 'Bay Window' – Endeavour crew sets up tower at ISS for arriving commercial spacecraft, *Aviation Week & Space Technology*, 22 February 2010.
122 National Aeronautic Association, 'NAA Announces the International Space Station as Winner of the 2009 Robert J. Collier Trophy', website, available at: https://naa.aero/userfiles/files/documents/Press%20Releases/Collier%20 2009%20PR.pdf
123 NASA, Heads of Agency International Space Station Joint Statement, Press Release 10-063, 11 March 2010.
124 NASA, 'STS-132', Shuttle Press Kit, May 2010.
125 NASA, 'STS-133', Shuttle Press Kit, February 2011.
126 NASA, 'STS-135', Shuttle Press Kit, July 2011.
127 Ibid.

Epilogue

1 Warner Brothers, *Gravity*, feature film, released October 2013, produced and directed by Alfonso Cuarón, starred Sandra Bullock and George Clooney.
2 NASA, 'NASA Congratulates 'Gravity' on Academy Award wins', 2 March 2014, website, available at: http://www.nasa.gov/content/nasa-congratulates-gravity-on-academy-award-wins/index.html#.UxcBL9y_ere.
3 NASA, 'Space Station's Benefits for Humanity in Plain Sight in New Video Feature', 25 November 2013, website, available at: http://www.nasa.gov/mission_pages/station/research/news/benefits_video/#.UxcDjty_erc.
4 NASA, 'International Space Station. Benefits for Humanity', NP-2012-02-003-JSC, Collaboration between the Canadian Space Agency, European Space Agency, Japan Aerospace Exploration Agency, NASA and the Russian Federal Space Agency.
5 NASA, 'ISS-RapidScat', NASA Facts, JPL 400-1569 9/14, Jet Propulsion Laboratory, 2014.
6 Ibid.
7 University of Calgary, 'NeuroArm Success, revolutionary procedure a world first', website, available at: http://www.ucalgary.ca/news/uofcpublications/oncampus/online/may29-08/neuroArm
8 NASA, 'MISSE: Testing materials in space', FS-2001-07-65-LaRC, July 2001, website, available at: http://www.nasa.gov/centers/langley/news/factsheets/MISSE.html
9-17 NASA, International Space Station, Experiment List, Alphabetical – 11.12.14, website, available at: http://www.nasa.gov/mission_pages/station/research/experiments/experiments_by_name.html
18 NASA, 'NASA Launches New Era of Earth Science from Space Station', Release 14-240, 8 September 2014, website, available at: http://www.nasa.gov/press/2014/september/nasa-launches-new-era-of-earth-science-from-space-station/index.html#.VA4KpEvAyec
19 European Space Agency, 'Video: Found at Sea – How the International Space Station can save lives on Earth', 5 September 2014, website, available at: http://www.esa.int/Our_Activities/Human_Spaceflight/VIDEO_Found_at_sea_How_the_International_space_station_can_save_lives_on_Earth
20 NASA, 'International Space Station. Benefits for Humanity', NP-2012-02-003-JSC, Collaboration between the Canadian Space Agency, European Space Agency, Japan Aerospace Exploration Agency, NASA and the Russian Federal Space Agency.
21 Uhran, ML, Assistant Associate Administrator, International Space Station, 'Positioning the International Space Station for the Utilization Era', Office of Space Exploration, NASA, Washington DC, AIAA, 2011.
22 ISS Program Science Forum, 'International Space Station Utilization Statistics. Expeditions 0-34. December 1998 –March 2013', draft, 21 June 2013.
23 Robinson, J, 'Update on Program Science Forum Activities and Utilization Products for MCB, 'Multilateral Coordination Board, NASA/OA, March 2014.
24 NASA, 'International Space Station Facilities. Research in Space. 2013 and Beyond', NP-2012-10-027-JSC, NASA ISS Program Science Office.
25 CERN, 'New results from the Alpha Magnetic Spectrometer', Press Release, AMS Collaboration, Geneva, 18 September 2014.
26 NASA, 'STS-134 – Final Flight of Endeavour', Shuttle Press Kit, April 2011.
27 CERN, 'New results from the Alpha Magnetic Spectrometer', Press Release, AMS Collaboration, Geneva, 18 September 2014.
28 Robinson, J, ISS Program Scientist, 'International Space Station NASA Research. NASA Outreach Seminar on the ISS at the United Nations', February 2011.
29 NASA, Radiogram No 5253nu. Form 24 for 04/07/2014, Flight plan timeline.
30 NASA, Radiogram No 5275u, Form 24 for 04/08/2014, Flight plan timeline.
31 NASA, Radiogram No 5278u, Form 24 for 04/09/2014, Flight plan timeline.
32 NASA, Radiogram No 5281u, Form 24 for 04/10/2014, Flight plan timeline.
33 NASA, Radiogram No 5300u, Form 24 for 04/11/2014, Flight plan timeline.
34 NASA, Radiogram No 5312u, Form 24 for 04/12/2014, Flight plan timeline.
35 NASA, Radiogram No 5328u, Form 24 for 04/13/2014, Flight plan timeline.
36 NASA, 'Reference Guide to the International Space Station. Assembly Complete edition 2010', NP-2010-09-682-HQ, NASA, Washington DC.
37 Cundieff, L, Barrios Technology Ltd (JSC-OP), written statement provided to author, April 2014.
38 Ibid.
39 Boucher, M, 'Canada's Dextre Becomes the First Robot to Repair Itself in Space', 22 May 2014, SpaceRef, website, available at: http://spaceref.ca/missions-and-programs/canadian-space-agency/canadas-dextre-becomes-the-first-robot-to-repair-itself-in-space.html
40 NASA, 'Reference Guide to the International Space Station. Assembly Complete edition 2010', NP-2010-09-682-HQ, NASA, Washington DC.
41 International Space Exploration Coordination Group, *The Global Exploration Roadmap*, published by the NASA, NP-2013-06-945-HQ, Washington DC, August 2013.
42 Ibid.
43 NASA, Exploration Forum to Showcase Human Path to Mars, 30 April 2014, NASA Headquarters, Washington DC, website, available at: http://spaceref.com/mars/nasa-hosts-exploration-forum-to-showcase-human-path-to-mars.html
44 NASA, 'The Joint US-Russian One-Year Mission: Establishing International Partnerships and Innovative Collaboration', 16 May 2013, website, available at: http://www.nasa.gov/exploration/humanresearch/Joint_US_Russian_One_Yr_Mission.html
45 University of Houston, 'Collaboration Aims to Reduce, Treat Vision Problems in Astronauts', 26 August 2014, website, available at: http://www.uh.edu/news-events/stories/2014/August/082614NASAvision.php
46 Carrasquillo, R, 'ISS Environmental Control and Life Support System (ECLSS) Future Development for Exploration', 2nd Annual ISS Research and Development Conference, 16-18 July, 2013.
47 Gatens, R, Broyan, J, Macatangay, A., Metcalf, J, Shull, S, Bagdigian, R, Stephan, R, 'NASA (NASA) Environmental Control and Life Support (ECLS) Technology Development and Maturation for Exploration 2013 to 2014 Overview', ICES-2014-019, 44th International Conference on Environmental Systems, Tucson, Arizona, 13-17 July 2014.
48 National Geographic News, 'Up on the Farm? Five Reasons NASA Needs Space Greenhouses', website, available at: http://news.nationalgeographic.com/news/2014/01/140121-space-greenhouses-plants-astronaut-mars/
49 NASA, 'Commercial Crew Program', NASA Facts, FS-2012-02-048-KSC, John F. Kennedy Space Center.

50 NASA, 'Commercial Crew Program Update. The Next Step in US Space Transportation', December 2012.
51 National Research Council, *Pathways to Exploration: Rationales and Approaches for a US Program of Human Space Exploration*, The National Academies Press, website, available at: http://www.nap.edu/catalog.php?record_id=18801
52 SpaceRef, 'NASA Selects SpaceX and Boeing to Ferry Astronauts to the Space Station', 17 September 2014, website, available at: http://spaceref.biz/nasa/nasa-selects-spacex-and-boeing-to-send-humans-to-space.html
53 Office of Audits and Office of Inspector General, 'Extending the Operational Life of the International Space Station Until 2024', IG-14-031 (A-13-021-00), NASA, 18 September 2014.
54 Ibid.
55 Gurrisi, C, Seidel, R et al., 'Space Station Control Moment Gyroscope Lessons Learned', NASA/CP-2010-216272, Proceedings of the 40th Aerospace Mechanisms Symposium, NASA Kennedy Space Center, May 2010.
56 Kluger, J, 'Mission Impossible', TIME, vol. 184, No 26-27, 29 December 2014 – 5 January 2015.
57 NASA, 'One-Year Mission & Twins Study, Human Research', website, available at: www.nasa.gov
58 NASA, 'Journey to Mars, Pioneering Next Steps in Space Exploration', NP-2015-08-2018-HQ, NASA Headquarters, Washington DC, October 2015.

Index

First published in 2016 by Circa Press

Circa Press
50 Great Portland Street
London W1W 7ND
+44 (020) 7637 0099
www.circapress.net

ISBN 978-0-9930721-3-0

Cataloguing-in-Publication Data for this book is available from the British Library

Editor: David Jenkins
Designer: Jean-Michel Dentand
Reproduction: DawkinsColour

Printed and bound in Italy
by Conti Tipocolor S.p.A.

All photographs courtesy of NASA, except where indicated otherwise.